Biomaterials for Tissue Engineering and Regeneration

Biomaterials for Tissue Engineering and Regeneration

Editors

Antonia Ressler
Inga Urlic

MDPI • Basel • Beijing • Wuhan • Barcelona • Belgrade • Manchester • Tokyo • Cluj • Tianjin

Editors
Antonia Ressler
Faculty of Engineering and
Natural Sciences
Tampere University
Tampere
Finland

Inga Urlic
Faculty of Science, Division of
Molecular Biology
University of Zagreb
Zagreb
Croatia

Editorial Office
MDPI
St. Alban-Anlage 66
4052 Basel, Switzerland

This is a reprint of articles from the Special Issue published online in the open access journal *Polymers* (ISSN 2073-4360) (available at: www.mdpi.com/journal/polymers/special_issues/Biomaterials_Tissue_Engineering_and_Regeneration).

For citation purposes, cite each article independently as indicated on the article page online and as indicated below:

LastName, A.A.; LastName, B.B.; LastName, C.C. Article Title. *Journal Name* **Year**, *Volume Number*, Page Range.

ISBN 978-3-0365-6361-9 (Hbk)
ISBN 978-3-0365-6360-2 (PDF)

© 2023 by the authors. Articles in this book are Open Access and distributed under the Creative Commons Attribution (CC BY) license, which allows users to download, copy and build upon published articles, as long as the author and publisher are properly credited, which ensures maximum dissemination and a wider impact of our publications.

The book as a whole is distributed by MDPI under the terms and conditions of the Creative Commons license CC BY-NC-ND.

Contents

S. Mary Stella, T. M. Sridhar, R. Ramprasath, Jolius Gimbun and U. Vijayalakshmi
Physio-Chemical and Biological Characterization of Novel HPC (Hydroxypropylcellulose): HAP (Hydroxyapatite): PLA (Poly Lactic Acid) Electrospun Nanofibers as Implantable Material for Bone Regenerative Application
Reprinted from: *Polymers* 2022, 15, 155, doi:10.3390/polym15010155 1

David Grijalva Garces, Carsten Philipp Radtke and Jürgen Hubbuch
A Novel Approach for the Manufacturing of Gelatin-Methacryloyl
Reprinted from: *Polymers* 2022, 14, 5424, doi:10.3390/polym14245424 21

Yuki Hachinohe, Masayuki Taira, Miki Hoshi, Wataru Hatakeyama, Tomofumi Sawada and Hisatomo Kondo
Bone Formation on Murine Cranial Bone by Injectable Cross-Linked Hyaluronic Acid Containing Nano-Hydroxyapatite and Bone Morphogenetic Protein
Reprinted from: *Polymers* 2022, 14, 5368, doi:10.3390/polym14245368 35

Siti Hajar Ahmad Shariff, Wan Khartini Wan Abdul Khodir, Shafida Abd Hamid, Muhammad Salahuddin Haris and Mohamad Wafiuddin Ismail
Poly(caprolactone)-*b*-poly(ethylene glycol)-Based Polymeric Micelles as Drug Carriers for Efficient Breast Cancer Therapy: A Systematic Review
Reprinted from: *Polymers* 2022, 14, 4847, doi:10.3390/polym14224847 61

Aniek Setiya Budiatin, Junaidi Khotib, Samirah Samirah, Chrismawan Ardianto, Maria Apriliani Gani and Bulan Rhea Kaulika Hadinar Putri et al.
Acceleration of Bone Fracture Healing through the Use of Bovine Hydroxyapatite or Calcium Lactate Oral and Implant Bovine Hydroxyapatite–Gelatin on Bone Defect Animal Model
Reprinted from: *Polymers* 2022, 14, 4812, doi:10.3390/polym14224812 87

Ana Chor, Christina Maeda Takiya, Marcos Lopes Dias, Raquel Pires Gonçalves, Tatiana Petithory and Jefferson Cypriano et al.
In Vitro and In Vivo Cell-Interactions with Electrospun Poly (Lactic-Co-Glycolic Acid) (PLGA): Morphological and Immune Response Analysis
Reprinted from: *Polymers* 2022, 14, 4460, doi:10.3390/polym14204460 105

Yao Fu, Sanne K. Both, Jacqueline R. M. Plass, Pieter J. Dijkstra, Bram Zoetebier and Marcel Karperien
Injectable Cell-Laden Polysaccharide Hydrogels: In Vivo Evaluation of Cartilage Regeneration
Reprinted from: *Polymers* 2022, 14, 4292, doi:10.3390/polym14204292 129

Ozgur Basal, Ozlem Ozmen and Aylin M. Deliormanlı
Bone Healing in Rat Segmental Femur Defects with Graphene-PCL-Coated Borate-Based Bioactive Glass Scaffolds
Reprinted from: *Polymers* 2022, 14, 3898, doi:10.3390/polym14183898 143

Antonia Ressler
Chitosan-Based Biomaterials for Bone Tissue Engineering Applications: A Short Review
Reprinted from: *Polymers* 2022, 14, 3430, doi:10.3390/polym14163430 159

Nohra E. Beltran-Vargas, Eduardo Peña-Mercado, Concepción Sánchez-Gómez, Mario Garcia-Lorenzana, Juan-Carlos Ruiz and Izlia Arroyo-Maya et al.
Sodium Alginate/Chitosan Scaffolds for Cardiac Tissue Engineering: The Influence of Its Three-Dimensional Material Preparation and the Use of Gold Nanoparticles
Reprinted from: *Polymers* 2022, 14, 3233, doi:10.3390/polym14163233 177

Georgina Carbajal-De la Torre, Nancy N. Zurita-Méndez, María de Lourdes Ballesteros-Almanza, Javier Ortiz-Ortiz, Miriam Estévez and Marco A. Espinosa-Medina
Characterization and Evaluation of Composite Biomaterial Bioactive Glass–Polylactic Acid for Bone Tissue Engineering Applications
Reprinted from: *Polymers* **2022**, *14*, 3034, doi:10.3390/polym14153034 197

Johannes J van den Heever, Christiaan J Jordaan, Angélique Lewies, Jacqueline Goedhals, Dreyer Bester and Lezelle Botes et al.
Impact of Three Different Processing Techniques on the Strength and Structure of Juvenile Ovine Pulmonary Homografts
Reprinted from: *Polymers* **2022**, *14*, 3036, doi:10.3390/polym14153036 219

Viviana R. Güiza-Argüello, Víctor A. Solarte-David, Angie V. Pinzón-Mora, Jhair E. Ávila-Quiroga and Silvia M. Becerra-Bayona
Current Advances in the Development of Hydrogel-Based Wound Dressings for Diabetic Foot Ulcer Treatment
Reprinted from: *Polymers* **2022**, *14*, 2764, doi:10.3390/polym14142764 233

Mika Suzuki, Tsuyoshi Kimura, Yukina Yoshida, Mako Kobayashi, Yoshihide Hashimoto and Hironobu Takahashi et al.
In Vitro Tissue Reconstruction Using Decellularized Pericardium Cultured with Cells for Ligament Regeneration
Reprinted from: *Polymers* **2022**, *14*, 2351, doi:10.3390/polym14122351 259

My Thi Ngoc Nguyen and Ha Le Bao Tran
In-Vitro Endothelialization Assessment of Heparinized Bovine Pericardial Scaffold for Cardiovascular Application
Reprinted from: *Polymers* **2022**, *14*, 2156, doi:10.3390/polym14112156 271

Kiran M. Ali, Yihan Huang, Alaowei Y. Amanah, Nasif Mahmood, Taylor C. Suh and Jessica M. Gluck
In Vitro Biocompatibility and Degradation Analysis of Mass-Produced Collagen Fibers
Reprinted from: *Polymers* **2022**, *14*, 2100, doi:10.3390/polym14102100 281

Maja Ledinski, Ivan Marić, Petra Peharec Štefanić, Iva Ladan, Katarina Caput Mihalić and Tanja Jurkin et al.
Synthesis and In Vitro Characterization of Ascorbyl Palmitate-Loaded Solid
Lipid Nanoparticles
Reprinted from: *Polymers* **2022**, *14*, 1751, doi:10.3390/polym14091751 295

Hongcai Wang, Xiuqiong Chen, Yanshi Wen, Dongze Li, Xiuying Sun and Zhaowen Liu et al.
A Study on the Correlation between the Oxidation Degree of Oxidized Sodium Alginate on Its Degradability and Gelation
Reprinted from: *Polymers* **2022**, *14*, 1679, doi:10.3390/polym14091679 307

Marika Faggioli, Arianna Moro, Salman Butt, Martina Todesco, Deborah Sandrin and Giulia Borile et al.
A New Decellularization Protocol of Porcine Aortic Valves Using Tergitol to Characterize the Scaffold with the Biocompatibility Profile Using Human Bone Marrow Mesenchymal Stem Cells
Reprinted from: *Polymers* **2022**, *14*, 1226, doi:10.3390/polym14061226 323

Laura Olariu, Brindusa Georgiana Dumitriu, Carmen Gaidau, Maria Stanca, Luiza Mariana Tanase and Manuela Diana Ene et al.
Bioactive Low Molecular Weight Keratin Hydrolysates for Improving Skin Wound Healing
Reprinted from: *Polymers* **2022**, *14*, 1125, doi:10.3390/polym14061125 341

Article

Physio-Chemical and Biological Characterization of Novel HPC (Hydroxypropylcellulose):HAP (Hydroxyapatite):PLA (Poly Lactic Acid) Electrospun Nanofibers as Implantable Material for Bone Regenerative Application

S. Mary Stella [1], T. M. Sridhar [2], R. Ramprasath [3], Jolius Gimbun [3] and U. Vijayalakshmi [1,*]

[1] Department of Chemistry, School of Advanced Sciences, Vellore Institute of Technology, Vellore 632014, India
[2] Department of Analytical Chemistry, Guindy Campus, University of Madras, Chennai 600025, India
[3] Department of Chemical Engineering, College of Engineering, Universiti Malaysia Pahang, Pekan 26600, Malaysia
* Correspondence: vijayalakshmi.u@vit.ac.in

Abstract: The research on extracellular matrix (ECM) is new and developing area that covers cell proliferation and differentiation and ensures improved cell viability for different biomedical applications. Extracellular matrix not only maintains biological functions but also exhibits properties such as tuned or natural material degradation within a given time period, active cell binding and cellular uptake for tissue engineering applications. The principal objective of this study is classified into two categories. The first phase is optimization of various electrospinning parameters with different concentrations of HAP-HPC/PLA(hydroxyapatite-hydroxypropylcellulose/poly lactic acid). The second phase is in vitro biological evaluation of the optimized mat using MTT (3-(4,5-dimethylthiazol-2-yl)-2,5-diphenyl tetrazolium bromide) assay for bone regeneration applications. Conductivity and dielectric constant were optimized for the production of thin fiber and bead free nanofibrous mat. With this optimization, the mechanical strength of all compositions was found to be enhanced, of which the ratio of 70:30 hit a maximum of 9.53 MPa (megapascal). Cytotoxicity analysis was completed for all the compositions on MG63 cell lines for various durations and showed maximum cell viability on 70:30 composition for more than 48 hrs. Hence, this investigation concludes that the optimized nanofibrous mat can be deployed as an ideal material for bone regenerative applications. In vivo study confirms the HAP-HPC-PLA sample shows more cells and bone formation at 8 weeks than 4 weeks.

Keywords: electrospinning; polymer; composite; XPS (X-ray photoelectron spectroscopy); tensile properties; in vitro study

1. Introduction

Bone tissue engineering is one of the most important approaches to repairing damaged and diseased tissue and it facilitates the complete recovery of the tissue itself as well as its function [1]. Bone serves several functions such as body support, organ protection, and storage of nutrients. Due to the applicability of bone, it is considered a complex tissue in the human body [2]. Moreover, it has an extremely anisotropic nature due to a variety of mechanical properties extending in all directions. The design of scaffolds as an extracellular matrix (ECM) and the process of regeneration is the major goals in tissue engineering [3]. ECM should be nontoxic and biocompatible as well as show the desired degradation rate and porosity with excellent mechanical properties. Additionally, ECM should not be the cause of foreign body reactions [4].

Alternatively, scaffolds are used as supporting materials and the results have been better cell growth, increased proliferation rate, and healthier ECM production [5]. The fabricated artificial material will be useful for the repair of damaged tissue and the regeneration of new tissues by cell proliferation and growth of their own ECM. Generally, several types of material sources, be it natural, synthetic, semi-synthetic or even composites are used to synthesize scaffolds used in bone tissue regeneration.

Cellulose fiber is a good choice for bone tissue engineering due to its high strength and better mechanical properties. Subsequently, the fabrication of cellulose composite scaffolds is attracting many researchers due to its highly applicable mechanical properties [6]. The main advantages of cellulose-based scaffolds are biodegradability in chemical as well as biological environments and mechanical stress factors, good mechanical properties in terms of tensile strength, negligible foreign body and inflammatory response reactivity within in vitro and in vivo applications, and abundant presence of this naturally occurring polymer [7,8]. In tissue engineering, cellulose has been used as a permanently implantable scaffold establishing the fact that it is stable for a long-time application in vivo [9,10].

Among the other cellulose fibers, hydroxypropyl cellulose (HPC) has the abovementioned desired properties and it can be used for the fabrication of HAP-reinforced fibrous mat. Structurally, hydroxypropyl groups substitute the hydroxyl groups present in cellulose, hence yielding better mechanical properties than cellulose [11]. The material has adequate biomedical application due to its biodegradability and biocompatibility [12].

In scaffold fabrication, polymers play a major role as they are responsible to mimic the organic nature of human native bone. Poly(lactic acid) (PLA) [13], polyglycolic acid (PGA) [14], poly(lactic-co-glycolic acid) (PLGA) [15], poly(vinyl alcohol) (PVA) [16], poly(vinyl pyrrolidone) (PVP) [17], poly(methyl methacrylate) [18] and polycaprolactone (PCL) [19] are some of the important polymers that may be deployed in fabricating scaffolds for bone tissue engineering applications. Among the above polymers, PLA has advantages such as biocompatibility, biodegradability and good modulus with adequate strength [20].

Certain polymeric materials are unable to accomplish the desired scaffold to provide valuable inputs in biomedical applications. To overcome these issues, the composite of natural and synthetic polymers can be employed. HPC polymer has good biocompatibility and fine particle grades are most favorable for faster hydration, and uniform dissolution of particles during mixing [21]. On the other hand, it has some disadvantages such as hydrophobicity and low molecular weight. Additionally, for electrospinning to occur, polymer chain entanglement is one of the important criteria to form nanofibers. Due to these reasons, HPC as such is not spinnable and hence the polymers such as PLA are mixed to provide sufficient chain entanglement in the solution, whereby electro-spinnability is feasible [22].

Hydroxyapatite [$Ca_{10}(PO_4)_6(OH)_2$] is one of the important materials which offer significant applications in bone therapy and tooth reconstruction. It exhibits the properties and mechanisms to directly bond to the natural host tissue and results in regeneration [23]. The major challenge in recent research is producing an apatite layer similar to human tissues to be confirmed by morphological and physiochemical analysis. It is necessary to understand the mechanism of HAP bonding and the biomineralization process leading to cell growth [24]. In biomedical aspects, scaffolds have been fabricated to renew or repair damaged organs by fracture or defects caused by disease (osteoporosis, osteoarthritis, osteosarcoma) or surgeries (tumor removal) or congenital disabilities. To figure out, the risk and demands, artificial scaffolds have been fabricated with the properties of biocompatibility, biodegradability, mineralization, mechanical resistance, and cell proliferation.

Electrospinning is an emerging technique for the fabrication of nanofibrous scaffolds which is useful for the production of ECM better than other bone scaffold fabrication techniques [25]. As we know, other fabrication techniques do not offer any ability to control porosity, whereas the electrospinning technique offers continuous fiber production with a controlled diameter. The diameter and alignment of the fiber can be easily controlled by adjusting the properties of the HAP–polymer composite solution and parameters of the electrospinning instrument [26].

In the present study, the novelty of the work is divided into two categories: Firstly, HAP-HPC/PLA nanofibrous mat has been fabricated using electrospinning techniques.

The composite of HAP-HPC/PLA prepared by 25% of HAP was gradually added into 5 wt% of HPC/PLA polymer solution. The different composition ratios such as 0:100, 40:60, 50:50, 60:40, and 70:30 of HAP-HPC/PLA have been used to optimize the composition in terms of mechanical strength, biodegradability, and bioactivity properties. Different physical, chemical, and mechanical characterization techniques were used for the analysis of composite nanofibers. Further, the optimized scaffold has been confirmed by in vitro biocompatibility study. In this study, the cytotoxicity analysis has been carried out using the MG63 (osteoblast) cell line to achieve the osteoconductive property in the form of new bone formation. Secondly, the optimized composite was further confirmed by in vivo performance using the calvarial defect on rats and qualitative analysis was carried out at 4 weeks and 8 weeks of implantation using radiological (X-Ray) evaluation and histological study. The objective of this work was to optimize composite scaffolds with respect to the different levels of cell growth (osteoblast and osteoclast etc.). Thus, the optimized study plays a favorable foundation for new bone formation both by in vitro and in vivo analysis.

2. Materials and Methods

2.1. Materials

PLA (molecular weight of 220 kDa), HPC (molecular weight of 100,000 kDa), aqueous ammonia, calcium nitrate tetrahydrate ($Ca(NO_3)_2 \cdot 4H_2O$) and ammonium dihydrogen orthophosphate ($NH_4H_2PO_4$) were purchased from the SD-Fine chemicals, Mumbai, India.

MG63 cell lines were obtained from NCCS, Pune. DMEM (Dulbecco's Modified Eagle Medium), fetal bovine serum (FBS), trypsin and 1× antibiotic solution were purchased from Hi-Media Laboratories, Mumbai, India. Methyl thiazolyl diphenyl-tetrazolium bromide (MTT) and dimethyl sulfoxide (DMSO) were purchased from Sigma-Aldrich Chemicals Company, Mumbai, India. The cells were maintained in Dulbecco's Modified Eagle's Medium (DMEM) supplemented with 10% FBS, penicillin (10,000 U/mL), and streptomycin (10 mg/mL) in a humidified atmosphere of 50 μg/mL CO_2 at 37 °C.

2.2. Methods

2.2.1. Preparation of Hydroxyapatite and Polymer Solution

1M solutions of calcium nitrate tetrahydrate and 0.6M of ammonium dihydrogen orthophosphate were prepared using double distilled water by adopting co-precipitation method. Both the solutions were mixed together under continuous stirring and pH of the solution was adjusted to 11 by adding aqueous ammonia [27]. Pure hydroxyapatite was obtained by aging the solution for 16 h, the raw powder was filtered and washed several times with distilled water followed by sintering the dry powder at 700 °C for 2 h. Nanofibrous scaffold was fabricated using 2.5 g of HAP powder dispersed in 10 mL of water under stirring conditions for 12 h. The solution of HPC/PLA was prepared by selecting 0.5 g of (5 wt %) HPC and 0.5 g of PLA in 10 mL of boiling water under stirring conditions. Finally, the resultant solution with the composition of HAP/HPC/PLA suspension was made for electrospinning.

The composite mixture was stirred for 3 h until a clear homogeneous solution was obtained. The homogenous solution is allowed to age for 12 h followed by ultra-sonication prior to electrospinning.

2.2.2. Viscosity and Electrical Conductivity Analysis

The obtained composite solutions were tested for their viscosity using Brookfield DV2 viscometer. The dielectric constant of the electrospun nanofibers was measured using a HIOKI 3532-50 LCR HITESTER meter at a temperature range of 35–125 °C at a frequency of 0–5 × 10^6 Hz.

2.2.3. Electrospinning

The nanofibrous mat was fabricated with different composite ratios (0:100, 40:60, 50:50, 60:40, and 70:30) of HAP and HPC/PLA. The composite mixture was stirred for 3 h until a clear homogeneous solution is obtained. The homogenous solution is allowed to age for 12 h. The composite solution to be electrospun was taken in a plastic syringe (5 mL) with a hypodermic needle and a flat-filed tip, with an internal diameter of 0.8 mm. The electrospun nanofibers were collected on aluminum foils on a rotating drum collector at a controlled relative humidity (20–25%) environment. The obtained nanofibrous mat was dried along with aluminum foil in a desiccator to reduce the effect of humidity on the nanofibers [28]. The solution preparation is tabulated in Table 1.

Table 1. The composite mixture and the volume of the solution.

S.No.	Sample Ratio	Solution Preparation
1	HPC/PLA (0:100)	(Without HAP)10 mL of HPC/PLA
2	HAP-HPC/PLA (40:60)	4 mL of HAP and 6 mL of HPC/PLA
3	HAP-HPC/PLA (50:50)	5 mL of HAP and 5 mL of HPC/PLA
4	HAP-HPC/PLA (60:40)	6 mL of HAP and 4 mL of HPC/PLA
5	HAP-HPC/PLA (70:30)	7 mL of HAP and 3 mL of HPC/PLA

2.2.4. Physio Chemical Characterization

FT-IR analysis was used to distinguish species, and analyze the functional groups, vibration modes, chemical interaction of HAP with polymers. Evaluation by using FT-IR was accomplished using SHIMADZU CROP IRAFFINITY-1 flourier transform infrared spectrophotometer within the variety 400–4000. X-ray diffraction is used to identify the nanoparticles such as HAP in composites. This characterization was performed for dried and finely tailored nano composite samples on XRD machine (BRUKER Germany with Cu K radiation; =1.5405 Å). Mechanical study used to measure the tensile strength and elastic modulus. Tensile strength for electrospun nanofibers was performed using an ASTM standard (D695) on the Tinus Olsen H5K5 universal testing machine. Nanofibers were cut into a rectangular shape of width 12 mm and placed at the height of 6 cm between two clamps bearing a 500 N load cell with velocity of 1mm/min. The average of three trials of tensile modulus were calculated from stress–strain response. X-ray photoelectron spectroscopy (XPS, ESCALAB 250Xi, Richardson, TX, USA) is used to analyze materials surface chemistry. Specimens were analyzed using a monochromatic Al Kα source (10 mA, 15 kV). Scanning Electron Microscope (ZISS-EVO18, Horn, Austria) is used to investigate the morphology of composite at excessive magnification and resolution by using lively electron beam.

2.2.5. Porosity

The solvent replacement method was used for the measurement of porosity of the HPC/PLA and HAP-HPC/PLA nanofibrous mat. The initial weight (Wi) of the nanofibers was measured after drying in desiccator. The nanofibrous mat was immersed in absolute ethanol and dried off immediately. The porosity of the electrospun nanofibrous mat was determined as follows:

$$\text{Porosity} = ((W_s - W_i))/W_s \times 100 \quad (1)$$

where Ws is the rehydrated nanofiber weight and Wi is the initial nanofiber weight.

2.2.6. Bioactivity Study
Bioactivity

The simulated body fluid (SBF) solution which is similar to the human physiological fluid was used for the in vitro bioactive study of the electrospun nanofibers. The procedure, which was suggested by Kokubo et al., 1990 [29] was used for the preparation of SBF solution. All chemicals were added together to 1 L of double distilled water, pH was adjusted to 7.35 ± 0.25, and stored in the refrigerator at −4 °C. For the analysis, 1 × 1 cm size of the electrospun nanofibrous mat (HPC/PLA and HAP-HPC/PLA) was immersed in the SBF solution (60 mL) at 37 °C for a different time duration of 7, 15, and 30 days, respectively. The contamination was avoided by the replacement of SBF solution with fresh solution for every 24 h. Carbonated hydroxyapatite (HCA) was formed in the process of immersion in the SBF solution and hence resulting in the formation of apatite layer. SEM and EDAX analysis were used for the confirmation of apatite layer formation on the surface of the electrospun nanofibers.

In Vitro Biodegradation Properties

Electrospun mat of dimensions 1 × 1 cm was immersed in SBF solution at a pH of 7.4 for different time intervals such as 7, 15, and 30 days, respectively. The samples were dried at room temperature after the respective time duration was completed [30]. The percentage degradation was calculated from the dried weight of the samples after degradation as follows:

$$\text{Percentage Degradation} = ((W_i - W_d))/W_i \times 100 \quad (2)$$

where W_i and W_d are initial and dried weight of the sample, respectively.

2.2.7. Cytotoxicity Analysis

Osteosarcoma (MG-63) cell lines were used for the cytotoxicity analysis of the fabricated mat. MG63 was passaged using Minimum Eagle's Medium (MEM), 10% fetal bovine serum, and 1% penicillin–streptomycin solution and maintained at 37 °C and 5% CO_2 concentration in a humidified incubator. Mats were sterilized by washing with 70% ethanol 2–3 times and exposing them to UV overnight (approximately for 12 h). The sterilized sample was soaked in cell culture medium for 2 h. The cells were harvested using 0.05% (v/v) trypsin–EDTA (ethylenediamine tetraacetic acid) and seeded on scaffolds followed by incubation at 37 °C in the presence of CO_2 environment. The medium was changed once in 2 days to supply the adequate amount of nutrients present in the culture plate. Further, 3-[4,5-dimethylthiazol-2-yl]-2,5-diphenyltetrazolium bromide (MTT) assay was used to check the toxicity effects of samples. The media were washed off with 0.1 M PBS solution at the end of 24 and 48 h incubation. Further, 300 μL of fresh media and 60 μL of MTT solution were mixed together, added to PBS-washed mats, and incubated at 37 °C for 3 h. About 200 μL of the incubated mixture was filled in 96-well plate and absorbance was measured at 570 nm using a Vmax Microplate reader. Triplicates were used for calculating the average of two sets of the assay [31]. The % cell viability was calculated using the following formula:

$$\%\text{cell viability} ((A570 \text{ of treated cells}))/((570 \text{ of control cells})) \times 100 \quad (3)$$

2.2.8. Statistical Analysis

Results were expressed as a mean and standard deviation. Comparative studies of means were performed using one-way analysis of variance (ANOVA). Significance was accepted with $p < 0.05$.

2.2.9. In Vivo Study

Sixteen 4–6 weeks-old Wistar albino rats weighing between 250 and 400 g were obtained from the Animal Center Laboratory of VIT University. All experimental rats were bred at the Animal Center Laboratory of VIT University, with a standard laboratory diet and environment. All animal experiments were approved and performed according to the regulations of the animal ethics committee of our university. VIT/IAEC/14/NOV5/47. The rats were anesthetized by intraperitoneal injection of pentobarbital (ketamine 0.2 mL and xylene 0.1 mL). Using sterile instruments and aseptic technique, a 1.0–1.5 cm sagittal incision was made on the scalp, and the calvarium was exposed by blunt dissection. A full-thickness defect (5 mm in diameter) was created in the central area of each parietal bone using a 5 mm electric trephine bur under constant irrigation with sterile 0.9% saline.

The defects were implanted randomly with the HAP-HPC-PLA scaffolds (n 1/4 12). After 4 and 8 weeks of implantation, the calvaria of the rats was harvested and immediately immersed in a 10% tempered solution of formalin and further analysis was studied such as R-ray and histopathological.

3. Results and Discussion

3.1. Preparation of HAP-HPC/PLA Nanofibers

The solution of HPC/PLA was prepared by HPC and was added to boiling water. Since HPC has a solubility issue, PLA with high viscosity was added to HPC to make the solution suitable for electrospinning. The different amounts of HAP-HPC/PLA were considered to fabricate nanofibers of different compositions.

3.2. Viscosity, Electrical Conductivity, and Dielectric Constant Analysis

HAP was blended with HPC/PLA polymer composite at different ratios such as 0:100 (HAP is absent here), 40:60, 50:50, 60:40, and 70:30 and these resulted in a viscosity of 387.6 2487, 1413, 1198, and 786.7 cP, respectively. Amongst the different composite solutions, 70:30 mixture had a viscosity of 786.7 cP, which is found to be the applicable viscosity, as it has to flow easily from the needle during electrospinning. Additionally, the viscosity parameter is significant, because the optimum viscosity can only yield a targeting diameter of nanofibers for various applications. The viscosity and conductivity of HAP-HPC/PLA composite solutions were measured and mentioned in Table 2.

Table 2. Optimizing different parameters and its blending properties.

S.No.	Sample	Viscosity(cP)	Conductivity (μS/cm)	Flow Rate (1/mL)	Collector Rotation Speed (RPM)	Tip to Collector Distance (cm)	Voltage (kV)
1	HPC	63	-	0.5	700	12	15
				0.7	1000	15	21
				1	1500	17	25
2	PLA	4519	-	0.5	700	12	15
				0.7	1000	15	21
				1	1500	17	25
3	HPC/PLA	3687	0.0018	0.5	700	12	15
				0.7	1000	15	21
				1	1500	17	25
4	40:60	2487	0.0023	0.5	700	12	15
				0.7	1000	15	21
				1	1500	17	25
5	50:50	1413	0.0024	0.5	700	12	15
				0.7	1000	15	21
				1	1500	17	25
6	60:40	1198	0.0036	0.5	700	12	15
				0.7	1000	15	21
				1	1500	17	25
7	70:30	786.7	0.0041	0.5	700	12	15
				0.7	1000	15	21
				1	1500	17	25

Conductivity and dielectric constants depend on the input temperature, frequencies, and viscosity of the solution. The overall parameters and conductivity depend on the viscosity of the sample preparation. Voltage, RPM (rotations per minute), and conductivity are the main parameters to form thinner fiber diameters. Conductivity remains constant at low temperatures, and it increases only when the temperature is increased. The dielectric constant and conductivity study of the nanofiber scaffolds is shown in Figure 1. In this study, the conductivity (σ) and dielectric constants (ε) depend on the concentration of the HAP at different temperature and the frequency was studied. The conductivity (σ) and dielectric constant (ε) were found to be increasing with an increase in temperature (35 to 125 °C) and with an increase in HAP concentration.

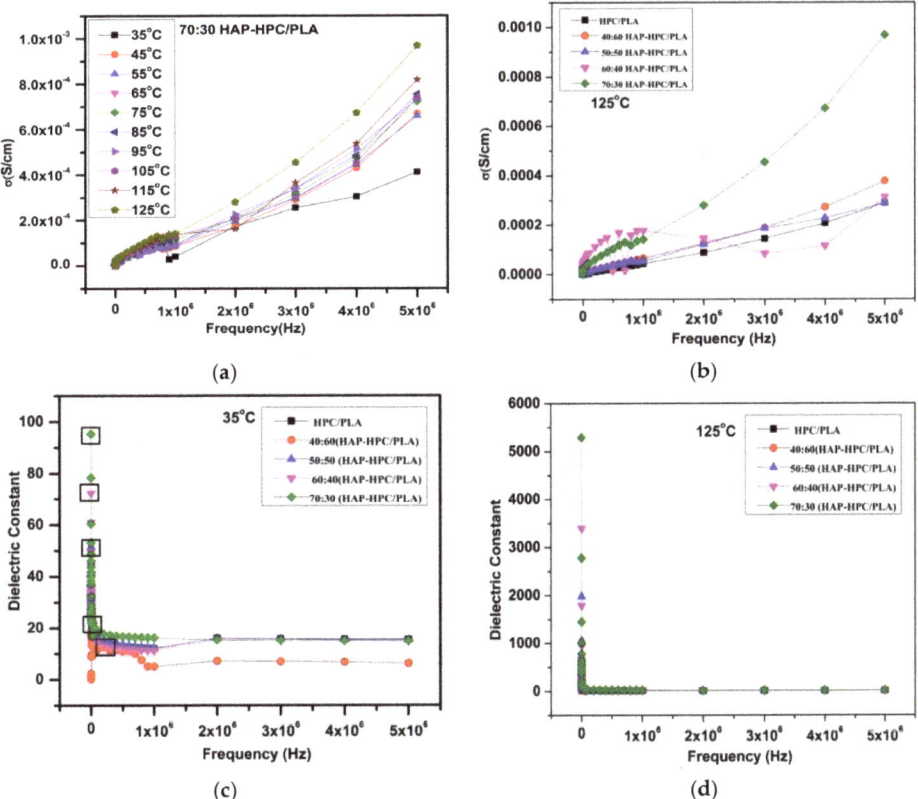

Figure 1. (a) Conductivity of HAP-HPC/PLA (70:30); (b) different ratios of HAP-HPC/PLA; (c) dielectric constant at 30 °C of 0% HAP and HAP-HPC/PLA; (d) dielectric constant at 125 °C of 0% HAP and HAP-HPC/PLA.

The dielectric constant decreases with increasing frequency and we obtain the maximum value of the dielectric constant 96 K for a 70:30 ratio of the nanofibrous mat at 5 kHz at room temperature. This is because of the fact that at lower frequency dipole moment of hydroxyl ions in HAP follow the variation of the field while at higher frequency those ions do not follow the variation of the field and the dielectric permittivity value decreases. Figure 1b shows the frequency dependence of the dielectric constant for different compositions at room temperature and at 125 °C. The dielectric constant increases with increasing the HAP content and we obtain the value of dielectric constant 5329 K at 5 kHz at 125 °C and at 89 K for RT which is due to the interfacial flow rate of the viscous solution during electrospinning. Similarly, the conductivity has a much stronger concentration dependence with respect to frequency. Hence, from the results observed, the values of both conductivity and dielectric constant for the 70:30 ratio were found to be increased in HAP-HPC/PLA composition. The percolation threshold limit of 70:30 ratio of HAP-HPC/PLA is flexible and hence applicable for the fabrication of mat.

3.3. FT-IR (Fourier Transform Infrared) Analysis of HAP-HPC/PLA Nanofibers

The FT-IR spectra of the fabricated HPC/PLA nanofibrous mat and HAP with various composite ratios of 40:60, 50:50, 60:40, and 70:30 have been illustrated in Figure 2A. The absorption band at 3351 cm^{-1} corresponds to a hydroxyl group in the pyranose unit of HPC. The absorption band at 2921 cm^{-1} appears due to the CH2 and CH stretching vibration. The absorption peaks at 1717 and 1422 cm^{-1} are attributed to the C=O stretch and –CH stretch, respectively. The absorption band appearing at 1079 cm^{-1} is formed due to C-O stretching vibration. The results confirm the presence of HPC in the composite nanofibrous mat.

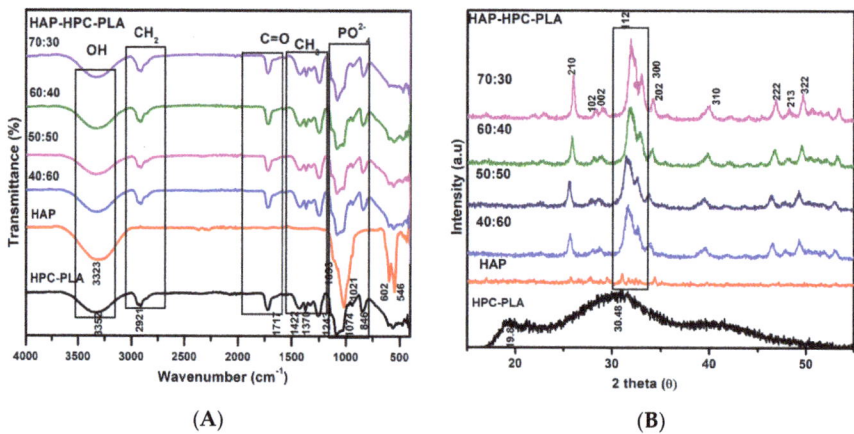

Figure 2. (**A**) FTIR and 2 (**B**) XRD (X-ray diffraction) analysis of HAP-HPC/PLA nanofibers of different ratios.

Figure 2A shows the broad peak at 3344 cm^{-1} corresponding to the OH stretch. Mainly, the prominent peaks appearing at 478 cm^{-1}, 555 cm^{-1}, 936 cm^{-1}, 1021 cm^{-1}, and 1081 cm^{-1} are characteristic bands assigned to PO_4^{3-} of HAP. The distinguishable peak at 936 cm^{-1} appears due to the asymmetric P-O stretching vibration of PO_4^{3-} bands. The sharp peaks at 478 cm^{-1} and 555 cm^{-1} correspond to the triply degenerate bending vibrations of PO_4^{3-} in HAP and these results further confirm the presence of HAP. Additionally, the characteristics peaks of HPC/PLA appear at 2931 cm^{-1} and 1437 cm^{-1} responsible for the asymmetric CH_2 and CH_3 stretch. The prominent peak at 1756 cm^{-1} identifies the carbonyl group. Eventually, FTIR spectroscopy results of Figure 2A confirmed that fabricated nanofibers contain HAP and HAP-PLA in all composite nanofibrous mats.

3.4. XRD Analysis of HAP-HPC/PLA Nanofibers

The phase analysis and purity of the fabricated HAP-HPC/PLA nanofibrous composite mats have been scanned using XRD studies which are shown in Figure 2B. The diffraction peaks at 19.8° and 30.48° are responsible for the HPC/PLA composite. Additionally, the diffraction peaks of HAP in HAP-HPC/PLA nanofibrous composite mats at 22.78°, 25.92°, 28.10°, 29.12°, 31.89°, 32.05°, 32.20°, 34.09°, 39.88°, 46.75°, 48.28°, 49.48°, 50.78°, and 51.23° are corresponding to (111), (002), (102), (210), (211), (112), (300), (202), (310), (222), (312), (213), (321), and (410) planes are confirmed. Hence both FTIR and XRD results confirmed the presence of pure HAP without any other secondary phases. Moreover, the triplet peak appearing at 31.89°, 32.05°, and 32.20° is due to an increase in the HAP concentration in the fabricated mats. The results suggest that the diffraction peak of HPC/PLA (Figure 2B) is slightly amorphous. Further, the intensity of the diffraction peaks is increasing with the addition of HAP to the HPC/PLA composite.

3.5. Mechanical Properties

As explained above, the smaller diameter of fibers with a definite orientation of the polymer chains has been observed in this study. The extended amorphous structures of the HPC/PLA and HAP-HPC/PLA nanofibers yield results as shown in Figure 3. The obtained results suggest that 70:30 ratio shows moderate tensile strength compared to the other ratios (Figure 3a). The results demonstrated that tensile strength and Young's modulus are found to be increased due to the alignment of the polymer along the axes of fibers, although the percentage of fracture is found to be decreasing, respectively. HAP-HPC/PLA nanofibrous mat with a composition ratio of 70:30 resulted in the high tensile strength of 9.53 MPa. The other composite ratios of 0:100, 40:60, 50:50, and 60:40 gave a tensile strength of 3.7 MPa, 4.88 MPa, 8.21 MPa, and 8.91 MPa, respectively. Further, Table 3 shows an increase in tensile strength as the percentage of HAP increases; hence maximum tensile strength was observed for "70:30" composition of HAP-HPC/PLA nanofibrous mat. Additionally, the percentage of fracture has decreased, and Young's modulus has increased, exhibiting high stiffness, regular chain orientation, and elevated mechanical strength to the nanofibers. The composite mat without HAP shows less Young's modulus and tensile strength. All these results clearly state that the HAP-HPC/PLA nanofibrous mat of composite ratio 70:30 shows an optimal increase in tensile strength (9.53 MPa) for a decreased diameter (110 ± 66 nm).

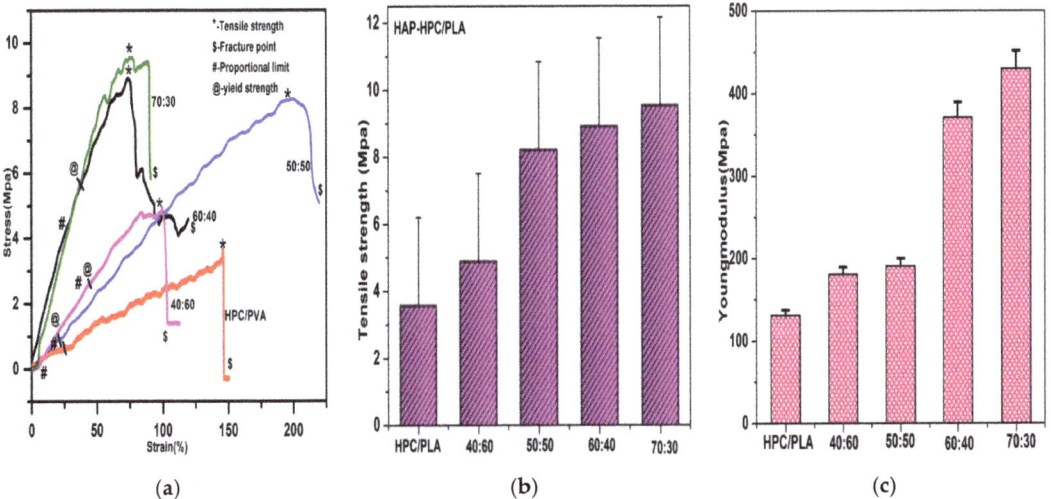

Figure 3. Correlation of (**a**) mechanical property (**b**) tensile strength (**c**) Young modulus.

Table 3. Mechanical properties of HAP-HPC/PLA.

S.No.	Sample	Proportional Limit (Pa)	Yield Strength (Mpa)	Tensile Strength (Mpa)	Young Modulus (MPa)	Fracture (%)
1	HPC/PLA	0.424	0.444	3.57	131.06	148
2	40:60	0.097	0.291	4.88	180.50	111
3	50:50	0.926	4.823	8.21	190.50	219
4	60:40	1.134	2.824	8.91	370.89	118
5	70:30	1.470	1.388	9.53	430.10	90

Structurally, all molecules of the nanofibers should be fully extended and aligned perfectly with the fiber axis. The secondary bonds of the molecular structure of the fibers help in the determination of the tensile strength. The high stiffness of the nanofibers can be obtained if the polymer chains are fully extended and oriented. Some polymer chains are defective due to the high molecular weight of the polymer leading to the high tenacity of the fiber. The mechanical strength of the nanofibrous mat is influenced by the size of the fiber diameter. This confinement effect is the main property for the improvement in the mechanical strength of the mat. Hence, the orientation and degree of alignment can be influenced by decreasing the diameter of the nanofiber [32,33].

3.6. X-ray Photoelectron Spectroscopy (XPS)

The results which are obtained from FT-IR, XRD and SEM (scanning electron microscopy) analysis suggest that HAP-HPC/PLA nanofibrous mat of composite ratio 70:30 is the desired and favorable composition, and the surface chemistry of the material was analyzed using XPS. The binding energy peaks which are available in the full survey spectrum confirm the presence of Ca2p (348.44), P2p (137.64), O1s (526.02) and C1s (287.22) in the fiber matrix (Figure 4).

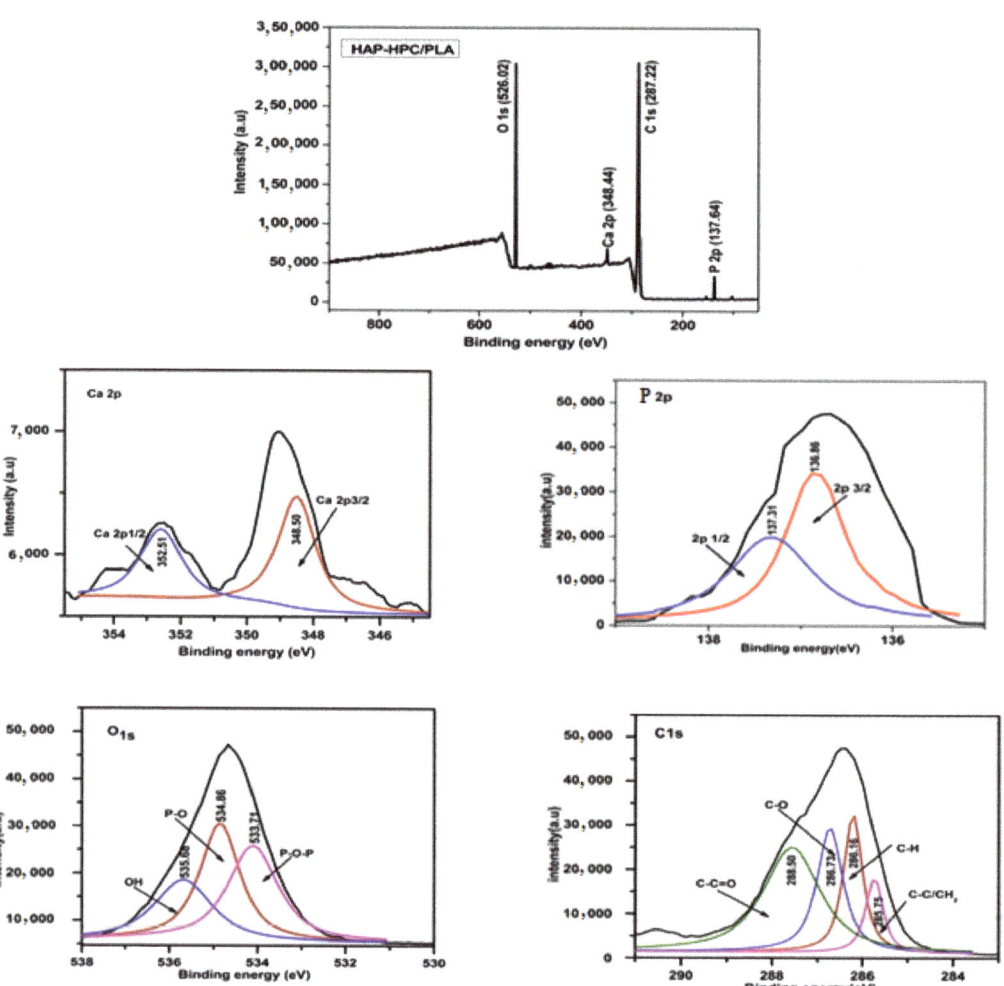

Figure 4. The XPS full survey for HAP-HPC/PLA nanofiber and core level spectrum of Ca 2p, P 2p, O 1s, and C1s.

In HAP, the main core level of Ca (2_P) appears at 348.44 eV region, and the two deconvoluted peaks such as Ca ($2_{P1/2}$) and Ca ($2_{P3/2}$) appear at 352.51 eV and 348.50 eV, respectively. Another core level of the P (2_P) exists at 137.31.1 eV and 136.86 eV which are responsible for P ($2_{P1/2}$) and $2(2_{P3/2}$), respectively [32,33]. The O1s appearing at 526.02 is deconvoluted into three elements which are present as P-O (535.68 eV), OH (534.86 eV), and P-O-P (533.71 eV), respectively. Additionally, Table 4 confirms the presence of C 1s of HPC and PLA appeared at 287.22 eV and its deconvoluted peak has four different elements at a different binding energy such as 288.50 eV (C-C=O), 286.73 eV (C-O), 286.16 eV (C-H), and 285.75 eV (C-CH$_3$), respectively. Therefore, XPS analysis confirms the nanofiber matrix surface chemistry and the Ca/P (1.67) ratio, which is specific to HAP [34].

Table 4. The binding energy of XPS spectra peak for HAP-HPC/PLA.

Sample	Ca 2P		P 2p		C1s	O1s	Ca/P Ratio
	$2p_{3/2}$	$2p_{1/2}$	$2p_{3/2}$	$2p_{1/2}$			
HAP-HPC/PLA	348.50	352.51	136.86	137.31	(a) 288.50-(C-C=O) (b) 286.73-(C-O) (c) 286.16-(C-H) (d) 285.75-(C-C/CH$_2$)	(a) 535.68-OH (b) 534.86-P-O (c) 533.71-P-O-P	1.66

3.7. SEM Morphology and EDAX

The morphology of the fabricated HAP-HPC/PLA electrospun nanofibrous mats was analyzed using SEM. Initially, the optimization study was performed with different parameters such as flow rate (0.5, 0.7, and 1 mL), collector rotation speed (700, 1000, and 1500 rpm), tip-to-collector distance (12, 15, and 17 cm), and voltage (15, 21, and 25 Kv (kilo volts)) as shown in Table 1. Accordingly, the samples were analyzed for the morphological studies and pore size measurements using SEM and image J software, respectively (Figure 5A).

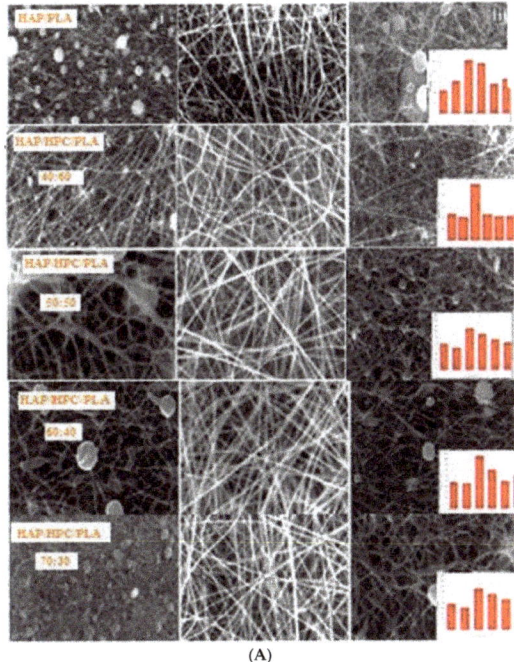

(A)

Figure 5. *Cont.*

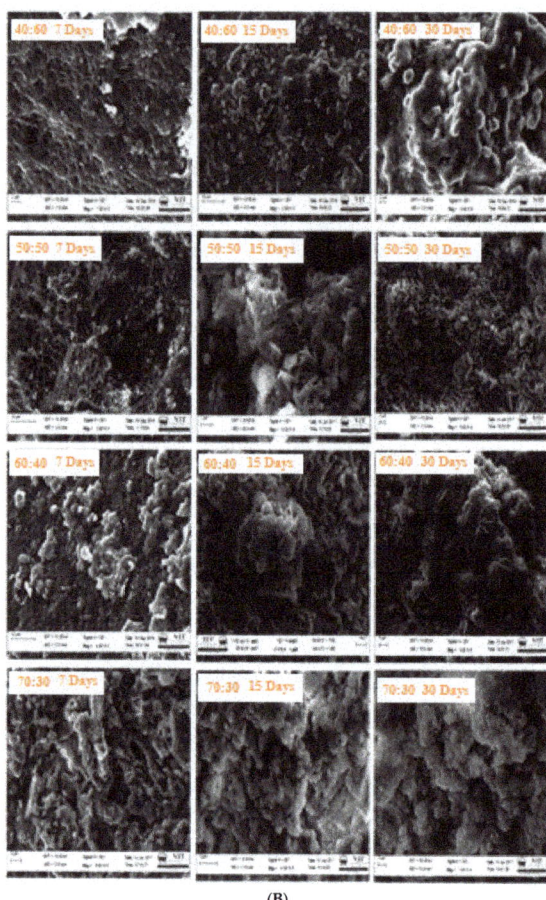

(B)

Figure 5. (**A**) SEM Morphology of electrospun nanofibers at various magnifications and fiber diameter (Insert) of various compositions. (**B**) SEM Morphology of SBF immersed electrospun nanofibers for various time intervals by in vitro bioactivity study.

Figure 5A represents the SEM micrograph for HAP-HPC/PLA composite nanofibrous mat of various compositions. The optimum parameters, i.e., a flow rate of 0.7 mL, a collector rotation speed of 1000 rpm, tip-to-collector distance of 15 cm, and voltage of 21 kV gave uniform nanofibers which were bead-free, porous, and non-woven amongst all the HAP-HPC/PLA blends. The other parameters resulted in the formation of improper HAP-HPC/PLA nanofibers. The Ca and P content in the HAP-HPC/PLA nanofibrous mat were determined using EDAX (energy dispersive analysis X-ray) analysis featured in Figure 6 and Table 5 respectively. The stoichiometric value of HAP is 1.67, which is on par with the stoichiometric Ca/P (calcium/phosphate) ratio of HAP. HAP-HPC/PLA nanofibrous mat of all ratios showed the same stoichiometric ratio of Ca and P content.

The HAP-HPC/PLA mat showed different diameter sizes for each composition ratio of HAP-HPC/PLA fabrication. The higher HAP-containing mat showed a smaller fiber diameter, while the mat with no HAP showed a higher fiber diameter. From Table 6, it was observed that the HAP-HPC/PLA nanofibrous mat of the composite ratio 70:30 showed a diameter of 110 ± 66 nm, whereas the mat of composite ratio 40:60 showed a fiber diameter of 277 ± 11 nm. Additionally, as stated earlier, the fiber diameter decreases with an increase in the concentration of HAP. The viscoelastic property may be the cause for the reduction in diameter size with an increase in HAP concentration since all parameters remained constant.

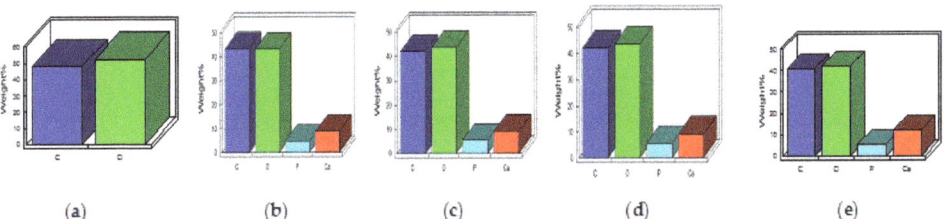

Figure 6. EDAX analysis of (**a**) HPC/PLA (**b**) 40:60 (**c**) 50:50 (**d**) 60:40 (**e**) 70:30 of HAP-HPC/PLA.

Table 5. The calculated Ca/P Ratio from EDAX Analysis.

S.No.	Sample	Ca/P Ratio
1	HPC/PLA	-
2	40:60	1.66
3	50:50	1.67
4	60:40	1.65
5	70:30	1.67

Table 6. The fiber pore size and diameter calculated from SEM images and solvent replacement method.

S.No.	Sample	Pore Size (μm)	Pore Diameter (nm)	Porosity (%)
1	HPC/PLA	4.54	355 ± 10	59.56
2	40:60 HAP-HPC/PLA	7.21	277 ± 11	62.97
3	50:50 HAP-HPC/PLA	9.23	133 ± 33	75.23
4	60:40 HAP-HPC/PLA	11.21	122 ± 33	89.45
5	70:30 HAP-HPC/PLA	15.54	110 ± 66	98.11

3.8. Pore Size and Porosity

It has been observed that the pore size of HAP-HPC/PLA nanofibrous mats increases with increasing HAP concentration owing to a decrease in fiber diameter. The nanofibrous mat of composite ratios 0:100, 40:60, 50:50, 60:40, and 70:30 ratios yielded pore sizes of 4.54, 7.21, 9.23, 11.21, and 15.54 μm, respectively. Favoring cell proliferation and migration, pore size plays a major role in tissue engineering applications [35]. Factors such as cell growth, migration, and nutrient supply completely depend on the pore size and their interconnectivity. Cell migration becomes limited if the pore size is less, and also, decreases in the surface area may lead to limited cell adhesion if the pore size is very high. The interconnection of pores results in providing space for vasculature, required to promote new bone formation. Prior studies suggest that microporosity provides bone growth on scaffolds and it can increase the acting surface area of the mat for protein adsorption, it may also provide osteoblast attachment points during biocompatibility evaluation [36]. The pore size of the mat obtained from SEM images was processed using image J software and Porosity was measured by solvent replacement method. The results obtained from both studies are tabulated in Table 6.

3.9. Bioactivity

The dissolution and precipitation process assists in the formation of an apatite layer on the surface of HAP which is essential for a good calcium-based biomaterial. HAP from the nanofibrous mat releases Ca^{2+} ions when it is immersed in the SBF solution, hence increasing apatite formation in the surrounding body fluid due to this ionic activity.

The release of Ca^{2+} ions may lead to an increase in positive charge on the surface of hydroxyapatite in the nanofibrous mat [37]. Further, the calcium-rich surface interacts with the PO_4^{3-} ions present in SBF. The calcium and phosphate ions migrate onto the surface and induce precipitation of apatite on the HAP surface of the nanofibrous mat. This formed apatite gets stabilized by the crystallization process and forms a bone analog. In addition, various ions play a key role in the formation of the apatite layer due to the ionic interaction taking place between the mat's surface and the ions present in SBF.

The apatite layer formation on the mat's surface was shown in Figure 5B. The apatite layer formation was obtained by immersing the Mat in SBF solution at different time intervals such as 7, 15, and 30 days. The SEM images exhibit the formation of apatite layers on the nanofibrous mat's surface as least after 7 days of immersion, and it was found that the quantity of apatite formation increases for 15 and 30 days of immersion. Further, the HAP-HPC/PLA nanofibrous mat with a composition ratio of 70:30 shows higher apatite layer formation after 30 days of immersion compared to the other composition ratios. The higher apatite layer formation is due to the high immersion time in SBF.

3.10. Degradation

The biodegradation of polymers attacks the anhydride, ester, amide groups, or even enzymatically cleaves the structural or functional bonds. The molecular weight and structure, composition, crystallinity, and presence of cross-linking in a polymer affect the polymer degradation. In addition, the degradation of a polymer is influenced by the diffusion coefficient of water in the polymer matrix, the hydrolysis rate constant of the ester bond, the diffusion coefficient of the chain fragments within the polymeric matrix, and the solubility of the degradation product [38]. The results are demonstrated in Figure 7a. The fast degradation is due to a higher concentration of polymers (40:60, 50:50) with low molecular weight and the hydrolytic reaction is high in the early stage of degradation. By increasing the concentration of HAP, the ionic exchange (ca^{2+} and PO_4^{3-}) is very high between the stimulated body fluid and nanofibrous mat when compared with a low concentration of HAP. When the ions entered into the pores, the mineralization process also occurred predominantly. Once the mineralization process occurred, the interchange of ions might be less, and the degradation was reduced.

Figure 7b, the FT-IR analysis confirms by increasing the days of immersion exhibits the polymer decrease that shows the degradation rate. After 4 days of degradation, the nanofiber breakage was obtained and there is no morphology change was observed. In 7 days of degradation, the result confirms that the surface of pore fibers was covered due to the swelling effect. In 15 and 30 days of immersion shows that interconnected fiber shape was merged together and melted during the degradation process. Figure 7c represents the SEM analysis of the SBF-immersed scaffolds at various time intervals. After 4 days of immersion, the nanofibrous scaffold starts to degrade and there was a slight change in the morphology was observed. At 7 days of degradation, the results further confirmed that the pore in the fibers was covered due to the swelling effect. At 15 and 30 days of immersion, the complete coverage of apatite was clearly visible on the surface of the scaffold. From the figure, it was clearly confirmed the process of degradation rate with respect to various time intervals. Hence from the observation, the stability of the scaffold for 4 weeks is confirmed.

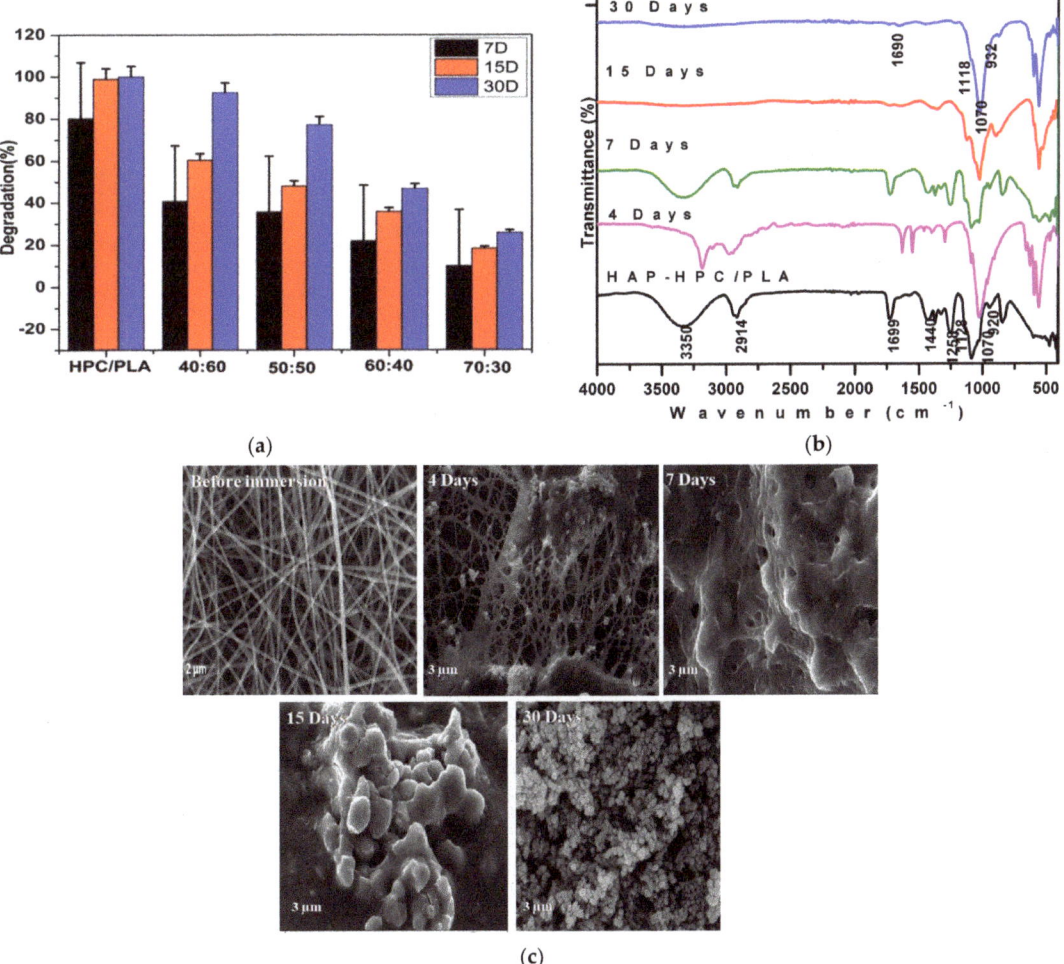

Figure 7. (**a**). Degradation rate of different composition ratios. (**b**). FTIR analysis of nanofibrous mat at different intervals of SBF immersion. (**c**). SEM analysis of nanofibrous mat at different intervals of SBF immersion.

3.11. Cytotoxic Analysis

The microscopic images of the osteosarcoma (MG-63) cells seeded in HAP-HPC/PLA nanofibrous mat of different ratios (40:60, 50:50, 60:40, and 70:30) are shown in Figure 8a–e. Different patterns of cell proliferation have been observed within two days of culture. Cell proliferation differs at 24 h and 48 h, meaning cell viability is high even after 48 h when compared to 24 h. HPC/PLA nanofibrous mat shows moderate results in cell viability at 48 h (Figure 8a*) than at 24 h (Figure 8a). Moreover, these microscopic images clearly conclude that complete cell proliferation was observed after 48 h for a 70:30 composition ratio (Figure 8a*). This may be due to the infiltration of HAP with HPC/PLA composite, hence showing better performance in cell proliferation. The proliferation rate of cells in HAP-HPC/PLA nanofibrous mat is much higher at a 70:30 ratio compared to 40:60, 50:50, and 60:40 composition ratios for both 24 and 48 h cultures, respectively.

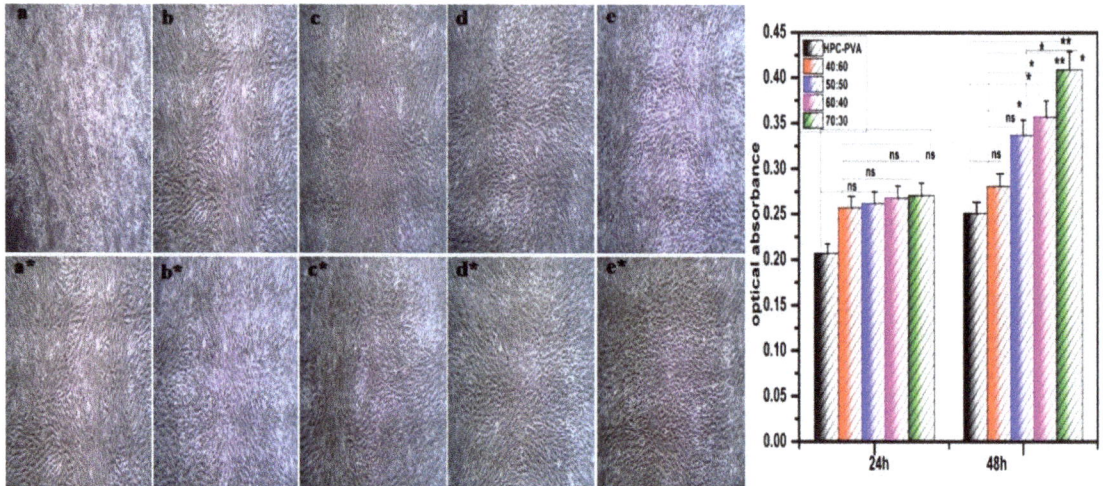

Figure 8. Cytotoxicity of electrospun nanofibrous (a–e) (24 h) and (a*–e*) (48 h) (a) HPC/PLA (b) 40:60 (c) 50:50 (d) 60:40 (e) 70:30 of HAP-HPC/PLA and its significant difference.

Hydrophobicity of the mat increases due to the cross-linking phenomenon and hence exhibiting cell adhesion as well as improved cellular functions. HPC and HAP usually show high biocompatibility and mechanical properties, therefore the cellulose derivatives are used as an important biomaterial for the fabrication of tissue engineering scaffolds. These biodegradable scaffolds are mainly used in tissue regeneration applications since the biodegradation rate of the scaffolds matches the biological process which takes place in tissue regeneration. Often, slow biodegradable scaffolds are preferred for regeneration applications due to minimal risks [39–41]. Interestingly, cellulose is an important biomaterial candidate for the design of tissue engineering scaffolds. The cells attach, grow and stimulate tissue growth depending on the stability of scaffolds in the body fluid; hence, scaffolds should be insoluble in water.

Figure 8 shows the results of the MTT assay at 24 h and 48 h at various composition ratios of 40:60, 50:50, 60:40, and 70:30 of HAP-HPC/PLA nanofibrous mat. The mat of ratio 70:30 is found to have the highest cell proliferation rate compared to other ratios at both 24 and 48 h. The electrospun nanofiber at 70:30 ratios showed a significant difference with $p < 0.01$ (Figure 8) levels in the cell proliferation for 48 h, respectively. The obtained results confirm that the cells were attached and proliferated on HAP-HPC/PLA nanofibrous mat. This may be due to the presence of β-glucose linkages in HPC. The β-glucose linkage of carbohydrate derivatives plays a major role in cell metabolism, and it activates the formation of HAP. High cell proliferation was achieved at the highest concentration of HAP in the HAP-HPC/PLA nanofibrous mat, hence proving the fact that this composition is non-toxic and significant material for regeneration and the rejuvenation of bone tissues.

3.12. In Vivo Study

3.12.1. X-Ray Radiology Results

Figure 9 shows the X-ray examination of with and without implantation of the HAP-HPC-PLA nanofibrous composite. Instead of new bone formation, the hollow void space was observed in the defected area and is clearly visible in Figure 9A,C. Further, the results confirmed that there was no new bone formation at 4 and 8 weeks of implantation. Figure 9B displays, no osteogenesis, and bone formation were observed at 4 weeks of HAP-HPC-PLA implantation. The border of the unfilled void with no osteogenesis was observed at 8 weeks of implantation with HAP-HPC-PLA nanofibrous mat (Figure 9D). When compared with

4 weeks of implantation, 8 weeks of implantation confirmed the new bone formation. Further study is in progress to study the effect of new bone formation with respect to different ratios of HAP in the scaffolds.

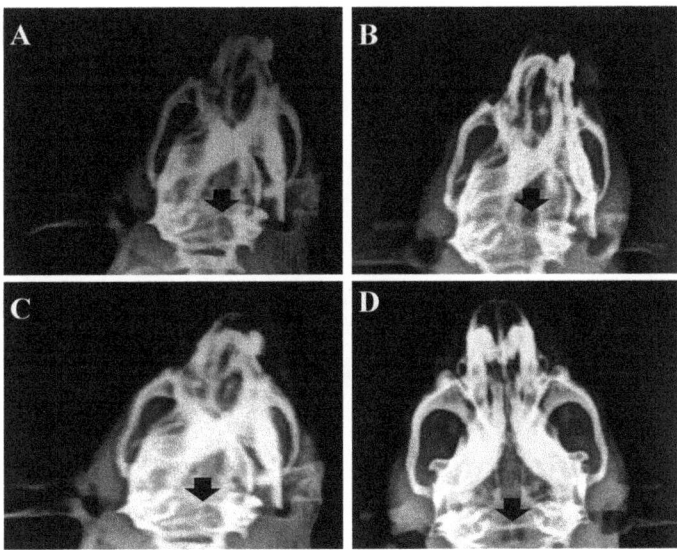

Figure 9. X-ray images of (**A**) without implantation at 4 weeks (**B**) HAP-HPC-PLA at 4 weeks (**C**) without implantation at 8 weeks (**D**) HAP-HPC-PLA at 8 weeks.

3.12.2. Histological Analysis

Figure 10 illustrates the highlights of the study of animal experiments. The histological section of in vivo studies with and without HAP-HPC-PLA polymer composite on rat skull defected area. The results demonstrated that the scaffolds promoted bone bonding activity at 4 and 8 weeks of implantation. The results of radiology and histology analysis indicated that this scaffold facilitated bone formation in the defects with excellent potential in bone defect repair.

Figure 10B shows the development of new cells and blood vessels at 4 weeks of HAP-HPC-PLA nanofiber implantation. When compared with 4 weeks intervals, 8 weeks of implantation of HAP-HPC-PLA nanofibers showed enhanced cell proliferation such as osteoblast, osteocyte with lacunae and osteoclast (Figure 10B1). The new bone formation was started to grow at 8 weeks of implantation. Architectural modification for scaffold such as pore size and fiber diameter also seems necessary for osteoblasts to favorably attach and grow on the hybrid scaffolds to substitute collagen sponge. Such scaffolds were reported to degrade in vivo after 2–6 months after their implantation [42]. Additionally, Pektok et al. discussed the use of scaffolds with better healing properties in vivo. They showed that faster extracellular matrix formation was achieved with the decomposition of nanofibers grafts [43]. So, these nanofibers with excellent healing properties can be applied for biomedical applications. Both in vitro and in vivo studies verified that these novel layered scaffolds can effectively deliver growth factors with better cell migration in a controlled manner for bone repair by promoting the healing process.

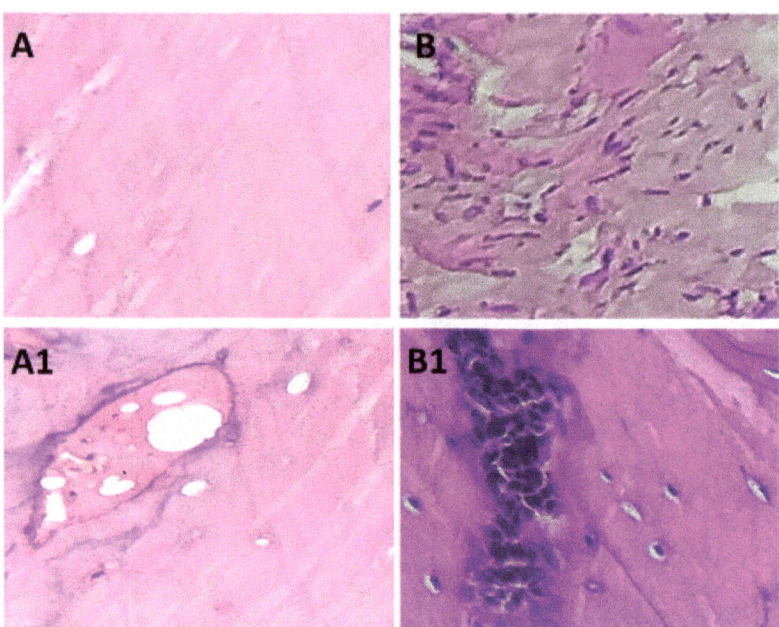

Figure 10. Histological images of (**A**) without implantation at 4 weeks (**B**) HAP-HPC-PLA at 4 weeks (**A1**) without implantation at 8 weeks (**B1**) HAP-HPC-PLA at 8 weeks.

4. Conclusions

Bone tissue engineering applications require fibrous mats, which have desired properties such as proper chemical integrity and crosslinking efficiency, and biodegradable properties to interact in the void space of the human native tissue without dissolving. Addressing porosity and optimum interconnectivity, optimum fiber diameter is required to ensure the necessary infiltration of cells and nutrients. The current study optimized the parameters required to produce the nanofibrous mat with the required porosity in terms of interconnectivity, mechanical property, bioactivity, and biocompatibility. Different concentrations of HAP were chosen to fabricate the nanofibrous mat resulting in the observation that the composite ratio of 70:30 exhibits all the desired properties required to be considered as an ideal biomaterial.

✓ The XRD results confirm the existence of HAP in the presence of a polymeric network and it was found that the triplet peak at 31.89°, 32.05°, and 32.20° increases with an increase in the HAP concentration in the fibrous mat.

✓ The mechanical property of 9.53 Mpa was obtained for the optimized composition with a high rate of HCA formation on SBF immersion, this may be due to the interconnected polymeric network and porosity of the sample which was confirmed favorable for cellular activity.

✓ The retention of the Ca/P ratio of HAP in the polymeric network was analyzed by XPS analysis. Finally, the biocompatibility evaluation on the MG-63 osteoblast cell line was conducted for 24 h and 48 h which deemed the material fit for biomedical applications. All the compositions revealed enhanced cell proliferation at 48 h of duration.

✓ In vivo animal study confirmed the effective bone formation at the 8th week of implantation of HAP-HPC-PLA grafted in the defective area with more cell differentiation when compared with the 4th week of implantation.

Hence, from the above study, the fabricated nanofibrous mat of this composition ratio is found to be the desirable type for bone tissue engineering applications such as bone void filling, repair of bone damage, and in vitro and in vivo bone disease modeling.

Author Contributions: Conceptualization, methodology and writing.; U.V. and S.M.S.; formal analysis and data curation.; T.M.S. resources, R.R. and J.G. All authors have read and agreed to the published version of the manuscript.

Funding: APC was funded by Vellore Institute of Technology.

Institutional Review Board Statement: The animal study protocol was approved by the Institutional Animal Ethical Committee of Vellore Institute of Technology (VIT/IAEC/14/Nov5/47 dated November 5, 2017)." for studies involving animals.

Acknowledgments: The authors would like to thank the management of VIT, Vellore, Tamil Nadu for rendering the necessary laboratory and characterization facilities and financial support for this manuscript.

Conflicts of Interest: All the authors declare no conflict of interest.

References

1. Hutmacher, D.W. Scaffolds in tissue engineering bone and cartilage. *Biomaterials* **2000**, *21*, 2529–2543. [CrossRef] [PubMed]
2. Zhu, B.; Lu, Q.; Yin, J.; Hu, J.; Wang, Z. Alignment of osteoblast-like cells and cell-produced collagen matrix induced by nanogrooves. *Tissue Eng.* **2005**, *11*, 825–834. [CrossRef] [PubMed]
3. Wahl, D.A.; Sachlos, E.; Liu, C.; Czernuszka, J. Controlling the processing of collagen-hydroxyapatite scaffolds for bone tissue engineering. *J. Mater. Sci. Mater. Med.* **2007**, *18*, 201–209. [CrossRef] [PubMed]
4. Teo, W.E.; Ramakrishna, S. A Review on Electrospinning Design and Nanofibre Assemblies. *Nanotechnology* **2006**, *17*, 89–106. [CrossRef]
5. Causa, F. A multi-functional scaffold for tissue regeneration: The need to engineer a tissue analogue. *Biomaterials* **2007**, *28*, 5093–5099. [CrossRef]
6. Chemar Huntley, J.; Kristy Crews, D.; Michael Curry, L. Chemical Functionalization and Characterization of Cellulose Extracted from Wheat Straw Using Acid Hydrolysis Methodologies. *Int. J. Polym. Sci.* **2015**, *2015*, 293981. [CrossRef]
7. Rodríguez, K.; Renneckar, S.; Gatenholm, P. Biomimetic calcium phosphate crystal mineralization on electrospun cellulose-based scaffolds. *ACS Appl. Mater. Interfaces* **2011**, *3*, 681–689. [CrossRef]
8. Samadian, H.; Salehi, M.; Farzamfar, S.; Vaez, A.; Ehterami, A.; Sahrapeyma, H.; Goodarzi, A.; Ghorbani, S. In vitro and in vivo evaluation of electrospun cellulose acetate/gelatin/hydroxyapatite nanocomposite mats for wound dressing applications. *Artif. Cells Nanomed. Biotechnol.* **2018**, *46*, 964–974. [CrossRef]
9. Bäckdahl, H.; Helenius, G.; Bodin, A.; Nannmark, U.; Johansson, B.R.; Risberg, B.; Gatenholm, P. Mechanical properties of bacterial cellulose and interactions with smooth muscle cells. *Biomaterials* **2006**, *27*, 2141–2149. [CrossRef]
10. Trivedi, M.K. Influence of Biofield Treatment on Physicochemical Properties of Hydroxyethyl Cellulose and Hydroxypropyl Cellulose. *J. Mol. Pharm. Org. Process. Res.* **2015**, *3*, 126. [CrossRef]
11. Sarode, A.L.; Malekar, S.A.; Cote, C.; Worthen, D. Hydroxypropyl cellulose stabilizes amorphous solid dispersions of the poorly water soluble drug felodipine. *Carbohydr. Polym.* **2014**, *4*, 512–519. [CrossRef] [PubMed]
12. Li, H.; Gong, G.; Gong, T. Permeability characterization of hydroxypropylcellulose/ polyacrylonitrile blend membranes. *e-Polymers* **2013**, *13*, 15. [CrossRef]
13. Gonzalez, E.; Shepherd, L.M.; Saunders, L.; Frey, M.W. Surface Functional Poly(lactic Acid) Electrospun Nanofibers for Biosensor Applications. *Materials* **2016**, *9*, 47. [CrossRef] [PubMed]
14. Sekiya, N.; Ichioka, S.; Terada, D.; Tsuchiya, S.; Kobayashi, H. Efficacy of a poly glycolic acid (PGA)/collagen composite nanofibre scaffold on cell migration and neovascularisation in vivo skin defect model. *J. Plast. Surg. Hand Surg.* **2013**, *47*, 498–502. [CrossRef] [PubMed]
15. Lao, L.; Wang, Y.; Zhu, Y.; Zhang, Y.; Gao, C. Poly(lactide-co-glycolide)/hydroxyapatite nanofibrous scaffolds fabricated by electrospinning for bone tissue engineering. *J. Mater. Sci. Mater. Med.* **2011**, *22*, 1873–1884. [CrossRef] [PubMed]
16. Dhivya, A.; Mary Stella, S.; Arunai Nambiraj, N.; Vijayalakshmi, U. Development of mechanically compliant 3D composite scaffolds for bone tissue engineering applications. *J. Biomed. Mater. Res. Part A* **2018**, *106*, 3267–3274. [CrossRef]
17. El-Arnaouty, M.B.; Eid, M.; Salah, M.; El-Sayed Hegazy, A. Preparation and Characterization of Poly Vinyl Alcohol/Poly Vinyl Pyrrolidone/Clay Based Nanocomposite by Gamma Irradiation. *J. Macromol. Sci. Part A* **2012**, *49*, 1041–1051. [CrossRef]
18. Kulkarni, M.B.; Mahanwar, P.A. Effect of Methyl Methacrylate–Acrylonitrile-Butadiene–Styrene (MABS) on the Mechanical and Thermal Properties of Poly (Methyl Methacrylate) (PMMA)-Fly Ash Cenospheres (FAC) Filled Composites. *J. Miner. Mater. Charact. Eng.* **2012**, *11*, 365–383. [CrossRef]
19. Uma Maheshwari, S.; Samuel, V.K.; Nagiah, N. Fabrication and evaluation of (PVA/HAp/PCL) bilayer composites as potential scaffolds for bone tissue regeneration application. *Ceram. Int.* **2014**, *40*, 8469–8477. [CrossRef]

20. Thanh, D.T.M.; Trang, P.T.T.; Huong, H.T.; Nam, P.T.; Phuong, N.T.; Trang, N.T.T.; Hoang, T. Fabrication of poly (lactic acid)/hydroxyapatite (PLA/HAp) porous nanocomposite for bone regeneration. *Int. J. Nanotechnol.* **2015**, *12*, 391. [CrossRef]
21. Martin-Pastor, M.; Stoyanov, E. Mechanism of interaction between hydroxypropyl cellulose and water in aqueous solutions: Importance of polymer chain length. *J. Appl. Polym. Sci.* **2020**, *58*, 1632–1641. [CrossRef]
22. Sundarrajan, S.; Ramakrishna, S. Green Processing of a Cationic Polyelectrolyte nanofibers in the Presence of Poly(vinyl alcohol). *Int. J. Green Nanotechnol.* **2011**, *3*, 244–249. [CrossRef]
23. Chuenjitkuntaworn, B.; Inrung, W.; Damrongsri, D.; Mekaapiruk, K.; Supaphol, P.; Pavasant, P. Polycaprolactone/hydroxyapatite composite scaffolds: Preparation, characterization, and in vitro and in vivo biological responses of human primary bone cells. *J. Biomed. Mater. Res. Part A* **2010**, *94*, 241–251. [CrossRef] [PubMed]
24. Priyadarshini, B.; Vijayalakshmi, U. Development of cerium and silicon co-doped hydroxyapatite nanopowder and its in vitro biological studies for bone regeneration applications. *Adv. Powder Technol.* **2018**, *29*, 2792–2803. [CrossRef]
25. Tighzert, W.; Habi, A.; Ajji, A.; Sadoun, T.; Daoud, F.B.-O. Fabrication and Characterization of Nanofibers Based on Poly(lactic acid)/Chitosan Blends by Electrospinning and Their Functionalization with Phospholipase A1. *Fibers Polym.* **2017**, *18*, 514–524. [CrossRef]
26. Mary Stella, S.; Vijayalakshmi, U. Influence of chemically modified Luffa on the preparation of nanofiber and its biological evaluation for biomedical applications. *J. Biomed. Mater. Res. Part A* **2018**, *107*, 610–620. [CrossRef]
27. Gopi, D.; Karthika, A.; Nithiya, S.; Kavitha, L. In vitro biological performance of minerals substituted hydroxyapatite coating by pulsed electrodeposition method. *Mater. Chem. Phys.* **2014**, *144*, 75–85. [CrossRef]
28. Rama, M.; Vijayalakshmi, U. Influence of silk fibroin on the preparation of nanofibrous scaffolds for the effective use in osteoregenerative applications. *J. Drug Deliv. Sci. Technol.* **2021**, *61*, 102182. [CrossRef]
29. Kokubo, T.; Kushitani, H.; Sakka, S.; Kitsugi, T.; Yamamuro, T. Solutions able to reproduce in vivo surface-structure changes in bioactive glass-ceramic A-W3. *J. Biomed. Mater. Res.* **1990**, *24*, 721–734. [CrossRef]
30. Yu, J.X. Preparation and *In Vitro* Degradation of PLLA Super Fiber Membrane. *Adv. Mater. Res.* **2013**, *750–752*, 221–223. [CrossRef]
31. Yin, Z.; Chen, X.; Song, H.; Hu, J.; Tang, Q.; Zhu, T.; Shen, W.; Chen, J.; Liu, H.; Heng, B.C.; et al. Electrospun scaffolds for multiple tissues regeneration in vivo through topography dependent induction of lineage specific differentiation. *Biomaterials* **2015**, *44*, 173–185. [CrossRef] [PubMed]
32. Parida, C.; Das, S.C.; Dash, S.K. Mechanical Analysis of Bio Nanocomposite Prepared from Luffa cylindrica. *Procedia Chem.* **2012**, *4*, 53–59. [CrossRef]
33. He, B.; Tan, D.; Liu, T.; Wang, Z.; Zhou, H. Study on the Preparation and Anisotropic Distribution of Mechanical Properties of Well-Aligned PMIA Nanofiber Mats Reinforced Composites. *J. Chem.* **2017**, *2017*, 8274024. [CrossRef]
34. Nagakane, K.; Yoshida, Y.; Hirata, I.; Fukuda, R.; Nakayama, Y.; Shirai, K.; Ogawa, T.; Suzuki, K.; Meerbeek, V.B.; Okazaki, M. Analysis of Chemical Interaction of 4-MET with Hydroxyapatite Using XPS. *Dent. Mater. J.* **2006**, *25*, 645–649. [CrossRef] [PubMed]
35. Salinas, A.J.; Vallet-Regı, M. Bioactive ceramics: From bone grafts to tissue engineering. *RSC Adv.* **2013**, *3*, 11116–11131. [CrossRef]
36. Mary Stella, S.; Vijayalakshmi, U. Influence of Polymer Based Scaffolds with Genipin Cross Linker for the Effective (Pore Size) Usage in Biomedical Applications. *Trends Biomater. Artif. Organs* **2018**, *32*, 83–92.
37. Dridi, A.; Riahi, K.Z.; Somrani, S. Mechanism of apatite formation on a poorly crystallized calcium phosphate in a simulated body fluid (SBF) at 37 °C. *J. Phys. Chem. Solids* **2021**, *156*, 110122. [CrossRef]
38. Dong, Y.; Yong, T.; Liao, S.; Chan, C.K.; Stevens, M.M.; Ramakrishna, S. Distinctive degradation behaviors of electrospun polyglycolide, poly(DL-lactide-co-glycolide), and poly(L-lactide-co-epsilon-caprolactone) nanofibers cultured with/without porcine smooth muscle cells. *Tissue Eng. Part A* **2010**, *16*, 283–298. [CrossRef]
39. Gomes, S.; Rodrigues, G.; Martins, G.G.; Roberto, M.A.; Mafra, M.; Henriques, C.M.R.; Silva, J.C. In vitro and in vivo evaluation of electrospun nanofibers of PCL, chitosan and gelatin: A comparative study. *Mater. Sci. Eng. C* **2015**, *46*, 348–358. [CrossRef]
40. Liu, T.; Ding, X.; Lai, D.; Chen, Y.; Zhang, R.; Chen, J.; Feng, X.; Chen, X.; Yang, X.; Zhao, R.; et al. Enhancing in vitro bioactivity and in vivo osteogenesis of organic–inorganic nanofibrous biocomposites with novel bioceramics. *J. Mater. Chem. B* **2014**, *2*, 6293–6305. [CrossRef]
41. Balusamy, B.; Senthamizhan, A.; Uyar, T. In vivo safety evaluations of electrospun nanofibers for biomedical applications. In *Electrospun Materials for Tissue Engineering and Biomedical Applications*, 1st ed.; Woodhead Publishing: Cambridge, UK, 2017; pp. 101–113. [CrossRef]
42. Lopresti, F.; Pavia, F.C.; Vitrano, I.; Kersaudy-Kerhoas, M.; Brucato, V.; Carrubba, V.L. Effect of hydroxyapatite concentration and size on morpho-mechanical properties of PLA-based randomly oriented and aligned electrospun nanofibrous mats. *J. Mech. Behav. Biomed. Mater.* **2020**, *101*, 103449. [CrossRef] [PubMed]
43. Pektok, E.; Nottelet, B.; Tille, J.-C.; Gurny, R.; Kalangos, A.; Moeller, M.; Walpoth, B.H. Degradation and Healing Characteristics of Small-Diameter Poly(ε-Caprolactone) Vascular Grafts in the Rat Systemic Arterial Circulation. *Circulation* **2008**, *118*, 2563–2570. [CrossRef] [PubMed]

Disclaimer/Publisher's Note: The statements, opinions and data contained in all publications are solely those of the individual author(s) and contributor(s) and not of MDPI and/or the editor(s). MDPI and/or the editor(s) disclaim responsibility for any injury to people or property resulting from any ideas, methods, instructions or products referred to in the content.

Article

A Novel Approach for the Manufacturing of Gelatin-Methacryloyl

David Grijalva Garces [1,2], Carsten Philipp Radtke [2] and Jürgen Hubbuch [1,2,*]

1. Institute of Functional Interfaces, Karlsruhe Institute of Technology, 76344 Eggenstein-Leopoldshafen, Germany
2. Institute of Process Engineering in Life Sciences Section IV: Biomolecular Separation Engineering, Karlsruhe Institute of Technology, 76131 Karlsruhe, Germany
* Correspondence: juergen.hubbuch@kit.edu

Abstract: Gelatin and its derivatives contain cell adhesion moieties as well as sites that enable proteolytic degradation, thus allowing cellular proliferation and migration. The processing of gelatin to its derivatives and/or gelatin-containing products is challenged by its gelation below 30 °C. In this study, a novel strategy was developed for the dissolution and subsequent modification of gelatin to its derivative gelatin-methacryloyl (GelMA). This approach was based on the presence of urea in the buffer media, which enabled the processing at room temperature, i.e., lower than the sol–gel transition point of the gelatin solutions. The degree of functionalization was controlled by the ratio of reactant volume to the gelatin concentration. Hydrogels with tailored mechanical properties were produced by variations of the GelMA concentration and its degree of functionalization. Moreover, the biocompatibility of hydrogels was assessed and compared to hydrogels formulated with GelMA produced by the conventional method. NIH 3T3 fibroblasts were seeded onto hydrogels and the viability showed no difference from the control after a three-day incubation period.

Keywords: biomaterials; cell culture; fibroblasts; gelatin; GelMA; hydrogel; tissue engineering

1. Introduction

Hydrogels are widely used as scaffolds for tissue engineering (TE) as the polymeric network resembles the extracellular matrix (ECM) within native tissue. For this purpose, hydrophilic polymers are covalently or physically crosslinked in order to create a water-swollen network [1,2]. The building blocks for hydrogel formulation can be purified from native tissue or produced synthetically. Naturally derived polymers show higher biocompatibility and can promote cellular adhesion and proliferation, whereas synthetic polymers have the advantage of low batch-to-batch variation compared to polymers extracted from natural sources [3,4]. The composition of the network influences cellular behavior not only by biochemical cues but also by the resulting physical structure and mechanical properties of the surrounding matrix [5,6].

Gelatin is derived from collagen, the most abundant protein in the native ECM, by acidic or alkaline treatment. The use of gelatin as a cell culture carrier is limited by the sol–gel transition of gelatin solutions at physiological temperatures [7]. In order to increase the stability of gelatin hydrogels at higher temperatures, the protein backbone can be chemically crosslinked. One approach for this purpose is the functionalization of gelatin with methacrylamide and methacrylate residues [8]. The resulting product, gelatin methacryloyl, also known as GelMA, retains sites for cell adhesion as well as for enzymatic degradation that are present in collagen and gelatin [9]. The use of photo-crosslinkable GelMA has gained popularity in the fields of TE and biofabrication [9]. The versatile GelMA has been used in cancer research [10], drug encapsulation and delivery [11], and in bioprinting. Therefore, bioinks containing solely GelMA are used [12], as well as bioink blends with other proteins like collagen [13] or in combination with polysaccharides such as alginate and gellan gum [14,15].

To meet the increasing demand for biomaterials required for TE applications, the manufacturing of hydrogel components requires a thorough understanding of the process parameters to reach a certain quality attribute [16]. In the case of GelMA, an aspect of particular interest is the degree of functionalization (DoF) of the biopolymer as the functional residues are relevant during photo-crosslinking. Since GelMA was first introduced by Van den Bulcke [8] over two decades ago, lab-scale studies have shown increasing reproducibility and efficiency during GelMA manufacturing, where aqueous gelatin in aqueous solution reacts with methacrylic anhydride (MAA). Lee [17] and Shirahama [18] studied the effect of the pH of the protein solution on the produced GelMA. As reaction by-products are acidic, the pH of the solution decreases, thus inhibiting the progress of the reaction. By introducing a carbonate-bicarbonate (CB) buffer at the isoelectric point (IEP) of type A gelatin, these methods showed a reduction of MAA excess over free amino groups from 30- to 2.2-fold. The study by Shirahama [18] also observed the effect of reaction temperature, which proved to have no influence on the DoF. This study was limited to the range of 35 to 50 °C as a lower temperature would allow the solution to form a physical gel, thus leading to inefficient mixing and distribution of reactants. To the best of our knowledge, no studies have reported the possibility of GelMA production at temperatures lower than 35 °C.

In this work, we propose a GelMA manufacturing process including a chaotropic salt, i.e., urea, as a buffer component at pH 9 that enables the processing of gelatin at room temperature. GelMA with various degrees of functionalization was produced and used as the basis for hydrogel formulation. Furthermore, the effect of polymer concentration and DoF on the elasticity and swelling behavior of hydrogels was characterized. Additionally, the suitability of GelMA hydrogels for cell culture was assessed by the cultivation of NIH 3T3 fibroblasts. Therefore, an automated image processing workflow was developed for the identification of single cells and the determination of cell viability. Fibroblasts were incubated on GelMA hydrogels produced by the presented method for a three-day period. GelMA produced by the conventional method was used as a control.

2. Materials and Methods

2.1. GelMA Manufacturing and Characterization

2.1.1. Synthesis and Purification

For the dissolution of gelatin and subsequent synthesis of GelMA, a 0.25 M carbonate-bicarbonate (CB) buffer was prepared according to the method of Lee [17] and Shirahama [18]. Buffer components were purchased from Merck (Darmstadt, Germany). CB buffer was composed of 0.023 M sodium carbonate and 0.227 M sodium bicarbonate, pH 9 was adjusted with 1 M sodium hydroxide (NaOH) or 1 M hydrochloridic acid (HCl). This buffer system was used for the dissolution of gelatin (Type A, 300 bloom strength, Sigma-Aldrich, St. Louis, MI, USA) to a concentration of 10% (w/v) at 40 °C under stirring. Subsequently, the reaction was performed at 40 °C by the addition of methacrylic anhydride (MAA, 94%, Sigma-Aldrich) dropwise with a syringe pump (Nemesys, CETONI, Korbussen, Germany) over a period of 100 min. The MAA-to-gelatin ratio was 100 µL g^{-1}. The reaction was carried out for additional 20 min. The process was terminated by 2-fold dilution with ultrapure water and pH adjustment to pH 7.4. The reaction mixture was purified with 3.5 kDa molecular weight cutoff dialysis tubing (Thermo Fisher Scientific, Waltham, MA, USA) in an ultrapure water reservoir for 4 days at 40 °C. The conductivity of the water in the reservoir was measured with a conductivity meter (CDM230, Radiometer Analytical SAS, Villeurbanne, France). By the end of the dialysis, the conductivity of the reservoir equaled the conductivity of fresh ultrapure water with a value of 5 µS cm^{-1}. This approach for the production of GelMA is referred to as the conventional approach throughout this manuscript. A second buffer system comprised of 0.25 M CB buffer and 4 M urea (Merck) was prepared. The pH of the solution was adjusted to pH 9. In the urea-containing CB buffer, gelatin was dissolved to 10% (w/v) at room temperature (RT) under stirring, no external heating sources were used. Subsequently, the reaction started by adding MAA dropwise to the appropriate amount. A summary of the produced samples within this study is provided in Table 1. The reaction mixture was then purified against ultrapure water in a tangential flow filtration (TFF) unit (Tandem 1082, Sartorius, Göttingen, Germany) equipped with a 2 kDa molecular weight cutoff membrane (Vivaflow® 200, Hydrosart®, Sartorius). The TFF process took place at 50 °C. During purification, the retentate pressure was set to 0.2 MPa, as recommended by the manufacturer as the maximal operating pressure. The purification was stopped after 10 diafiltration volumes. The conductivity of the retentate was equal to the conductivity of ultrapure water by the end of the purification. After each run, the TFF membrane was cleaned with ultrapure water, a 0.1 M sodium hydroxide solution, and a 15% (v/v) ethanol (Merck) solution. The GelMA solutions produced by both methods were frozen at −80 °C and lyophilized at −55 °C and 0.66 Pa for 4 days. Solid GelMA was stored at RT until further use. The methodology used for the synthesis and purification of GelMA in this study is schematically presented in Figure 1.

Table 1. Overview of the experimental set-up for the synthesis of gelatin methacryloyl (GelMA), and the sample nomenclature used throughout this manuscript. CB: carbonate-bicarbonate, DoF: degree of functionalization RT: room temperature.

Synthesis Buffer	MAA-to-Gelatin Ratio µL/g	Reaction Temperature	DoF -	Nomenclature
0.25 M CB	100	50 °C	0.926	50C100MA
0.25 M CB, 4 M urea	100	RT	0.963	RT100MA
0.25 M CB, 4 M urea	50	RT	0.657	RT50MA
0.25 M CB, 4 M urea	25	RT	0.176	RT25MA
0.25 M CB, 4 M urea	12.5	RT	0.044	RT12.5MA

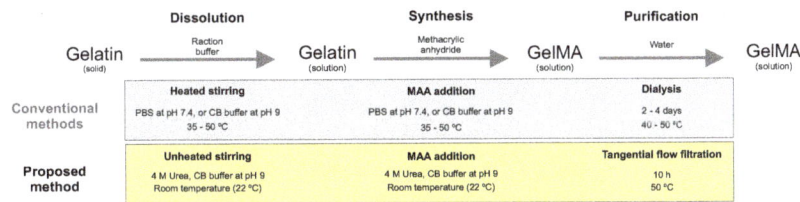

Figure 1. Synthesis and purification of gelatin methacryloyl (GelMA). Conventional methods require the heating of the phosphate-buffered saline (PBS) or the carbonate-bicarbonate (CB) buffer for the dissolution of gelatin. Similarly, the synthesis of GelMA requires heated stirring of the biopolymer solution in order to ensure homogeneous mixing and distribution of the added methacrylic anhydride (MAA). Reaction by-products as well as buffer salts are removed using a dialysis membrane at temperatures above the gelation temperature of GelMA [8,17,18]. The presented method allows for the dissolution of gelatin and the synthesis of GelMA at room temperature under stirring as the urea-containing CB buffer inhibits the formation of a physical gel. Subsequently, GelMA is purified in a tangential flow filtration unit at 50 °C, thereby reducing the processing time to hours.

2.1.2. DoF Determination

The degree of functionalization (DoF) was determined based on the trinitrobenzene-sulfonic (TNBS) acid (Sigma-Aldrich) method by Habeeb [19]. A 0.1 M CB Buffer (0.009 M sodium carbonate, 0.091 M sodium bicarbonate) was prepared and used as a reaction buffer containing 0.01% (w/v) TNBS. Glycine (Sigma-Aldrich) standards, a gelatin reference, and GelMA samples were dissolved in ultrapure water. A volume of 250 µL of each sample was mixed with an equal volume of the TNBS reagent solution and incubated for 2 h at 40 °C. The reaction was stopped by addition of 250 µL of a 10% (w/v) sodium dodecyl sulfate (Sigma-Aldrich) solution and 125 µL of a 1 M HCl solution. The absorbance of each sample was measured at 335 nm using a microplate reader (infiniteM200, Tecan Group, Männedorf, Switzerland). The gelatin and GelMA samples were prepared at 0.1, 0.3, 0.5 and 0.8 mg mL^{-1}. The concentration of primary amino groups in the samples was determined in comparison to a glycine standard curve, which was measured at 3, 5, 8, 10 and 20 µg mL^{-1}, and normalized to the respective protein concentration. The DoF was estimated as the difference between the number of free amines present in gelatin ($c_{NH_2,gelatin}$), i.e., before the functionalization, and the amount in the produced GelMA ($c_{NH_2,GelMA}$), i.e., after the reaction, divided by the number of free amines in the raw gelatin, as shown in Equation (1).

$$\text{Degree of Functionalization}/\text{-} = \frac{c_{NH_2,gelatin} - c_{NH_2,GelMA}}{c_{NH_2,gelatin}} \quad (1)$$

The experimental setup for the characterization of the synthesized GelMA samples consisted of three experimental runs at each MAA-to-gelatin ratio. The absorbance measurements were performed for each independently synthesized batch.

2.2. Hydrogel Characterization

2.2.1. Preparation

Prior to the addition of GelMA, the photo-initiator lithium phenyl-2,4,6-trimethylbenzoylphosphinate (LAP, Sigma-Aldrich) was dissolved to the final concentration of 0.5% (w/v) in Dulbecco's phosphate buffered saline (DPBS, without calcium and magnesium, 1x, pH 7.4, Thermo Fisher Scientific). GelMA synthesized at different MAA/gelatin ratios was used for hydrogel preparation by dissolving the lyophilized material to 5, 10 and 15% (w/v). The resulting hydrogel precursor solution was incubated at 37 °C until complete dissolution of GelMA. A volume of 235 µL GelMA solution was transferred to cylindrical polytetrafluoroethylene (PTFE) molds (diameter 10 mm, height 3 mm). The samples were polymerized by exposure to a UV lamp (365 nm, OSRAM, Munich, Germany) with an irradiation dose of 2850 mJ cm^{-2}. Polymerized GelMA hydrogels were incubated in DPBS until further analysis.

2.2.2. Physical Characterization

The rheological and swelling behavior of the hydrogels were characterized in this study. The viscoelastic properties of cell-free hydrogels were characterized based on their storage and loss modulus as a function of frequency. The cylindrical hydrogels were placed between the plate-plate geometry (diameter 10 mm) of a rotational rheometer (Physica MCR301, Anton Paar, Graz, Austria). The gap height was adjusted to 2.5 mm. All conditions were tested within the linear viscoelastic (LVE) regime covering the frequency range of 0.5 to 50 rad s^{-1} at a stress amplitude of 0.5 Pa.

The swelling behavior of the crosslinked hydrogels was characterized by the ratio of the weight of the hydrogel in the swollen state ($m_{swollen}$) to the weight in the dry state (m_{dry}), as shown in Equation 2. The weight in the swollen state was determined after the equilibration of hydrogels in DPBS. In order to weigh the samples in the dry state, GelMA hydrogels were frozen at −80 °C overnight and lyophilized for 24 h.

$$\text{Swelling Ratio}/\text{-} = \frac{m_{swollen}}{m_{dry}} \quad (2)$$

Data for the physical characterization study shown below were acquired from measurements performed with three experimental runs with a set of three samples (i.e., technical replicates) each. For each experimental run, GelMA hydrogels from independently synthesized batches were prepared.

2.3. Biocompatibility Assessment

2.3.1. Cell Culture

Culture media and supplements were purchased from Gibco™ (Thermo Fisher Scientific). NIH 3T3 mouse fibroblasts (CLS Cell Lines Service, Eppelheim, Germany) were cultured in Dulbecco's Modified Eagle Medium (DMEM, high glucose, GlutaMAX™) supplemented with 10% (v/v) FBS (fetal bovine serum), 50 U mL^{-1} penicillin, and 50 µg mL^{-1} streptomycin. Cells were seeded in tissue culture flasks and maintained at 37 °C in a humidified, 5% CO_2 atmosphere. Subcultivation proceeded at 70 to 80% confluence.

2.3.2. Cell Seeding on Hydrogel Coated Well Plates

The biocompatibility of GelMA hydrogels manufactured at room temperature was investigated. Hydrogels formulated with GelMA produced by the conventional method were used as a control. For these experiments, hydrogels were prepared with GelMA at a reactant ratio of 100 µL MAA per gram gelatin. Lyophilized GelMA was dissolved in ultrapure water to a 2% (w/v) solution at 50 °C. Warm GelMA solutions were sterile filtered using 0.2 µm polyethersulfone (PES) filters (diameter 50 mm, Merck) in a laminar flow cabinet. The sterile solutions were frozen at −80 °C and lyophilized as described in Section 2.1.1. Hydrogel precursor solutions were then prepared at a concentration of 10% (w/v) as mentioned in Section 2.2.1. The bottoms of 12-well plates (Thermo Fisher) with a surface area of 3.5 cm^2 were coated with GelMA by transferring a volume of 350 µL to each well and a subsequent crosslinking under UV exposure. Cells were detached from culture flasks with Trypsin/ethyleneaminetetraacetic acid (Gibco), centrifuged, and resuspended in fresh media. The seeding density on the hydrogel-coated plates was set to 80×10^3 cells per well within a total media volume of 2 mL. Cell-laden samples were kept at incubation conditions until further analysis.

2.3.3. Cell Staining and Imaging

The biocompatibility was assessed after one and after three days of incubation. Staining compounds were purchased from Thermo Fisher (Invitrogen, Waltham, MA, USA). The cell-laden samples were washed with DPBS prior to staining. The samples were covered with 2 mL of staining solution comprised of calcein-AM, propidium iodide, and Hoechst 33342 with concentrations 0.1, 1.5 and 1.66 µM, respectively. Incubation followed for 15 min at 37 °C. Imaging was performed immediately after staining using an inverted microscope (Axio Observer.Z1, Carl Zeiss, Oberkochen, Germany).

2.3.4. Cell Counting and Viability

In order to obtain an objective determination of the number of cells, an image processing workflow was developed using Matlab® R2021b (TheMathWorks, Natick, MA, USA). The acquired green signal originated from calcein retained within the cellular membrane, whereas the red signal arose from stained DNA of cells with compromised membranes. The blue signal of stained nuclear DNA is present in all cells, both viable and dead, as Hoechst 33342 is membrane permeable. The three signals were used as input for the image processing workflow. The preprocessing of the images consisted of individual binarisation of each signal. The identification of single nuclear regions on the binary image of the blue channel was performed by the watershed segmentation method. Binary images from the green and red signals were used as masks on the blue signal. Nuclei behind the green mask and behind the red mask were considered nuclei of viable and dead cells, respectively. The image processing workflow is shown schematically in Figure 2. Cell viability was calculated according to Equation (3), where the number of viable cells (N_{viable}) was divided by the total number of cells consisting of both viable and dead cells (N_{dead}). The experimental setup consisted of three experimental runs (i.e., biological replicates). For each biological replicate, GelMA precursor solutions from independently synthesized batches were used.

Two independent hydrogel samples (i.e., technical replicates) and at least 6 images were recorded of each sample.

$$\text{Cell viability}/\% = 100 * \frac{N_{viable}}{N_{viable} + N_{dead}} \quad (3)$$

2.4. Data Handling and Statistical Analysis

Data evaluation, image processing, statistical analysis, and data visualization were done with Matlab® R2021b. Results are shown as a mean ± standard deviation. The normal distribution of data sets was verified using the Jarque–Bera test with the α-value set to 0.05. A one-way analysis of variance (ANOVA) was performed in order to find significant differences. A p-value below 0.05 was classified as statistically significant.

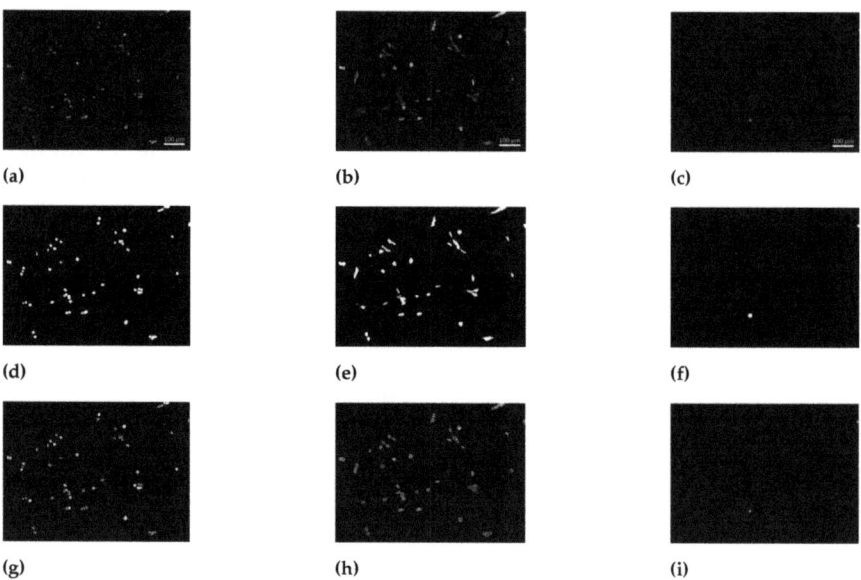

Figure 2. Image processing workflow for cell counting. (**a–c**) Single signal images used as inputs for the image processing workflow. Scale bar: 100 µm. (**a**) Nuclear DNA stained using membrane-permeable Hoechst 33342, (**b**) calcein-AM converted to calcein and retained in the cytoplasm, and (**c**) DNA stained using membrane-impermeable propidium iodide. The input images are preprocessed to binary data. (**d–f**) Resulting binary images of the regions of interest. (**d**) Nuclear region within all cells, (**e**) cytoplasmic region within viable cells, and (**f**) nuclear region within dead cells. (**g**) Single nuclear regions are identified with a watershed segmentation algorithm. The identified nuclei behind the produced mask from the calcein stain are considered viable, whereas nuclei behind the mask from the propidium iodide stain are considered dead cells. (**h**) Outline of nuclei of cells identified as viable shown in green as a composite with the raw image. (**i**) Outline of nuclei of cells identified as dead shown in red as a composite with the raw image.

3. Results and Discussion

3.1. GelMA Synthesis and Characterization

In order to use the biopolymer as a cell culture carrier, gelatin was modified to GelMA. Type A Gelatin was used as starting material for the production of GelMA. The protein was dissolved in a CB buffer solution at pH 9. After the complete dissolution of the gelatin, MAA was added continuously. Both steps were performed under thorough stirring at 50 °C. After the reaction was stopped, the purification of the GelMA took place using dialysis tubing in an ultrapure water reservoir at 40 °C. A second approach for the dissolution and synthesis of GelMA was performed. For this purpose, gelatin was dissolved at RT in a CB buffer at pH 9 that contained 4 M urea. The dissolution of gelatin in the urea-containing buffer solely required stirring. The presence of urea in the buffer also allowed the efficient mixing of MAA during the synthesis, which was also performed at RT. In contrast, the purification set-up for this approach did require the heating of the TFF unit to 50 °C. The DoF of GelMA produced at MAA-to-gelatin ratio of 100 µL g^{-1} was determined. GelMA synthesized in the CB buffer (50C100MA) exhibited a DoF of 0.926 ± 0.057 and did not significantly differ from GelMA produced in the urea containing CB buffer (RT100MA), which showed a DoF of 0.963 ± 0.027, as presented on Table 1 and Figure 3.

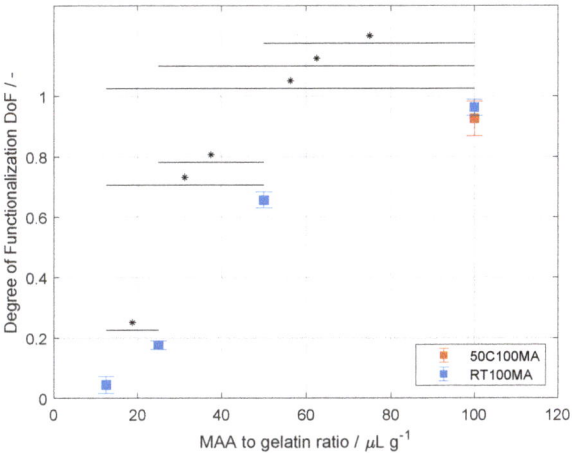

Figure 3. Degree of functionalization (DoF) of produced gelatin methacryloyl (GelMA) batches the DoF was determined by the TNBS method [19]. No difference was found between samples produced a room temperature and samples produced at 50 °C at 100 µL g^{-1}. The DoF increased significantly with increasing the methacrylic anhydride (MAA)-to-gelatin ratio. Asterisks denote a significant difference between samples ($p < 0.05$).

As MAA reacts with neutral free amine groups, a pH value higher than the isoelectric point of gelatin had to be maintained during the reaction. The typical isoelectric point of type A gelatin lies between pH 7 and pH 9. Therefore, the functionalization process was performed at pH 9 in a buffered solution, similar to the methods by Lee [17] and Shirahama [18]. These and previous studies report the processing of gelatin at temperatures above the gelation point of gelatin solutions [8,20,21]. A brief overview of the synthesis parameters found in the literature is schematically shown in Figure 1. Gelatin undergoes a sol–gel transition due to inter- and intramolecular interactions of the biopolymer as the protein backbone forms helical structures similar to those found in collagen [22,23]. In order to process gelatin at RT, urea was used in the reaction buffer. Urea causes the unfolding of proteins as the hydrophobic interaction between protein chains is disrupted [24–26]. This effect was used in order to ensure the sol state at RT, thus enabling the homogeneous mixing of MAA during the synthesis of GelMA. The purification of GelMA produced at RT required heating of the complete TFF set-up, GelMA, and water reservoirs, as well as the membrane. The decreasing concentration of urea during purification allows the re-folding of GelMA protein coils to helical structures, i.e., the transition from a solution to a gel [8]. Nevertheless, the purification time was reduced to about ten hours compared to the several days required by the dialysis process. It is noteworthy to mention that the purity of the samples should be analyzed using other methodologies in future studies, as in the presented study the purification was indirectly controlled by measuring the conductivity of the liquid used during purification, and no direct quantification of the purity of the sample was performed. In addition, there were no significant differences between the DoF of GelMA produced by both methods, and the values are in the same range as previous studies [17,18,20]. Additionally, Shirahama [18] also showed that the produced GelMA was not influenced by the reaction temperature in the range from 35 to 50 °C. The DoF of GelMA produced at higher temperatures is comparable to the DoF of GelMA synthesized by the presented approach at RT; thus, this finding could indicate that the reaction kinetics is controlled by the molecular weight and molecular weight distribution of the used gelatin, rather than by the reaction temperature. Differences in molecular weight can arise not only from the species of the gelatin source but also from the bloom strength and even from batch-to-batch variations of the same gelatin product. The DoF of type B GelMA produced with bovine gelatin has been studied in the literature showing a higher value than that of type A GelMA produced with porcine gelatin [27]. This is due to the fact that the pH of the buffered solution remains above the IEP of bovine gelatin during the reaction. The presented method of reaction buffer containing urea should also be implemented in future studies for the comparison of the properties of GelMA from different species as well as varying bloom strength.

The effect of the MAA-to-gelatin ratio was also studied for the developed synthesis at RT and is shown in Figure 3. All synthesized batches are summarized in Table 1. The ratio was varied from 12.5 to 100 $\mu L\,g^{-1}$, where the increasing ratio leads to higher DoF ranging from 0.044 ± 0.028 to 0.963 ± 0.027. Statistically significant differences between the DoF values were found for all data sets ($p < 0.05$). The effect of the increasing ratio is in accordance with other studies [17,18,20,21]. Based on the reactant ratio, the resulting GelMA product can be controlled in a reproducible manner as required for the formulation of hydrogels.

3.2. Hydrogel Characterization

GelMA hydrogels can be covalently crosslinked to generate hydrogels with structural properties according to the required application. In cell culture, the physical properties of the matrix environment are known to influence cellular migration, proliferation, and differentiation [28–30]. Hydrogels were prepared as described previously and the resulting storage modulus was determined by oscillatory frequency sweeps on a rheometer. Holding the DoF constant, the elastic modulus showed a significant increase with higher polymer concentration ($p < 0.05$), as shown in Figure 4. The moduli of RT100MA hydrogels at 5, 10 and 15% (w/v) were 1.4, 11.2 and 29.4 kPa, respectively. The same trend was observed with all used samples, i.e., RT50MA, RT25MA, and RT12.5MA. The increase of the elastic modulus corresponds to a higher crosslink density in the polymeric network [31]. The higher GelMA content leads to an increased availability of crosslinking sites as well as an increasing amount of physical entanglements. This behavior is comparable with other studies [27,32,33].

The elasticity of the hydrogels was influenced by the DoF, this effect is presented in Figure 4 as well. Hydrogels prepared from the samples RT50MA, RT25MA, and RT12.5MA with DoFs of 0.657, 0.176 and 0.044, respectively, show a significant increase of elasticity with increasing DoF at a constant GelMA concentration ($p < 0.05$). At a concentration of 15% (w/v), the moduli increased from 1.8 to 29.4 kPa with increasing DoFs. The elasticity of hydrogels produced from RT100MA and RT50MA was in the same range and did not differ significantly at any tested GelMA concentration, even though the DoFs of both samples differed significantly. The DoF of GelMA RT100MA was 0.963, while the DoF of the RT50MA sample was 0.657.

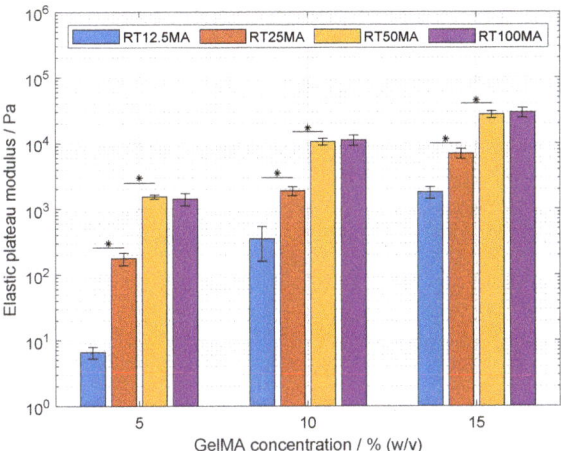

Figure 4. Elastic plateau modulus of gelatin methacryloyl (GelMA) hydrogels at various concentrations. The elasticity of hydrogels increased significantly with higher GelMA concentrations of the same synthesized sample ($p < 0.05$). These differences are not shown for the purpose of clarity. Significant differences between the elastic modulus of samples at a constant GelMA concentration are denoted with an asterisk ($p < 0.05$).

The results observed from the samples RT50MA, RT25MA, and RT12.5MA are in accordance with other studies [27,32,33]. The increasing elasticity of the hydrogels at a constant polymer concentration is attributed to the higher crosslink density proceeding from the higher amount of methacrylamide and methacrylate residues. The missing difference between the samples prepared with RT50MA and RT100MA could arise from the crosslinking condition used in this study. The irradiation dose was set to 2850 mJ cm^{-2} and the concentration of the photo-initiator was set to 0.5% (w/v). The irradiance dose is higher than those used in similar studies by Van Den Bulcke [8], Nichol [32], and Pepelanova [20] which were set to 10, 6.9 and 1200 mJ cm^{-2}, respectively. The concentration of the used photo-initiator was also higher than the concentration presented by Van Den Bulcke (0.006% (w/v)) [8], Schuurman (0.05% (w/v)) [12], and Lee (0.1% (w/v)) [27]. In free radical polymerization, the reaction rate is proportional to the irradiance and photo-initiator concentration, as studied by O'Connell [33]. The fast generation of free radicals, which initiates the chain polymerization, is opposed by the increasing elasticity of the matrix, i.e., the decreasing mobility of the polymeric network and the lower diffusivity of free radicals. Thus, reaching a limit of the created crosslinks.

The swelling behavior of hydrogels reflects the ability of the polymeric network to bind and retain aqueous media. This property influences the diffusion of nutrients to the cells and metabolic by-products away from the cells [34]. As shown in Figure 5, it was observed that the swelling ratio of hydrogels was influenced by the used GelMA concentration as well as the DoF. Maintaining a constant DoF, an increasing amount of GelMA significantly reduced the swelling capacity of the network ($p < 0.05$). Furthermore, a higher DoF led to lower swelling ratios at a constant concentration. This effect was significant between the samples RT50MA, RT25MA, and RT12.5MA ($p < 0.05$). Similar to the observation during the characterization of the mechanical properties, the swelling ratios of hydrogels produced with GelMA RT50MA and RT100MA did not differ significantly. The effect of GelMA concentration and DoF on the swelling capacity of GelMA hydrogels have also been demonstrated by other studies [11,32,35,36] and are in agreement with the observations presented in this study. The driving mechanism of swelling is the osmotic pressure difference between the fluid within the network and the outer solution, which is opposed by the elasticity of the network preventing the dissolution due to its crosslinked structure [34,37]. Therefore, more elastic hydrogels produced with higher GelMA concentrations or higher DoFs have a lower swelling capability due to the higher crosslink density. As previously described, the similarity between the swelling ratio of the hydrogel samples RT50MA and RT100MA could be explained by the crosslinking conditions. It has to be kept in mind that the physical properties of a crosslinked GelMA hydrogel also depend on the properties of the used gelatin before functionalization such as the animal source, the bloom strength, and the crosslinking conditions [20,27,38]. By variation of the gelatin bloom strength, and, therefore, variation of the molecular weight of the biopolymer, the range of elasticity of the produced hydrogels could be expanded in further investigations. The GelMA hydrogels produced in this study proved to be suitable for mimicking the physical properties of tissue. Such properties can be adjusted in a controllable manner over the concentration of biopolymer and/or DoF in order to adjust the hydrogel to the specific cell type requirements.

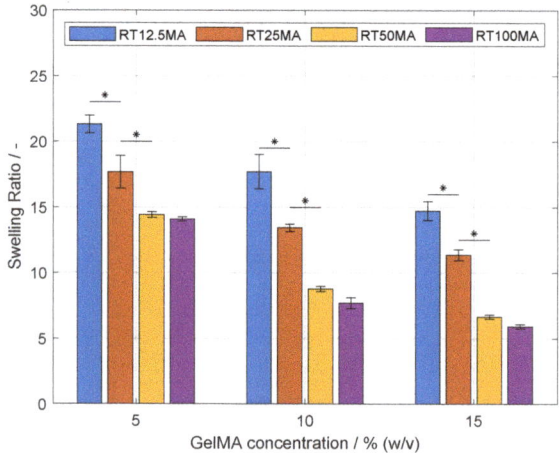

Figure 5. Equilibrium swelling ratio of gelatin methacryloyl (GelMA) hydrogels at various concentrations. The swelling behavior increased significantly with higher GelMA concentrations of the same synthesized sample ($p < 0.05$). These differences are not shown for the purpose of clarity. Significant differences between the swelling ratio of samples at a constant GelMA concentration are denoted with an asterisk ($p < 0.05$).

3.3. Biocompatibility Assessment

GelMA hydrogels were used as carriers for NIH 3T3 fibroblasts in culture. The biocompatibility of synthesized GelMA as a hydrogel was determined via quantification of the cell viability after one, and after three days in cultivation. For this purpose, fluorescent staining and subsequent imaging of the samples was performed. The acquired images were imported into Matlab® and evaluated using the image processing workflow, as described in Section 2.3.4. During processing, the detected signals were binarized separately. Single nuclei in contiguous regions were identified using a watershed segmentation tool. Nuclei of cells under the binary mask proceeding from the green channel were counted as live cells, whereas nuclei under the red mask were classified as dead cells. The image processing workflow enabled the objective and automated analysis of the gained frames. The developed tool is intended to reduce the time required for the analysis of images. Moreover, the automated identification of cells is also advantageous as it increases reproducibility and reduces observer-dependent errors [39–41]. The used image processing workflow allowed the analysis of relatively large data sets, which consisted of at least 40 frames with more than 1500 identified cells for each tested material. In contrast, similar studies in the field of tissue engineering and biofabrication are limited to a relatively low number of counted cells [27,42] and/or a low number of acquired images by microscopy [12,43].

Figure 6 shows the viability of cells seeded onto 10% (*w/v*) GelMA hydrogels. Cells growing on GelMA synthesized according to the conventional method at 50 °C (50C100MA) showed viabilities of 94.8% and 94.5% after one and three days, respectively. The viability of cells on GelMA produced at room temperature (RT100MA) was quantified to a value of 95.8% after one day of incubation, and a value of 93.3% after three days.

The viability of cells after one day did not significantly differ from the viability after three days for either hydrogel. Hence, no significant difference in viability was observed between cells growing on GelMA 50C100MA and on GelMA RT100MA. This approach of GelMA production at room temperature enabled the synthesis of a biocompatible material for hydrogel formulation. The high viability is in accordance with similar studies that use GelMA produced following conventional methods [27,32,38,42]. In the field of tissue engineering, the applications of hydrogels also include cell growth within the produced scaffold as well as cellular invasion in scaffolds. Such studies should be performed with GelMA produced according to the presented methodology.

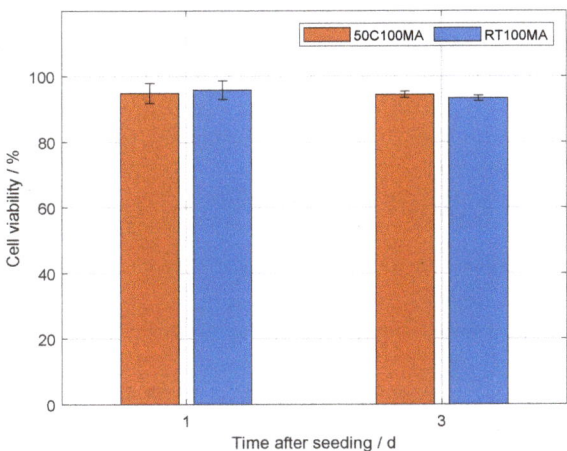

Figure 6. Cell viability of NIH 3T3 fibroblasts on gelatin methacryloyl (GelMA) coated well plates. GelMA produced similar to the method of Shirahama [18] at 50 °C (50C100MA) was used as a control in comparison with GelMA produced by the novel approach presented in this study, i.e., at room temperature (RT100MA). The viability of cells seeded onto both hydrogels samples did not differ significantly after one and three days.

4. Conclusions

In the field of tissue engineering, the use of GelMA-based hydrogels is well established due to the biocompatible nature of the biomaterial. The current manufacturing of products containing gelatin and gelatin derivatives requires heating, as gelatin solutions undergo a transition to a gel state at lower temperatures. This study presents a novel approach to functionalize gelatin to GelMA at room temperature. This process was possible due to the presence of urea in the synthesis buffer, as urea decreases the protein–protein interaction, thereby inhibiting the gelation of the solution and allowing efficient mixing of the reactants. GelMA with different degrees of functionalization was produced in a controllable manner by variation of the reactant ratio. By variation of the concentration of GelMA and its DoF, hydrogels were prepared with elastic properties in the range of 1.8 to 24 kPa, thus enabling the precise adaptation to specific cell type requirements. Moreover, the determination of cell viability was performed by imaging and a subsequent image processing workflow that allowed the time-saving analysis of several frames. GelMA produced at room temperature proved to be suitable for cell culture applications with the cell line NIH 3T3 and no difference was observed in comparison to GelMA produced by the conventional method at elevated temperatures, i.e., 50 °C. The possibility to manufacture GelMA at room temperature shows several advantages. First, gelatin solutions are prone to the growth of microorganisms and would require cooling during longer transition periods between operations such as gelatin dissolution and GelMA synthesis. The presence of urea in the solution can be helpful during mentioned time intervals as urea is known to inhibit microbial growth. Second, the growing demand for biomaterials and related products such as bioinks requires suitable means to produce them efficiently in large quantities. Processing at room temperature, i.e., without heating, could facilitate the transition to large-scale production as undesired temperature profiles in reactors can lead to poor mixing and improper distribution of reactants, thus leading to non-reproducible processes. Thirdly, and of significant importance, the production of GelMA for clinical stages might face conditions imposed by regulatory agencies. These require detailed information on the range of operating conditions that will result in the production of materials meeting quality criteria. In the case of GelMA, the presented approach widens the operating range in terms of temperature in order to achieve the aimed degree of functionalization, and, therefore, the robustness of the synthesis process is enhanced.

Author Contributions: Conceptualization, D.G.G.; methodology, D.G.G.; formal analysis, D.G.G. and C.P.R.; data curation, D.G.G.; software, D.G.G.; writing—original draft preparation, D.G.G.; writing—review and editing, C.P.R. and J.H.; visualization, D.G.G.; supervision, J.H.; project administration, J.H.; funding acquisition, J.H. All authors have read and agreed to the published version of the manuscript.

Funding: This work was funded by the German Federal Ministry of Education and Research (BMBF) as project SOP-Bioprint under contract number 13XP5071B.

Institutional Review Board Statement: Not applicable

Data Availability Statement: The raw data supporting the conclusions of this article as well as the written codes for Matlab® will be made available on request. Inquiries can be directed to the corresponding author.

Acknowledgments: We would like to thank Amelie Wirth and Johanna Ossmann for their valuable contributions in the form of experimental work for this project. We acknowledge support by the KIT-Publication Fund of the Karlsruhe Institute of Technology.

Conflicts of Interest: The authors declare that they have no known competing financial interest or personal relationship that could have appeared to influence the work reported in this paper.

Abbreviations

The following abbreviations are used in this manuscript:

CAM	Calcein-AM
DoF	Degree of functionalization
ECM	Extracellular matrix
GelMA	Gelatin-methacryloyl
IEP	Isoelectric point
MAA	Methacrylic anhydride
PI	Propidium iodide
TE	Tissue engineering
TFF	Tangential flow filtration
TNBS	2,4,6-Trinitrobenzenesulfonic acid solution

References

1. Slaughter, B.V.; Khurshid, S.S.; Fisher, O.Z.; Khademhosseini, A.; Peppas, N.A. Hydrogels in Regenerative Medicine. *Adv. Mater.* **2009**, *21*, 3307–3329. [CrossRef] [PubMed]
2. Tibbitt, M.W.; Anseth, K.S. Hydrogels as extracellular matrix mimics for 3D cell culture. *Biotechnol. Bioeng.* **2009**, *103*, 655–663. [CrossRef] [PubMed]
3. Wade, R.J.; Burdick, J.A. Engineering ECM signals into biomaterials. *Mater. Today* **2012**, *15*, 454–459. [CrossRef]
4. Ruedinger, F.; Lavrentieva, A.; Blume, C.; Pepelanova, I.; Scheper, T. Hydrogels for 3D mammalian cell culture: A starting guide for laboratory practice. *Appl. Microbiol. Biotechnol.* **2015**, *99*, 623–636. [CrossRef]
5. Ingber, D.E. Cellular mechanotransduction: Putting all the pieces together again. *FASEB J.* **2006**, *20*, 811–827. [CrossRef]
6. Jansen, K.A.; Donato, D.M.; Balcioglu, H.E.; Schmidt, T.; Danen, E.H.; Koenderink, G.H. A guide to mechanobiology: Where biology and physics meet. *Biochim. Et Biophys. Acta Mol. Cell Res.* **2015**, *1853*, 3043–3052. [CrossRef]
7. Lee, K.Y.; Mooney, D.J. Hydrogels for Tissue Engineering. *Chem. Rev.* **2001**, *101*, 1869–1880. [CrossRef]
8. Van Den Bulcke, A.I.; Bogdanov, B.; De Rooze, N.; Schacht, E.H.; Cornelissen, M.; Berghmans, H. Structural and rheological properties of methacrylamide modified gelatin hydrogels. *Biomacromolecules* **2000**, *1*, 31–38. [CrossRef]
9. Klotz, B.J.; Gawlitta, D.; Rosenberg, A.J.; Malda, J.; Melchels, F.P. Gelatin-Methacryloyl Hydrogels: Towards Biofabrication-Based Tissue Repair. *Trends Biotechnol.* **2016**, *34*, 394–407. [CrossRef]
10. Kaemmerer, E.; Melchels, F.P.; Holzapfel, B.M.; Meckel, T.; Hutmacher, D.W.; Loessner, D. Gelatine methacrylamide-based hydrogels: An alternative three-dimensional cancer cell culture system. *Acta Biomater.* **2014**, *10*, 2551–2562. [CrossRef]
11. Vigata, M.; Meinert, C.; Bock, N.; Dargaville, B.L.; Hutmacher, D.W. Deciphering the molecular mechanism of water interaction with gelatin methacryloyl hydrogels: Role of ionic strength, ph, drug loading and hydrogel network characteristics. *Biomedicines* **2021**, *9*, 574. [CrossRef] [PubMed]
12. Schuurman, W.; Levett, P.A.; Pot, M.W.; van Weeren, P.R.; Dhert, W.J.A.; Hutmacher, D.W.; Melchels, F.P.W.; Klein, T.J.; Malda, J. Gelatin-Methacrylamide Hydrogels as Potential Biomaterials for Fabrication of Tissue-Engineered Cartilage Constructs. *Macromol. Biosci.* **2013**, *13*, 551–561. [CrossRef] [PubMed]
13. Stratesteffen, H.; Köpf, M.; Kreimendahl, F.; Blaeser, A.; Jockenhoevel, S.; Fischer, H. GelMA-collagen blends enable drop-on-demand 3D printablility and promote angiogenesis. *Biofabrication* **2017**, *9*, 045002. [CrossRef] [PubMed]
14. Aldana, A.A.; Valente, F.; Dilley, R.; Doyle, B. Development of 3D bioprinted GelMA-alginate hydrogels with tunable mechanical properties. *Bioprinting* **2021**, *21*, e00105. [CrossRef]
15. Mouser, V.H.; Melchels, F.P.; Visser, J.; Dhert, W.J.; Gawlitta, D.; Malda, J. Yield stress determines bioprintability of hydrogels based on gelatin-methacryloyl and gellan gum for cartilage bioprinting. *Biofabrication* **2016**, *8*, 035003. [CrossRef]
16. Chandra, P.; Yoo, J.J.; Lee, S.J. *Biomaterials in Regenerative Medicine: Challenges in Technology Transfer from Science to Process Development*; Elsevier Inc.: Amsterdam, The Netherlands, 2015; pp. 151–167. [CrossRef]
17. Lee, B.H.; Shirahama, H.; Cho, N.J.; Tan, L.P. Efficient and controllable synthesis of highly substituted gelatin methacrylamide for mechanically stiff hydrogels. *RSC Adv.* **2015**, *5*, 106094–106097. [CrossRef]
18. Shirahama, H.; Lee, B.H.; Tan, L.P.; Cho, N.J. Precise Tuning of Facile One-Pot Gelatin Methacryloyl (GelMA) Synthesis. *Sci. Rep.* **2016**, *6*, 31036. [CrossRef]
19. Habeeb, A.F. Determination of free amino groups in proteins by trinitrobenzenesulfonic acid. *Anal. Biochem.* **1966**, *14*, 328–336. [CrossRef]
20. Pepelanova, I.; Kruppa, K.; Scheper, T.; Lavrentieva, A. Gelatin-Methacryloyl (GelMA) Hydrogels with Defined Degree of Functionalization as a Versatile Toolkit for 3D Cell Culture and Extrusion Bioprinting. *Bioengineering* **2018**, *5*, 55. [CrossRef]
21. Zhu, M.; Wang, Y.; Ferracci, G.; Zheng, J.; Cho, N.J.; Lee, B.H. Gelatin methacryloyl and its hydrogels with an exceptional degree of controllability and batch-to-batch consistency. *Sci. Rep.* **2019**, *9*, 1–13. [CrossRef]
22. Pezron, I.; Djabourov, M.; Leblond, J. Conformation of gelatin chains in aqueous solutions: 1. A light and small-angle neutron scattering study. *Polymer* **1991**, *32*, 3201–3210. [CrossRef]
23. Ross-Murphy, S.B. Structure and rheology of gelatin gels: Recent progress. *Polymer* **1992**, *33*, 2622–2627. [CrossRef]
24. Zou, Q.; Habermann-Rottinghaus, S.M.; Murphy, K.P. Urea effects on protein stability: Hydrogen bonding and the hydrophobic effect. *Proteins Struct. Funct. Genet.* **1998**, *31*, 107–115. [CrossRef]
25. Stumpe, M.C.; Grubmüller, H. Interaction of urea with amino acids: Implications for urea-induced protein denaturation. *J. Am. Chem. Soc.* **2007**, *129*, 16126–16131. [CrossRef]
26. Das, A.; Mukhopadhyay, C. Urea-mediated protein denaturation: A consensus view. *J. Phys. Chem. B* **2009**, *113*, 12816–12824. [CrossRef]
27. Lee, B.H.; Lum, N.; Seow, L.Y.; Lim, P.Q.; Tan, L.P. Synthesis and characterization of types A and B gelatin methacryloyl for bioink applications. *Materials* **2016**, *9*, 797. [CrossRef]
28. Lo, C.M.; Wang, H.B.; Dembo, M.; Wang, Y.l. Cell Movement Is Guided by the Rigidity of the Substrate. *Biophys. J.* **2000**, *79*, 144–152. [CrossRef]
29. Hadjipanayi, E.; Mudera, V.; Brown, R.A. Close dependence of fibroblast proliferation on collagen scaffold matrix stiffness. *J. Tissue Eng. Regen. Med.* **2009**, *3*, 77–84. [CrossRef]
30. Marklein, R.A.; Burdick, J.A. Controlling stem cell fate with material design. *Adv. Mater.* **2010**, *22*, 175–189. [CrossRef]

31. Anseth, K.S.; Bowman, C.N.; Brannon-Peppas, L. Mechanical properties of hydrogels and their experimental determination. *Biomaterials* **1996**, *17*, 1647–1657. [CrossRef]
32. Nichol, J.W.; Koshy, S.T.; Bae, H.; Hwang, C.M.; Yamanlar, S.; Khademhosseini, A. Cell-laden microengineered gelatin methacrylate hydrogels. *Biomaterials* **2010**, *31*, 5536–5544. [CrossRef] [PubMed]
33. O'Connell, C.D.; Zhang, B.; Onofrillo, C.; Duchi, S.; Blanchard, R.; Quigley, A.; Bourke, J.; Gambhir, S.; Kapsa, R.; Di Bella, C.; et al. Tailoring the mechanical properties of gelatin methacryloyl hydrogels through manipulation of the photocrosslinking conditions. *Soft Matter* **2018**, *14*, 2142–2151. [CrossRef] [PubMed]
34. Peppas, N.A.; Hilt, J.Z.; Khademhosseini, A.; Langer, R. Hydrogels in biology and medicine: From molecular principles to bionanotechnology. *Adv. Mater.* **2006**, *18*, 1345–1360. [CrossRef]
35. Kirsch, M.; Birnstein, L.; Pepelanova, I.; Handke, W.; Rach, J.; Seltsam, A.; Scheper, T.; Lavrentieva, A. Gelatin-Methacryloyl (GelMA) Formulated with Human Platelet Lysate Supports Mesenchymal Stem Cell Proliferation and Differentiation and Enhances the Hydrogel's Mechanical Properties. *Bioengineering* **2019**, *6*, 76. [CrossRef]
36. Krishnamoorthy, S.; Noorani, B.; Xu, C. Effects of Encapsulated Cells on the Physical–Mechanical Properties and Microstructure of Gelatin Methacrylate Hydrogels. *Int. J. Mol. Sci.* **2019**, *20*, 5061. [CrossRef]
37. Rička, J.; Tanaka, T. Swelling of Ionic Gels: Quantitative Performance of the Donnan Theory. *Macromolecules* **1984**, *17*, 2916–2921. [CrossRef]
38. Shie, M.Y.; Lee, J.J.; Ho, C.C.; Yen, S.Y.; Ng, H.Y.; Chen, Y.W. Effects of gelatin methacrylate bio-ink concentration on mechanophysical properties and human dermal fibroblast behavior. *Polymers* **2020**, *12*, 1930. [CrossRef]
39. Malpica, N.; De Solórzano, C.O.; Vaquero, J.J.; Santos, A.; Vallcorba, I.; García-Sagredo, J.M.; Del Pozo, F. Applying watershed algorithms to the segmentation of clustered nuclei. *Cytometry* **1997**, *28*, 289–297. [CrossRef]
40. Wiesmann, V.; Franz, D.; Held, C.; Münzenmayer, C.; Palmisano, R.; Wittenberg, T. Review of free software tools for image analysis of fluorescence cell micrographs. *J. Microsc.* **2015**, *257*, 39–53. [CrossRef]
41. Eggert, S.; Hutmacher, D.W. In vitro disease models 4.0 via automation and high-throughput processing. *Biofabrication* **2019**, *11*, 043002. [CrossRef]
42. Wang, Z.; Tian, Z.; Menard, F.; Kim, K. Comparative study of gelatin methacrylate hydrogels from different sources for biofabrication applications. *Biofabrication* **2017**, *9*, 044101. [CrossRef] [PubMed]
43. Sharifi, S.; Sharifi, H.; Akbari, A.; Chodosh, J. Systematic optimization of visible light-induced crosslinking conditions of gelatin methacryloyl (GelMA). *Sci. Rep.* **2021**, *11*, 1–12. [CrossRef] [PubMed]

Article

Bone Formation on Murine Cranial Bone by Injectable Cross-Linked Hyaluronic Acid Containing Nano-Hydroxyapatite and Bone Morphogenetic Protein

Yuki Hachinohe [1], Masayuki Taira [2,*], Miki Hoshi [1], Wataru Hatakeyama [1], Tomofumi Sawada [2] and Hisatomo Kondo [1]

[1] Department of Prosthodontics and Oral Implantology, School of Dentistry, Iwate Medical University, 19-1 Uchimaru, Morioka 020-8505, Japan
[2] Department of Biomedical Engineering, Iwate Medical University, 1-1-1 Idaidori, Yahaba-cho 028-3694, Japan
* Correspondence: mtaira@iwate-med.ac.jp; Tel.: +81-19-651-5110

Abstract: New injection-type bone-forming materials are desired in dental implantology. In this study, we added nano-hydroxyapatite (nHAp) and bone morphogenetic protein (BMP) to cross-linkable thiol-modified hyaluronic acid (tHyA) and evaluated its usefulness as an osteoinductive injectable material using an animal model. The sol (ux-tHyA) was changed to a gel (x-tHyA) by mixing with a cross-linker. We prepared two sol–gel (SG) material series, that is, x-tHyA + BMP with and without nHAp (SG I) and x-tHyA + nHAp with and without BMP (SG II). SG I materials in the sol stage were injected into the cranial subcutaneous connective tissues of mice, followed by in vivo gelation, while SG II materials gelled in Teflon rings were surgically placed directly on the cranial bones of rats. The animals were sacrificed 8 weeks after implantation, followed by X-ray analysis and histological examination. The results revealed that bone formation occurred at a high rate (>70%), mainly as ectopic bone in the SG I tests in mouse cranial connective tissues, and largely as bone augmentation in rat cranial bones in the SG II experiments when x-tHyA contained both nHAp and BMP. The prepared x-tHyA + nHAp + BMP SG material can be used as an injection-type osteoinductive bone-forming material. Sub-periosteum injection was expected.

Keywords: cross-linked hyaluronic acid; nano hydroxyapatite; bone morphogenetic protein; injection-type bone forming material; ectopic bone formation; bone augmentation

1. Introduction

In dental implantology, bone formation is often desired in patients whose implants cannot be firmly placed due to shallow and narrow jaw bones [1]. Sinus lift or socket lift with autogenous bones and/or alloplasts (granules) is often performed to enlarge the areas of the bone that receive implants [2]. Treatment with an alloplast without bone collection is preferred as a remedy for patients [3]. Incisions and sutures are inevitable when placing granules of beta-tricalcium phosphate [4] and apatite [5,6]. However, alloplastic granules often spill out of implanted areas [7], causing infection problems. Therefore, the use of less invasive injection-type bone forming materials without suturing is expected [8].

In dental implantology, injection-type bone substitute materials are rarely used. Meanwhile, in orthopedic surgery, self-setting apatite-based bone cement has been used to treat vertebral compression fractures [9]. Hydrogels—such as polyethylene glycol [10], chitosan [11], alginate [12], hyaluronic acid (HyA) [13–15], and gelatin [16,17]—which are often coupled with calcium phosphate (i.e., apatite or tricalcium phosphate) [18] and growth factors, such as bone morphogenetic protein (BMP) [19], have been studied as injectable bone forming materials. However, most have not been used routinely in dental practice. New injectable biomaterial systems are expected to be developed in implantology.

HyA is a natural polysaccharide composed of D-glucuronic acid and N-acetylglucosamine [20] and is a component of the extracellular matrix of most connective tissues that exhibit excellent biocompatibility when applied to the human body [21]. Depending on the processing method, HyA materials can be prepared in the form of sponges, hydrogels, or injectable gels [20]. In cosmetics, HyA is often used to eliminate nasolabial folds [22]. HyA is also used for joint fluid supplementation [23], eye operation [24], wound recovery [25], and soft tissue restoration [26].

Due to its chemical structure, HyA is a hydrophilic polymer and can be characterized by a fast degradation rate (e.g., for 3–5 days) [20]. HyA-based materials have been intensively assessed for biomedical applications due to their excellent biocompatibility, biodegradability, and chemical modification [20]. HyA requires a chemical cross-link for more than a month in vivo [20,27]. HyA can be cross-linked by chemical modification and the use of an appropriate cross-linker, while HyA has been chemically modified with hydrazide [28], amino or aldehyde functional groups [29], and methacrylate groups [30] to form stable cross-link networks [20,27]. Another important approach is the thiol modification of the side chains of HyA (Figure 1a) and cross-linking by a di-functional cross-linker (Figure 1b) [31,32]. Hystem®—a cross-linkable thiol-modified hyaluronic acid (tHyA)—was developed in the USA for biomedical research, and is claimed to be capable of transplanting cells and/or slowly releasing growth factors [33–37]. This material has been followed by several rival clinical products, such as Restylane Lyft® [33,38], and has not been thoroughly examined as an injection-type bone-forming material [39].

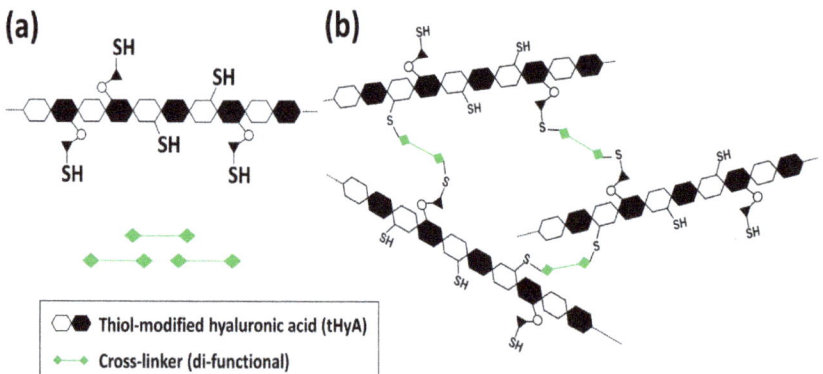

Figure 1. (a) Thiol modification of side chains of HyA, (b) cross-linking of tHyA by a di-functional cross-linker.

HyA is not osteoconductive, while BMP is a strong bone-forming growth factor [40]. Adding a growth factor and its carriers can render HyA-based materials osteoinductive and osteoconductive [39,40]. Nano-hydroxyapatite (nHAp) has been reported to be an osteoconductive, bio-absorbable, and carrier material, while larger hydroxyapatite blocks and granules are more inert, less bio-absorbable, and less protein adsorbed [41]. BMP can be bound to and slowly released from nHAp, sustaining long-term bone-forming activity [40,42]. We previously reported that injected x-tHyA + nHAp + BMP sol–gel (SG) successfully caused ectopic bone formation in the back subcutaneous and thigh muscles of mice by endochondral ossification [39]. As a next step, we believed that it was necessary for clinical application to check the bone-forming capability of x-tHyA + nHAp + BMP SG in the cranial osseous area of living animals.

The materials considered for bone augmentation—namely, bone grafts and substitute materials—have wide variations in the type and use method. Briefly, these materials can be classified into naturally derived materials (autografts, allografts, and xenografts), synthetic materials (hydroxyapatite, beta-tricalcium phosphate, calcium phosphate, bioactive glasses [43,44], metals, and polymers), composite materials (e.g., HyA/nHAp), growth factor-based materials, and materials with infused living osteogenic cells. Xenografts contain bovine bone, collagen, HyA, and silk [45]. Materials can also be classified based on several attributes. According to the source, they can be classified into two groups: biological (e.g., HyA) or non-biological. By chemical composition, they are metals and alloys, ceramics, or polymers (e.g., natural polymers, such as collagen and HyA, and synthetic polymers, such as polyetherether-ketone, polyethylene glycol, polylactide, and polycaprolactone). Due to their material consistency, they are three-dimensional (3D) printable, implantable solids, injectable (e.g., HyA), or adhesive. Additives may contain stem cells and bioactive agents (e.g., BMP). They may be composed of nanografts in the form of nanosized tubes, fibers, sheets, crystals, and cages [46]. The base material studied (x-tHyA) is an injectable natural-origin polymeric SG material.

First, we characterized un-cross-linked and cross-linked thiol-modified HyA (ux-tHyA and x-tHyA, respectively) by Fourier transform infrared spectroscopy (FTIR), scanning electron microscopy (SEM), hyaluronidase dissolution tests, and thermogravimetry (TG) coupled with differential scanning calorimetry (DSC). We performed two experiments to investigate the use of nHAp. We observed direct binding between nHAp and protein using confocal laser scanning microscopy and performed protein release tests in saline solution from a mixed gel of x-tHyA and nHAp.

Second, the main purpose of this investigation was to prepare an injection-type boneforming material using x-tHyA, nHAp, and BMP and examine its usefulness in animal experiments. We examined the bone-forming ability of x-tHyA + nHAp + BMP (i) using x-tHyA + BMP with and without nHAp (SG I series) in mouse cranial subcutaneous connective tissues, and (ii) employing x-tHyA + nHAp with and without BMP (SG II series) on rat cranial bones by X-ray approaches and histological observations. The final objective of this study was the rapid development of a novel injectable bone-forming material system using existing HyA material (x-tHyA), nHAp, and BMP and to search for information on the technology and supporting materials necessary to realize this original purpose. The novelty of this study lies in the open publication of the bone-forming capability by the combined use of nHAp and BMP in x-tHyA, which could lead to its future direct-wide therapeutic use in dentistry and medicine. In particular, we have examined the usefulness of nHAp in biomedical applications [47,48]. The final intended use of the investigated materials is its direct subperiosteal injection to achieve alveolar bone augmentation for dental implant placement and insertion of denture.

2. Materials and Methods
2.1. Material

A commercial cross-linkable hyaluronic acid kit (Hystem® Kit, 12.5 mL, Part Number GS1004, Sigma-Aldrich, St. Louis, MO, USA) consisting of hyaluronic acid possessing –SH functional groups (Glycosil®) (Catalog Number GS220, ESI BIO, Alameda, CA, USA), thiol-reactive polyethylene glycol diacrylate (PEGDA) cross-linker (Extralink Lite®, Catalog Number GS3008, ESI BIO), and degassed (DG) water (Calatog Number GS241, ESI BIO) (Figure 2). Other drugs and materials used were commercial recombinant human/mouse/rat (CHO cell-derived) bone morphogenetic protein-2 (BMP) (R&D Systems, Catalog Number 355-BM, Minneapolis, MN, USA), nHAp with a mean diameter of 40 nm (nano-SHAp, SofSera, Tokyo, Japan), and a microorganism-derived HyA (HYALURONSAN HA-SHY, average molecular weight = 1,500,000–3,900,000, Kewpie Co., Tokyo, Japan) (HyA control). The nHAp particles were autoclaved before mixing.

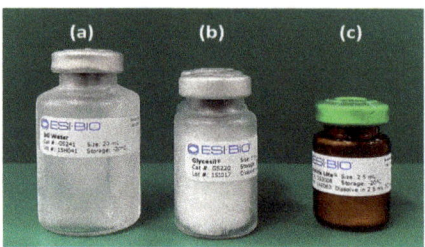

Figure 2. A cross-linkable hyaluronic acid kit, Hystem®, consisting of: (**a**) degassed (DG) water; (**b**) hyaluronic acids possessing –SH functional groups (Glycosil®); and (**c**) thiol-reactive polyethylene glycol diacrylate (PEGDA) cross-linker (Extralink Lite®).

2.2. Material Constitution and Experimental Design

Gycosil® was re-constituted with DG water to form an uncross-linked sol (ux-tHyA). The Hystem® hydrogel (x-tHyA) was prepared by adding the cross-linker PEGDA (Extralink Lite®) with DG water to the ux-tHyA sol, following the manufacturer's instructions. Tables 1 and 2 show the material composition and experimental design of this study, respectively. The raw x-tHyA material used was Gycosil® (sponge) (tHyA). The addition of water to Gycosil® produced sol (ux-tHyA). The addition of a cross-linker to ux-tHyA created x-tHyA. x-tHyA was both sol and gel, depending on the timing of cross-linker mixing completion. Before and after 20 min of mixing, x-tHyA was a sol and a gel, respectively. Therefore, x-tHyA is called a SG material. During the sol stage, it can be injected and gelled in vivo over time (Table 1). In this study, three types of experiments were carried out: (a) material characterization of HyA; (b) characterization of the use of nHAp; and (c) animal experiments using x-tHyA (Table 2).

Table 1. Materials explanation.

Materials	Material Constituent	State
HyA control	HYALURONSAN HA-SHY	Powder
tHyA	Glycosil®	Sponge body
ux-tHyA	Glycosil® + DG water	Sol
At this time, BMP, and nHAp were added to ux-tHyA		
x-tHyA	Glycosil® + DG water + Extralink Lite®	Sol (up to 20 min) Gel (after 20 min)
Dried HyA control	Freeze-dried (HyA control + water)	Sponge body
Dried ux-tHyA	Freeze-dried ux-tHyA	Sponge body
Dried x-tHyA	Freeze-dried x-tHyA	Sponge body

Table 2. Design of materials and experiments.

	(a)			
Dried Samples	Material Characterization of HyA			
	FTIR	SEM	Hyaluronidase Dissolution Tests	TG/DSC
HyA control	•	•	•	•
ux-tHyA	•	•	•	•
x-tHyA	•	•	•	•

Table 2. Cont.

	(b)	
	Characterization of the Use of nHAp	
Samples	Observation of Protein Binding	Accelerated Protein Release Tests
nHAp*FITC-Collagen (±)	●	
x-tHyA SG*nHAp (±)		●

	(c)	
	Animal Experiments	
Samples	Direct Injection into Cranial Area	Set in a Teflon Ring and Placed on Cranial Bone
SG I*nHAp (±)	●	
SG II*BMP (±)		●

Note: sol–gel (SG). ● means the experiment conducted. * means "and".

2.3. Material Characterization of Control HyA, ux-tHyA, and x-tHyA

For material studies, the HyA control was dissolved in distilled water at 0.25 wt% concentration. Samples of HyA control sol, ux-tHyA sol, and x-tHyA gel were frozen at −80 °C and freeze-dried for 12 h. To differentiate the chemical and physical properties of the three HyA materials, four in vitro experiments were performed using dried samples.

2.3.1. FTIR

FTIR equipped with an attenuated total reflectance attachment (Nicolet6700, Thermo Fisher Scientific, Waltham, MA, USA) (using a single reflection diamond, a refractive index of 2.38 at 1000 cm^{-1} and angle of incidence of 45°) was used to characterize the chemical structures of the dried HyA control, ux-tHyA, x-tHyA, and cross-linker PEGDA. During the measurement, the resolution was 4 cm^{-1}, the wavenumber range was 4000–400 cm^{-1}, and the number of scans was 10. The OMNIC software (Thermo Fisher Scientific, Waltham, MA, USA) was used to collect and process the IR spectra. For all recorded FT-IR spectra, corrections for noise from the diamond attachment and CO_2 were performed manually.

2.3.2. SEM

SEM (SU8010, Hitachi High-Tech Corp., Tokyo, Japan) was used at 15 kV to morphologically compare dried HyA control, ux-tHyA, and x-tHyA. The dried HyA samples were glued to carbon tape, placed on an aluminum stub and plasma coated with OsO_4 using an OPC60A (Filgen, Nagoya, Japan). The thickness of OsO_4 was 30 nm.

2.3.3. Hyaluronidase Dissolution Tests

In the hyaluronidase dissolution tests, each sample (1.0 mg) (n = 6) of dried HyA control, ux-tHyA, and x-tHyA samples were dissolved in 0.01 wt% hyaluronidase solution (Code 18240-36, Nacalai Tesque, Kyoto, Japan) diluted in distilled water (0.5 mL) in a 1.5 mL microtube, and had been placed in a constant temperature bath, and kept at 37 °C. The dissolution condition was visually inspected every 6 h, and the time to complete disappearance (min) was recorded.

2.3.4. TG/DSC Thermal Analyses

TG/DSC was performed on each 1 mg sample (dried HyA control, ux-tHyA, and x-tHyA) (n = 1), using specialized equipment (STA409C, Netzsch, Selk, Germany) so that the thermal stability of x-tHyA could be scaled with reference to those of HyA control and ux-tHyA. The experimental conditions for TG/DSC were as follows: atmospheric gas, nitrogen; gas flow rate (sample), 50 mL/min; gas flow rate (reference), 20 mL/min;

temperature range, 20 °C to 550 °C; heating rate, 10 °C/min; sample holder, open aluminum crucible; reference, alumina (6.8 mg); TG resolution, 5 µg; and DSC resolution, <1 µW.

2.4. Characterization of the Use of nHAp

2.4.1. Observation of Binding between nHAp and Protein

Fluorescein isothiocyanate (FITC)-labeled bovine type I collagen (1 mg/mL) (#4001, Chondrex, Inc., Redmond, WA, USA) (0.5 mL) was mixed with nHAp (0.6 mg) containing PBS (−) x2 buffered solution (0.5 mL) in a 1.5 mL microtube and held at 4 °C for 12 h (nHAp*FITC-Collagen (+)). The same solution containing nHAp particles was prepared by mixing 0.01 M acetic solution (0.5 mL) and PBS (−) ×2 buffered solution (0.5 mL) without FITC-labeled collagen (nHAp*FITC-Collagen (−)). Both solutions with nHAp were centrifuged at 56× g (rotation radius = 50 mm, rotation speed = 1000 rpm) for 1 min. The supernatants were discarded, and the bottom pellets were resuspended in PBS (−) solution (300 µL). The solutions were then transferred to glass dishes (25 mm in diameter), stood still for 1 h, and the powders on the bottoms of the two glasses were observed with a confocal laser scanning microscope (A1RHD25, Nikon Co., Tokyo, Japan). The measurement conditions were an excitation wavelength of 488 nm and an emission wavelength of 500–550 nm.

2.4.2. Accelerated Protein Release Tests from x-tHyA Containing nHAp

For accelerated protein release tests of x-tHyA with nHAp in an aqueous environment, x-tHyA sol (6.25 mL) was produced using Gycosil® (50 mg), DG water (4 mL), bovine serum albumin standard (BSA) (2 mg/mL) (1 mL) (Thermo Scientific, Rockford, IL, USA), and PEGDA cross-linker (1.25 mL) with nHAp powder (10 mg) (x-tHyA*nHAp (+)), and each sol was poured into four 1.5 mL microtubes, followed by gelation. Sols without nHAp were prepared in the same proportion and separated into four tubes (x-tHyA*nHAp (−)). Phosphate buffered saline solution (−) at a volume of 1 mL made from PBS tablets (#T900, Takara Bio, Kusatsu, Shiga, Japan) was added to gel samples in tubes after gelation, which had been stored at 37 °C in a constant temperature bath and the solution was collected 1, 3, 5, and 7 days after gelation and stored at −20 °C until measurements while new saline solutions (1 mL) were added to gels 1, 3, and 5 days later. The quantities of BSA eluted in solution were measured using a Pierce BCA protein assay kit (Thermo Scientific, Rockford, IL, USA) with four samples with two repetition measurements (n = 4 × 2) so that the protein release kinetics of x-tHyA and the protein binding/releasing trend of nHAp in x-tHyA could be visualized in a time-dependent manner.

2.5. Preparation of SG I and SG II Materials

Material preparation was performed aseptically on a clean bench.

2.5.1. SG I Sample

The preparation protocol for the SG I samples (x-tHyA + BMP ± nHAp) was as follows. First, BMP (50 µg) was re-constituted with a 4 mM HCl solution (0.25 mL in total) with 0.5 wt% bovine fetal albumin standard (fraction V) (Production no. DK59769, Thermo Scientific Pierce, Waltham, MA, USA) as adjuvant and diluted in DG water (5 mL in total). Second, freeze-dried Glycosil® (50 mg) was re-constituted in the sol state with DG water and BMP (5 mL) on a vibrating mixer (Mild Mixer PR-12, Tokyo, Japan) for 12 h at 20 °C (Liquid A). The Extralink Lite® was diluted with DG water (1.25 mL) (Liquid B). Third, Liquid B was mixed with Liquid A to obtain a viscous solution (Figure 3a). Membrane filtration (0.22 µm) was used for sterilization. Half of the sol was manually mixed with nHAp (50 µg) with a plastic spatula in a 35 mm culture dish (test samples; SG I*nHAp (+) = x-tHyA + BMP + nHAp), while the other half was unmixed (control samples; SG I*nHAp (−) = x-tHyA + BMP). Both SGs were injected with a needle and syringe (Figure 3b), followed by gelation for approximately 20 min at 37 °C. The injection volume of the SG I samples was set at 200 µL in the sol stage. The BMP content of each injected SG I*nHAp (+) was 1.6 µg.

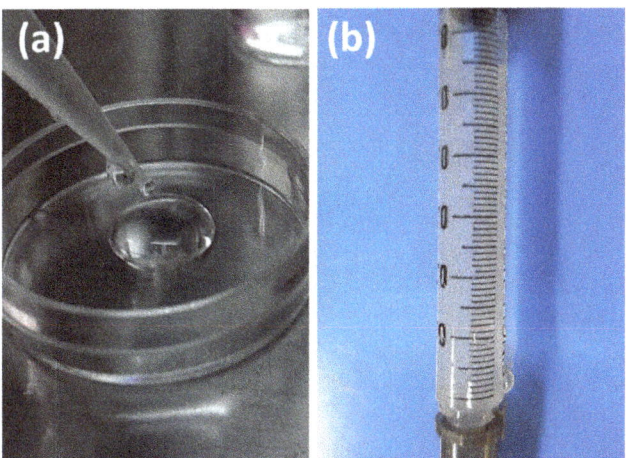

Figure 3. (a) SG I*nHAp (−) = x-tHyA + BMP in sol stage; (b) SG I*nHAp (+) = x-tHyA + BMP + nHAp in sol stage in a syringe.

2.5.2. SG II Sample

In the case of the SG II samples (tHyA + nHAp ± BMP), a mixture of ux-tHyA and nHAp was first prepared, followed by the addition of a BMP-containing HCL solution and a cross-linker solution to form the test samples (SG II*BMP (+) = x-tHyA + nHAp + BMP)—using mixing proportions of x-tHyA, nHAp, and BMP—similar to those of SG I samples. Control samples (SGII*BMP (−) = x-tHyA + nHAp) were also produced using the HCL solution without BMP. Before animal studies, test and control SG II in sol stage (50 µL each) were poured into a Teflon ring (inner hole diameter = 4 mm, outer hole diameter = 6 mm, and thickness = 2 mm) placed on a glass slide and set at 20 °C. The amount of BMP in each SG II*BMP (+) gel sample was 0.4 µg.

2.6. Animal Experiments

The study protocol was approved by the Ethics Committee on Animal Research of Iwate Medical University (#30-001).

2.6.1. SG I Sample

Twenty 10-week-old male BALB/cAJcl mice (CLEA Japan) were used. Groups of two to three mice were housed in separate cages and provided with a standard diet and water ad libitum. Before injection, the skull hairs of the mice were removed mainly using an electric shaver. Under anesthesia with a mixture of isoflurane (3 vol%) and oxygen (0.5 L/min) gas generated by a carburetor (IV-ANE; Olympus, Tokyo, Japan), the test and control SG I samples were injected in sol stages (0.2 mL) (SG I*nHAp (+) and SG I*nHAp (−), respectively) into the cranial subcutaneous tissue of mice (Figure 4), respectively (n = 10 each) with the use of a 24-gauge needle, and fed for a duration of 8 weeks. The animals were sacrificed by CO_2 inhalation.

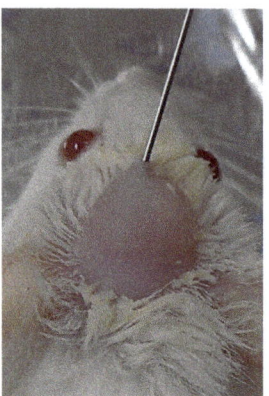

Figure 4. Subcutaneous injection of SG I*nHAp (+) = x-tHyA + BMP + nHAp sample in sol stage into the mouse cranial area.

2.6.2. SG II Sample

Four male Wistar rats weighing 340 ± 16 g (mean ± SD) were used. All rats were housed in separate cages (two rats per cage) with a standard diet and water ad libitum. Under anesthesia with a mixture of isoflurane and oxygen, the centers of the rat calvariae were shaved and sterilized with 10% povidone iodine, followed by a local injection of anesthetic (0.2 mL, 2% lidocaine with 1:80,000 epinephrine). Then, the periosteum flaps were elevated and the cranial bone was exposed with a scalpel and bone forceps. Two rats were used for test and control SG II gels, respectively. Three tests and control SG II gels (SG II*BMP (+) and SG II*BMP (−), respectively) in Teflon rings (Figure 5a) were placed directly on the cranial bones of a rat (Figure 5b) and tightly closed with soft nylon (Softretch 4-0; GC, Tokyo, Japan). Most of the periosteum was detached from the cranial bone during the operation. Eight weeks after surgery, all rats were sacrificed by CO_2 inhalation.

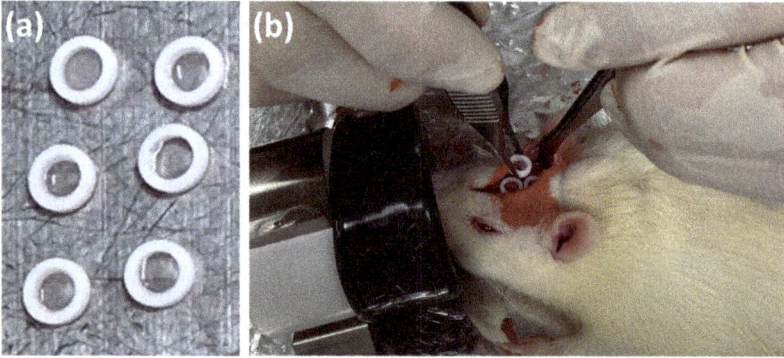

Figure 5. (a) Prepared SG II*BMP (+) = x-tHyA + nHAp + BMP samples gelled in Teflon rings; and (b) implantation of SG II*BMP (+) samples on exposed rat cranial bone.

2.6.3. X-ray Analyses

We evaluated new bone formation by SG I samples in the cranial subcutaneous tissues of mice using a 3D microcomputed tomography (micro-CT) system (eXplore Locus; GE Healthcare, Wilmington, MA, USA). The ossification trends in the operation areas of the cranial bones of rats using SG II samples were evaluated using a soft X-ray apparatus (M60, Softex, Tokyo, Japan).

2.6.4. Histological Observations

Skulls or cranial skins of rats used for SG I and SG II samples were collected after feeding for 8 weeks with a diamond saw (MC-201 Microcutter; Maruto, Tokyo, Japan) or scissors, fixed in 10% neutral buffered formaldehyde equivalent (Mildform, Wako Chemical, Osaka, Japan) for 4 weeks at 4 °C, and decalcified in 0.5 wt% ethylene diamine tetra-acetate solution (Decalcifying solution B, Wako Chemical, Osaka, Japan) for 4 weeks at 4 °C. The cranial regions were cut from the skulls, treated with graded alcohol and xylene, and embedded in wax. The wax specimens were then cut into five μm sections using a microtome (IVS-410, Sakura Finetek, Tokyo, Japan). The sections of the slides were stained with hematoxylin and eosin (HE), followed by histological observations using fluorescence microscopy (All-in-one BZ-9000; Keyence, Osaka, Japan).

In Figure 6, the evaluation method of ossification of cranial bone for SG II samples based on line measurements is schematically illustrated. In Figure 6a, the newly formed bone by SG II is indicated in the boxed area on the left side. The dotted box area in Figure 6a is enlarged in Figure 6b. Bone formation activities were scaled by the length of the bone zone on multiple perpendicular lines. For example, the bone length in line 4 was determined by the sum of 4a, 4b, and 4c. The bone-forming activities in the Teflon ring (4 mm wide) were assessed using lines at 100 μm intervals (40 lines). As a control, the boxed area of the sham bones was selected (Figure 6c), and the bone length in lines at 100 μm intervals (40 lines) inside the boxed area in Figure 6c was measured (Figure 6d). The bone-forming activities of the SG II samples were evaluated with respect to those of the sham control bones.

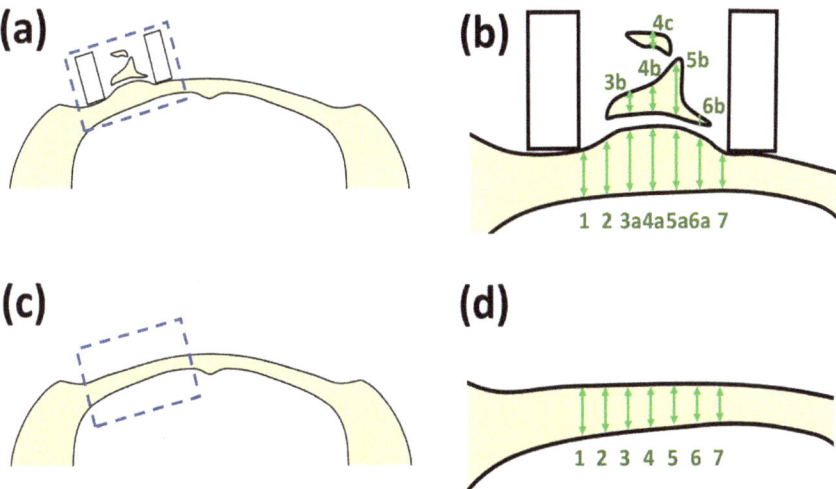

Figure 6. (a) Newly formed bone by SG II sample on the left side cranial bone; (b) enlarged bone area of the dotted box area in Figure 6a. Note: bone formation activities were scaled by the length of the bone zone on multiple perpendicular lines; (c) the sham rat cranial bone; and (d) the enlarged bone area of the dotted box area in Figure 6c. Note: Bone thickness was scaled by the length of the bone zone on multiple perpendicular lines.

2.7. Statistical Analyses

Free statistical software (EZR version 1.55, Saitama Medical Center, Jichi Medical University, Saitama, Japan) [49] was used for nonparametric tests, such as Fisher's exact, Kruskal–Wallis and Mann–Whitney U tests. The null hypothesis was rejected at $p < 0.05$.

3. Results

3.1. Material Characterization of HyA Control, ux-tHyA, and x-tHyA

3.1.1. FTIR

Figure 7a shows the FTIR charts of dried (i) HyA control, (ii) ux-tHyA, (iii) x-tHyA, and (iv) PEGDA cross-linker. Figure 7b shows the chemical structures of ux-tHyA with detailed side chains [35], PEGDA cross-linker, and x-tHyA. The three materials (HyA control, ux-tHyA, and x-tHyA) had similar IR absorption peaks, such as OH or NH peaks at approximately 3300 cm^{-1}, C=O peak at approximately 1600 cm^{-1}, and C-O peak at 1045 cm^{-1} due to the common basic HyA structures. In contrast, two test tHyA samples (ux-tHyA and x-tHyA) possessed a weak SH stretching peak at approximately 2600 cm^{-1} while control HyA lacked this peak, which reflected thiol modification of both ux-tHyA and x-tHyA. The effect of PEGDA cross-linking on x-tHyA is ambiguous. The relative peak height intensity at approximately 2900 cm^{-1} due to C-H of x-tHyA was higher than that of ux-tHyA, resulting from adding PEGDA elements with abundant C-H bonds to ux-tHyA.

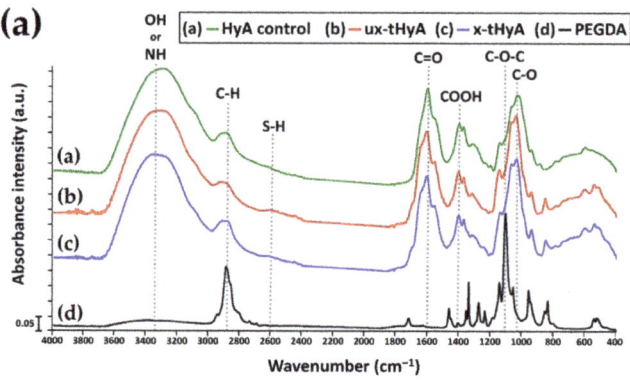

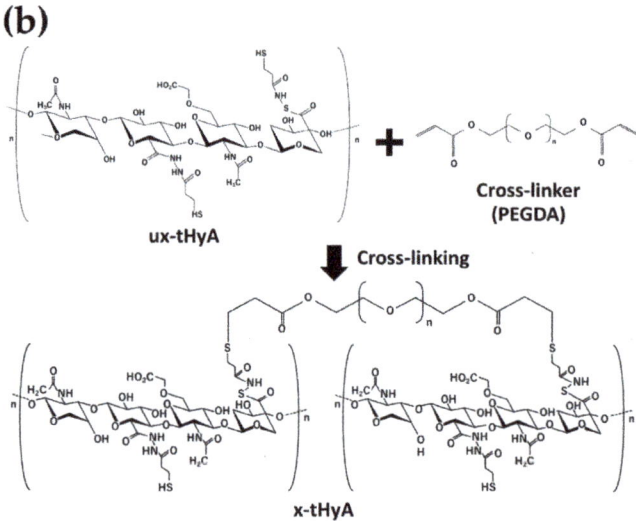

Figure 7. (a) FTIR charts of dried (i) HyA control, (ii) ux-tHyA, (iii) x-tHyA, and (iv) PEGDA cross-linker. (b) Chemical structures of ux-tHyA with detailed side chains, PEGDA cross-linker and x-tHyA.

3.1.2. SEM Observations

Figure 8a shows an SEM photomicrograph of a dried HyA control. It was highly porous and fibrous. Figure 8b,c show those of the dried ux-tHyA and x-tHyA, respectively. Microscopically, ux-tHyA had a loose and porous structure, while x-tHyA had a denser and flat structure.

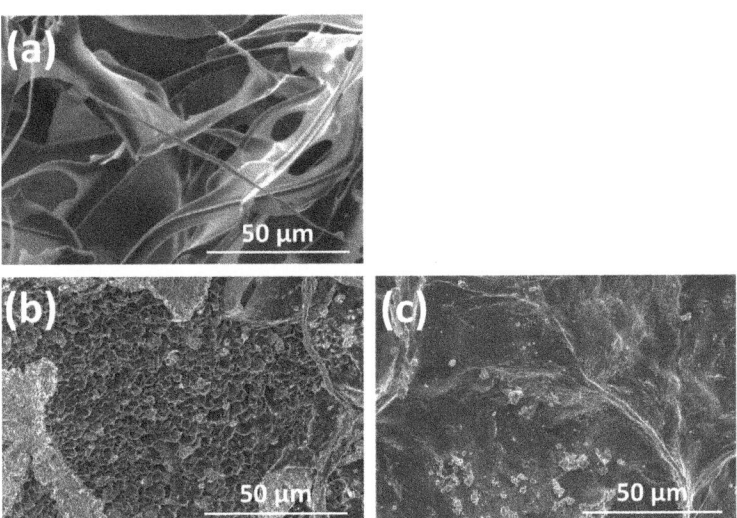

Figure 8. SEM photo-micrographs of dried (**a**) HyA control, (**b**) ux-tHyA, and (**c**) x-tHyA.

3.1.3. Hyaluronidase Dissolution Tests

Figure 9 shows the results of the hyaluronidase dissolution tests of three dried HyA samples. The mean dissolution time of x-tHyA was 2763 min, significantly longer than those of the HyA control and ux-tHyA of < 110 min ($p < 0.05$). This means that cross-linking of x-tHyA resulted in a significant increase in hyaluronidase dissolution time compared with that of uncross-linked ux-tHyA. The dissolution time of ux-tHyA (104 min) was longer than that of the HyA control (17 min) ($p < 0.05$).

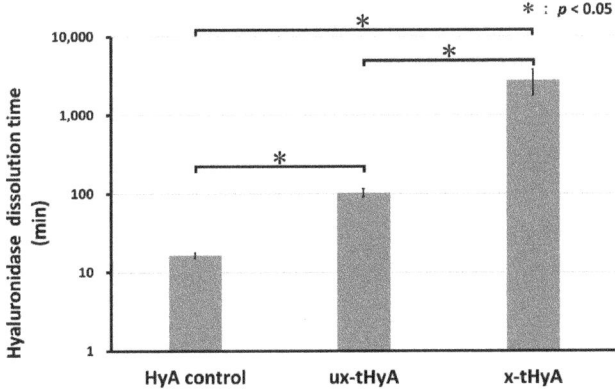

Figure 9. Results of the hyaluronidase dissolution test of three dried HyA samples—HyA control, ux-tHyA, and x-tHyA. Standard deviation error bars were added to graphs. The statistical analysis between two graphs combined by horizontal bar was carried out by Kruskal–Wallis test. *: $p < 0.05$.

3.1.4. TG/DSC

Figure 10a,b show the TG and DSC curves of the TG/DSC thermal analyses of the dried HyA control, ux-tHyA, and x-tHyA, respectively. The TG curves revealed that all three HyL samples maintained their weight up to approximately 220 °C, followed by gradual weight loss, while the loss rates of ux-tHyA and x-tHyA were smaller than that of the HyA control (Figure 10a). The DSC curves clarified that both ux-tHyA and x-tHyA had broad distinctive endothermic peaks in the temperature range of 200–400 °C, while the HyA control lacked a peak (Figure 10b). The two peak endothermic temperatures of the DSC curves are indicated by the arrows.

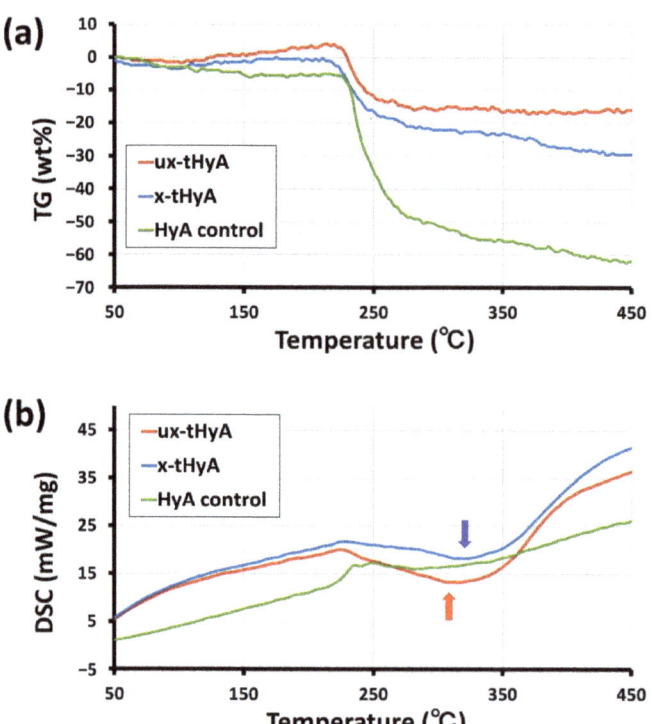

Figure 10. (a) TG and (b) DSC charts of TG/DSC thermal analyses of dried HyA control, ux-tHyA, and x-tHyA. Note: Arrows indicated peak temperatures of DSC endothermic curves of two samples.

3.2. Characterization of the Use of nHAp

3.2.1. Observation of Direct Binding between nHAp and Protein

Figure 11a–c show the bright field, fluorescence, and overlay images of nHAp particles without FITC-labeled collagen (nHAp*FITC-Collagen (−)), respectively. It was confirmed that the nHAp itself was not fluorescent and that the nHAp particles tended to agglomerate in the saline solution. Figure 11d–f show those of nHAp particles mixed with FITC-labeled collagen(nHAp*FITC-Collagen (+)). It became evident that FITC-labeled type I collagen was strongly bound to nHAp powders, causing strong fluorescent reflections. This is a direct proof of the binding between nHAp and the protein. In the latter case, nHAp agglomeration appeared to be suppressed by the presence of collagen.

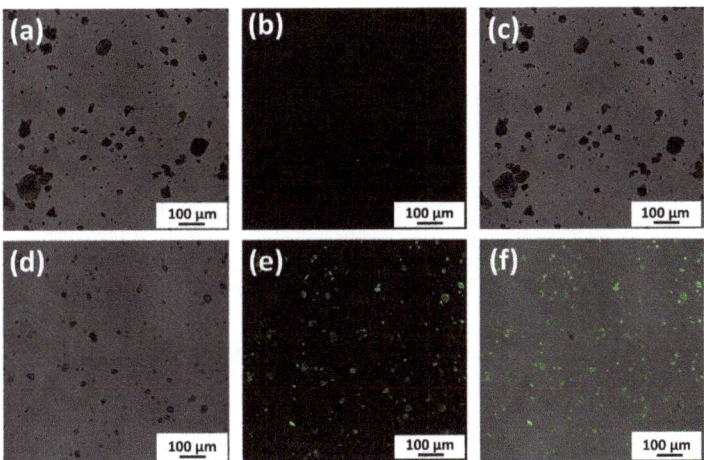

Figure 11. (**a**) The bright field, (**b**) fluorescence, and (**c**) overlay images of nHAp particles without FITC-labeled collagen (nHAp*FITC-Collagen (−)); (**d**) bright field, (**e**) fluorescence, and (**f**) overlay images of nHAp particles mixed with FITC-labeled collagen (nHAp*FITC-Collagen (+)).

3.2.2. Accelerated Protein Release Test

Figure 12 indicates the results of the accelerated protein release test of x-tHyA with and without nHAp (x-tHyA*nHAp (−) and x-tHyA*nHAp (+)) in a saline solution. The cumulative protein release amounts versus the original BSA quantity (wt%) in the x-tHyA gels were plotted as a function of the elution period (days). Figure 12 presents two important observations. The BSA protein was retained but slowly released from x-tHyA without nHAp (x-tHyA*nHAp (−)) in a time-dependent manner. However, the amount of protein eluted (wt%) from x-tHyA*nHAp (−) reached 100% after incubation for 7 days. In the case of the addition of nHAp to x-tHyA (x-tHyA*nHAp (+)), the amount of BSA protein released slightly decreased in all elution periods compared to those of x-tHyA*nHAp(−) ($p < 0.05$; Mann–Whitney U test). These results implied that nHAp was strongly bound to the BSA protein in x-tHyA and retarded the release of protein from x-tHyA into a saline solution, and x-tHyA*nHAp(+) could release the protein for more than 7 days.

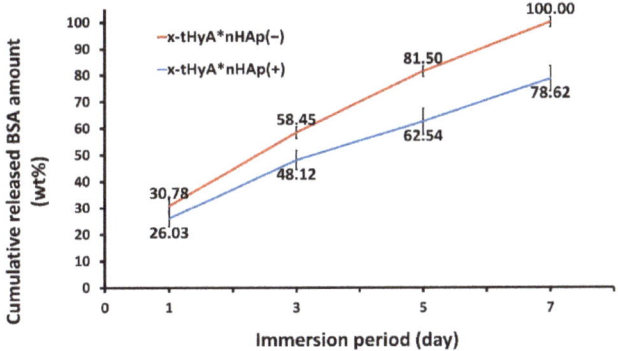

Figure 12. Results of the accelerated protein (BSA) release test of x-tHyA with and without nHAp (x-tHyA*nHAp (−) and x-tHyA*nHAp (+)) in saline solution.

3.3. Animal Studies

3.3.1. Animal Experiments with SG I

When using the SG I test material (SG I*nHAp (+)), cranial ossification was observed in subcutaneous connective tissue in 7 out of 10 mice, while the control SG I materials (SG I*nHAp (−)) did not form bone in 10 mice, and most Teflon rings were dropped. According to Fisher's exact test, it can be stated that test SG I caused statistically significant ossification in the cranial subcutaneous connective tissues (success rate = 70%) of mice compared to the control SG I (success rate = 0%) ($p < 0.05$) (Table 3).

Table 3. Bone-forming numbers in the cranial subcutaneous connective tissues of mice 8 weeks after injection of 10 control SG I*nHAp (−) and 10 test SG I*nHAp (+) samples, respectively.

Sample	Number of Ossification Confirmed	Number of Ossification Not Confirmed
Control SG I*nHAp (−) [#] (x-tHyA + BMP)	0	10
Test SG I*nHAp (+) [#] (x-tHyA + BMP + nHAp)	7	3

* means "and". [#] $p < 0.05$ by Fisher's exact test.

The ossification of SG I (SG I*nHAp (+)) is unique. Figure 13a,b show the front and side micro-CT images of one mouse cranial bone 8 weeks after SG I*nHAp (+) was injected into the cranial connective tissue. Cranial bone formation was evident. Figure 13c shows four sliced rectangles inside the newly formed bone on the existing cranial bone shown in Figure 13b. Figure 13d schematically indicates the state of bone existence on four slices of Figure 13c. Most of the ossification occurred inside subcutaneous connective tissues such as ectopic bone, while bone augmentation and connection with preexisting cranial bone were quite limited and partial (e.g., cases (ii) and (iii) of Figure 13d). Figure 13e shows a low-magnification HE-stained image of the cranial subcutaneous tissue of a mouse 8 weeks after injection of SG I (SG I*nHAp (+)) material. Figure 13f shows a high-magnification HE image of the yellow rectangle area in Figure 13e. The residual material (RM) appears to be light purple. The periosteum (PS) originated in the deep interior of the cranial bone and thickened. Island-like small new bone (NB) fragments exist above the existing cranial bone (EB), both of which are anchored to the periosteum (PS). This continuation of bone between EB and NB bridged by PS was evidence of partial bone augmentation of preexisting bone.

In contrast, the control SG I (SG I*nHAp (−)) material did not induce new bone (NB) formation in mouse cranial bone. Figure 14a shows a front micro-CT image of one mouse cranial bone 8 weeks after control SG I was injected into the cranial connective tissue. Figure 14b shows a low-magnification HE-stained image of the cranial subcutaneous tissue of a mouse 8 weeks after injection of SG I control material (SG I*nHAp (−)). Figure 14c shows a high-magnification HE image of the yellow rectangle area in Figure 14b. Residual material (RM) was present above the cranial bone, but did not cause bone formation.

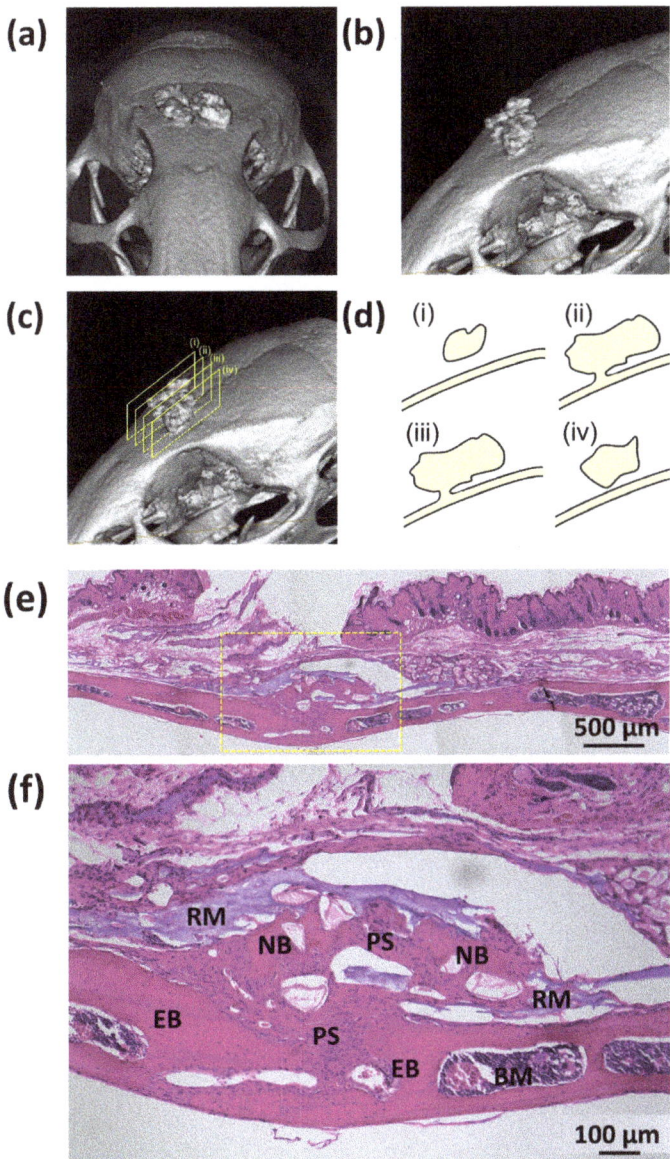

Figure 13. (**a**) Front and (**b**) side micro-CT images of one mouse cranial bone 8 weeks after test SG I (SG I*nHAp (+)) was injected into the cranial connective tissue; (**c**) four sliced rectangles inside the newly formed bone on the cranial existing bone of Figure 13b; (**d**) schematical indication of bone existence states on four slices of Figure 13c; (**e**) the low magnified HE-stained image of the cranial subcutaneous tissue of a mouse 8 weeks after injection of SG I (SG I*nHAp (+)) material; (**f**) the high-magnified HE image of the yellow rectangle area of Figure 13e. Note: RM, residual material; PS, periosteum; NB, new bone; EB, existing bone; BM, bone marrow.

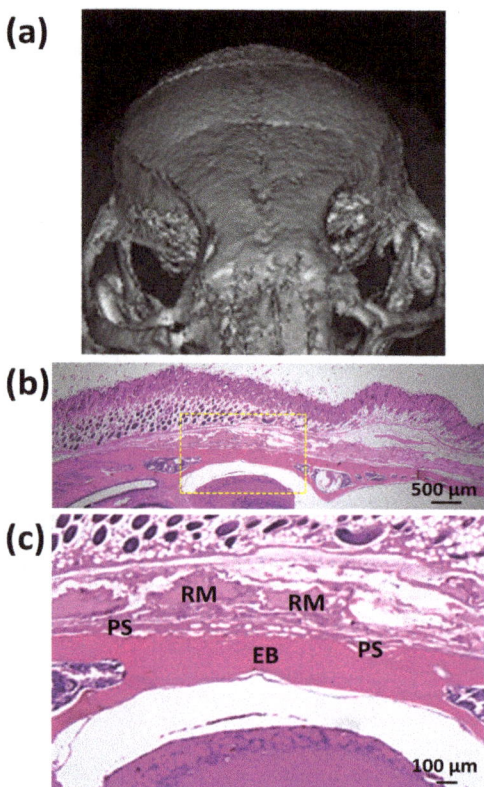

Figure 14. (**a**) Front micro-CT image of one mouse cranial bone 8 weeks after control SG I (SG I*nHAp (−)) was injected into the cranial connective tissue; (**b**) the low magnified HE-stained image of the cranial subcutaneous tissue of a mouse 8 weeks after injection of SG I (SG I*nHAp (−)) material; and (**c**) the high-magnified HE image of the yellow rectangle area of Figure 14b. Note: RM, residual material; PS, periosteum; EB, existing bone.

3.3.2. Animal Experiments with SG II

When using test SG II (SG II*BMP (+)) in six Teflon rings on rat cranial bones, one ring fell off and five rings remained (83% survival and bone formation rate). Figure 15a,b show three test SG II samples inside Teflon rings placed on one rat cranial bone after 8 weeks of placement and the corresponding soft X-ray images, respectively. Test SG II induced additional bone formation inside the Teflon ring. Figure 15c shows a low-magnification HE-stained image of the cranial bone of a rat 8 weeks after the placement of test SG II (SG II*BMP (+)). Figure 15d,e show higher magnified HE images the sites of which are indicated by yellow rectangles (i) and (ii) in Figure 15c, respectively. Test SG II (SG II*BMP (+)) produced NB both on the preexisting cranial bone (EB) (Figure 15c,d) and around the RM that appears in light purple (Figure 15c,e). The boundary between the NB and EB was observed, as indicated by the * marks in Figure 15d. This case of bone formation can be termed bone augmentation. It was also characteristic that the periosteum (PS) thickened and appeared to prelude to NB. Around the residual test SG II*BMP (+) (RM), both the NB and PS existed as island forms in Figure 15e far away from the preexisting bone. This type of bone may be called ectopic bone in connective tissue. Figure 16a shows a low-magnification HE image of the sham rat cranial bone. Figure 16b shows a high-magnification HE image of the yellow rectangle area shown in Figure 16a. No bone increment was observed above the preexisting bone. Figure 16c shows the tool-box graph of multiple line-scaled cranial bone lengths using test SG II (SG II*BMP (+)) and sham cranial bones. In the test SG II material, the bone length in the cranial region was enlarged with a statistically significant difference compared to the sham bones ($p < 0.05$).

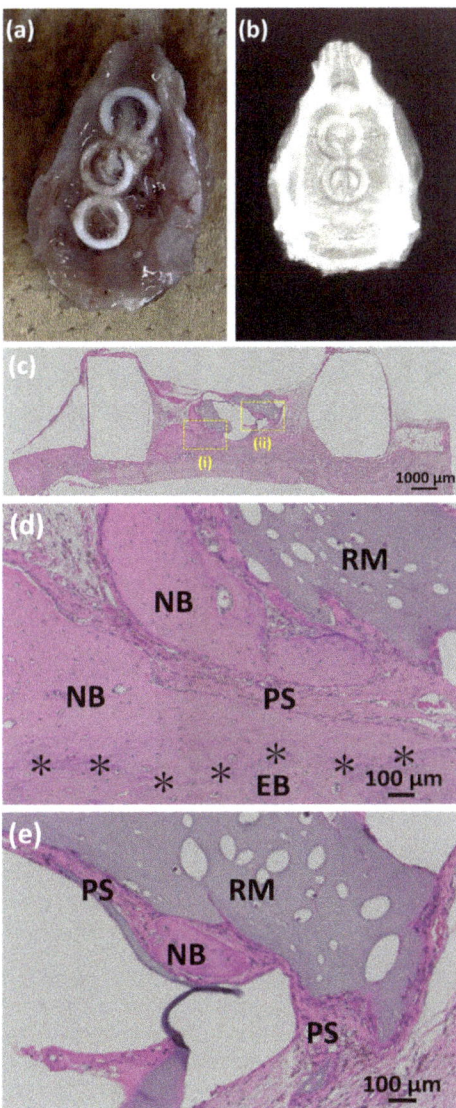

Figure 15. (**a**) Photo and (**b**) soft X-ray image of three test SG II samples (SG II*BMP (+)) inside Teflon rings placed on one rat cranial bone after 8 weeks of placement; (**c**) the low magnified HE-stained image of the cranial bone of a rat 8 weeks after the placement of test SG II (SG II*BMP (+)); (**d**) higher magnified HE images of the yellow rectangle (i) of Figure 15c; and (**d**,**e**) higher magnified HE images of the yellow rectangle (ii) of Figure 15c. Note: RM, residual material; PS, periosteum; NB, new bone; EB, existing bone. Note: * indicates the border between NB and EB.

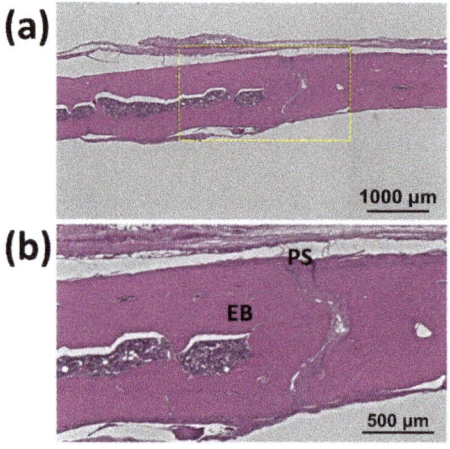
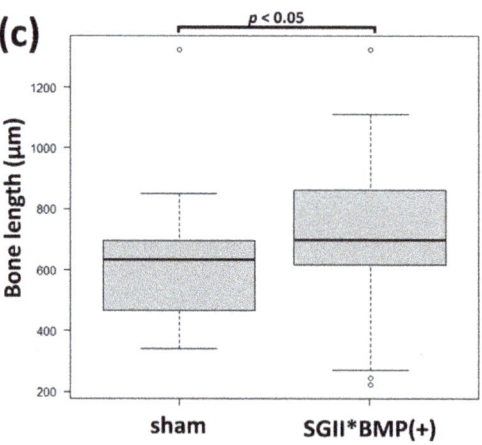

Figure 16. (a) The low-magnified HE image of the sham rat cranial bone, (b) the high-magnified HE image of the yellow rectangle area in Figure 16a, and (c) the tool-box graph of multiple line-scaled bone length of cranial area bone using test SG II (SG II*BMP (+) = x-tHyA + nHAp + BMP) and sham cranial bones. The statistical analysis was performed by Mann–Whitney U test.

When using the control SG II (SG II*BMP (−)), two rings were lost; two rings were filled with dermal tissue and a slight ossification was found inside the two rings on the cranial bone. Control SG II did not induce stable cranial bone formation within Teflon rings. Figure 17a,b show the skin of a rat with control II gel (SG II*BMP (−)) in two Teflon rings after 8 weeks of placement and the corresponding soft X-ray image, respectively. The Teflon rings were separated from the cranial bone and embedded in the cranial skin. NB did not form inside the rings, while weak X-ray opacity was observed inside the rings, analogous to the surrounding skin tissues (Figure 17b). Figure 17c shows a low-magnification HE-stained image of the interior of the Teflon ring 8 weeks after placement of control gel II (SG II*BMP (−)). Figure 17d shows the magnified HE-stained image of the yellow-dotted rectangle in Figure 17c. Inside the ring, dermal tissues (DM) were filtered and filled the inner space of the Teflon ring, while residual material (control gel II) (RM) was minimally observed. Inside the dermal tissues, inflammatory cells were widely infiltrated, blood vessels existed, and no bony structures were found.

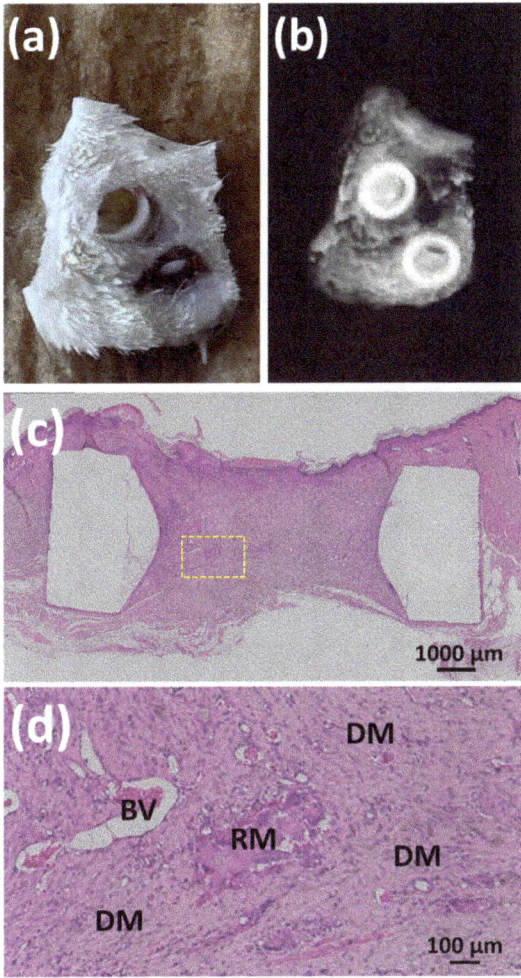

Figure 17. (**a**,**b**) Photograph of the skin of a rat with control II gels (SG II*BMP (−)) in two Teflon rings after 8 weeks of placement and the corresponding soft X-ray image, respectively; (**c**) low-magnification HE-stained image of the interior of the Teflon ring 8 weeks after the placement of control gel II (SG II*BMP (−)); and (**d**) magnified HE-stained image of the yellow-dotted rectangle in Figure 17c. Note: RM, residual material; DM, dermal tissues; BV, blood vessels.

4. Discussion

Cross-linkable HyA (x-tHyA) was confirmed by FTIR to be thiol-modified and cross-linked using a PEGDA cross-linker (Figure 7). The physical and chemical properties of uncross-linked and cross-linked thiol-modified HyA (ux-tHyA and x-tHyA) appear to be caused by the characteristic base structure and polymerization reaction. Glycosil® and Extralink Lite® covalently bond to each other like Lego blocks. When mixed, Extralink's acrylate reacts with the thiol groups of the Glycosil® components by click chemistry (Michael addition reaction). Crosslinks form in trans (e.g., Glycosil molecules can link to neighboring Glycosil molecules). In addition, given Glycosil's large molecular weight and ability to adopt semiflexible random coil configurations, it is likely to loop back and bind to the cis. The final clear, transparent, viscoelastic hydrogel formed at physiological pH and temperature in approximately 20 min and was greater than 98% water. This timeframe allows an investigator to customize the hydrogel with drugs to load and deliver the mixture through a cannula [33]. The molecular weight of ux-tHyA must be significantly greater and more hydrophobic than that of the HyA control. Therefore, ux-tHyA absorbed less water, resulting in a porous but denser structure compared to the HyA control after freeze-drying (Figure 8); and it was relatively heat resistant, comparable to x-tHyA (Figure 10). It also became evident that by cross-linking, x-tHyA became very resistant to hyaluronidase (Figure 9), had a dense and flat surface upon freeze-drying (Figure 8), and was relatively heat-durable when heated at temperatures higher than 200 °C (Figure 10). This plain surface might be attributed to the formation of stronger film-like structures of x-tHyA after freeze-drying. The freeze-dried structures of HyA samples might reflect the molecular and cross-linked structures of gels with abundant water, while an increase in molecular weight leads to a decline in water content [50]. Water evaporation from the lower molecular structures of HyA materials (e.g., HyA control) during freeze-drying might create fibrous and porous freeze-dried structures. However, the fibrous and porous structures of the HyA control and ux-tHyA were not used in the animal studies of this study. Gels (x-tHyA) prepared by mixing ux-tHyA and water were used as base materials and applied to the cranial area of murines.

It was confirmed that the protein (collagen type I labeled with FITC) was directly bound to nHAp (Figure 11), which made the study to consider adding nHAp and BMP to x-tHyA meaningful. Two experimental results using test SG I and SG II indicate that x-tHyA + nHAp + BMP SG material was osteoinductive in the murine cranial areas (Figures 13, 15 and 16). The dual use of nHAp and BMP is a prerequisite for successful bone formation using x-tHYA (Table 3). The lack of nHAp or BMP led to the failure of reliable bone formation in both SG I and SG II studies. For bone formation using x-tHyA, BMP is primarily necessary for direct bone induction [51], and nHAp is also indispensable as a carrier to absorb and slowly release BMP to maintain the subsequent time required bone formation [52–55]. Proteins have been reported to be absorbed and slowly released by nHAp through electrostatic interactions [56]. It was reported that x-tHyA maintained and could release BMP for up to 4 weeks [33], but our experimental results (Figure 12) contradict this report. x-tHyA itself (x-tHyA*nHAp(−)) retained the BSA protein for only 7 days (1 week). In contrast, with the addition of nHAp, x-tHyA (x-tHyA*nHAp(+)) retained BSA protein over 7 days, and its dissolution was capable of continuing for 2–3 weeks (Figure 12). This delayed protein release has been considered beneficial for bone regeneration [52,53], which generally requires a long period of time (more than a month) [57]. Before animal studies, it is important to unveil these material characteristics of x-tHyA because few studies have been published [33].

Although the protein type and quantities between BSA in the elution tests and BMP used in the animal studies differed considerably, the phenomena observed in the previous test results (Figure 12)—especially the delayed protein elution trend by nHAp—might be applicable to the latter animal studies. We used three mixed proteins due to their availability and cost performance. The approximate molecular weights of BMP, BSA, and collagen type I are 26, 66, and 300 kDa, respectively [58]. The size of the protein is a key element for apatite binding. Although apatite can bind to all three proteins, a protein with a lower molecular weight is considered to bind more easily to nHAp. Such proteins with lower molecular weights could be loaded and released from nHAp for a longer period. Increased molecular weight of the protein may present a 3D conformational hindrance [59] to bind with nHAp. The surface charge may be another important factor [60]. HAp has positive and negative charges, which are beneficial for binding to oppositely charged proteins (including almost all charged proteins). Therefore, the findings observed in BSA and type I collagen for nHAp could be applicable to BMP for nHAp with intensified levels. The use of BSA instead of BMP for slow-release studies of carrier materials has been common and is considered a standard protocol [61]. BSA has been loaded as a protein model drug in many studies [62].

The consideration of animal studies is as follows. As mentioned, x-tHyA was biocompatible and largely bio-absorbed but remained in vivo for 8 weeks and did not cause adverse effects such as inflammation, immunological reactions, or fibrous tissue encapsulation [63].

We used mice and rats for the SG I and SG II experiments, respectively, because mice were more cost-effective and easier to handle for simple injection, and Teflon ring insertion was practically limited to the size of the rat cranial zone. During feeding, the animals actively moved and touched the injected sites, leading to movement and loss of the injected material in vivo. Consequently, bone formation at the target location may be hindered. The use of the cap and band could stabilize and protect the injected materials in future studies [64].

Martinez-Sanz et al. [19] reported a successful bone augmentation in mandibular bone in rats by injection of gels consisting of self-formulated cross-link-type HyA, nHAp, and BMP. They precisely injected their biomaterials subperiosteally into the innate mandibular diastema (an inactive site) and increased bone volume proportional to the dose of BMP applied. However, subperiosteal injection of gel into the cranial bone is generally quite difficult [65], and success reports have been rare [66].

In this study, we performed the injection of test SG I into the subcutaneous tissues above the cranial bone of mice and placement of SG II directly on the exposed cranial bone of rats, lacking full cover of the periosteum, respectively, and achieved partial success in additional cranial bone formation. During the injection of test SG I, the tip of the needle might slightly break the periosteum, leading to limited contact between the SG and the cranial bone, and most of the injected substances were located within the cranial connective soft tissue. Ectopic bone formation [39] occurred predominantly in cranial connective tissues, and cranial bone augmentation of preexisting bone by the membranous ossification mechanism [67] occurred in limited amounts (Figure 13). The former type of bone formation is undesirable for future clinical use. The size and morphology of the newly formed bones by test SG I varied between samples. We report a case of ectopic bone formation by SG I in Figure 13. Ectopic bone formation has been reported to significantly alter the size and morphology, making morphological analysis quite difficult [68,69]. When placed directly on the cranial bone using test SG II inside the Teflon ring, the periosteum was detached from the cranial bone, and most folded to the periphery during the first operation. During the healing process, the periosteum might recover cranial bone, while test SG II was in the process of biodegradation. We adopted a new morphological evaluation method for bone formation using SG II materials containing BMP based on line measurements (Figure 6). The bone formation level of the test SG II materials was not compared with that of the control SG II materials because the latter did not produce reliable numerical data. We used the heights of the sham bones as a control for comparison with test SG II and confirmed that the test SG II materials induced considerable new bone formation (Figure 16). Two mixed modes of bone formation were observed. One major part was bone augmentation from preexisting bone, while the other minor part was ectopic bone in the connective tissue around the remaining material above the preexisting bone (Figure 15). The latter bone was also undesirable.

It should be stated that the BMP application site determines bone quality, such as in ectopic and orthotopic models [70]. When applied to an osseous site, the osseous bone may be formed. However, when applied to soft tissue, such as connective tissues and muscle, BMP might cause ectopic bone formation. Prudent site selection is necessary for tissue engineering when using injectable scaffolds with BMP for bone augmentation. The appropriate dose of BMP is also an important factor for successful bone augmentation. The amount of BMP used in this study was comparable to that used in other studies [17,19,40,52,68,70]. Precise subperiosteal injection of SG in sol stage containing moderate dose of BMP is anticipated by developing a new surgical technique to achieve 100% bone augmentation from preexisting bone [71].

An alternative approach could be to fill box-type bone defects with SG material coupled with a covering of membranous material [72], although it deviates from injection-only treatment. Bone formation using only injected sols is a fascinating technique for future clinical dentistry due to its ease of handling and less invasive treatment, as mentioned previously [8,73]. In situ gelation of the poured sol is desirable for long-term position stability [74]. It is hoped that the set x-tHyA gel that is currently still soft and viscous will be hardened to increase the position stability.

5. Conclusions

Within the limitations of this study, using an SG material—thiol-modified hyaluronic acid (x-tHyA)—the following conclusions were obtained:

(1) Instrumental analyses (FTIR, TG/DSC, and SEM) confirmed the presence of thiol modifications and characteristic cross-linking reactions. The cross-linked structure appeared to result in a considerably long hyaluronidase dissolution time and a denser/plain microstructure upon freeze-drying.

(2) We confirmed that nHAp is directly bound to proteins and could play an important role as a growth factor carrier to hold and release proteins for a long period, which is beneficial for sustaining the time required bone formation.

(3) BMP is a strong bone-forming growth factor. However, its application must be localized to osseous sites for clinically significant bone augmentation. BMP easily causes undesirable ectopic bone formation when applied to connective soft tissues. It is highly anticipated that a technique will be developed for the precise subperiosteal injection of bone-forming SG materials.

(4) The combined use of nHAp and BMP for x-tHyA succeeded in bone formation at the rate of 70–83%, achieving a relatively high positive record. The dual use of nHAp and BMP is vital for successful bone formation.

(5) Because the materials used (x-HyA, nHAp, and BMP) are widely available, it is highly desirable to develop a new injectable bone-forming system consisting of these materials for widespread clinical use soon.

Author Contributions: Conceptualization, M.T. and H.K.; Methodology, M.T.; Validation and investigation, Y.H., M.T., M.H., W.H., T.S. and H.K.; Formal analysis, Y.H., M.T., M.H. and T.S.; Writing—original draft preparation, Y.H., M.T. and M.H.; Writing—review and editing, W.H. and T.S.; Supervision and project administration, H.K. All authors have read and agreed to the published version of the manuscript.

Funding: This study was funded in part by supported by JSPS KAKENHI grant nos. 21K09984 and 21K17070.

Institutional Review Board Statement: The animal study protocol was approved by the Institutional Ethics Committee of Iwate Medical University on 16 April 2018 (approval number: 30-001).

Data Availability Statement: All data are included in the manuscript.

Acknowledgments: We thank Tomohito Hanasaka of Technical Support Center for Life Science Research, Iwate Medical University for assistance on obtaining SEM photo-micrographs and fluorescence bio-images.

Conflicts of Interest: The authors declare no conflict of interest.

Abbreviations

BMP	Bone morphogenetic protein
BV	blood vessels
DG water	Degassed water
DM	Dermal tissues
DSC	Differential scanning calorimetry
EB	Existing bone

FTIR	Fourier-transformed infrared spectroscopy
HAp	Hydroxyapatite
HE	Hematoxylin and eosin
HyA	Hyaluronic acid
micro-CT	Micro-computed tomography
NB	New bone
nHAp	Nano-hydroxyapatite
PEGDA	Polyethylene glycol diacryalate
PS	Periosteum
RM	Residual material
SEM	Scanning electron microscopy
TG	Thermogravimetric analysis
ux-tHyA	Un-cross-linked thiol-modified hyaluronic acid
x-tHyA	Cross-linked thiol-modified hyaluronic acid

References

1. Zhang, W.; Tullis, J.; Weltman, R. Cone Beam Computerized Tomography Measurement of Alveolar Ridge at Posterior Mandible for Implant Graft Estimation. *J. Oral Implant.* **2015**, *41*, e231–e237. [CrossRef] [PubMed]
2. Yamada, M.; Egusa, H. Current bone substitutes for implant dentistry. *J. Prosthodont. Res.* **2018**, *62*, 152–161. [CrossRef] [PubMed]
3. Sheikh, Z.; Sima, C.; Glogauer, M. Bone Replacement Materials and Techniques Used for Achieving Vertical Alveolar Bone Augmentation. *Materials* **2015**, *8*, 2953–2993. [CrossRef]
4. Roca-Millan, E.; Jané-Salas, E.; Marí-Roig, A.; Jiménez-Guerra, A.; Ortiz-García, I.; Velasco-Ortega, E.; López-López, J.; Monsalve-Guil, L. The Application of Beta-Tricalcium Phosphate in Implant Dentistry: A Systematic Evaluation of Clinical Studies. *Materials* **2022**, *15*, 655. [CrossRef] [PubMed]
5. El Deeb, M.E.; Tompach, P.C.; Morstad, A.T. Porous hydroxylapatite granules and blocks as alveolar ridge augmentation materials: A preliminary report. *J. Oral Maxillofac. Surg.* **1988**, *46*, 955–970. [CrossRef]
6. Ishikawa, K.; Hayashi, K. Carbonate apatite artificial bone. *Sci. Technol. Adv. Mater.* **2021**, *22*, 683–694. [CrossRef]
7. El Deeb, M. Comparison of three methods of stabilization of particulate hydroxylapatite for augmentation of the mandibular ridge. *J. Oral Maxillofac. Surg.* **1988**, *46*, 758–766. [CrossRef] [PubMed]
8. Tomas, M.; Čandrlić, M.; Juzbašić, M.; Ivanišević, Z.; Matijević, N.; Včev, A.; Peloza, O.C.; Matijević, M.; Kačarević, P. Synthetic Injectable Biomaterials for Alveolar Bone Regeneration in Animal and Human Studies. *Materials* **2021**, *14*, 2858. [CrossRef]
9. Yang, H.; Zou, J. Filling Materials Used in Kyphoplasty and Vertebroplasty for Vertebral Compression Fracture: A Literature Review. *Artif. Cells Blood Substit. Biotechnol.* **2010**, *39*, 87–91. [CrossRef]
10. Shi, J.; Yu, L.; Ding, J. PEG-based thermosensitive and biodegradable hydrogels. *Acta Biomater.* **2021**, *128*, 42–59. [CrossRef]
11. Peers, S.; Montembault, A.; Ladavière, C. Chitosan hydrogels incorporating colloids for sustained drug delivery. *Carbohydr. Polym.* **2021**, *275*, 118689. [CrossRef] [PubMed]
12. Xu, M.; Qin, M.; Cheng, Y.; Niu, X.; Kong, J.; Zhang, X.; Huang, D.; Wang, H. Alginate microgels as delivery vehicles for cell-based therapies in tissue engineering and regenerative medicine. *Carbohydr. Polym.* **2021**, *266*, 118128. [CrossRef] [PubMed]
13. Feng, J.; Mineda, K.; Wu, S.-H.; Mashiko, T.; Doi, K.; Kuno, S.; Kinoshita, K.; Kanayama, K.; Asahi, R.; Sunaga, A.; et al. An injectable non-cross-linked hyaluronic-acid gel containing therapeutic spheroids of human adipose-derived stem cells. *Sci. Rep.* **2017**, *7*, 1548. [CrossRef] [PubMed]
14. Fujioka-Kobayashi, M.; Schaller, B.; Kobayashi, E.; Hernandez, M.; Zhang, Y.; Miron, R.J. Hyaluronic Acid Gel-Based Scaffolds as Potential Carrier for Growth Factors: An In Vitro Bioassay on Its Osteogenic Potential. *J. Clin. Med.* **2016**, *5*, 112. [CrossRef] [PubMed]
15. Kim, D.-S.; Kim, J.H.; Baek, S.-W.; Lee, J.-K.; Park, S.-Y.; Choi, B.; Kim, T.-H.; Min, K.; Han, D.K. Controlled vitamin D delivery with injectable hyaluronic acid-based hydrogel for restoration of tendinopathy. *J. Tissue Eng.* **2022**, *13*, 20417314221122089. [CrossRef] [PubMed]
16. Kimura, A.; Kabasawa, Y.; Tabata, Y.; Aoki, K.; Ohya, K.; Omura, K. Gelatin Hydrogel as a Carrier of Recombinant Human Fibroblast Growth Factor-2 During Rat Mandibular Distraction. *J. Oral Maxillofac. Surg.* **2014**, *72*, 2015–2031. [CrossRef]
17. Uehara, T.; Mise-Omata, S.; Matsui, M.; Tabata, Y.; Murali, R.; Miyashin, M.; Aoki, K. Delivery of RANKL-Binding Peptide OP3-4 Promotes BMP-2–Induced Maxillary Bone Regeneration. *J. Dent. Res.* **2016**, *95*, 665–672. [CrossRef]
18. Lorenz, J.; Barbeck, M.; Kirkpatrick, C.; Sader, R.; Lerner, H.; Ghanaati, S. Injectable Bone Substitute Material on the Basis of β-TCP and Hyaluronan Achieves Complete Bone Regeneration While Undergoing Nearly Complete Degradation. *Int. J. Oral Maxillofac. Implant.* **2018**, *33*, 636–644. [CrossRef]
19. Martínez-Sanz, E.; Varghese, O.P.; Kisiel, M.; Engstrand, T.; Reich, K.M.; Bohner, M.; Jonsson, K.B.; Kohler, T.; Müller, R.; Ossipov, D.A.; et al. Minimally invasive mandibular bone augmentation using injectable hydrogels. *J. Tissue Eng. Regen. Med.* **2012**, *6* (Suppl. S3), s15–s23. [CrossRef]

20. Chircov, C.; Grumezescu, A.M.; Bejenaru, L.E. Hyaluronic acid-based scaffolds for tissue engineering. *Rom. J. Morphol. Embryol.* **2018**, *59*, 71–76. Available online: https://rjme.ro/RJME/resources/files/590118071076.pdf (accessed on 10 November 2022).
21. Salwowska, N.M.; Bebenek, K.A.; Żądło, D.A.; Wcisło-Dziadecka, D.L. Physiochemical properties and application of hyaluronic acid: A systematic review. *J. Cosmet. Dermatol.* **2016**, *15*, 520–526. [CrossRef] [PubMed]
22. Peng, T.; Hong, W.; Fang, J.; Luo, S. The selection of hyaluronic acid when treating with the nasolabial fold: A meta-analysis. *J. Cosmet. Dermatol.* **2022**, *21*, 571–579. [CrossRef] [PubMed]
23. Wu, Y.; Stoddart, M.; Wuertz-Kozak, K.; Grad, S.; Alini, M.; Ferguson, S. Hyaluronan supplementation as a mechanical regulator of cartilage tissue development under joint-kinematic-mimicking loading. *J. R. Soc. Interface* **2017**, *14*, 20170255. [CrossRef] [PubMed]
24. Durrie, D.S.; Wolsey, D.; Thompson, V.; Assang, C.; Mann, B.; Wirostko, B. Ability of a new crosslinked polymer ocular bandage gel to accelerate reepithelialization after photorefractive keratectomy. *J. Cataract. Refract. Surg.* **2018**, *44*, 369–375. [CrossRef] [PubMed]
25. Jiang, D.; Liang, J.; Noble, P.W. Hyaluronan in Tissue Injury and Repair. *Annu. Rev. Cell Dev. Biol.* **2007**, *23*, 435–461. [CrossRef]
26. Canciani, E.; Sirello, R.; Pellegrini, G.; Henin, D.; Perrotta, M.; Toma, M.; Khomchyna, N.; Dellavia, C. Effects of Vitamin and Amino Acid-Enriched Hyaluronic Acid Gel on the Healing of Oral Mucosa: In Vivo and In Vitro Study. *Medicina* **2021**, *57*, 285. [CrossRef] [PubMed]
27. Buckley, C.; Murphy, E.J.; Montgomery, T.R.; Major, I. Hyaluronic Acid: A Review of the Drug Delivery Capabilities of This Naturally Occurring Polysaccharide. *Polymers* **2022**, *14*, 3442. [CrossRef]
28. Martínez-Sanz, E.; Ossipov, D.A.; Hilborn, J.; Larsson, S.; Jonsson, K.B.; Varghese, O.P. Bone reservoir: Injectable hyaluronic acid hydrogel for minimal invasive bone augmentation. *J. Control. Release* **2011**, *152*, 232–240. [CrossRef]
29. Hulsart-Billström, G.; Bergman, K.; Andersson, B.; Hilborn, J.; Larsson, S.; Jonsson, K.B. A uni-cortical femoral defect model in the rat: Evaluation using injectable hyaluronan hydrogel as a carrier for bone morphogenetic protein-2. *J. Tissue Eng. Regen. Med.* **2012**, *9*, 799–807. [CrossRef]
30. Patterson, J.; Siew, R.; Herring, S.W.; Lin, A.S.; Guldberg, R.; Stayton, P. Hyaluronic acid hydrogels with controlled degradation properties for oriented bone regeneration. *Biomaterials* **2010**, *31*, 6772–6781. [CrossRef]
31. Pike, D.B.; Cai, S.; Pomraning, K.R.; Firpo, M.A.; Fisher, R.J.; Shu, X.Z.; Prestwich, G.D.; Peattie, R.A. Heparin-regulated release of growth factors in vitro and angiogenic response in vivo in implanted hyaluronan hydrogels containing VEGF and bFGF. *Biomaterials* **2006**, *27*, 5242–5251. [CrossRef] [PubMed]
32. Summonte, S.; Racaniello, G.F.; Lopedota, A.; Denora, N.; Bernkop-Schnürch, A. Thiolated polymeric hydrogels for biomedical application: Cross-linking mechanisms. *J. Control. Release* **2020**, *330*, 470–482. [CrossRef] [PubMed]
33. Zarembinski, T.I.; Skardal, A. HyStem®: A Unique Clinical Grade Hydrogel for Present and Future Medical Applications. In *Hydrogels-Smart Materials for Biomedical Applications*; Popa, L., Ghica, M.V., Dinu-Pîrvu, C., Eds.; IntechOpen: London, UK, 2019. [CrossRef]
34. Ogasawara, T.; Okano, S.; Ichimura, H.; Kadota, S.; Tanaka, Y.; Minami, I.; Uesugi, M.; Wada, Y.; Saito, N.; Okada, K.; et al. Impact of extracellular matrix on engraftment and maturation of pluripotent stem cell-derived cardiomyocytes in a rat myocardial infarct model. *Sci. Rep.* **2017**, *7*, 1–8. [CrossRef] [PubMed]
35. Maloney, E.; Clark, C.; Sivakumar, H.; Yoo, K.; Aleman, J.; Rajan, S.A.P.; Forsythe, S.; Mazzocchi, A.; Laxton, A.W.; Tatter, S.B.; et al. Immersion Bioprinting of Tumor Organoids in Multi-Well Plates for Increasing Chemotherapy Screening Throughput. *Micromachines* **2020**, *11*, 208. [CrossRef] [PubMed]
36. Glickman, R.D.; Onorato, M.; Campos, M.M.; O'Boyle, M.P.; Singh, R.K.; Zarembinski, T.I.; Binette, F.; Nasonkin, I.O. Intraocular Injection of HyStem Hydrogel Is Tolerated Well in the Rabbit Eye. *J. Ocul. Pharmacol. Ther.* **2021**, *37*, 60–71. [CrossRef] [PubMed]
37. Ravina, K.; Briggs, D.I.; Kislal, S.; Warraich, Z.; Nguyen, T.; Lam, R.K.; Zarembinski, T.I.; Shamloo, M. Intracerebral Delivery of Brain-Derived Neurotrophic Factor Using HyStem®-C Hydrogel Implants Improves Functional Recovery and Reduces Neuroinflammation in a Rat Model of Ischemic Stroke. *Int. J. Mol. Sci.* **2018**, *19*, 3782. [CrossRef]
38. AlHowaish, N.A.; AlSudani, D.I.; AlMuraikhi, N.A. Evaluation of a hyaluronic acid hydrogel (Restylane Lyft) as a scaffold for dental pulp regeneration in a regenerative endodontic organotype model. *Odontology* **2022**, *110*, 726–734. [CrossRef]
39. Ikeda, K.; Sugawara, S.; Taira, M.; Sato, H.; Hatakeyama, W.; Kihara, H.; Kondo, H. Ectopic bone formation in muscles using injectable bone-forming material consisting of cross-linked hyaluronic acid, bone morphogenetic protein, and nano-hydroxyapatite. *Nano Biomed.* **2019**, *11*, 11–20. [CrossRef]
40. Zhou, P.; Wu, J.; Xia, Y.; Yuan, Y.; Zhang, H.; Xu, S.; Lin, K. Loading BMP-2 on nanostructured hydroxyapatite microspheres for rapid bone regeneration. *Int. J. Nanomed.* **2018**, *13*, 4083–4092. [CrossRef]
41. Watari, F.; Takashi, N.; Yokoyama, A.; Uo, M.; Akasaka, T.; Sato, Y.; Abe, S.; Totsuka, Y.; Tohji, K. Material nanosizing effect on living organisms: Non-specific, biointeractive, physical size effects. *J. R. Soc. Interface* **2009**, *6*, S371–S388. [CrossRef]
42. Kim, B.-S.; Yang, S.-S.; Kim, C.S. Incorporation of BMP-2 nanoparticles on the surface of a 3D-printed hydroxyapatite scaffold using an ε-polycaprolactone polymer emulsion coating method for bone tissue engineering. *Colloids Surf. B Biointerfaces* **2018**, *170*, 421–429. [CrossRef] [PubMed]
43. Beltrán, A.M.; Alcudia, A.; Begines, B.; Rodríguez-Ortiz, J.A.; Torres, Y. Porous titanium substrates coated with a bilayer of bioactive glasses. *J. Non-Cryst. Solids* **2020**, *544*, 120206. [CrossRef]

44. Jones, J.R.; Lin, S.; Yue, S.; Lee, P.; Hanna, J.V.; Smith, M.E.; Newport, R. Bioactive glass scaffolds for bone regeneration and their hierarchical characterisation. *Proc. Inst. Mech. Eng. Part H J. Eng. Med.* **2010**, *224*, 1373–1387. [CrossRef] [PubMed]
45. Zhao, R.; Yang, R.; Cooper, P.; Khurshid, Z.; Shavandi, A.; Ratnayake, J. Bone Grafts and Substitutes in Dentistry: A Review of Current Trends and Developments. *Molecules* **2021**, *26*, 3007. [CrossRef]
46. Wickramasinghe, M.L.; Dias, G.J.; Premadasa, K.M.G.P. A novel classification of bone graft materials. *J. Biomed. Mater. Res. Part B Appl. Biomater.* **2022**, *110*, 1724–1749. [CrossRef]
47. Hatakeyama, W.; Taira, M.; Sawada, T.; Hoshi, M.; Hachinohe, Y.; Sato, H.; Takafuji, K.; Kihara, H.; Takemoto, S.; Kondo, H. Bone Regeneration of Critical-Size Calvarial Defects in Rats Using Highly Pressed Nano-Apatite/Collagen Composites. *Materials* **2022**, *15*, 3376. [CrossRef]
48. Hatakeyama, W.; Taira, M.; Chosa, N.; Kihara, H.; Ishisaki, A.; Kondo, H. Effects of apatite particle size in two apatite/collagen composites on the osteogenic differentiation profile of osteoblastic cells. *Int. J. Mol. Med.* **2013**, *32*, 1255–1261. [CrossRef]
49. Kanda, Y. Investigation of the freely available easy-to-use software 'EZR' for medical statistics. *Bone Marrow Transplant.* **2013**, *48*, 452–458. [CrossRef]
50. Alonso, J.; del Olmo, J.A.; Gonzalez, R.P.; Saez-Martinez, V. Injectable Hydrogels: From Laboratory to Industrialization. *Polymers* **2021**, *13*, 650. [CrossRef]
51. Wozney, J.M. The bone morphogenetic protein family: Multifunctional cellular regulators in the embryo and adult. *Eur. J. Oral Sci.* **1998**, *106* (Suppl. S1), 160–166. [CrossRef]
52. Xu, X.; Jha, A.K.; Duncan, R.L.; Jia, X. Heparin-decorated, hyaluronic acid-based hydrogel particles for the controlled release of bone morphogenetic protein 2. *Acta Biomater.* **2011**, *7*, 3050–3059. [CrossRef] [PubMed]
53. Holloway, J.L.; Ma, H.; Rai, R.; Burdick, J.A. Modulating hydrogel crosslink density and degradation to control bone morphogenetic protein delivery and in vivo bone formation. *J. Control. Release* **2014**, *191*, 63–70. [CrossRef] [PubMed]
54. Munir, M.U.; Salman, S.; Javed, I.; Bukhari, S.N.A.; Ahmad, N.; Shad, N.A.; Aziz, F. Nano-hydroxyapatite as a delivery system: Overview and advancements. *Artif. Cells Nanomed. Biotechnol.* **2021**, *49*, 717–727. [CrossRef] [PubMed]
55. Zaffarin, A.S.M.; Ng, S.-F.; Ng, M.H.; Hassan, H.; Alias, E. Nano-Hydroxyapatite as a Delivery System for Promoting Bone Regeneration In Vivo: A Systematic Review. *Nanomaterials* **2021**, *11*, 2569. [CrossRef] [PubMed]
56. Duanis-Assaf, T.; Hu, T.; Lavie, M.; Zhang, Z.; Reches, M. Understanding the Adhesion Mechanism of Hydroxyapatite-Binding Peptide. *Langmuir* **2022**, *38*, 968–978. [CrossRef]
57. Piskounova, S.; Gedda, L.; Hulsart-Billström, G.; Hilborn, J.; Bowden, T. Characterization of recombinant human bone morphogenetic protein-2 delivery from injectable hyaluronan-based hydrogels by means of125I-radiolabelling. *J. Tissue Eng. Regen. Med.* **2012**, *8*, 821–830. [CrossRef] [PubMed]
58. Lehnfeld, J.; Dukashin, Y.; Mark, J.; White, G.; Wu, S.; Katzur, V.; Müller, R.; Ruhl, S. Saliva and Serum Protein Adsorption on Chemically Modified Silica Surfaces. *J. Dent. Res.* **2021**, *100*, 1047–1054. [CrossRef] [PubMed]
59. Georgieva, E.R. Protein Conformational Dynamics upon Association with the Surfaces of Lipid Membranes and Engineered Nanoparticles: Insights from Electron Paramagnetic Resonance Spectroscopy. *Molecules* **2020**, *25*, 5393. [CrossRef]
60. Bystrov, V.; Bystrova, A.; Dekhtyar, Y. HAP nanoparticle and substrate surface electrical potential towards bone cells adhesion: Recent results review. *Adv. Colloid Interface Sci.* **2017**, *249*, 213–219. [CrossRef]
61. Levingstone, T.; Ali, B.; Kearney, C.; Dunne, N. Hydroxyapatite sonosensitization of ultrasound-triggered, thermally responsive hydrogels: An on-demand delivery system for bone repair applications. *J. Biomed. Mater. Res. Part B Appl. Biomater.* **2021**, *109*, 1622–1633. [CrossRef]
62. Liu, J.; Sun, H.; Peng, Y.; Chen, L.; Xu, W.; Shao, R. Preparation and Characterization of Natural Silk Fibroin Hydrogel for Protein Drug Delivery. *Molecules* **2022**, *27*, 3418. [CrossRef] [PubMed]
63. Isık, S.; Taşkapılıoğlu, M.; Atalay, F.O.; Dogan, S. Effects of cross-linked high-molecular-weight hyaluronic acid on epidural fibrosis: Experimental study. *J. Neurosurg. Spine* **2015**, *22*, 94–100. [CrossRef] [PubMed]
64. Van Loock, K.; Menovsky, T.; Kamerling, N.; De Ridder, D. Cranial Bone Flap Fixation using a New Device (Cranial LoopTM). *min-Minim. Invasive Neurosurg.* **2011**, *54*, 119–124. [CrossRef]
65. Pavicic, T.; Yankova, M.; Schenck, T.L.; Frank, K.; Freytag, D.L.; Sykes, J.; Green, J.B.; Hamade, H.; Casabona, G.; Cotofana, S. Subperiosteal injections during facial soft tissue filler injections—Is it possible? *J. Cosmet. Dermatol.* **2019**, *19*, 590–595. [CrossRef]
66. Kisiel, M.; Klar, A.S.; Martino, M.; Ventura, M.; Hilborn, J. Evaluation of Injectable Constructs for Bone Repair with a Subperiosteal Cranial Model in the Rat. *PLoS ONE* **2013**, *8*, e71503. [CrossRef]
67. Zhang, W.; Wang, N.; Yang, M.; Sun, T.; Zhang, J.; Zhao, Y.; Huo, N.; Li, Z. Periosteum and development of the tissue-engineered periosteum for guided bone regeneration. *J. Orthop. Transl.* **2022**, *33*, 41–54. [CrossRef] [PubMed]
68. Tachi, K.; Takami, M.; Sato, H.; Mochizuki, A.; Zhao, B.; Miyamoto, Y.; Tsukasaki, H.; Inoue, T.; Shintani, S.; Koike, T.; et al. Enhancement of Bone Morphogenetic Protein-2-Induced Ectopic Bone Formation by Transforming Growth Factor-β1. *Tissue Eng. Part A* **2011**, *17*, 597–606. [CrossRef]
69. Cappato, S.; Gamberale, R.; Bocciardi, R.; Brunelli, S. Genetic and Acquired HeterotopicOssification: A Translational Tale of Mice and Men. *Biomedicines* **2020**, *8*, 611. [CrossRef]
70. Mumcuoglu, D.; Fahmy-Garcia, S.; Ridwan, Y.; Nicke, J.; Farrell, E.; Kluijtmans, S.G.; Van Osch, G.J. Injectable BMP-2 delivery system based on collagen-derived microspheres and alginate induced bone formation in a time- and dose-dependent manner. *Eur. Cells Mater.* **2018**, *35*, 242–254. [CrossRef]

71. Abrahamsson, P.; Isaksson, S.; Andersson, G. Guided bone generation in a rabbit mandible model after periosteal expansion with an osmotic tissue expander. *Clin. Oral Implant. Res.* **2011**, *22*, 1282–1288. [CrossRef]
72. Abid, W.K.; AL Mukhtar, Y.H. Repair of surgical bone defects grafted with hydroxylapatite + β-TCP combined with hyaluronic acid and collagen membrane in rabbits: A histological study. *J. Taibah Univ. Med. Sci.* **2019**, *14*, 14–24. [CrossRef] [PubMed]
73. Kim, J.; Kim, I.S.; Cho, T.H.; Kim, H.C.; Yoon, S.J.; Choi, J.; Park, Y.; Sun, K.; Hwang, S.J. In vivo evaluation of MMP sensitive high-molecular weight HA-based hydrogels for bone tissue engineering. *J. Biomed. Mater. Res. Part A* **2010**, *95*, 673–681. [CrossRef] [PubMed]
74. Fricain, J.; Aid, R.; Lanouar, S.; Maurel, D.; Le Nihouannen, D.; Delmond, S.; Letourneur, D.; Vilamitjana, J.A.; Catros, S. In-vitro and in-vivo design and validation of an injectable polysaccharide-hydroxyapatite composite material for sinus floor augmentation. *Dent. Mater.* **2018**, *34*, 1024–1035. [CrossRef] [PubMed]

Review

Poly(caprolactone)-*b*-poly(ethylene glycol)-Based Polymeric Micelles as Drug Carriers for Efficient Breast Cancer Therapy: A Systematic Review

Siti Hajar Ahmad Shariff [1], Wan Khartini Wan Abdul Khodir [1], Shafida Abd Hamid [1], Muhammad Salahuddin Haris [2] and Mohamad Wafiuddin Ismail [1,*]

[1] Department of Chemistry, Kulliyyah of Science, International Islamic University Malaysia, Kuantan 25200, Pahang Darul Makmur, Malaysia
[2] Department of Pharmaceutical Technology, Kulliyyah of Pharmacy, International Islamic University Malaysia, Kuantan 25200, Pahang Darul Makmur, Malaysia
* Correspondence: wafisnj@iium.edu.my

Abstract: Recently, drug delivery systems based on nanoparticles for cancer treatment have become the centre of attention for researchers to design and fabricate drug carriers for anti-cancer drugs due to the lack of tumour-targeting activity in conventional pharmaceuticals. Poly(caprolactone)-*b*-poly(ethylene glycol) (PCL-PEG)-based micelles have attracted significant attention as a potential drug carrier intended for human use. Since their first discovery, the Food and Drug Administration (FDA)-approved polymers have been studied extensively for various biomedical applications, specifically cancer therapy. The application of PCL-PEG micelles in different cancer therapies has been recorded in countless research studies for their efficacy as drug cargos. However, systematic studies on the effectiveness of PCL-PEG micelles of specific cancers for pharmaceutical applications are still lacking. As breast cancer is reported as the most prevalent cancer worldwide, we aim to systematically review all available literature that has published research findings on the PCL-PEG-based micelles as drug cargo for therapy. We further discussed the preparation method and the anti-tumour efficacy of the micelles. Using a prearranged search string, Scopus and Science Direct were selected as the databases for the systematic searching strategy. Only eight of the 314 articles met the inclusion requirements and were used for data synthesis. From the review, all studies reported the efficiency of PCL-PEG-based micelles, which act as drug cargo for breast cancer therapy.

Keywords: PCL-PEG; polymer micelle; drug cargo; drug delivery; breast cancer

1. Introduction

The statistics from the International Agency for Research on Cancer in 2020 have estimated around 19.3 million new cases and almost 10.0 million deaths worldwide, with female breast cancer as the most commonly diagnosed cancer [1]. The World Health Organization in 2019 reported cancer as the first or second most significant cause of mortality among people before the age of 70 in 112 of 183 countries [2]. According to the National Cancer Institute, some cancer treatments include surgery, chemotherapy, immunotherapy, stem cell transplant, and precision medicine [3]. Among the many treatments, chemotherapy is one of the most commonly used methods for cancer therapy [4].

Nevertheless, several drawbacks have limited the benefits of chemotherapy treatment, such as the poor water solubility of some anti-cancer drugs that lower their efficiency. In addition, anti-cancer drugs also possess a high toxicity that results in severe side effects which cannot be reduced using traditional pharmaceutical dosage forms [5]. While conventional pharmaceutical formulations are lacking in the activity that targets tumours, resulting in only a limited amount of drugs that enter systemic circulation to target the tumour tissues. Consequently, the drug uptake of conventional pharmaceuticals is lowered [6]. As such, drug delivery systems based on nanoparticles for cancer treatment have become the centre of attention for researchers to design and fabricate drug carriers for anti-cancer drugs, such as vesicles, liposomes, nanogels, and polymer micelles [5,7,8]. These nanodrug delivery methods can lower drug toxicity while increasing bioavailability since the drugs are dissolved, adsorbed, and covalently bound to the surface of the nanocarriers. The surface of the carriers can also be modified to direct the drug toward the tumour, minimising drug transport to healthy tissues and increasing treatment safety. The discovery of this cancer-targeting technology based on nanodrug delivery has resurrected the medicinal use of many powerful anti-cancer drugs that previously contained a high toxicity [6,7,9].

Researchers highly sought polymer micelles in breast cancer treatment due to their flexibility in designing and modifying their structures and compositions [10]. As a result, the development of polymer micelles as breast anti-cancer drug carriers has been ongoing even up to this day. Polymer micelles are usually spherical, nano-sized, amphiphilic copolymers that self-assembled spontaneously in aqueous media to form micelles above the critical micelle concentration (CMC), comprised of hydrophobic and hydrophilic block domains. The hydrophobic core dissolves the hydrophobic drugs, improving their solubility and biostability. At the same time, the hydrophilic outer shell provides micelles with compatibility in the aqueous media and shields the drugs in the core from interactions with the blood components [11].

One of the most commonly used polymeric micelles is based on a polyester-polyether, poly(caprolactone)-b-poly(ethylene glycol) (PCL-PEG) block copolymer due to their amphiphilicity, high biocompatibility, controlled biodegradability, and self-assembling ability to produce polymeric micelles in aqueous media [12–15]. Polyester cores were reported to have higher hydrophobic anti-cancer drug loading than polyether [16]. PCL has a very low glass transition temperature, making it suitable for developing drug delivery systems based on nanoparticles, in addition to its biodegradability, biocompatibility, and FDA approval [17]. Compared to other hydrophobic polyesters such as poly(lactic-co-glycolic acid) (PLGA) and polylactic acid (PLA), PCL displayed a relatively weak acidic environment, lessening the biological inflammation. Furthermore, PCL is less expensive and more stable in the body than PLGA and degrades rather slowly [18]. PCL is the most promising polyester for the development of novel, commercial medical devices. This capability is related to PCL's unique physicochemical features, relatively harmless biodegradation behaviour, and the ability to fine-tune and make significant chemical alterations [19].

On its own, pure PCL has poor water stability as it easily aggregates. Therefore, the addition of PEG chains is used to address this issue. PEG has been used as a therapeutic agent for a long time as an FDA-approved hydrophilic constituent of polymeric micelles because it can prevent micelle uptake in the reticuloendothelial system (RES), hence lengthening the blood circulation time of the drug in the polymer micelles [20,21]. PEGylation of the hydrophobic PCL results in amphiphilicity properties and a controllable degradation speed and drug release profile, enhancing its biocompatibility and circulation time [17]. Apart from that, the addition of PEG to PCL increased the degree of crystallinity of PCL, resulting in more refined PCL crystals. In fact, even a substantial amount of PEG resulted in a significant decrease in the degradation temperature, crystallinity, time of crystallisation, and an increase in the crystal's average size [22]. The synergistic effects of PCL-PEG copolymers make them attractive in the anti-cancer drug delivery system. Over the years, the application of PCL-PEG-based polymeric micelles in the area of cancer therapeutics has been studied numerous times as drug carriers for various types of cancer, such as breast cancer [23,24], colorectal cancer [25], lung cancer [26], colon cancer [27,28], prostate cancer [29], and others. PCL-PEG based nanoparticles as nanocarriers for chemotherapeutic drugs such as paclitaxel, campothecin, and doxorubicin confirmed increased cellular internalisation, sustained drug release, and lower cytotoxic effects compared to free drugs [30].

These studies focus on the efficacies and capabilities of the amphiphilic polymeric micelles to act as anti-cancer drug cargo. However, there is still a limited number of existing studies that have been reviewed systematically, specifically on breast cancer therapy. Hence, this review was conducted to systematically review past studies in detail. The fundamental question that led to this review is: what is the efficiency of PCL-PEG-based micelles as drug carriers for breast cancer therapy? This study aims to narrow the gap by systematically examining past related studies (2016–2021) to acquire better knowledge on the efficiency of the PCL-PEG-based micelles as drug carriers for breast cancer treatment.

2. Methods

In this section, the reviewers discussed the strategy used to identify papers about poly(caprolactone)-*b*-poly(ethylene glycol)-based polymeric micelles as drug carriers for efficient breast cancer therapy. The method used for this systemic review was PRISMA, which included Scopus and ScienceDirect as resources.

2.1. The Review Protocol—PRISMA

This systematic review uses the Preferred Reporting Items for Systematic Reviews and Meta-Analyses Protocols (PRISMA) guideline. PRISMA is a published standard that guides researchers on how to conduct a systematic literature review. It is widely used in medical research and can identify the inclusion and exclusion criteria of the study [31].

2.2. Formulation of the Research Question

The research question for this review was formulated based on PICO. PICO is a method that helps construct a relevant research question for systematic literature reviews. The four main concepts in PICO are Population or Problem, Intervention or Experimental Variables, Control Variable, and Outcome [32]. Based on the concepts, the researchers outlined four main aspects in the review: PCL-PEG (population), micelle (Intervention/experimental variables), drug carrier (control variable), and breast cancer (outcome), which then guided the formulation of the leading research question: what is the efficiency of the PCL-PEG-based micelles as drug carriers for breast cancer therapy?

2.3. Resources

Two electronic databases were used as sources for this study: Scopus and ScienceDirect. These databases are relevant and provide high-impact factor publication [33]. The researchers analysed the titles and abstracts of the published articles according to the inclusion criteria in this study.

2.4. Systematic Searching Strategies

The systematic search process for selecting relevant articles for this review was done in three stages: identification, screening, and eligibility.

2.4.1. Identification

In the first stage, the keywords were identified and expanded by looking for similar or relevant terms in dictionaries, thesauruses, and previous research. To make the search process more accessible and to limit the results to relevant articles, a combination of symbols and coding, such as field codes, Boolean operators (AND, OR), wildcards, and truncation, were used to connect the keywords. The search strings were developed and used on Scopus and ScienceDirect after all keywords were determined (Table 1). Different search strings were used between Scopus and ScienceDirect due to some of the characters not being accepted as keywords.

Table 1. The search strings.

Database	Search String
Scopus	TITLE-ABS-KEY ("PCL-PEG" OR "polycaprolactone polyethylene glycol" OR "PCLPEG") AND ("micelle*s" OR "micellar") AND ("drug delivery" OR "drug cargo" OR "drug carrier") AND ("breast cancer")
ScienceDirect	("PCL-PEG" OR "polycaprolactone polyethylene glycol" OR "PCL PEG") AND ("micelle" OR "micellar") AND ("drug delivery" OR "drug cargo" OR "drug carrier") AND ("breast cancer")

2.4.2. Screening

A total of 314 articles were automatically screened using the sorting function available in the databases by selecting the predefined inclusion and exclusion criteria (Table 2). The first criterion decided was the article category, with the researchers agreeing to focus solely on research articles because they are classified as primary sources and provide actual data [31,34]. As a result, publications other than research articles were excluded from the current evaluation, including systematic reviews, review papers, meta-analyses, meta-syntheses, proceedings, books, book chapters, and book series. Aside from that, the current study solely looked at articles written in English. Therefore, publications in other languages were not considered. Moreover, articles' acceptable timeline to be included in the review was six years (2016–2021). There were 240 publications removed from the study because they did not fit the criteria. Five articles were identified to be duplicated during the screening process and thus were removed. The remaining 69 articles were found and prepared for the next step in the process: eligibility.

Table 2. The inclusion and exclusion criteria.

Criteria	Inclusion	Exclusion
Literature type	Journal article (empirical data)	Systematic reviews, review papers, meta-analyses, meta-syntheses, proceedings, books, book chapters, book series
Language	English	Non-English
Timeline	2016–2021	<2016

2.4.3. Eligibility

The researchers manually examined the remaining 69 articles to ensure they were fit to be included in the present study to achieve the study's objectives by thoroughly reading the articles' titles and abstracts. A total of 61 articles were excluded because the articles focus on other types of cancers rather than breast cancer, other types of copolymers rather than PCL-PEG copolymers, and polymersomes and nanoparticles rather than micelles. As a result, eight articles were selected for the next process: quality appraisal.

2.5. Quality Appraisal

Quality appraisal was conducted to assess the quality of the articles' content. Two authours independently examined eight papers and categorised them as high, medium, or poor quality based on the pre-set criteria. The criteria were developed in response to the systematic review's research questions. Mutual agreement between authours was practiced during the rating process to eliminate bias. As a result, data abstraction and analysis were carried out in eight papers.

2.6. Data Abstraction and Analysis

The data abstraction was done in response to the formulated research question that had been developed. Any data from the reviewed researchers that could be used to answer the research question was retrieved and tabulated. Thematic analysis was then used to find the themes and sub-themes from the abstracted data based on patterns, similarities, and linkages. The creation of topics was the initial step in the thematic analysis. The researchers attempted to find the patterns that appeared to connect the abstracted data from all eight pieces of study in this step. Any related or comparable data was organised, and five themes were eventually produced. The researchers repeated the process for each new theme, yielding six sub-themes in all (Table 3). The researchers reviewed the data's accuracy and discussed any anomalies in the resulting themes and sub-themes with one another to verify the data's relevance and reliability. The researchers then named the five themes: synthesis and characterisation of PCL-PEG, preparation of micelles, characterisation of micelles, drug delivery study, and anti-tumour efficacy (Table 3).

Table 3. The themes and sub-themes.

Authors	Synthesis and Characterisation of PCL-PEG	Preparation of Micelles	Characterisation of Micelles				Drug Release Study	Anti-Tumour Activity	
			Morphology	Particle Size Distribution and Zeta Potential	Drug Loading and Encapsulation Efficiency	DSC Analysis		Cell Viability and Cytotoxicity Study	Anti-Tumour Efficacy
Kheiri Manjili et al., (2016) [35]	/	/	/	/	/	/	/	/	/
Kheiri Manjili et al., (2017a) [36]	/	/	/	/	/	x	/	/	/
Kheiri Manjil et al., (2017b) [37]	/	/	/	/	/	/	/	/	/
Mahdaviani et al., (2017) [38]	/	/	/	/	/	x	x	/	/
Hu et al., (2017) [39]	/	/	/	/	/	/	x	x	/
Zamani et al., (2018) [40]	/	/	/	/	/	/	/	/	/
Kheiri Manjili et al., (2018) [41]	/	/	/	/	/	/	/	/	/
Peng et al., (2019) [42]	x	/	/	/	/	x	/	/	/

3. Results

3.1. Selected Articles' Background

Eight selected articles were successfully obtained for the current review (Figure 1). Five themes were developed based on the thematic analysis: synthesis and characterisation of PCL-PEG, preparation of micelles, characterisation of micelles, drug release study, and anti-tumour efficacy. Six sub-themes resulted from further study of the themes. From the eight selected papers, one was published in 2016 [35], four were published in 2017 [36–39], two were published in 2018 [40,41], and one was published in 2019 [42].

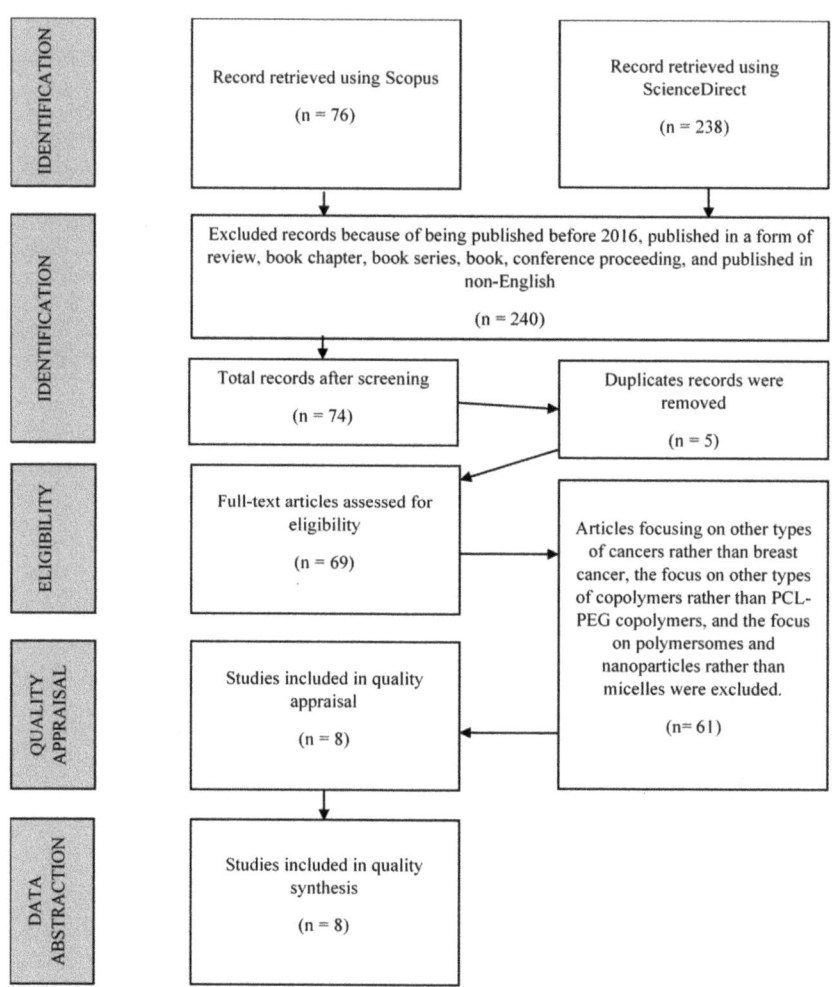

Figure 1. The flow diagram of the systematic literature review (adapted from Mohamed Shaffril et al. [31]).

3.2. Themes and Sub-Themes

3.2.1. Synthesis and Characterisation of PCL-PEG Copolymers

Seven studies synthesised the PCL-PEG copolymers, except for Peng et al., which used pre-synthesized PCL-PEG copolymers. The copolymers were synthesised by using ring-opening polymerisation (ROP) of ε-caprolactone (ε-CL) [35–41]. In most studies, the PEG with the hydroxyl end group was used to initiate the ring opening of ε-CL in the presence of the catalyst stannous octoate (Sn(Oct)$_2$) and under a nitrogen atmosphere (Figure 2) [35–37,39–41]. However, Mahdaviani et al., initiated the ROP of ε-CL by utilising N-protected 3-aminopropan-1-ol, and PCL-PEG copolymer was synthesised via an amidation reaction between the -COOH groups of PCL and the -NH3 end of PEG [38]. In addition, Zamani et al., added folic acid (FA) as a molecular probe to the synthesised PCL-PEG copolymers for targeted cancer treatment in the form of folate-lysine-PCL-PEG (FA-L-PEG-PCL) (Figure 3). To ensure sufficient conversion of ε-CL to PCL, the reaction was heated at 110 °C for 24 h [38]. However, most studies conducted the reaction at 120 °C to shorten the reaction period to 12 h. Various molecular weight copolymers were produced by manipulating the ε-CL: PEG ratio. The summary of the reaction condition is tabulated in Table 4.

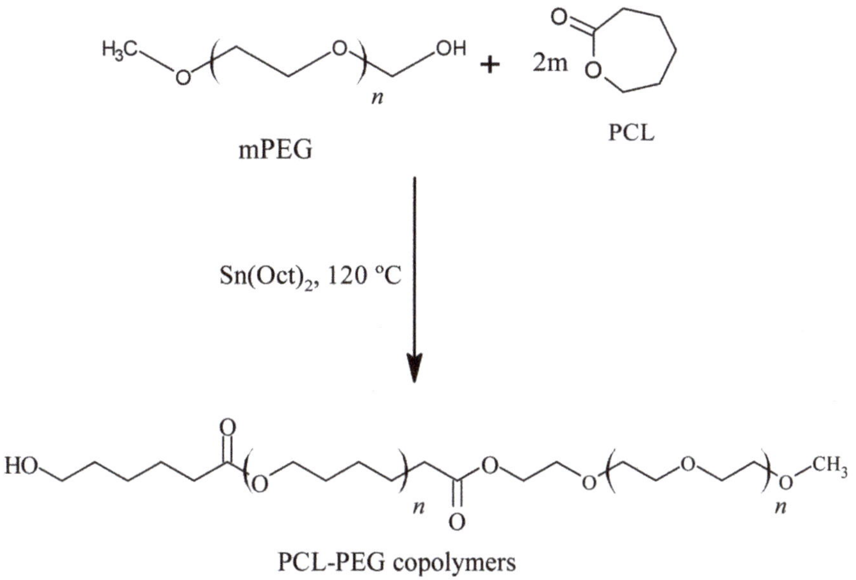

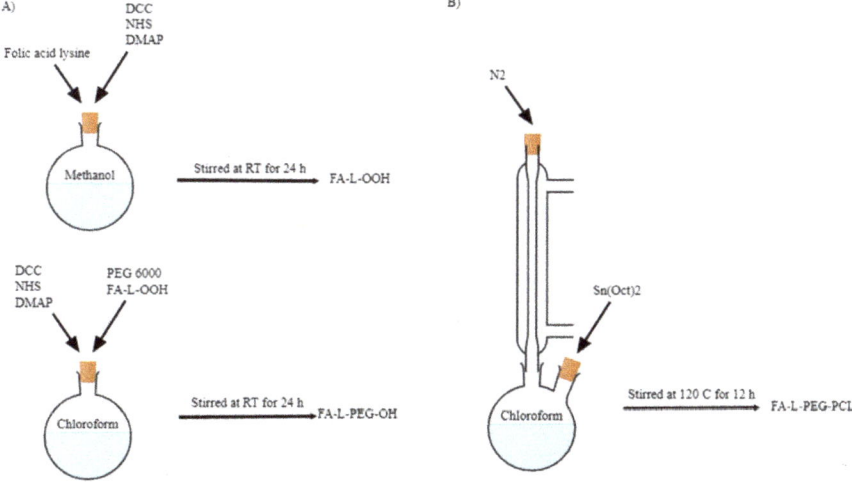

Figure 2. Schematic route for the preparation of PCL-PEG copolymers.

Figure 3. Preparation of (**A**) FA-L-PEG-OH and (**B**) FA-L-PEG-PCL. Reprinted/adapted with permission from Elsevier [40].

Table 4. Summary of data obtained from the synthesis and characterisation of PCL-PEG copolymers.

Authors	Amount of ε-CL	Amount of PEG	Catalyst	Reaction Condition	Drying Condition	Products Synthesised	GPC Analysis			DSC Analysis
							Mn (Da)	Mw (Da)	PDI	Melting Temperature (°C)
Kheiri Manjili et al., (2016) [35]	0.5, 1, 2, 4, 5, and 6 g	1 g	Sn(Oct)$_2$, 0.01 mol	120 °C, 12 h, oil bath	23 °C, 24 h	mPEG-PCL di-block copolymer	9342–20,543	10,231–21,932	1.04–1.09	55.00
Kheiri Manjili et al., (2017a) [36]	1, 2, 4, 8, and 10 g	2 g	Sn(Oct)$_2$, 0.01 mol	120 °C, 12 h, oil bath	23 °C, 24 h	PCL-PEG-PCL tri-block copolymer	8940–18,976	9731–21,321	1.05–1.12	54.72
Kheiri Manjil et al., (2017b) [37]	0.5, 1, and 2 g	1 g	Sn(Oct)$_2$, 0.01 mol	120 °C, 12 h, oil bath	23 °C, 24 h	PCL-PEG-PCL tri-block copolymer	8722–11,257	10,200–15,961	1.16–1.41	52.34
Kheiri Manjili et al., (2018) [41]	2 g	1 g	Sn(Oct)$_2$, 0.01 mol	120 °C, 12 h, oil bath	23 °C, 24 h	PCL-PEG-PCL tri-block copolymer	16,987	18,765	1.10	62.84
Mahdaviani et al., (2017) [38]	Not mentioned	Not mentioned	1.2 mmol EDC and 2.5 mmol NHS	Room temperature, 24 h	At reduced pressure	PCL-PEG di-block copolymer	Not stated	45,000	Not stated	-
Hu et al., (2017) [39]	Not mentioned	Not mentioned	Sn(Oct)$_2$	120 °C, 48 h, oil bath	Vacuum-dried	PCL-PEG-PCL tri-block copolymer	12,875–27,844	13,615–33,084	1.057–1.218	Not stated
Zamani et al., (2018) [40]	3 g	1 g	Sn(Oct)$_2$, 0.01 mol	120 °C, 12 h	Room temperature, vacuum-dried	mPEG-PCL di-block copolymer	-	-	-	57.0

Each study employed Fourier transform infrared spectroscopy (FTIR), proton nuclear magnetic resonance spectroscopy (1H NMR), and gel permeation chromatography (GPC) to characterise the synthesised copolymers. The successful formation of PCL-PEG copolymers in the studies was indicated by the sharp and intense bands around 1720.0 cm^{-1} and 1100.0 cm^{-1} of the FTIR analysis, representing the carboxylic ester (C=O) and ether (C–O) groups of the PCL-PEG linkage. The NMR analysis from the studies revealed that the peaks around 2.2–2.5 ppm of methylene proton –OCCH$_2$- and 4.06 ppm of methylene proton –CH$_2$OOC– belongs to the linkage between the PCL and PEG, which confirmed the formation of PCL-PEG copolymers. Mahdaviani et al., found that the synthesised copolymers had a 1:1 ratio of PCL and PEG by assessing the peak area associated with the PEG methylene group to that of PCL [38]. The ratio of the copolymers can be determined by equating the peak integration area at ~4.06 ppm (PCL block) to that at ~3.6 ppm (PEG block) [39]. Meanwhile, the number average molecular weight (M_n) of the PCL-PEG copolymers can be determined based on the equation:

$$M_n = (1 + R) \times PEG$$

where R = the ratio of the integration area of PCL:PEG and PEG = the number average molecular weight of PEG.

GPC analysis from all studies showed a polydispersity index (PDI) closer to one (<1.4), indicating the uniform distribution of the polymerisation of PCL-PEG copolymers. Hu et al., also stated that the study's unimodal curve of the GPC chromatogram suggested no impurities in the polymers [39].

Thermal analysis of the synthesised PCL-PEG was done using differential scanning calorimetry (DSC) analysis. The DSC thermograms in each study recorded an endothermic peak for the melting point of the copolymers. The melting transition temperatures of PCL and PEG homopolymers that were greater than those of respective blocks in copolymers indicated the interaction between the PCL and PEG blocks [39]. In addition, the formation of two peaks during the cooling process in the thermogram was most likely due to the length of the PCL block, which was half of the PEG block. The PEG block's higher crystallisation temperature compared to the PCL block resulted in the appearance of two peaks in the thermogram. However, when the PEG block's molecular weight and the PCL block's molecular weight were close enough, their peaks merged and became exothermic peaks. The fact that the crystallisation temperature of PEG was more significant when the block length was longer supports the conclusion that crystallisation temperature is positively related to the block length [39]. The same pattern was observed by Kheiri Manjili et al. [35–37,41] where the DSC thermogram copolymers showed two merged endothermic peaks of PCL and PEG (Figure 4). The characterisation of PCL-PEG copolymers is summarised in Table 4.

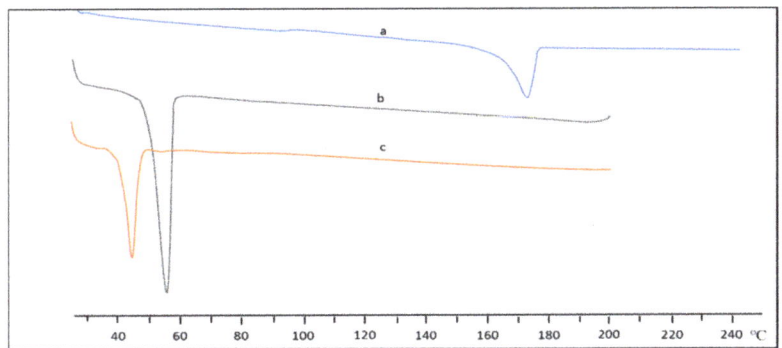

Figure 4. DSC thermogram of (a) CUR (b) mPEG-PCL, and (c) CUR/mPEG-PCL micelles. Reprinted/adapted with permission from Elsevier [35].

3.2.2. Preparation of Micelles

The selected studies adopted four types of micelle preparation methods: nanoprecipitation, thin-film hydration, co-solvent evaporation, and emulsion solvent evaporation. The preparation methods of the micelles are summarised in Table 5. The suitable techniques were chosen based on various variables such as the particle's size, the active agent's stability, and the finished product's toxic effect [38]. In addition, PCL-PEG copolymers can self-assemble in appropriate solvent conditions because of their duality and are able to transform into different types of structures, such as spherical, cylindrical, and others, depending on the optimisation of the solvent conditions [41]. Solvents such as acetone, methylene dichloride, and chloroform were used to dissolve both PCL-PEG copolymers and hydrophobic drug models before adding water as the aqueous phase for the copolymers to self-assemble into micelles. Meanwhile, Mahdaviani et al., optimised the polymeric micelles' preparation method regarding the carrier size and encapsulation efficacy by modifying the organic: aqueous phase ratio and the sequence in which the phases were added. They found that the optimised ratio was 10% (w/w) of the organic to aqueous phase [38]. In contrast, Peng et al., optimised the PCL-PEG chain length to encapsulate the hydrophobic drug into the semi-crystalline PCL core and to promote self-assembly into specific morphology. The weight ratio of 3:7 of PCL-PEG copolymers was reported as the optimised matrix to encapsulate the drug and self-assemble it into worm-like micelles. The ratio was chosen as the ideal formulation due to the micelles' smaller and homogeneous size distribution [42].

Two studies adopted functionalised conjugations to the micelles for the targeted delivery of drugs. Mahdaviani et al., used a peptide-functionalized ligand, a cyclic ten amino acid peptide (GCGNVVRQGC), which was a tumour metastasis-targeting (TMT) peptide to promote the effective targeting of the anti-tumour CBZ. The TMT peptide was conjugated onto PCL-PEG micelles via the carboxyl group PCL-PEG copolymer's covalent bonds. The Moc-protected TMT peptide was linked to PCL-PEG using EDC and NHS [38]. In another study, Peng et al., used Herceptin (HER) as antibody-conjugated nanoparticles (ACN) to bind extracellularly to the p185 glycoprotein domain of the HER2-positive breast cancer receptor to cause apoptosis in tumour cells or to stop the cell cycle progression. The drug-loaded PCL-PEG-HER was prepared by performing a Schiff base reaction between the aldehyde group of CHO-PCL-PEG anchored at the surface of the drug-loaded micelles with the amine group of HER using a molar ratio of 5:1 aldehyde to amine. Then, the -C=N- was reduced to -C-N- by using CH3BNNa [42].

Table 5. Summary of the preparation and characterisation of micelles.

Authour	Method Preparation	Type of Drug	Solvent Used	Selected Micelles in the Study	Size (nm)	DL %	EE %	Zeta Potential	Melting Temperature (°C) Drug	Melting Temperature (°C) Micelles
Kheiri Manjili et al., (2016) [35]	Nanoprecipitation	Curcumin (CUR)	Acetone	0.25 (CUR/copolymer mass ratio)					173.56	45.92
Kheiri Manjili et al., (2017a) [36]	Nanoprecipitation	Sulforaphane (SF)	Acetone	0.25 (SF/copolymer mass ratio)	81.70	20.65	89.32	−11.5	-	-
Kheiri Manjil et al., (2017b) [37]	Nanoprecipitation	Artemisinin (ART)	Acetone	0.25 (ART/copolymer mass ratio)	114.00	19.33	87.1	−7.57	154.35	52.34
Kheiri Manjili et al., (2018) [41]	Nanoprecipitation	Atorvastatin and rosuvastatin	Acetone	Atorvastatin-loaded	83.22	18.62	89.23	−15.45	167.73	55.48
Mahdaviani et al., (2017) [38]	Cosolvent evaporation	Cabazitaxel (CBZ)	Acetone	Rosuvastatin-loaded	55.66	20.0	88.19	−7.72	75.51	51.12
Hu et al., (2017) [39]	Thin-film hydration and ultrasonic dispersion	Paclitaxel (PTX)	Methylene chloride	CBZ-loaded	53.72	13.21	59.01	−2.99	-	-
Zamani et al., (2018) [40]	Cosolvent evaporation	Tamoxifen (TMX)	Acetone	PTX-loaded	110.00	8.5	82.5	Not mentioned	-	-
Peng et al., (2019) [42]	Emulsion solvent evaporation	Paclitaxel	Chloroform	1:6 (TMX/copolymer mass ratio)	255.80	8.87	87.97	−17.9	142.00	55.00

3.2.3. Characterisation of Micelles

The determination of the micelles' morphology was common under this theme, followed by the distribution of particle size and zeta potential, drug loading and encapsulation efficiency, and thermal analysis. Table 5 summarised the characterisation of the selected micelles.

Morphology of Micelles

In each study, the morphological characterisation of the micelles was examined by utilising atomic force microscopy (AFM) [35–37,40,41], transmission electron microscopy (TEM) [39,42], and scanning electron microscopy (SEM) [38]. The PCL-PEG micelles demonstrated a uniform spherical shape when observed using atomic force microscopy (AFM). TEM results revealed that most nanoparticles had a distinct spherical shape and uniform size [39]. Hu et al., reported that when the PCL-PEG block ratio was 1, the copolymers formed polymeric micelles. Meanwhile, Peng et al., found the self-assembled spherical amphiphilic PCL-PEG micelles became worm-like micelles when PTX was loaded into the PCL-PEG micelles in an aqueous solution (Figure 5). After conjugation with a particular number of HER molecules, the worm-like structure of the micelles was preserved [42]. Meanwhile, the SEM image in the study showed that CBZ polymeric micelles had a uniform spherical morphology, with a particle size of around 100 nm and a small particle size range [38].

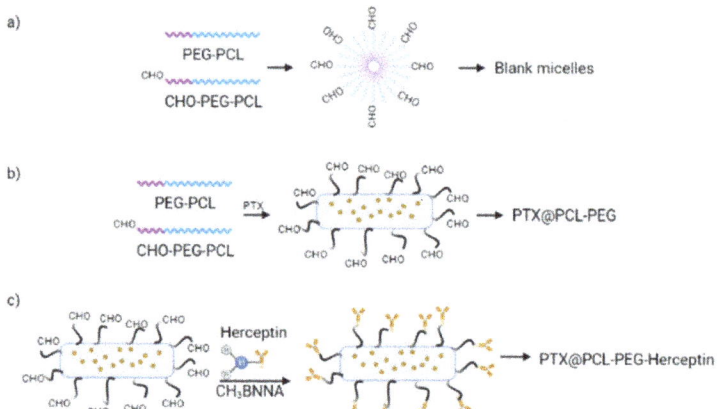

Figure 5. The formulation strategy of (**a**) Blank micelles, (**b**) PTX/CL-PEG, and (**c**) PTX/PCL-PEG-Herceptin. Reprinted/adapted with permission from Elsevier [42].

Particle Size and Zeta Potential

In each study, the size of the micelle observed by AFM, TEM, and SEM was somewhat smaller compared to DLS because the diameter calculated by DLS represents the hydrodynamics diameter. In contrast, the diameter acquired by AFM, SEM, and TEM was measured after the evaporation of water. Kheiri Manjili et al., found that the CUR/PCL-PEG micelles' stability was reduced when the particle size and the polydispersity index (PDI) were increased [35]. In another study, the micelles were unstable in water when the ART/PCL-PEG mass ratio was 1 due to ART's hydrophobicity, guiding more drugs at high concentrations to be adsorbed onto the micelle's exterior [37]. Hu et al., found that increasing the length of the hydrophobic PCL block increased the size of the particles for two types of nanoparticles with a fixed PEG block [39], while for a sequence of nanoparticles with fixed PCL:PEG ratios, the diameters of nanoparticles increased when the molecular weight of the copolymers increased. Meanwhile, Mahdaviani et al., found that the particle size of the copolymers had a unimodal distribution, which was advantageous in terms of providing a more extended pharmacological profile in vivo [38]. Furthermore, Peng et al., reported that the PTX wrapping significantly increased the micelles' average diameter, and HER conjugation onto the surfaces expanded the micelles' size marginally [42]. Additionally, all studies reported a slightly negative zeta potential of the copolymeric micelles. Kheiri Manjili et al., discovered that a minor negative charge surface of the SF/PCL–PEG–PCL micelles can increase the drug's circulation time [36]. Similarly, Peng et al., discovered that the slightly negatively charged PTX/PCL-PEG-HER was more ideal for blood persistence and nano complexes contacting the cell surface than their negatively charged counterparts [42].

Drug Loading (DL) and Encapsulation Efficiency (EE)

The two methods used to measure the DL and EE were high-performance liquid chromatography (HPLC) and UV-Vis spectrophotometry. The various parameters for the HPLC analysis are summarised in Table 6. The wavelength of 420 nm was used for CUR. In contrast, the wavelengths of 284 nm and 244 nm were used for atorvastatin and rosuvastatin, respectively, and the wavelength of 278 nm was used for TMX in the UV-Vis analysis [35,40,41].

The DL and EE were calculated using the following equations:

$$DL\ (\%) = \frac{\text{weight of drug in micelles}}{\text{weight of micelles}} \times 100$$

$$EE\ (\%) = \frac{\text{weight of drug in micelles}}{\text{weight of initial drug}} \times 100$$

Different mass feed ratios of drug/PCL-PEG were tested to improve the development parameters and investigate the influence of the drug/copolymer ratio on the drug loading and encapsulation efficiency of CUR, SF, and ART [35–37]. Kheiri Manjili et al. [35–37] found that different drug/PCL-PEG ratios were found to give other micelles stability. For example, increasing the particle size and polydispersity index (PDI) reduced the strength of the drug/PCL-PEG micelles. Besides, the mass ratio of 0.75 and 1 of drug/PCL-PEG resulted in instability of the micelles in water due to aggregation of the micelles. Thus, the drug/PCL-PEG mass ratio at 0.25 was preferred.

Additionally, the effects of the CL:PEG ratio on the DL and EE were also investigated to develop an ideally high DL or EE. The DL and EE of micelles with fixed PEG length were found to increase when the PCL length increased [35–37,39]. This observation was attributed to the encapsulation of the hydrophobic drug into the centre via the hydrophobic interaction of PCL and the drugs and the hydrophilic interaction between water and PEG. The DL and EE of the hydrophobic drug in PCL-PEG copolymers increased due to the strong hydrophobic interaction between the longer PCL chain and the drug. This outcome was anticipated since the encapsulation of the drugs into the hydrophobic centre increases the micelles' volume.

Kheiri Manjili et al. [41] revealed that the drug loading of atorvastatin was greater than rosuvastatin because of its hydrophobic nature, which is preferred by the micelles. Hence, the rosuvastatin-loaded micelles appeared smaller than atorvastatin due to their increased size during drug loading. Whereas Mahdaviani et al., discovered that by encapsulating the CBZ into the micelle, the solubility of the drug in an aqueous solution was increased hundreds-fold, which, in turn, increased the drug substance at the site of action due to a high encapsulation efficiency [38].

Table 6. HPLC parameters for drug loading and encapsulation efficiency analysis.

Authors	Special Solvent	Mobile Phase	Column	Temperature	Flow Rate	Sample Injection Volume	Sample Detection
Kheiri Manjili et al., (2017a) [36]	-	Acetonitrile and water (45:55, v/v)	C18 analytical column (250 mm × 4.6 mm, particle size 5 μm)	Not mentioned	1.0 mL/min	20 μL	λ max = 241 nm, SF
Kheiri Manjil et al., (2017b) [37]	-	Methanol and 5% (w/v) acetic acid (70:30, v/v)	C18 analytical column (150 mm × 4.6 mm, particle size 5 μm)	Not mentioned	1.0 mL/min	20 μL	λ max = 420 nm, ART
Mahdaviani et al., (2017) [38]	Acetonitrile (to dissolve CBZ)	Methanol	Agilent ZORBAX Eclipse Plus C18 column (5 μm, 4.6 mm × 150 mm)	Not mentioned	1.0 mL/min	Not mentioned	λ max = 248 nm, CBZ
Hu et al., (2017) [39]	Acetonitrile (to dissolve PTX)	Acetonitrile and water (50:50, v/v)	Reverse-phase column (Symmetry, 150 mm × 4.6 mm, five μm)	40 °C	1.0 mL/min	Not mentioned	λ max = 227 nm, PTX
Peng et al., (2019) [42]	Acetonitrile (to dissolve PTX)	Acetonitrile and water (50:50, v/v)	C18 column (5 μm, 4.6 × 150 mm)	30 °C	1.0 mL/min	Not mentioned	λ max = 227 nm, PTX

Thermal Analysis

Thermal analysis of the copolymeric micelles was done using differential scanning calorimetry (DSC). In all studies, the DSC thermogram demonstrated endothermic peaks for the PCL-PEG copolymeric micelles. Zamani et al., suggested that the single peak shown on the DSC curve indicates the absence of impurities in the copolymers. A sharp endothermic peak at 142 °C (Figure 6) belongs to the melting point of free TMX, which indicates its crystallinity. After the drug and copolymers are combined, the peak vanishes as the crystalline phase changes to amorphous, indicating successful drug incorporation in the copolymers [40]. Correspondingly, Kheiri Manjili et al. [35,37,41] reported a higher thermogram display of the melting crystalline PCL block of the copolymers' peak compared to the melting temperature of their micelles, confirming the physical interaction between the drug and PCL-PEG copolymers.

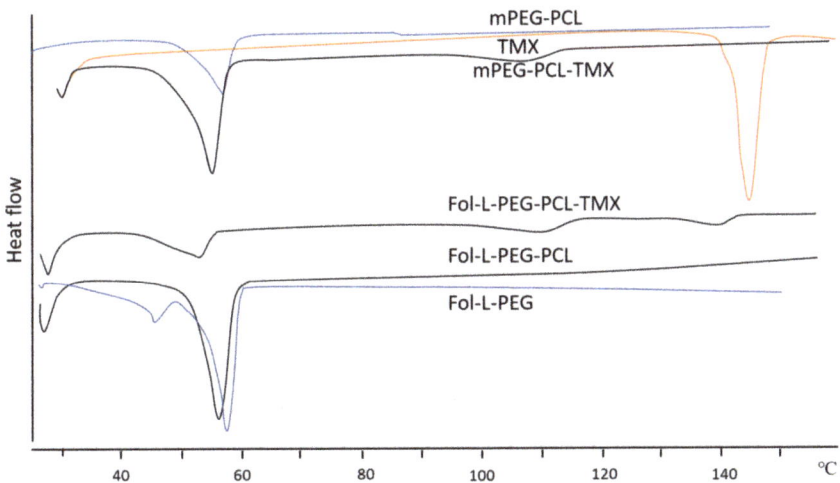

Figure 6. DSC curves of copolymers and drug-loaded copolymers. Reprinted/adapted with permission from Elsevier [40].

3.2.4. Drug Release Study

The dialysis method was adopted in the drug release study [35–37,40–42] and no considerable initial burst of drug release was seen from the micelles. The prolonged release of the hydrophobic drugs was due to the incorporation of the drugs in the core of the micelles. The summary of the release study was tabulated in Table 7.

Table 7. Drug release study of drug-loaded PCL-PEG copolymers.

Authors	Release Medium	pH	Molecular Weight Cut-Off	Incubation Temperature	Shaking Speed	Method of Drug Concentration Analysis	Cumulative Drug Release
Kheiri Manjili et al., (2016) [35]	PBS with 5% (v/v) Tween 80	7.4 5.5 Human plasma	120,000 Da	37 °C	Not mentioned	UV-Vis	~45.32% ~76.8% ~63.21%
Kheiri Manjili et al., (2017a) [36]	PBS	7.4 5.5 Human plasma	120,000 Da	37 °C	Not mentioned	HPLC	~56.75% ~65.75% ~63.21%
Kheiri Manjili et al., (2017b) [37]	PBS with 2% (v/v) tween 80	7.4 5.5 Human plasma	120,000 Da	37 °C	Not mentioned	HPLC	~38.0% ~50.0% ~42.0%
Kheiri Manjili et al., (2018) [41]	PBS with 5% (v/v) Tween 80	7.4 5.5	12,000 Da	37 °C	Not mentioned	UV-Vis	~55.76 % ~60.12 %
Zamani et al., (2018) [40]	PBS	7.4 5.5	140,000 Da	37 °C	100 rpm	UV-Vis	~25.0% ~55.0%
Peng et al., (2019) [42]	PBS with 1% Tween 80	7.4 6.5	10,000 Da	37 °C	120 rpm	HPLC	~60.0% ~63.5%

A pH-sensitive release was observed in each study. Kheiri Manjili et al. [35–37,41] performed the drug release study on drug-loaded micelles at neutral pH (pH 7.4), acidified PBS solution (pH 5.5), and human plasma. Freshly prepared human plasma was collected from a volunteer to investigate the influence of chemical and biological parameters on the hydrophobic drug release from the micelles (Figure 7). The free drug release was done as a control to ensure that drug molecules' diffusion over the dialysis membrane was not the rate-limiting step. They discovered that the percentage of the drugs released from the micelles increased from a pH value of 7.4 to 5.5 due to the hydrolysis degradation of copolymers in acidic conditions. Similar action was observed in the plasma medium when hydrolysis of the esoteric link of the copolymer occurred due to the enhancement of certain enzymes present in human plasma.

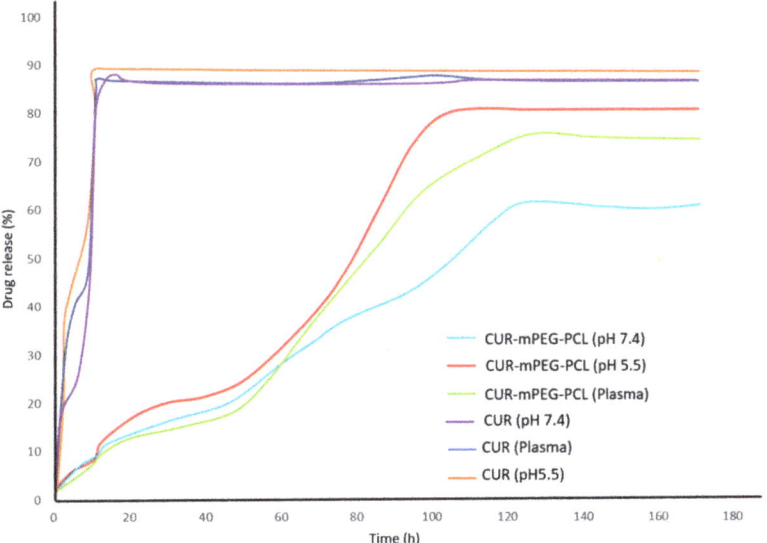

Figure 7. The release profile of CUR from CUR/MPEG-PCL micelles in different release media. Reprinted/adapted with permission from Elsevier [35].

Comparatively, Peng et al., found that the cumulative release rate of PTX in TAXOL® was higher than that of PTX/PCL-PEG and PTX/PCL-PEG-HER at pH 7.4 but showed a similar release rate for all at pH 6.5. This proved that PTX/PCL-PEG and PTX/PCL-PEG-HER had a higher stability than PTX/TAXOL® in physiological conditions and that Herceptin conjugation has no influence on the release rates of PTX/PCL-PEG-HER when compared to PTX/PCL-PEG [42].

In another study, Zamani et al., found no initial burst release observed for the first three hours for the TMX/PCL-PEG micelles at pH 2. Since food and drugs pass the gastrointestinal tract in less than 2 h, this drug-loaded copolymer could be an excellent nanodrug carrier for oral application [40].

Bioavailability of Drugs

The bioavailability of the drugs was assessed using a pharmacokinetic study via oral [31,32] and intravenous routes [37,42] and this was analysed using HPLC and LC-MS. The pharmacokinetic study of the drugs is summarized in Table 8.

Table 8. Pharmacokinetic study of drug-loaded PCL-PEG copolymeric micelles.

Authors	Solution	Dose	Values of Area under the Plasma Concentration—Time Curve (AUC0-∞)	Apparent Volume of Distribution (Vd)
Kheiri Manjili et al., (2016) [35]	CUR aqueous solution	50 mg/kg	8.734 ± 1.09 h ng/mL	8769.132 ± 1321.3 L/kg
	CUR-loaded micelles		488.17 ± 1.23 h ng/mL	807.123 ± 342.9 L/kg
Kheiri Manjili et al., (2017a) [36]	SF aqueous solution	30 mg/kg	8.3425 ± 1.564 h ng/mL	9420.132 ± 2221.3 L/kg
	SF-loaded micelles		465.87 ± 34.2 h ng/mL	909.123 ± 252.1 L/kg
Kheiri Manjil et al., (2017b) [37]	ART aqueous solution	25 mg/kg	320 ± 4.02 h ng/mL	-
	ART-loaded micelles		5234 ± 1.13 h ng/mL	-
Peng et al., (2019) [42]	PTX-PCL-PEG-Her	5 mg/kg	1922.78 μg * h/L	6.59 L/kg
	PTX-PCL-PEG		1874.83 μg * h/L	11.93 L/kg
	TAXOL®		5591.67 μg * h/L	3.32 L/kg

Kheiri Manjili et al., reported an increase in the bioavailability of drugs loaded in micelles compared to the free drug solution. The micelles were effective in increasing drug absorption and delaying the drug release, indicating sustained release of the drugs [35–37]. Meanwhile, Peng et al., found that worm-like micelles enter into the peripheral tissues more quickly than the spherical micelles based on the apparent volume of distribution (V_d). This showed that PTX-PCL-PEG-Her possessed an improved targeting capability compared to PTX-PCL-PEG [42].

3.2.5. Anti-Tumour Activity

Cell Viability and Cytotoxicity Study

The cancerous cell viability and cytotoxicity after treatment using drug-loaded PCL-PEG micelles were done via MTT assay [35–38,40–42]. The cell toxicity of the drug/PCL-PEG micelles was reported to be directly proportional to the micelles' concentration. There was no significant difference in the anti-cancer effect of the CUR-loaded and ART-loaded copolymer micelles at different treatment times in all of the cancer lines tested (mice breast adenocarcinoma, 4T1 and human breast adenocarcinoma, MCF-7), indicating the remarkable anti-cancer effect of the drug-loaded PCL-PEG micelles for breast cancer lines [35,37]. No toxicity was recorded for the drug and blank micelles at different concentrations. However, Kheiri Manjili et al. [37] observed that the MCF-7 cell line was more susceptible to ART treatment compared to the 4T1 cell line. Meanwhile, MTT data statistics showed that, except in MCF10A cells, SF/PCL-PEG micelles significantly ($p < 0.05$) decreased cell viability at each concentration when compared to their copolymers in MCF-7 and 4T1 cancerous cells [36]. The MTT assay also showed that SF-loaded micelles enhanced the cytotoxic effect of SF in MCF-7 and 4T1 cell lines. The blank micelles' biocompatibility as nanocarriers was evaluated by performing the MTT assay against HFF-2 (normal cell line) and MCF-7. At 48 and 72 h, minimal cytotoxic effects were seen on MCF-7 and HFF2 cell lines at the highest tested concentrations of blank polymer. All statin-loaded micelles on normal cells (HFF2) had a minimal cytotoxic effect compared to free statins.

A study by Mahdaviani et al., showed that the CBZ encapsulated in TMT-PCL-PEG showed a greater efficacy in killing highly metastatic cancer cells (MDA-MB-231) than the one encapsulated in PCL-PEG micelles alone and recorded no substantial increase in the cytotoxic effect compared to non-targeted micelles in MCF-7 cells. There was also no significant distinction in the percentage of cell viability between the blank micelles and the control (Figure 8) [38]. The same pattern was reported by Zamani et al., where TMX-loaded FA-L-PCL-PEG micelles decreased the cell viability of MCF-7 cell lines up to 53% compared to the PCL-PEG micelles alone [40], and Peng et al., where PTX-loaded PCL-PEG-Herceptin showed the highest anti-cancer effects compared to PCL-PEG and TAXOL® [42]. Peng et al., also found that the typical antibody dose for in vivo anti-tumour experiments was

10 mg/kg, while Herceptin used in each formulation studied was less than 1.3 mg/kg. The results showed that Herceptin had neither cytotoxic nor therapeutic effects at the utilised dose [42].

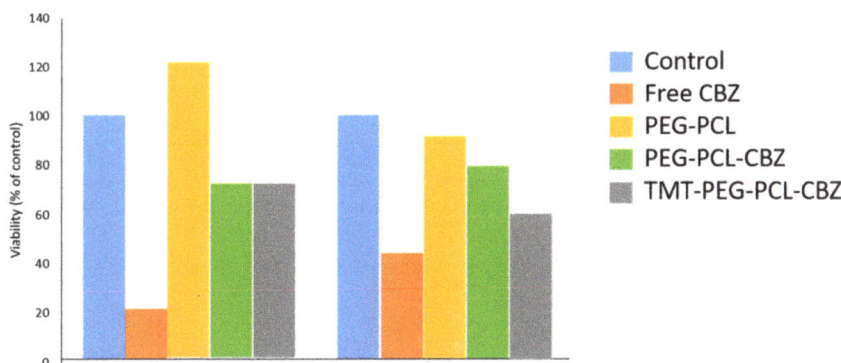

Figure 8. In vitro cell cytotoxicity of free CBZ and CBZ-loaded block copolymer micelles against MCF-7 and MDA-MB 231 after a 72 h incubation. Reprinted/adapted with permission from Elsevier [38].

Anti-Tumour Efficacy

An in vivo experiment by Kheiri Manjili et al. [35–37] was done by measuring the tumour growth profile of 4T1 mice breast cancer as a function of time. The mean tumour volumes of the mice treated with a low concentration of CUR/PCL-PEG micelles after 25 days (1075 ± 195 mm^3) was comparable to the one treated with higher concentrations of free CUR (960 ± 158 mm^3), which corresponded to 45.7% and 40% tumour volume reductions, respectively. Whereas the mean tumour volumes of ART/micelle-treated mice reached below 27 mm^3 compared to the free ART, which was about 500 mm^3 after 20 days, and the tumour volume of the mice treated with SF-loaded micelle decreased up to 78.5% compared to the free SF, which decreased to 49.5% only. These results showed that the drug-loaded micelle had a higher anti-tumour efficacy and more therapeutic effects than the free drug. Additionally, weight variations in all mouse groups were also measured to observe the side effects of the drugs. Mice who were given a high intravenous dose of the free drug lost significantly more weight than the control, saline-treated mice. The significant weight loss observed following high-dose treatment demonstrates the severe systemic toxicity of the free drug. In contrast, mice treated with drug/PCL-PEG micelles showed no signs of severe toxicity, although they had a similar in vivo anti-tumour efficacy to high-dose free drugs. The endurance of animals treated with drug/PCL-PEG micelles or high-dose free drugs was also found to increase significantly.

The anti-tumour efficacy of PTX/PCL-PEG-HER was studied by Peng et al., using the SKBR-3 xenograft model via in vivo experiment. The tumour growth curve revealed that PTX/PCL-PEG-HER was 2-fold and 3-fold more efficient in suppressing tumour development than PTX/TAXOL® and PTX/PCL-PEG within the respective days, and 3-fold, 2.4-fold, and 3.4-fold more effective at stopping tumour development than saline, Herceptin, and blank micelles, respectively, over 48 days. The median survival time of the mice in the TAXOL® group was 36 days. However, all mice in the PTX/PCL-PEG-HER group were still thriving after those days, suggesting the capabilities of PTX/PCL-PEG-HER for the suppression of tumour growth and increases in the longevity of the cancerous mice [42].

Meanwhile, an in vitro experiment was done by Kheiri Manjili et al. [41] to investigate the anti-tumour efficacy of rosuvastatin and atorvastatin and statin-loaded PCL-PEG-PCL micelles on MCF-7 cells. Typically, the presence of statins in nanoparticle form decreased MCF-7 cancer cell viability. Compared to free statins, statin-loaded micelles at the lowest tested doses were observed to generate more cytostatic effect in the MCF-7 cell line. The anti-tumour effect of atorvastatin-loaded micelles was slightly lower compared to rosuvastatin-loaded micelles. The anti-tumour effect of rosuvastatin-loaded micelles was significantly greater than the atorvastatin when observed using IC50 at a concentration of 28.5 μM (48 h) and 17.5 μM (72 h) on the MCF-7 cell line. In comparison, free rosuvastatin suppressed MCF-7 cell growth at a concentration of 30.2 μM (48 h) and 20.7 μM (72 h), respectively. These showed that lower concentrations of drug-loaded micelles were needed compared to the free drug to exhibit the tumour-suppressive effect.

Mahdaviani et al., investigated the anti-cancer efficacy by analysing the apoptotic and necrotic induction effect of CBZ-loaded copolymer micelles in MCF-7 and MDA-MB-231 cell lines using flow cytometry. The study revealed significant apoptosis (16%) and necrosis (65%) in MDA-MB-231 cells, as well as mildly elevated apoptosis (2%) and necrosis (33%) in MDA-MB-231 cells treated with CBZ-loaded PCL-PEG. Compared to the control group, there was an overall increase in the percentages of early and late apoptotic MCF-7 cells, although this increase was practically the same in the targeted and non-targeted groups. However, treatment with the targeted copolymeric micelles demonstrated better apoptosis and necrosis induction than the non-targeted copolymeric micelles for the MDA-MB-231 cell line [38].

Additionally, Hu et al., researched the efficiency of the PCL-PEG polymeric micelles as drug cargo for breast cancer cells via cellular uptake experiments. The internalisation of the drug-loaded micelles into the EMT-6 breast cancer cells was examined using confocal laser scanning microscopy (CLSM). They observed that the internalised polymeric micelles (red) successfully bypassed the EMT-6 cancer cell's nucleus (blue) in the cytoplasm after 4 h of incubation and increased in number after 24 h (refer to Figure 2 from the previous study [39]), indicating the high cellular uptake efficiency of the copolymeric micelles towards breast cancer cells [39].

4. Discussion

Five themes and six sub-themes were developed from the thematic analyses. Further discussion of the developed themes is presented in this section. Except for Mahdaviani et al. [38], who utilised N-protected-3-aminopropan-1-ol as the initiator for the ROP of ε-CL to PCL, the synthesis process of PCL-PEG copolymers of other studies was done using PEG as the initiator in the presence of stannous octoate as the catalyst [35–37,39–41]. The reaction had to be conducted at 110 °C to ensure complete conversion of ε-CL to PCL. The ratio of ε-CL:PEG was controlled in the studies to produce copolymers with different molecular weights.

The micelles were prepared by nanoprecipitation, thin film hydration, co-solvent evaporation, and emulsion evaporation methods. The nanoprecipitation method was chosen due to its simplicity and its ease of scaling up [35–37,41]. The PCL-PEG micelles produced were biocompatible, biodegradable, stable in blood, and small in size, making them an exceptional choice for drug delivery systems. The co-solvent evaporation method has the advantage of scalability and minimal drug loss during the encapsulation process compared to the dialysis method [38]. The three main aspects of optimising the size of the polymeric micelles and drug encapsulation efficiency in the co-solvent evaporation method are the type of organic co-solvent, the volume fraction of the organic: aqueous phase during the production, and the sequence to add the organic and aqueous phase. Mahdaviani et al., stated that the best organic:aqueous phase ratio for the polymeric micellar carrier to function optimally is 10% w/w. Additionally, PCL-PEG chain length can be optimised to stimulate the micelles to self-assemble into specific morphology [38]. Peng et al., reported the optimised matrix of the chain length of PCL-PEG copolymers at a weight ratio of 3:7 to encourage the hydrophobic drug to self-assemble and wrap into worm-like micelles. Interestingly, they found that the micelles' morphology changed from spherical to worm-like structures after 6.3% of the drug was encapsulated into the micelles, which indicates that the drug can influence the morphology of the micelles [42].

Apart from that, micelles conjugated with functionalised ligands can be prepared for drug delivery to a specific target. Mahdaviani et al., found that the conjugated metastatic cancer-specific TMT ligand polymeric micelles displayed effective in vitro anti-tumour activities in highly metastatic MBA-MD-23 breast cancer cell lines but not in non-metastatic MCF-7 cell lines [38]. Correspondingly, the conjugation of Herceptin ACN to the PCL-PEG copolymers [42] modified the antibodies' spectra from a beta-tum to a polypro II helix confirmation but still retained the targeting ability to bind to overexpressed HER2 receptors on the surface of SKBR-3 human breast cancer cell lines [42].

PCL-PEG copolymers can self-assemble into different structures depending on certain conditions. Hence, the morphology determination of the micelles was essential to observe the specific morphology the micelles assembled into. The critical parameter that affects the morphology of the self-assembled polymeric micelles is the mass or volume fraction of the PEG fraction of the PCL-PEG copolymers. Hu et al., expressed that if the PEG fraction fell between 45% and 55%, the cylindrical micelles were likely to form, while if the PEG fraction was between 55% and 70%, then spherical micelles were more likely to develop [39]. All studies exhibited spherical morphology of the polymeric micelles when observed using AFM, TEM, and SEM. A visible core-shell structure for the polymeric micelles was reported, where the PCL is the inner hydrophobic part acting as the core, encircled by the hydrophilic PEG corona as the shell of the structure. The surface of the micelles is closed since both terminals of the PEG block are attached at the core/shell interfaces, limiting the potential of opsonisation. Finally, the hydrophobic drugs are encapsulated into the PCL's hydrophobic core through hydrophobic interactions. Additionally, Peng et al., observed worm-like micelles when the PCL-PEG micelles encapsulated the drug. As a drug cargo, the poorly water soluble PCL-PEG worm-like micelles were reported to be more efficient than spherical micelles at exiting the body's mononuclear phagocyte system, thus loading and delivering a more significant amount of drugs to tumours and reaching farther into tumours with a higher penetration ratio, leading to possible tumour shrinkage [42].

The particle size of the micelles observed using the DLS technique showed a slightly bigger diameter than the one observed via AFM, TEM, and SEM. The surface charge of the micelles showed that the PCL-PEG micelles possess a slightly negative charge. The surface charge influences the behaviour of the oppositely charged cell membrane, whether they cluster in the blood flow, stick to, or interact with it. Because plasma and blood cells are constantly negatively charged, nanoparticles with a slightly negative surface charge might reduce nonspecific contact with them via electrostatic forces.

The capacity of drug loading and encapsulation efficiency are crucial variables for nanoparticles. A decent drug delivery system must always have a high capacity for drug loading (DL) and encapsulation efficiency (EE) [39]. Due to their hydrophobic interaction, the PCL-PEG micelles showed a high capacity for hydrophobic DL and EE. Increasing the PCL length of the PCL-PEG copolymers will improve the DL and EE of the hydrophobic drug because of the stronger hydrophobic interaction between the longer PCL chain and the drug. The incorporation of the drugs into the micelles can be observed by increasing the volume of the micelles. It is also worth noting that the PCL-PEG micelles favoured hydrophobic drugs more than hydrophilic drugs, and the encapsulation of the drugs into the PCL-PEG micelles resulted in the increase in the solubility of the drug at the site of action.

The in vitro drug release study revealed that the PCL-PEG micelles can slowly release the drug into the medium, indicating the ability of the drug-loaded PCL-PEG micelles to increase in vivo systemic circulation, i.e., the t1/2 of the drug. The prolonged drug release observed from the copolymeric micelles was because of the diffusion of the drug from the micelles and the degradation or hydrolysis of the micelles. The drug released from the PCL-PEG micelle was pH-sensitive. The drug was released faster in acidic pH than in neutral pH due to the protonation of the polymer, making the polymer matrix swell and degrade. This feature is highly desirable in numerous applications, particularly in anti-cancer drug delivery, since the microenvironments of tumour extracellular spaces, intracellular lysosomes, and endosomes are acidic; hence, the drug release from micelles may be facilitated. The resulting copolymeric micelles are highly appealing to the time-controlled administration of hydrophobic drugs for various medicinal purposes.

Cell viability and cytotoxicity studies were done to examine the in vitro anti-tumour effects of the drug-loaded PCL-PEG micelles against selected breast cancer cells. The drug-loaded PCL-PEG micelles' increased cytotoxicity compared to the free drug can be attributed to their differential cellular absorption. This was because the internalisation of the free drug was diffusion-mediated, which was restricted after reaching saturation in the cytoplasm, resulting in minimal drug uptake compared to drugs loaded in PCL-PEG micelles. The endocytosis pathways were likely involved in the increased absorption of drug-loaded micelles and subsequent long-term drug release [35–37]. On the contrary, Mahdaviani et al., listed three pieces of information derived from the cytotoxicity results of the drug loaded in the conjugated polymeric micelles. First, there was no substantial change in the viability percentage between the blank micelles and the control group, indicating that the blank micelles were biocompatible. Second, the free drug interacts directly with the cells without experiencing the release process, making them more cytotoxic than the drug encapsulated in polymeric micelles. This was because the highly hydrophobic free drug can readily penetrate the lipid membranes and subsequently diffuse into the cells, resulting in a more significant cellular build-up and, consequently, increasing the cytotoxicity. Since the efficient delivery of the drug was often hindered by its hydrophobicity, using amphiphilic PCL-PEG micelles as a potent drug delivery vehicle is essential due to their biocompatible and anti-biofouling characteristics. Third, the conjugation of the cancer-specific ligand to the polymeric micelles resulted in an efficient in vitro anti-tumour effect in the specific cancer cell line [38].

Lastly, a different mode of analysis was done in each study to examine the breast anti-tumour efficiency of the drug-loaded PCL-PEG micelles. The much greater anti-tumour efficiency of drug/PCL-PEG micelles compared to the same dose of the free drug was attributed to the more extended drug circulation in the plasma and the enhanced drug accumulation in the tumour. The sensitive hydrophobic drug-loaded PCL-PEG micelles have the benefits of more extended blood circulation, prevention of RES absorption, and passive targeting of polymeric micelles to tumour tissues via the EPR effect. Also, signs of severe toxicity, such as weight loss, were absent because of the extended half-life in blood circulation and improved tumour localisation of conjugate-associated drugs [35–37]. Meanwhile, the flow cytometry analysis by Mahdaviani et al., showed that the active targeted group (CBZ-TMT-PCL-PEG) induced a greater percentage of apoptosis than the passive targeted group (CBZ-PCL-PEG) for the same cell line. This was due to polymeric micelles with tumour cell-targeting ligands that can facilitate drug binding and internalisation into tumour cells, which increased the drug accumulation in the tumour and improved the anti-cancer effect [38]. Moreover, Hu et al., stated that the high-efficiency cellular uptake of the PCL-PEG copolymeric micelles into the breast cancer cell line was due to the ability of the micelles to harness a nonspecific endocytosis pathway to internalise via lipid bilayer cellular membranes, depending on their shape and size [39].

5. Recommendation

This systematic literature review gives rise to several recommendations to guide researchers in their future studies. First, future studies should focus on the synergistic anti-cancer activity of conjugated PCL-PEG copolymers with drugs to clarify whether the conjugation functions only target specific cancerous cells in the anti-cancer treatment. Second, the effects of various morphologies, such as the worm-like structure of the micelles, should be further studied to optimise their efficacy as drug cargo. Next, more types of anti-cancer drugs and other types of cancer cell lines should be investigated to determine their anti-cancer effect and cytotoxicity on the cancer cell line since the current study only focused on a breast cancer cell line and a few drugs such as curcumin, sulforaphane, artemisinin, atorvastatin, rosuvastatin, cabazitaxel, paclitaxel, and tamoxifen.

To advance the therapy provided in the clinic, the micelles' interactions with the complex environment and barriers that determine the fate of the micelles' drug carrier in the body and the final therapeutic efficiency must be studied. This is because, once they enter the blood circulation, the micelles face and interact with complex biological environments such as blood component, phagocytosis, and biodistribution before arriving at the lesion site, which might modify their capability to target and transport the anti-cancer drug [43]. Additionally, micellar stability is significant in determining the delivery efficacy since micelles become unstable when facing blood shear stress, resulting in the release of the drug before reaching the target site [44]. The most effective strategy to increase the stability of the micelles' formulation is to increase the hydrophobicity of the micellar core, which, in turn, decreases the CMC value. Other strategies include introducing covalent cross linking in the micellar core, stereocomplexation, and using polymers possessing ester or amide bonds [43]. Another strategy to consider is the synthesis of PCL-PEG copolymers via sequential nucleophilic substitution reactions. This method produced copolymers that self-assembled and stayed as stable polymeric micelles at pH 7.4 and showed good encapsulation ability of both hydrophobic and hydrophilic drugs [45,46]. Therefore, to design micelles for drug delivery, consideration should be given to the following elements: appropriate effect on blood components to ensure hemocompatibility, modulation of protein adsorption to reduce subsequent phagocytosis, biodistribution, and increased micellar stability in blood and extended circulation time [47].

Most studies in this review relied on electronic keyword searches, which are often regarded as the best way to conduct a systematic review [31]. However, researchers can use several complementary techniques to enhance their search efforts. First, citation tracking, which refers to efforts to find related publications based on papers that cite the article under evaluation, can augment the results since it may find other publications that ordinary database searches would otherwise miss due to the vocabulary limitation in search strategies or bibliographic records [48]. Second, reference searching, where the reference lists in the selected articles are examined for other related articles, can also be conducted. This strategy can lower the risk of missing related information when researchers encounter challenges finding related material [49]. Third, a snowballing approach, including forward and backward snowballing, can be helpful. These two methods are similar to the first two strategies mentioned. However, these three searching strategies can sometimes get out of control due to the increased number of articles being retrieved and the possibility of needing to manually appraise each article [50]. Researchers can also consider contacting experts if the specialist literature is vaguely specified [51].

6. Conclusions

In conclusion, the PCL-PEG copolymeric micelles with different formulations showed a remarkable performance as drug cargo for the hydrophobic anti-breast cancer drug. Copolymeric micelles conjugated with functionalised ligands enable the targeting of specific cancer cell lines. The PCL-PEG copolymer micelles showed a high drug loading and encapsulation efficiency. Additionally, the drug release of drug-loaded PCL-PEG copolymer micelles was a sustained release. Cell viability and anti-cancer efficacy studies showed that the drug-loaded copolymeric micelles have a lower toxicity and can inhibit breast cancer cell lines.

Author Contributions: Conceptualization, S.H.A.S. and M.W.I.; validation, S.A.H., M.S.H. and W.K.W.A.K.; formal analysis, S.H.A.S., W.K.W.A.K. and M.S.H.; investigation, S.H.A.S. and M.W.I.; writing—original draft preparation, S.H.A.S. and M.W.I.; writing—review and editing, S.H.A.S., M.W.I. and S.A.H.; supervision, M.W.I.; project administration, M.W.I.; funding acquisition, M.W.I. All authors have read and agreed to the published version of the manuscript.

Funding: This research was funded by Ministry of Higher Education (grant number FRGS/1/2019/STG01/UIAM/02/1).

Acknowledgments: The Ministry of Higher Education Malaysia financially supported this research under the FRGS grant [FRGS/1/2019/STG01/UIAM/02/1].

Conflicts of Interest: The authors declare no conflict of interest.

References

1. Sung, H.; Ferlay, J.; Siegel, R.L.; Laversanne, M.; Soerjomataram, I.; Jemal, A.; Bray, F. Global Cancer Statistics 2020: GLOBOCAN Estimates of Incidence and Mortality Worldwide for 36 Cancers in 185 Countries. *CA Cancer J. Clin.* **2021**, *71*, 209–249. [CrossRef] [PubMed]
2. World Health Organization (WHO). *Global Health Estimates 2020: Deaths by Cause, Age, Sex, by Country and by Region 2000–2019*; WHO: Geneva, Switzerland, 2020.
3. National Cancer Institute. *Types of Cancer Treatment*; National Institutes of Health (NIH): Rockville, MD, USA. Available online: https://www.cancer.gov/about-cancer/treatment/types (accessed on 20 June 2021).
4. Bernstein, S. *Chemotherapy: How the Drugs That Treat Cancer Work*; WebMD LLC: New York, NY, USA, 2020; Available online: https://www.webmd.com/cancer/how-chemo-works (accessed on 10 August 2021).
5. Wang, X.; Wu, Z.; Li, J.; Pan, G.; Shi, D.; Ren, J. Preparation, characterization, biotoxicity, and biodistribution of thermo-responsive magnetic complex micelles formed by Mn0.6Zn0.4Fe2O4 and a PCL/PEG analogue copolymer for controlled drug delivery. *J. Mater. Chem. B* **2016**, *5*, 296–306. [CrossRef] [PubMed]
6. Fang, X.; Cao, J.; Shen, A. Advances in anti-breast cancer drugs and the application of nano-drug delivery systems in breast cancer therapy. *J. Drug Deliv. Sci. Technol.* **2020**, *57*, 101662. [CrossRef]
7. Deng, C.; Jiang, Y.; Cheng, R.; Meng, F.; Zhong, Z. Biodegradable polymeric micelles for targeted and controlled anticancer drug delivery: Promises, progress and prospects. *Nano Today* **2012**, *7*, 467–480. [CrossRef]
8. Feng, C.; Huang, X. Polymer Brushes: Efficient Synthesis and Applications. *Accounts Chem. Res.* **2018**, *51*, 2314–2323. [CrossRef] [PubMed]
9. Yao, Y.; Zhou, Y.; Liu, L.; Xu, Y.; Chen, Q.; Wang, Y.; Wu, S.; Deng, Y.; Zhang, J.; Shao, A. Nanoparticle-Based Drug Delivery in Cancer Therapy and Its Role in Overcoming Drug Resistance. *Front. Mol. Biosci.* **2020**, *7*, 193. [CrossRef]
10. Shen, Y.; Leng, M.; Yu, H.; Zhang, Q.; Luo, X.; Gregersen, H.; Wang, G.; Liu, X. Effect of Amphiphilic PCL-PEG Nano-Micelles on HepG2 Cell Migration. *Macromol. Biosci.* **2014**, *15*, 372–384. [CrossRef] [PubMed]
11. Biswas, S.; Kumari, P.; Lakhani, P.M.; Ghosh, B. Recent advances in polymeric micelles for anti-cancer drug delivery. *Eur. J. Pharm. Sci.* **2016**, *83*, 184–202. [CrossRef]
12. Khoee, S.; Kavand, A. Preparation, co-assembling and interfacial crosslinking of photocurable and folate-conjugated amphiphilic block copolymers for controlled and targeted drug delivery: Smart armored nanocarriers. *Eur. J. Med. Chem.* **2014**, *73*, 18–29. [CrossRef]
13. Grossen, P.; Witzigmann, D.; Sieber, S.; Huwyler, J. PEG-PCL-based nanomedicines: A biodegradable drug delivery system and its application. *J. Control. Release* **2017**, *260*, 46–60. [CrossRef]
14. Kang, L.; Gao, Z.; Huang, W.; Jin, M.; Wang, Q. Nanocarrier-mediated co-delivery of chemotherapeutic drugs and gene agents for cancer treatment. *Acta Pharm. Sin. B* **2015**, *5*, 169–175. [CrossRef] [PubMed]
15. Nair, P.R. Delivering Combination Chemotherapies and Targeting Oncogenic Pathways via Polymeric Drug Delivery Systems. *Polymers* **2019**, *11*, 630. [CrossRef] [PubMed]
16. Cabral, H.; Miyata, K.; Osada, K.; Kataoka, K. Block Copolymer Micelles in Nanomedicine Applications. *Chem. Rev.* **2018**, *118*, 6844–6892. [CrossRef]
17. Gou, M.; Wei, X.; Men, K.; Wang, B.; Luo, F.; Zhao, X.; Wei, Y.; Qian, Z. PCL/PEG Copolymeric Nanoparticles: Potential Nanoplatforms for Anticancer Agent Delivery. *Curr. Drug Targets* **2011**, *12*, 1131–1150. [CrossRef] [PubMed]
18. Chang, S.H.; Lee, H.J.; Park, S.; Kim, Y.; Jeong, B. Fast Degradable Polycaprolactone for Drug Delivery. *Biomacromolecules* **2018**, *19*, 2302–2307. [CrossRef]
19. Manavitehrani, I.; Fathi, A.; Badr, H.; Daly, S.; Negahi Shirazi, A.; Dehghani, F. Biomedical Applications of Biodegradable Polyesters. *Polymers* **2016**, *8*, 20. [CrossRef] [PubMed]
20. Cai, M.; Cao, J.; Wu, Z.; Cheng, F.; Chen, Y.; Luo, X. In vitro and in vivo anti-tumor efficiency comparison of phosphorylcholine micelles with PEG micelles. *Colloids Surf. B Biointerfaces* **2017**, *157*, 268–279. [CrossRef] [PubMed]
21. Senevirathne, S.A.; Washington, K.E.; Biewer, M.C.; Stefan, M.C. PEG based anti-cancer drug conjugated prodrug micelles for the delivery of anti-cancer agents. *J. Mater. Chem. B* **2015**, *4*, 360–370. [CrossRef]
22. Douglas, P.; Albadarin, A.; Sajjia, M.; Mangwandi, C.; Kuhs, M.; Collins, M.N.; Walker, G. Effect of poly ethylene glycol on the mechanical and thermal properties of bioactive poly(ε-caprolactone) melt extrudates for pharmaceutical applications. *Int. J. Pharm.* **2016**, *500*, 179–186. [CrossRef]
23. Zhou, X.; Qin, X.; Gong, T.; Zhang, Z.-R.; Fu, Y. d-Fructose Modification Enhanced Internalization of Mixed Micelles in Breast Cancer Cells via GLUT5 Transporters. *Macromol. Biosci.* **2017**, *17*, 1600529. [CrossRef]
24. Pourjavadi, A.; Dastanpour, L.; Tehrani, Z.M. Magnetic micellar nanocarrier based on pH-sensitive PEG-PCL-PEG triblock copolymer: A potential carrier for hydrophobic anticancer drugs. *J. Nanopart. Res.* **2018**, *20*, 282. [CrossRef]

25. Yu, T.; Huang, X.; Liu, J.; Fu, Q.; Wang, B.; Qian, Z. Polymeric nanoparticles encapsulating α-mangostin inhibit the growth and metastasis in colorectal cancer. *Appl. Mater. Today* **2019**, *16*, 351–366. [CrossRef]
26. Yang, L.; Zhang, Z.; Hou, J.; Jin, X.; Ke, Z.; Liu, D.; Du, M.; Jia, X.; Lv, H. Targeted delivery of ginsenoside compound K using TPGS/PEG-PCL mixed micelles for effective treatment of lung cancer. *Int. J. Nanomed.* **2017**, *12*, 7653–7667. [CrossRef] [PubMed]
27. Gao, X.; Gou, M.; Huang, M.; Huang, N.; Qian, Z.; Wei, X.; Wang, B.; Yang, B.; Men, K.; Rao, W.; et al. Preparation, characterization and application of star-shaped PCL/PEG micelles for the delivery of doxorubicin in the treatment of colon cancer. *Int. J. Nanomed.* **2013**, *8*, 971–982. [CrossRef]
28. Tazehkand, A.P.; Salehi, R.; Velaei, K.; Samadi, N. The potential impact of trigonelline loaded micelles on Nrf2 suppression to overcome oxaliplatin resistance in colon cancer cells. *Mol. Biol. Rep.* **2020**, *47*, 5817–5829. [CrossRef] [PubMed]
29. Jin, J.; Sui, B.; Gou, J.; Liu, J.; Tang, X.; Xu, H.; Zhang, Y.; Jin, X. PSMA Ligand Conjugated PCL-PEG Polymeric Micelles Targeted to Prostate Cancer Cells. *PLoS ONE* **2014**, *9*, e112200. [CrossRef]
30. Avramović, N.; Mandić, B.; Savić-Radojević, A.; Simić, T. Polymeric Nanocarriers of Drug Delivery Systems in Cancer Therapy. *Pharmaceutics* **2020**, *12*, 298. [CrossRef]
31. Shaffril, H.A.M.; Krauss, S.E.; Samsuddin, S.F. A systematic review on Asian's farmers' adaptation practices towards climate change. *Sci. Total Environ.* **2018**, *644*, 683–695. [CrossRef]
32. Mckenzie, J.E.; Brennan, S.E.; Ryan, R.E.; Thomson, H.J.; Johnston, R.V.; Thomas, J. Chapter 3: Defining the criteria for including studies and how they will be grouped for the synthesis. In *Cochrane Handbook for Systematic Reviews of Interventions Version 6.2*; Higgins, J., Thomas, J., Chandler, J., Cumpston, M., Li, T., Page, M., Welch, V., Eds.; Cochrane: London, UK, 2021.
33. Busalim, A.H.; Hussin, A.R.C. Understanding social commerce: A systematic literature review and directions for further research. *Int. J. Inf. Manag.* **2016**, *36*, 1075–1088. [CrossRef]
34. Siddaway, A.P.; Wood, A.M.; Hedges, L.V. How to Do a Systematic Review: A Best Practice Guide for Conducting and Reporting Narrative Reviews, Meta-Analyses, and Meta-Syntheses. *Annu. Rev. Psychol.* **2018**, *70*, 703–718. [CrossRef]
35. Kheiri Manjili, H.; Ghasemi, P.; Malvandi, H.; Mousavi, M.S.; Attari, E.; Danafar, H. Pharmacokinetics and in vivo delivery of curcumin by copolymeric mPEG-PCL micelles. *Eur. J. Pharm. Biopharm.* **2017**, *116*, 17–30. [CrossRef] [PubMed]
36. Manjili, H.K.; Sharafi, A.; Attari, E.; Danafar, H. Pharmacokinetics and in vitro and in vivo delivery of sulforaphane by PCL–PEG–PCL copolymeric-based micelles. *Artif. Cells Nanomed. Biotechnol.* **2017**, *45*, 1728–1739. [CrossRef] [PubMed]
37. Manjili, H.K.; Malvandi, H.; Mousavi, M.S.; Attari, E.; Danafar, H. In vitro and in vivo delivery of artemisinin loaded PCL–PEG–PCL micelles and its pharmacokinetic study. *Artif. Cells Nanomed. Biotechnol.* **2017**, *46*, 926–936. [CrossRef] [PubMed]
38. Mahdaviani, P.; Bahadorikhalili, S.; Navaei-Nigjeh, M.; Vafaei, S.Y.; Esfandyari-Manesh, M.; Abdolghaffari, A.H.; Daman, Z.; Atyabi, F.; Ghahremani, M.H.; Amini, M.; et al. Peptide functionalized poly ethylene glycol-poly caprolactone nanomicelles for specific cabazitaxel delivery to metastatic breast cancer cells. *Mater. Sci. Eng. C* **2017**, *80*, 301–312. [CrossRef] [PubMed]
39. Hu, C.; Chen, Z.; Wu, S.; Han, Y.; Wang, H.; Sun, H.; Kong, D.; Leng, X.; Wang, C.; Zhang, L.; et al. Micelle or polymersome formation by PCL-PEG-PCL copolymers as drug delivery systems. *Chin. Chem. Lett.* **2017**, *28*, 1905–1909. [CrossRef]
40. Zamani, M.; Rostamizadeh, K.; Manjili, H.K.; Danafar, H. In vitro and in vivo biocompatibility study of folate-lysine-PEG-PCL as nanocarrier for targeted breast cancer drug delivery. *Eur. Polym. J.* **2018**, *103*, 260–270. [CrossRef]
41. Kheiri, M.H.; Alimohammadi, N.; Danafar, H. Preparation of biocompatible copolymeric micelles as a carrier of atorvastatin and rosuvastatin for potential anticancer activity study. *Pharm. Dev. Technol.* **2018**, *24*, 303–313. [CrossRef]
42. Peng, J.; Chen, J.; Xie, F.; Bao, W.; Xu, H.; Wang, H.; Xu, Y.; Du, Z. Herceptin-conjugated paclitaxel loaded PCL-PEG worm-like nanocrystal micelles for the combinatorial treatment of HER2-positive breast cancer. *Biomaterials* **2019**, *222*, 119420. [CrossRef]
43. Majumder, N.; Das, N.G.; Das, S.K. Polymeric micelles for anticancer drug delivery. *Ther. Deliv.* **2020**, *11*, 613–635. [CrossRef]
44. Sun, X.; Wang, G.; Zhang, H.; Hu, S.; Liu, X.; Tang, J.; Shen, Y. The Blood Clearance Kinetics and Pathway of Polymeric Micelles in Cancer Drug Delivery. *ACS Nano* **2018**, *12*, 6179–6192. [CrossRef]
45. Nutan, B.; Chandel, A.K.S.; Jewrajka, S.K. Synthesis and Multi-Responsive Self-Assembly of Cationic Poly(caprolactone)-Poly(ethylene glycol) Multiblock Copolymers. *Chem.-A Eur. J.* **2017**, *23*, 8166–8170. [CrossRef] [PubMed]
46. Nutan, B.; Chandel, A.K.S.; Jewrajka, S.K. Liquid Prepolymer-Based in Situ Formation of Degradable Poly(ethylene glycol)-Linked-Poly(caprolactone)-Linked-Poly(2-dimethylaminoethyl)methacrylate Amphiphilic Conetwork Gels Showing Polarity Driven Gelation and Bioadhesion. *ACS Appl. Bio Mater.* **2018**, *1*, 1606–1619. [CrossRef] [PubMed]
47. Hou, Z.; Zhou, W.; Guo, X.; Zhong, R.; Wang, A.; Li, J.; Cen, Y.; You, C.; Tan, H.; Tian, M. Poly(ε-Caprolactone)-Methoxypolyethylene Glycol (PCL-MPEG)-Based Micelles for Drug-Delivery: The Effect of PCL Chain Length on Blood Components, Phagocytosis, and Biodistribution. *Int. J. Nanomed.* **2022**, *17*, 1613–1632. [CrossRef] [PubMed]
48. Wright, K.; Golder, S.; Rodriguez-Lopez, R. Citation searching: A systematic review case study of multiple risk behaviour interventions. *BMC Med. Res. Methodol.* **2014**, *14*, 73. [CrossRef]
49. Horsley, T.; Dingwall, O.; Sampson, M. Checking reference lists to find additional studies for systematic reviews. *Cochrane Database Syst. Rev.* **2011**, *2011*, MR000026. [CrossRef]
50. Tsafnat, G.; Glasziou, P.; Choong, M.K.; Dunn, A.; Galgani, F.; Coiera, E. Systematic review automation technologies. *Syst. Rev.* **2014**, *3*, 74. [CrossRef]
51. Gøtzsche, P.C.; Ioannidis, J.P.A. Content area experts as authors: Helpful or harmful for systematic reviews and meta-analyses? *BMJ* **2012**, *345*, e7031. [CrossRef]

Article

Acceleration of Bone Fracture Healing through the Use of Bovine Hydroxyapatite or Calcium Lactate Oral and Implant Bovine Hydroxyapatite–Gelatin on Bone Defect Animal Model

Aniek Setiya Budiatin [1,*], Junaidi Khotib [1], Samirah Samirah [1], Chrismawan Ardianto [1], Maria Apriliani Gani [1], Bulan Rhea Kaulika Hadinar Putri [1], Huzaifah Arofik [1], Rizka Nanda Sadiwa [1], Indri Lestari [1], Yusuf Alif Pratama [1], Erreza Rahadiansyah [2] and Imam Susilo [3]

1 Department of Pharmacy Practice, Faculty of Pharmacy, Universitas Airlangga, Surabaya 60115, Indonesia
2 Department of Orthopaedics and Traumatology, Faculty of Medicine, Universitas Airlangga, Surabaya 60131, Indonesia
3 Department of Anatomical Pathology, Faculty of Medicine, Universitas Airlangga, Surabaya 60131, Indonesia
* Correspondence: anieksb@yahoo.co.id

Abstract: Bone grafts a commonly used therapeutic technique for the reconstruction and facilitation of bone regeneration due to fractures. BHA–GEL (bovine hydroxyapatite–gelatin) pellet implants have been shown to be able accelerate the process of bone repair by looking at the percentage of new bone, and the contact between the composite and bone. Based on these results, a study was conducted by placing BHA–GEL (9:1) pellet implants in rabbit femoral bone defects, accompanied by 500 mg oral supplement of BHA or calcium lactate to determine the effectiveness of addition supplements. The research model used was a burr hole defect model with a diameter of 4.2 mm in the cortical part of the rabbit femur. On the 7th, 14th and 28th days after treatment, a total of 48 New Zealand rabbits were divided into four groups, namely defect (control), implant, implant + oral BHA, and implant + oral calcium lactate. Animal tests were terminated and evaluated based on X-ray radiology results, *Hematoxylin-Eosin* staining, vascular endothelial growth Factor (VEGF), osteocalcin, and enzyme-linked immunosorbent assay (ELISA) for bone alkaline phosphatase (BALP) and calcium levels. From this research can be concluded that Oral BHA supplementation with BHA–GEL pellet implants showed faster healing of bone defects compared to oral calcium lactate with BHA–GEL pellet implants.

Keywords: defect; bone remodeling; bovine hydroxyapatite; calcium lactate; BHA–GEL pellet

1. Introduction

Bone is a special connective tissue that hardens via the process of mineralization by calcium phosphate in the form of hydroxyapatite [1]. Various kinds of bone, joint, and muscle diseases in humans include open fractures, closed fractures, osteoporosis, osteoarthritis, osteomalacia, osteomyelitis, rheumatic polymyalgia, gouty arthritis, rheumatoid arthritis and others, with fractures being the most common large organ traumatic injury in humans [2]. A fracture is neuromuscular damage due to trauma to the tissue [3] and results in a gap in the bone. Fracture repair can generally restore damaged skeletal organs to their preinjury cellular composition, structure, and biomechanical function, but about 10% of fractures will not heal normally [2]. In 2011, the World Health Organization (WHO) recorded more than 1.3 million people suffering from fractures due to accidents. Accident cases that have a fairly high prevalence, amongst which are lower extremity fractures, reprsent 40% of the accidents that occur. Bone graft therapy, with a surgical procedure that places new bone or replacement material (composite matrix) into the space around the fracture or hole in the damaged bone (defect) to help speed up the healing process [3], is commonly used in fracture management [4].

Bovine hydroxyapatite (BHA) is an inorganic bovine bone material used as an alternative composite component, consisting of 93% hydroxyapatite ($Ca_{10}(PO_4)_6(OH)_2$) and 7% β-tricalcium phosphate ($Ca_3(PO_4)_2$). As a result, it is more porous and can only absorb antibiotics, hormones or growth factors [5]. BHA has properties similar to hydroxyapatite in human bone and is a scaffold that is more osteoconductive than other synthetic hydroxyapatites, is biocompatible and has high porosity. The high porosity of BHA accelerates the process of colonization of osteoblast cells and becomes a medium for osteoblast cells to stick to [6]. On the other hand, BHA is brittle as a new bone-forming material, so gelatin is added as an adhesive and smoothing agent [5]. Gelatin (GEL) is a macromolecule produced by partial hydrolysis of collagen from skin, white connective tissue and animal bones of amino acid residues [7]. GEL is commonly called type one collagen which together with osteoblasts forms osteoids (soft callus). BHA–GEL composites that resemble mineral components in humans are able to form new bone and fill bone gaps due to fractures [5] with high biocompatible properties, osteoconductive, osteoinductive, biodegradable, bioresorbable, and non-toxic [8,9]. The addition of gelatin can also control the degradation of pellet time and increase the synthesis of new bone in the defect area [10].

BHA–GEL composite implants have been shown to be able to accelerate the bone repair process in fractured rabbit femurs within 28 to 42 days, and have good biomaterials for bone filling [11]. However, this period of time is still relatively long when considering the effect of pain felt by patients with fractured bones, and can affect the patient's psychological (anxiety) level [12]. Repair of bone mineralization alone is not sufficient to meet demand and can affect skeletons with inadequate bone tissue function. It is necessary to increase the supply of calcium. Simple calcium preparations are generally administered orally as an adjuvant to treatment, in combination with more specific drugs [13]. For this reason, a study was carried out on bone defects due to fractures with BHA–GEL pellet implants and the addition of BHA or calcium lactate orally as calcium supplements. Calcium intake plays an important role in maintaining bone health, namely to achieve peak bone mass and prevent loss of bone mass [14]. The purpose of supplementation in this study was to prove that the period of bone growth around the defect could be shortened. Osteoblast proliferation is expected to increase with the addition of oral BHA supplements to form more osteoids (soft callus), which are then converted into osteocytes (hard callus) and can increase bone stability around the fracture.

BHA is known to have carbonate substitutions like that of human hydroxyapatite and that can be found in synthetic biomimetic hydroxyapatites [15,16]. The carbonate group increases the proliferation of osteoblasts, thereby accelerating the formation of new bone [17]. Hydroxyapatite raises blood calcium levels less than calcium carbonate and calcium citrate. This indicates that hydroxyapatite is more effective at entering bone cells [18]. BHA bioavailability is better than calcium carbonate [14] and absorbed about 42.5% [19]. A European study showed that hydroxyapatite was more effective than calcium carbonate in slowing bone loss of the peripheral trabeculae of the distal tibia and distal radius [14]. The role of oral BHA is to increase the proliferation of osteoblasts by stimulating mesenchymal cells. It is shown by the calcium deposition from hydroxyapatite will interact with the collagen fibers along with type I collagen, a substance from the degradation of gelatin. During the mineralization process, the ends of the bone fragments are covered by a fusiform mass filled with woven bone. The more minerals deposited, the harder the callus formed [20]. While calcium lactate which is a salt form of lactic acid in the form of white powder, crystals or grains [21], which is also given orally because it can increase the number of osteoblasts [22]. Calcium lactate intake can increase extracellular calcium and intracellular calcium levels, stimulating bone formation significantly and increasing the proliferation or chemotaxis of osteoblasts [22]. Calcium lactate also significantly reduces bone resorption [23]. In addition, calcium lactate in the body is easily converted into calcium bicarbonate, and only about 25% is absorbed in the small intestine from calcium intake through passive diffusion and active transport [24]. A study shows a comparison of 500 mg calcium lactate, 500 mg carbonate and 500 mg gluconate (it is known that each calcium content is different), the results of which show the absorption, AUC and excretion of calcium lactate to be better than others [25]. Calcium lactate absorption is better than calcium phosphate and calcium in milk, and stimulates bone activity more than calcium carbonate or calcium citrate in experimental rats. As a consequence, it is very effective, especially in bone metabolism [22]. However, based on the research of previous results, it is known that BHA has more osteoinductive properties that other materials do not have [26]. For this reason, it is possible that oral administration of BHA on BHA–GEL pellet implants is more effective in terms of bone growth than oral administration of calcium lactate on BHA–GEL pellet implants.

2. Materials and Methods

2.1. Ethical Approval

The submission of an ethical feasibility proposal was addressed to the Research Ethics Commission of the Faculty of Veterinary Medicine, Universitas Airlangga (Animal Care and Use Committee/ACUC) and has been declared ethically eligible via Ethical Clearence No. 2.KEH.075.05.2022.

The research was conducted in the laboratory of the Faculty of Pharmacy, Universitas Airlangga, Surabaya in a true experimental manner with a posttest-only control group study design using 48 New Zealand rabbits aged 4–8 months, weighing 1.5–2.5 kg, healthy and without bone disorders in femur. The rabbits were randomly divided into defect (control groups), BHA–GEL pellet implant group, BHA–GEL pellet implant group with oral BHA and BHA–GEL pellet implant group with oral calcium lactate. Rabbits were adapted for one week with adequate food provided during the study. Making a rabbit fracture model was made by generating a defect in the femur, followed by implanting a BHA–GEL pellet implant according to the group division. Furthermore, oral supplements of BHA or calcium lactate were administered until termination was carried out on days 7, 14 and 28, according to group division. After termination, the femur bone that was treated as a sample was taken; this was followed by evaluation through X-ray radiology, and then bone decalcification, to evaluate the number and distribution of bone cells through HE (*Hematoxylin-Eosin*) staining, evaluation of anti-vascular endothelial growth factor (VEGF) and osteocalcin levels using the immunohistochemical (IHC) method.

2.2. Materials

In terms of test materials, bovine hydroxyapatite (Universitas Airlangga, Indonesia), gelatin, sodium carboxymethyl cellulose powder, aquadest, ketamine, xylazine, gentamicin ointment, ampicillin, 70% alcohol (pharmaceutical grade), cotton balls, povidone iodine, savlon, cotton bud, sterile gauze, hypafix, handsaplast, calcium lactate 500 mg tablet, thrombophob gel, water for injection, 10% formalin buffer, and a Calcium Colorimetric Assay Kit, ELISA Kit (Cat No. MAK022) (Sigma-aldrich, St. Louis, MO, USA) were acquired. In terms of equipment, a Carver manual pellet press, punch and die (4 mm diameter), 1 cc and 3 cc syringe, bone drill with 4.2 mm drill bit, surgical blade, forceps, needle holder, needle circle or surgical needles, Silk no.3 surgical thread, gillette razor, shaver, tweezers, scissors, leukoplast, vacutainer gel separator, water bath, mortar, stamper, granule sieve, oven, feeding tube, oral catheter, X-ray machine, ELISA reader, light microscope, histology slides, object glass and cover glass were used.

2.3. BHA–GEL Pellet Preparation

Weigh and put 9 g Bovine hydroxyapatite (BHA) powder into a mortar and then reduce the particle size using a stamper. Heat 6 mL of distilled water in a glass beaker at 40 °C, then add 2 g of gelatin and stir until the gelatin dissolves. Put 3 mL of dissolved gelatin into a mortar containing BHA and stir until the mass is formed. Next, sieve the mass to obtain a uniform particle size and dry in an oven at 40 °C for 24 h. Weigh the granules as large as 100 mg and compress with a load of 1 ton with a diameter of 4 mm to form BHA–GEL pellets, and continue with UV sterilization for ±3 h.

Implantation of BHA–GEL implant and oral administration of BHA or oral calcium lactate is performed by injecting a combination anesthetic ketamine 50 mg/kgBW and xylazine 5 mg/kgBW intramuscularly in experimental animals. Clean the rabbit's thighs with 70% alcohol and shave. Disinfect the shaved area using betadine then make an incision in the required area of about 1.5 cm. The defect was made using a 4.2 mm drill bit and followed by implantation of BHA–GEL pellets. Next, suture the wound, disinfect it with 70% alcohol and then betadine, and then apply gentamicin ointment as an antibiotic. Cover the wound with sterile gauze and dressing retention tape, then administer injection of Ampicillin intramuscularly at a dose of 25 mg/kgBW as an antibiotic. During the recovery period, the wound was treated with betadine and tape replacement until the surgical wound was dry. In addition, the rabbits were treated with 1 mL oral BHA or calcium lactate according to the group division.

2.4. Characterizations of Pellet

The characterization of the prepared pellets was carried out using Fourier Transform Infrared (FT-IR) Spectroscopy (Perkin Elmer, MA, USA). The BHA, Gelatin, and BHA–GELatin that have been made were then mixed with potassium bromide to make pellets and measured at a wave number of 400–4000 cm^{-1} with only one scanning, while the size and morphology of the particles were observed using scanning electron microscopy (Inspect S-50, FEI, Japan). The sputter coating of SEM used the ultra-thin coating of gold.

2.5. Blood Sampling Technique

Xylol was applied to the rabbit's ear on the marginal vein and 3 mL of blood were using a disposable syringe one hour from the time of taking the drug; this was carried out on the 7th, 14th, and 28th days before termination. Applying heparin gel to the area around the injection to prevent blood clots. Inject blood into a vacutainer containing a gel separator (serum separator tube) (OneMed, Krian, Indonesia) slowly to prevent hemolysis. Continue to centrifuge at 4000 rpm for 15 min to obtain a serum. Serum was stored in the freezer at −80 °C. BALP levels were using ELISA microplate reader IMark series No. 12096 (Bio-Rad, Hercules, CA, USA), while calcium levels using a microplate reader Biochrom EZ Read 2000 serial num-ber 135247 (Biochrom Ltd, Cambridge, United Kingdom).

2.6. Bone Sampling Technique

Termination of the experimental rabbits on the 7th, 14th and 28th days after treatment with oral drugs was performed in the instance of both BHA and calcium lactate. A bone sample of the femur in a 10% formalin was followed followed by observation of the process of closing the bone gap by X-ray radiology. Then decalcification of bone in 10% EDTA solution for observation of bone cell development through HE staining (*Hematoxylin-Eosin*), examination of VEGF and osteocalcin levels using the immunohistochemical (IHC) method.

2.7. Radiology Examination

An evaluation of bone integrity was performed using X-ray radiography and clarified with *ImageJ V1.44p*, before then compared being with the initial diameter of the bone defect (4.2 mm). This was followed by macroscopic observations to see the percentage of callus growth around the bone defect and calculations using Lane–Sandhu Scoring (Table 1).

Table 1. Lane–Sandhu Scoring Criteria [27].

Criteria	Score	Characteristics
No callus	0	no callus tissue, fracture line clear
Minimal callus	1	25% callus tissue, fracture line still clearly visible
Callus evident but healing incomplete	2	50% callus tissue, fracture line blurred
Callus evident with stability expected	3	75% callus tissue, fracture line barely visible
Complete healing with bone remodeling	4	100% callus tissue, no remaining fracture line visible

2.8. Haematoxylin and Eosin Staining

The histological examination started with processing the paraffin blocks by dehydrating with alcohol concentration 70% to 100% for 60 min each. This was followed by 3 clarifications of xylene, for 15 min each time. After that, in an incubator at 60 °C, the permeation treatment with paraffin solution was carried out three times for 60 min each. The tissue was then immersed in liquid paraffin and brought to room temperature. Each paraffin block was then cut to a thickness of 4–6 m using a microtome. Cell morphology was determined with *hematoxylin* and *eosin*. The slides were dipped in xylene three times for 5 min each and hydrated with alcohol (96% to 70% alcohol) for 2 min each. The slides were then rinsed under running water for 10 min, placed in *Mayer's hematoxylin* for 15 min, rinsed with running water and examined microscopically. The slides were then placed in 1% eosin solution for 30 s, dried, washed, and mounted with an EZ mount. This was performed on five visual fields, with 400× magnification around the defect or implant area. The results obtained are the number of each bone cell, including osteoblasts, osteoclasts and osteocytes.

2.9. Immunohistochemistry

The immunohistochemical technique was used to stain VEGF and osteocalcin immunopositive cells. Slides that have been paraffinized are soaked with an antigen retrieval decloaking chamber, cooled for 20 min, and washed with PBS for 3 min. Sniper blocking followed for 15 min. The slides were then incubated with rabbit anti-rat VEGF primer (cat. no. PA1-21796, Thermo Fisher Scientific, 1:100 dilution) (Waltham, Massachusetts, United States) and washed in PBS for 3 min. After that, the universal link was performed for 20 min, and the slide was washed in PBS for 3 min. The Trecavidin-HRP Label was then applied for 10 min and washed with PBS for 3 min. The slides were then reacted with Chromogen DAB + Buffer Substrate for 2–5 min, followed by rinsing for 5 min with running water. The slides were then stained with *hematoxylin* for 1–2 min followed by rinsing for 5 min with running water (twice). The slides were then dehydrated with alcohol (70% absolute alcohol) for 5 min each, followed by three xylol washes for 5 min each. Finally, they were mounted on a slide (Ecomount) and covered with a cover glass.

2.10. Bone Alkaline Phosphatase

Examination of BALP levels was carried out using the Rabbit Bone-Specific Alkaline Phosphatase ELISA Kit (Cat. No. BZ-08173140-EB) (Bioenzy, Jakarta, Indonesia) with the following steps: prepare all reagents, standard solutions and samples; bring all reagents to room temperature before use. Add 50 µL standard to standard well; add 40 µL sample to sample wells and add 10 µL anti-BAP antibody to sample wells; then add 50 µL streptavidin-HRP to sample wells and standard wells (not blank control well) and mix well. Cover the plate with a sealer. Incubate for 60 min at 37 °C. After that, remove the sealer and wash the plate 5 times with a wash buffer. Soak wells with at least 0.35 mL wash buffer for 30 s to 1 min for each wash. Blot the plate onto paper towels or other absorbent material. Then, add 50 µL substrate solution A to each well and then add 50 µL substrate solution B to each well. Incubate plate covered with a new sealer for 10 min at 37 °C in the dark. Add 50 µL Stop Solution to each well, and the blue color will change into yellow immediately. Determine the optical density (OD value) of each well immediately, using a microplate reader set to 450 nm within 10 min after adding the stop solution.

2.11. Calcium Concentration

Examination of calcium levels in the blood is carried out using the Calcium Colorimetric Assay Kit (Cat. No. MAK022) (Sigma-aldrich, St. Louis, MO, USA). Serum samples can be used directly in this assay. Add 90 mL of the Chromogenic Reagent (Sigma-aldrich, St. Louis, MO, USA) to each well containing standards, samples, or controls and mix gently. Then, add 60 mL of Calcium Assay Buffer (Sigma-aldrich, St. Louis, MO, USA) to each well and mix gently. After that, incubate the reaction for 5–10 min at room temperature.

Protect the plate from light during incubation. Measure the absorbance at 575 nm (A575) before assay.

2.12. Statistical Analysis

Evaluation of the size of the bone gap that was closed radiologically was analyzed used the Shapiro–Wilk test to determine the normality of the data and homogeneity test using the Levene test. If the data is normally distributed, then the one-way ANOVA test is continued. If the results show a difference in meaning, then it is continued with the LSD post hoc test. Meanwhile, if the data is not normally distributed; then, the Kruskal–Wallis test is carried out and then the Mann–Whitney test. This also applies to the evaluation of number of osteoblasts, osteocytes, and osteoclasts by HE staining, BALP and calcium levels in the blood. The evaluation percentage of callus growth around bone defects was assessed by means of Lane–Sandhu scoring. Evaluation of VEGF and osteocalcin expression using immunohistochemistry was performed semi-quantitatively via the Remmele method. The three tests were analyzed using the Kruskal–Wallis test, and assessment was continued with the Mann–Whitney test if there were significant differences between groups.

3. Results

3.1. Characterizations of Pellet

The results of the SEM identification show that the resulting pellet has a hexagonal particle shape with a particle size mean of 1.350 ± 0.243 µm (Mean $\pm$ SD) (Figure 1a). The results of the FT-IR from BHA describe the percentage of transmission at the specific wave numbers $PO_4 = 1048.57$ cm^{-1}, $OH = 3571.03$ cm^{-1}, and $CO_3 = 1460.65$ cm^{-1}, which are the harsh characteristics of bovine hydroxyapatite. Meanwhile, the result of the assessment in gelatin describe the emergence of the percentage of transmission in OH; NH_2; = 3500–3000 cm^{-1}; C=O and NH = 1655–1540 cm^{-1}; COO = 1450–1240 cm^{-1} (Figure 1b–d).

3.2. Radiology Examination

A study on the effect of oral BHA or calcium lactate on the repair of bone defects, implanted with BHA–GEL pellets, was carried out using a burr hole defect model in the cortical bone of the rabbit femur. Defects with implants appear circular, with higher intensity than their surroundings. The radiographic results showed that the implant group experienced accelerated bone growth in the area of the defect compared to the group without implants. The pellet intensity in the implant + oral BHA group was known to be fainter than that in the implant + oral calcium lactate group (Figure 2a,b), indicating the union of bone with the composite. The results of the evaluation of bone gap closure-obtained data that were not normally distributed ($p < 0.05$) nor homogeneous ($p < 0.05$) in the termination group on days 7, 14 and 28. Furthermore, using the Kruskal–Wallis, it was found that there was no difference which was significant ($p > 0.05$) between groups for each termination period (Figure 2c). The development of bone regeneration can also be known based on callus formation, which is characterized by changes in the intensity and transformation of the implants implanted. The longer treatment times showed the intensity of the pellets was fading, but the radiological results did not describe the percentage of callus growth so that it was continued with macroscopic observations. Lane–Sandhu scoring results (macroscopic observation) showed that the longer the treatment time, the higher the percentage of callus growth. The data generated from the scoring process includes non-parametric data so that the test carried out is Kruskal–Wallis and shows a sig. value ($p > 0.05$), or there is no significant difference between groups in each termination period (Figure 2d).

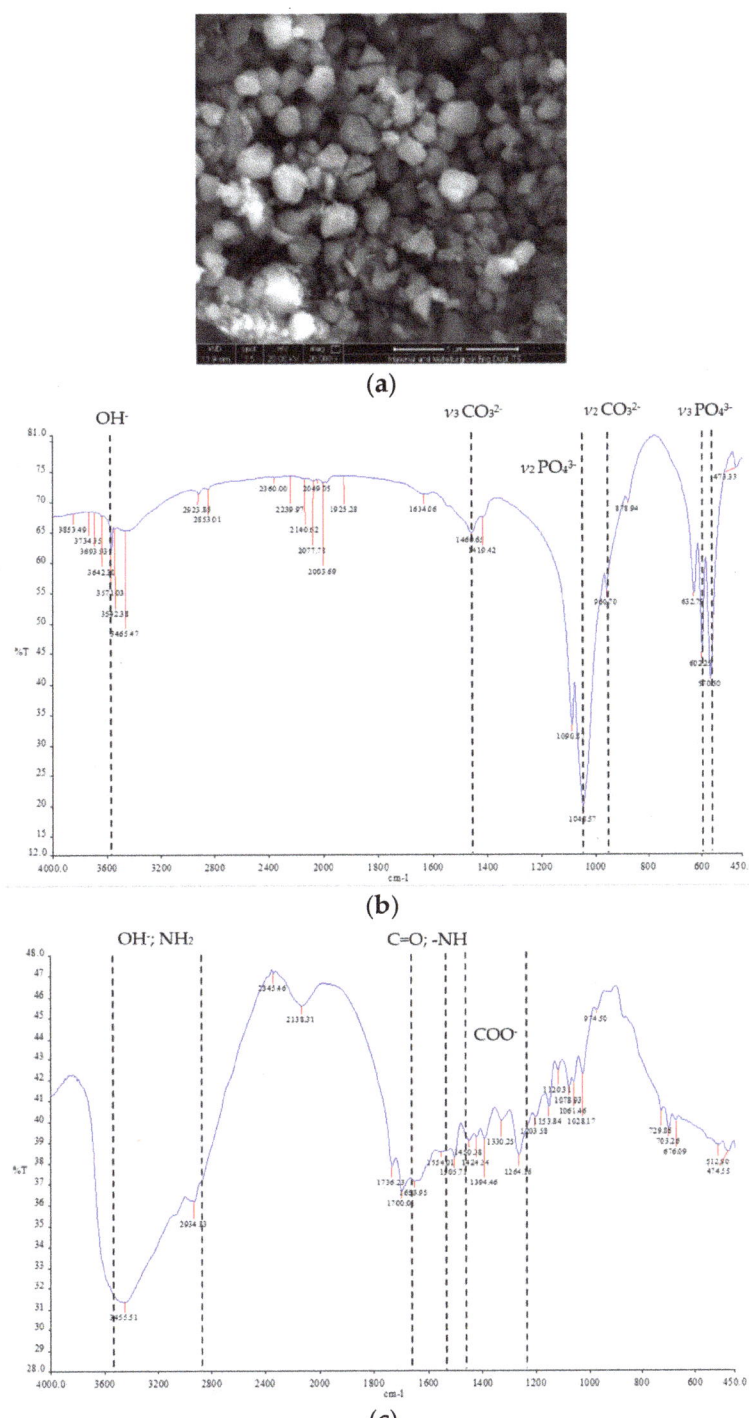

Figure 1. Cont.

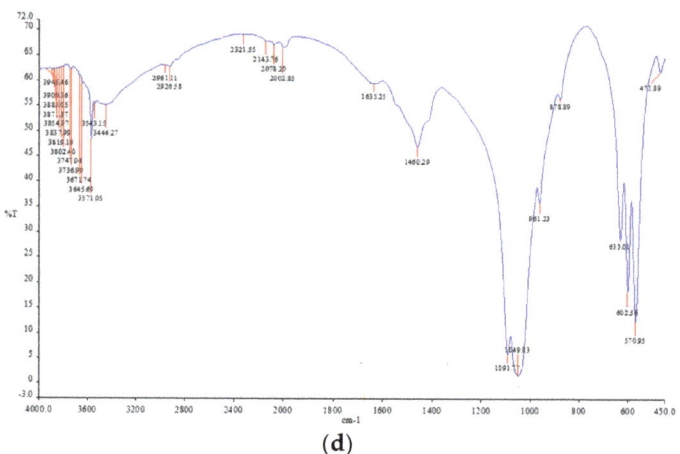

Figure 1. (**a**) Representative of SEM Scanning Result of BHA–GEL Pellet; (**b**) FT-IR Spectra Profile of BHA (**c**) FT-IR Sepctra Profile of Gelatin; (**d**) FT-IR Spectra Profile of BHA–GEL Pellet.

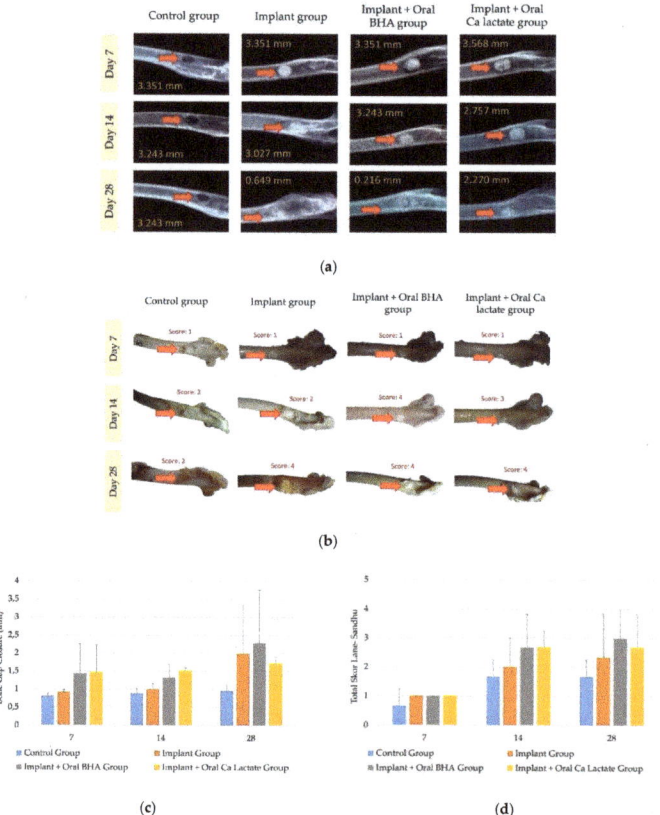

Figure 2. (**a**) X-ray radiology results of rabbit femur (1024 × 1024 pixel); (**b**) rabbit femur bone macroscopic; (**c**) result of measurement of bone cleft closure; (**d**) calculation results with Lane–Sandhu Scoring. The red arrow in subfigures (**a**,**b**) indicates the location of the defect. The implanted group showed accelerated bone growth in the defect area.

3.3. Examination of the number of Osteoblasts, Osteoclasts and Osteocytes through Hematoxylin-Eosin Staining

The data from the observation of osteoblast cells (Figure 3a) obtained showed that the data were normally distributed ($p > 0.05$) and homogeneous ($p > 0.05$) for each termination period. On days 7 and 28, there was no significant difference (one-way ANOVA; $p > 0.05$) between groups. Meanwhile, on the 14th day, it was known that the implant + oral BHA and implant + oral calcium lactate groups were significantly different (one-way ANOVA post hoc LSD; $p < 0.05$) compared to the defect group (control) (Figure 3b).

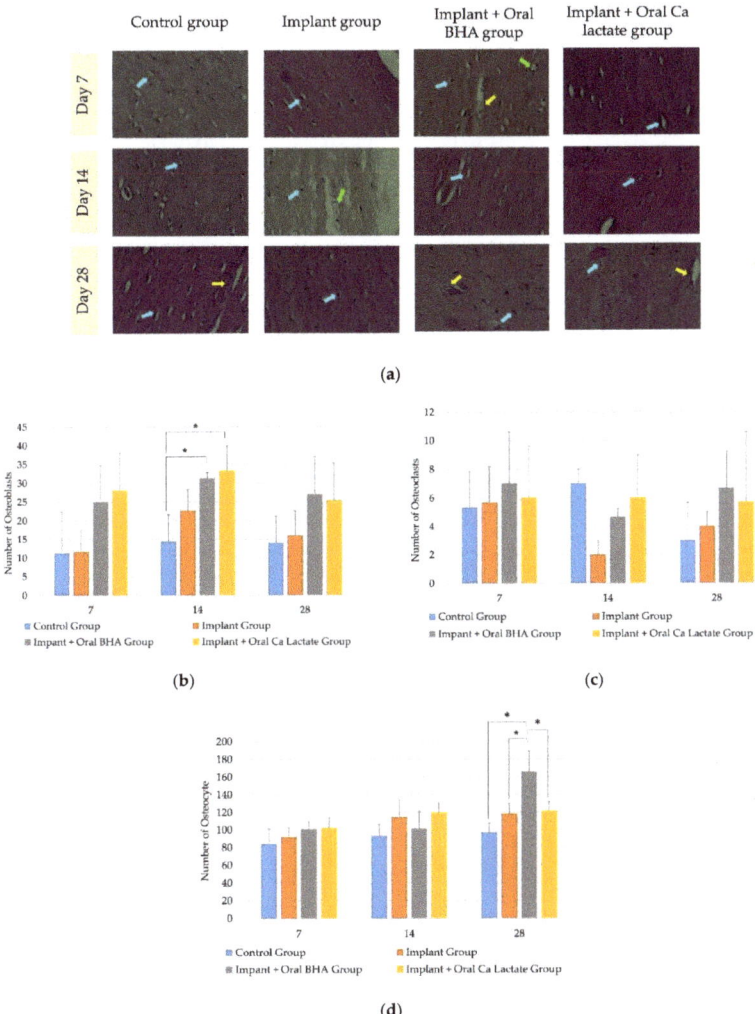

Figure 3. (a) Observation of bone cells with *Hematoxylin-Eosin* staining, magnification 400×. Osteoblast (yellow arrows), osteoclast (green arrows), and osteocyte (blue arrows); (b) observation of the number of osteoblast cells; (c) observation of the number of osteoclast cells; (d) observation of the number of osteocyte cells. The sign (*) indicates $p < 0.05$ with one-way ANOVA post hoc LSD test.

The data from the observation of osteoclast cells (Figure 3a) on days 7 and 28 were normally distributed ($p > 0.05$) and homogeneous ($p > 0.05$), and there was no significant difference (one-way ANOVA; $p > 0.05$) between groups. On day 14 the data were not normally distributed ($p < 0.05$) and there was no significant difference (Kruskal–Wallis; $p > 0.05$) between groups (Figure 3c).

The data from the observation of osteocyte cells (Figure 3a) obtained data that were normally distributed ($p > 0.05$) and homogeneous ($p > 0.05$) for each termination period. On days 7 and 14 there was no significant difference (one-way ANOVA; $p > 0.05$) between groups. Meanwhile, on day 28, it was found that the implant + oral BHA group was significantly different from the defect group (control), the implant group and the implant + oral calcium lactate group (one-way ANOVA post hoc LSD; $p < 0.05$). This indicates the effect of the addition of oral BHA on the number of osteocyte cells (Figure 3d).

3.4. Examination of the Amount and Distribution of VEGF through Immunohistochemistry

Immunohistochemical examination with anti-VEGF was performed to determine the vascularity of bone tissue in each treatment group. The results of the observation of VEGF expression using the Kruskal–Wallis test analysis in the negative control group, the implant group, the implant + oral BHA group, and the implant + oral calcium lactate group at 7, 14, and 28 days (Figure 4b) showed different values. VEGF expression was calculated ($p < 0.05$). When the results of VEGF expression in the implant and oral BHA group and the implant and oral calcium lactate group on days 7, 14, and 28, there was no difference in VEGF expression value ($p > 0.05$) with the Mann–Whitney test.

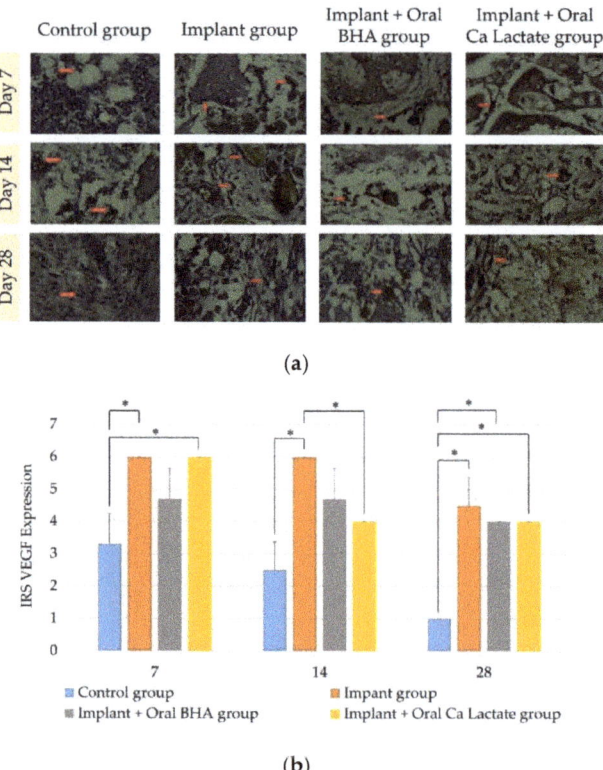

Figure 4. (a) Immunohistochemistry of VEGF expression (400× magnification). VEGF-positive cells are indicated by brown osteoblasts (red arrows); (b) differences in IRS values of VEGF expression, each bar graph represents IRS ± SD. The sign (*) indicates $p < 0.05$ with the Mann–Whitney Test.

3.5. Examination of the Amount and Distribution of Osteocalcin through Immunohistochemistry

Immuno-histochemical examination with anti-osteocalcin was performed to determine the number of osteoblasts which are markers of mineral deposition and growth of mature callus in each treatment group. The results of observations of osteocalcin expression in the negative control group, implant group, implant + oral BHA group, and implant + oral calcium lactate group at 7, 14, and 28 days (Figure 5b) showed differences in the value of osteocalcin expression ($p < 0.05$) with the Kruskal–Wallis test. On days 14 and 28, the implant + oral BHA group showed higher IRS scores than the implant + calcium lactate group (Mann–Whitney; $p < 0.05$).

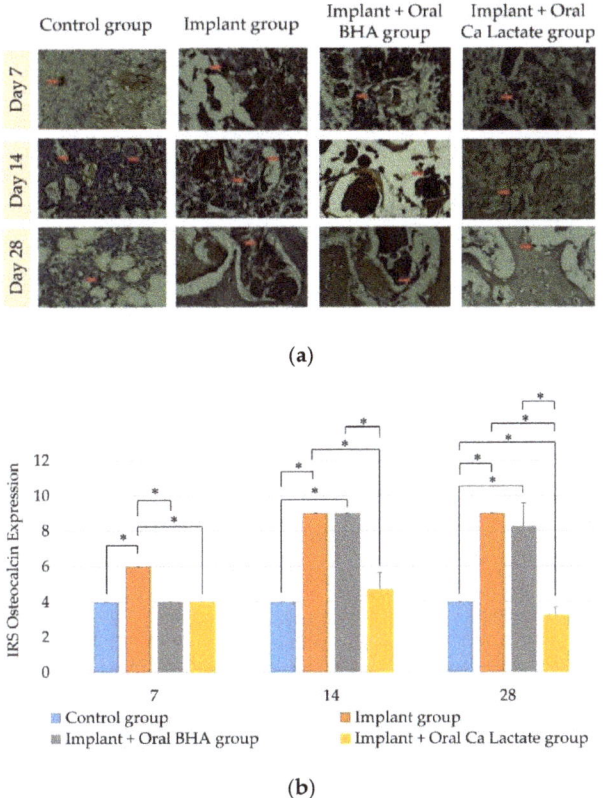

Figure 5. (a) Immunohistochemistry of osteocalcin expression (400× magnification). Osteocalcin-positive cells are indicated by brown osteoblasts (red arrows); (b) differences in IRS values of osteocalcin expression, each bar graph representing IRS ± SD. The sign (*) indicates $p < 0.05$ with the Mann–Whitney Test.

3.6. Examination of BALP Levels through ELISA

The results of measurements of BALP levels obtained data that were normally distributed ($p > 0.05$) and homogeneous ($p > 0.05$). There was a significant difference between the implant + oral BHA group and the implant + oral calcium lactate compared to the defect and implant groups on day 7. Meanwhile, on day 14, there was a significant difference between the defect group with an implant, an implant + oral BHA, and an implant + oral calcium lactate (One-way ANOVA post hoc LSD; $p < 0.05$) (Figure 6a).

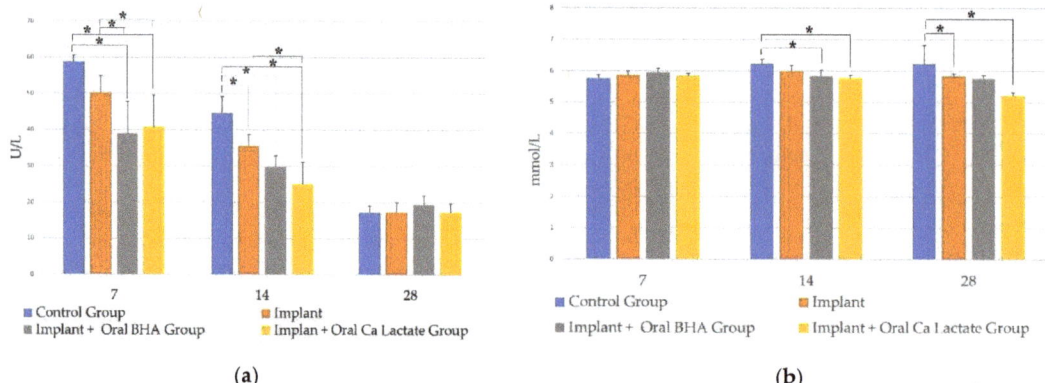

Figure 6. (a) BALP level examination results; (b) results of examination of calcium levels sign (*) indicates $p < 0.05$ in post hoc LSD test.

3.7. Examination of Calcium Levels in the Blood through the Calcium Colorimetric Assay Kit

The results of the measurement of calcium levels in the blood obtained data that were normally distributed ($p > 0.05$) and homogeneous ($p > 0.05$). There was a significant difference between the defect group with implant + oral BHA, implant + oral calcium lactate on day 14. Meanwhile, on day 28 there was a difference between the defect group with implant and implant + oral calcium lactate (one-way ANOVA post hoc LSD; $p < 0.05$) (Figure 6b).

4. Discussion

An in vivo study to test the effectiveness of oral BHA and calcium lactate in conjunction with BHA–GEL implants was performed on femoral bone defects in rabbits. Based on previous studies, BHA–GEL implants showed accelerated bone growth in terms of the percentage of new bone and bone-composite contact [10]. Castelo et al. [28] stated that the ossein hydroxyapatite complex (OHC) consisting of ossein, a protein that forms the organic matrix of bone and hydroxyapatite ($Ca_5[PO_4]_3OH$) has been shown to be effective in maintaining bone mineral density (BMD) and has a strong osteogenic effect which is stronger than calcium supplements. Other studies have shown that OHC stimulates bone metabolism by stimulating osteoblast differentiation, activity, and proliferation [29]. OHC has been shown to reduce the rate of bone resorption and stimulate ossification. Administration of OHC as monotherapy for 1–4 weeks in fracture patients has shown a reduction in the time required for healing by stimulating callus formation, thereby accelerating consolidation and clinical improvement [30].

From the characterizations of the pellet, the morphological shape of the particle was hexagonal, with a particle size mean of 1.350 ± 0.243 μm (mean ± SD). This is in line with previous studies of natural hydroxyapatite, one of which shows BHA possess a hexagonal particle shape [31,32]. Several factors that can influence the morphology of the particles are changes in the calcination temperature, as described by Khoo et al. [33]. Then, the FTIR spectra showed the specific wave number of functional group of the structure as well as the BHA–GEL pellet. The band at 1048 cm^{-1} is attributed to an asymmetric stretching vibration mode v_3 of PO_4^{3-}, while two sharp peaks at 570 nd 602 cm^{-1} are attributed to asymmetric bending vibration mode v_3, and symmetric bending vibration mode v_2 of PO_4^{3-} group. The bands observed at 962 and 1458 cm^{-1} at the FTIR spectra indicate the presence of B-type carbonate CO_3^{2-} bands [34].

Radiological results revealed the development of bone growth; after measuring bone gap closure, it was found that the implant + oral BHA group (the average diameter on day 28 is 1.910 mm) had a smaller final diameter of the bone gap than the implant + oral calcium lactate group (the average diameter on day 28 is 2.487 mm). This is supported by the observation of callus growth in bone defects, where changes in transformation and pellet intensity were more faded in the implant + oral BHA group. This indicated the growth of new tissue that was fused (union) with the bone due to the penetration of tissue cells around the bone into the implant.

The reparative process of bone fractures involves a series of events that include migration, proliferation, differentiation and activation of several cell types. These include mesenchymal and hematopoietic stem cells that ultimately lead to bone formation and remodeling [35]. These stem cells will differentiate into several types of bone cells, including osteoblasts and osteoclasts, while osteocytes are transformed osteoblasts that are embedded in the bone matrix. All three cells were expressed by *Hematoxylin-Eosin* staining. On day 14, it was found that the osteoblast cells in the implant + oral BHA group were significantly different from the defect group (control). In accordance with the previous theory, where BHA has a carbonate group which is known to increase the proliferation and differentiation of osteoblasts as well as mesenchymal stem cells which are osteoprogenitor cells that produce osteoblasts, thus bone matrix synthesis occurs more quickly [36]. Furthermore, there was also a significant difference between the implant + oral calcium lactate group and the defect group (control). In previous studies, it was stated that calcium lactate can increase extracellular calcium and intracellular calcium levels, thereby stimulating bone formation significantly, as well as increasing osteoblast proliferation or chemotaxis [22]. The same thing was observed in the observation of the number of osteocytes, where an increase in osteoblast proliferation would result in more mineralized osteoid so that the number of osteocytes also increased.

Osteoclasts on day 14 of the defect group (control) increased, indicating that the activity of mature osteoclasts was maximal in resorption of bone in the second week. However, in the group with the implant, the number of osteoclasts was lower than in the defect group (control). This may have been because the group with the implant treatment or the addition of calcium supplements had accelerated fracture healing. Substitution of carbonate contained in BHA causes bioresorption to occur earlier [37]. Increased osteoclastogenesis accelerates cartilage resorption and increased osteoblastogenesis during fracture healing promotes bone fusion, whereas inhibition of osteoclast or osteoblast differentiation has been reported to delay bone healing [38]. The processes of resorption by osteoclasts and bone formation by osteoblasts proceed sequentially but overlap substantially, so that there is a change in the cell population that indicates regeneration in the tissue [2]. This can be seen through the results of research where when there is an increase in the number of osteoblasts, the number of osteoclasts decreases, or vice versa.

Parameters measured by the IHC method were vascular endothelial growth factor (VEGF) and osteocalcin. VEGF is a protein that plays an important role in endothelial proliferation, migration, and activation [39]. VEGF can be detected in the first week after injury, namely in the inflammatory phase [40]. The results of the examination of VEGF expression in the implant + oral BHA group compared to the implant + oral calcium lactate group on days 7, 14, and 28 showed no significant difference. This is probably because VEGF expression in the implant + oral BHA group was seen on day 3, and then peaked on day 5 [41]. In the implant + oral BHA group, there may have been a decrease in the number of macrophages, so that VEGF expression decreased. This decrease was influenced by a decrease in the number of blood vessels damaged in the wound; the extracellular matrix begins to fill the missing area and stable blood vessels begin to form [42]. Based on this study, the implant + oral BHA group was found to be effective in increasing VEGF expression on day 14 compared to the implant + oral calcium lactate group based on the IRS value of VEGF expression.

Osteocalcin is a non-collagenous protein that has an important role in the mineralization process and calcium ion homeostasis [43,44]. The results of the examination of osteocalcin expression in this study were all groups experienced an increase in osteocalcin expression since day 14. This is in line with the research of Jafary et al. [45], which states that an increase in osteocalcin can be detected at week 2, which is day 14. Osteocalcin can be detected in mature callus [4]. Mature callus begins to form from the repair phase to the remodeling phase [20]. Osteoblast maturation was demonstrated by increasing the value of osteocalcin expression [43]. Osteocalcin plays an important role in regulating mineral nucleation by binding to hydroxyapatite. Based on the study, the value of osteocalcin expression on days 14 and 28 showed a higher value of osteocalcin expression in the implant + oral BHA group than the implant + oral calcium lactate group. This is because BHA increases the proliferation of hydroxyapatite by stimulating mesenchymal cells, resulting in faster hard callus formation [18].

These results were also similar to the results of measuring BALP levels, where the group receiving oral BHA had higher BALP levels than the oral calcium lactate group. Bone alkaline phosphatase (BALP) is a metalloenzyme that is produced when osteoblast cells work [46]. High expression of BALP indicates better activity where BALP is useful in the synthesis of collagen fibers and in bone mineralization. On the 7th day, BALP levels were higher than the 14th and 28th days which could be due to the process of differentiation of osteoblasts. This is also in accordance with the theory which states that bone alkaline phosphatase (BALP) is a metalloenzyme that is produced when osteoblast cells work [46,47]. Osteoblasts will express collagen, which plays a role in the process of bone mineralization to a form soft callus. This is also in accordance with the statement of Einhorn and Gerstenfeld [2], and on the 7th day it enters the proliferative or endochondral phase where the soft callus begins to form. Soft callus growth also occurs because the gelatin contained in the implant is degraded to form type 1 collagen which is the main constituent of bone [8]. On the 14th day BALP levels decreased compared to the 7th day; this is because at this stage the process of fibrous tissue formation has occurred and soft callus maturation has begun, which has occurred on the 7th day [2]. On day 14, the mean number of osteoblasts also decreased as, according to the matrix maturation phase, osteoblasts lost their function and differentiated into osteoid in the bone [48]. On day 28, BALP also decreased, which could be due to the near-complete fracture healing process [49]. Birmingham et al. [50] also stated that the differentiation of osteoblast cells occurred at the beginning of day 5 to 14, after which there was a decrease in BALP expression. In addition, the results of the examination of higher blood calcium levels in the implant + oral BHA group compared to the implant + oral calcium lactate group will also increase the proliferation and/or chemotaxis of osteoblasts, thereby stimulating bone formation significantly.

Calcium lactate is known to be highly soluble and has high bioavailability; however, hydroxyapatite, which is generally considered insoluble, shows absorption values one-fourth to one-third as good as the most soluble preparations [51]. Hydroxyapatite raises blood calcium levels less than calcium carbonate and calcium citrate. This indicates that hydroxyapatite is more effective in entering bone cells [18]. In addition, in vitro examination of bone cell cultures showed that the organic part of OHC contains factors that influence osteoblast proliferation [13]. This causes the bone growth of the implant + oral BHA group to be better than the implant + oral calcium lactate group. BHA also has a carbonate group that is naturally present in human bone and that can be found in synthetic biomimetic hydroxyapatites. Carbonated hydroxyapatite is known to increase protein adsorption and increase adhesion, proliferation, and osteogenic differentiation of mesenchymal stem cell tissue. In addition, carbonated hydroxyapatite is also known to increase the proliferation and differentiation of osteoblasts, thereby increasing bone matrix synthesis [36]. The same thing happened in the study by Castelo et al. [29], based on histological observations of bone using a fluorescent microscope showed that oral administration of ossein hydroxyapatite could increase bone formation compared to the group given hydroxyapatite alone or with calcium carbonate. In addition, this is influenced by the absorption to the bioavailability of the two supplements given. It is known that the absorption of microcrystalline hydroxyapatite compound (MCHC) is about 42.5% [19], and it is higher than the absorption of calcium lactate which is around 25% [24].The results showed that the bioavailability of calcium lactate in rats was 8.9 + 1.4% [52]. Meanwhile, hydroxyapatite is reported to have better bioavailability than calcium carbonate [14], and as good as or better than calcium gluconate [53].

5. Conclusions

BHA supplements, given orally together with BHA–GEL pellet implants, showed faster healing of bone defects compared to oral calcium lactate with BHA–GEL pellet implants.

Author Contributions: Conceptualization, A.S.B., J.K. and S.S.; Data curation, C.A. and I.S.; Formal analysis, S.S., M.A.G. and I.S.; Investigation, B.R.K.H.P., H.A., R.N.S. and I.L.; Methodology, A.S.B. and J.K.; Project administration, A.S.B. and M.A.G.; Resources, S.S., M.A.G. and E.R.; Supervision, A.S.B., J.K. and C.A.; Validation, S.S., C.A., M.A.G. and E.R.; Visualization, B.R.K.H.P., H.A., R.N.S., I.L. and Y.A.P.; Writing—original draft, B.R.K.H.P., H.A., R.N.S., I.L. and Y.A.P.; Writing—review & editing, A.S.B., J.K. and C.A. All authors have read and agreed to the published version of the manuscript.

Funding: This research was funded by Lembaga Pengelola Dana Pendidikan Ministry of Finance of the Republic of Indonesia via Pendanaan Riset Inovatif Produktif (RISPRO) Batch 1 Tahun 2019 No. KEP-51/LPDP/2019.

Institutional Review Board Statement: The animal study protocol was approved by the Institutional Review Board (or Ethics Committee) of Faculty of Veterinary Medicine, Universitas Airlangga (Animal Care and Committee/ACUC) (protocol code No. 2.KEH.075.05.2022 by 5 May 2022).

Data Availability Statement: Not applicable.

Acknowledgments: The authors thank to the Lembaga Pengelola Dana Pendidikan Ministry of Finance of the Republic of Indonesia, BPBRIN Universitas Airlangga, PT. INOBI Indonesia, and Faculty of Pharmacy Universitas Airlangga.

Conflicts of Interest: The authors declare no conflict of interest.

References

1. Barrett, K.E.; Barman, S.M.; Boitano, S.; Brooks, H. *Ganong's Review of Medical Physiology*; McGraw-Hill Education: New York City, NY, USA, 2012; 768p.
2. Einhorn, T.A.; Gerstenfeld, L.C. Fracture Healing: Mechanisms and Interventions. *Nat. Rev. Rheumatol.* **2015**, *11*, 45–54. [CrossRef] [PubMed]
3. Noorisa, R.; Apriliwati, D.; Aziz, A.; Bayusentono, S. The Characteristic of Patients With Femoral Fracture In Department Of Orthopaedic And Traumatology RSUD Dr. Soetomo Surabaya 2013–2016. *J. Orthop. Traumatol. Surabaya* **2017**, *6*, 1–11. [CrossRef]
4. Ferdiansyah Mahyudin, N. *Graft Tulang & Material Pengganti Tulang (Karaktersitik Dan Strategi Aplikasi Klinis)*; Airlangga University Press: Surabaya, Indonesia, 2018.
5. Budiatin, A.S.; Zainuddin, M.; Khotib, J.; Ferdiansyah. Pelepasan Gentamisin Dari Pelet Bovine-Hydroxyapatite-Gelatin Sebagai Ssitem Penghantaran Obat Dan Pengisi Tulang. *J. Farm. Dan Ilmu Kefarmasian Indones.* **2014**, *1*, 10–15.
6. Ardhiyanto, H. Peran Hidroksiapatit Sebagai Material Bone Graft Dalam Menstimulasi Kepadatan Kolagen Tipe L Pada Proses Penyembuhan Tulang. *Stomatognatic-J. Kedokt. Gigi* **2015**, *9*, 16–18.
7. Marfil, P.H.M.; Anhê, A.C.B.M.; Telis, V. Texture and Microstructure of Gelatin/Corn Starch-Based Gummy Confections. *Food Biophys.* **2012**, *7*, 236–243. [CrossRef]
8. Budiatin, A.S.; ZAINUDDIN, M.; KHOTIB, J. Biocompatible Composite as Gentamicin Delivery System for Osteomyelitis and Bone Regeneration. *Int. J. Pharm. Pharm. Sci.* **2014**, *6*, 4.
9. Chao, W.W.; Lin, B.F. Isolation and Identification of Bioactive Compounds in Andrographis Paniculata (Chuanxinlian). *Chin. Med.* **2010**, *5*, 17. [CrossRef]
10. Gani, M.A. *Studi Komparasi Osteokonduktivitas Natural Dan Synthetic Hydroxyapatite Serta Keterlibatan Polaritas Makrofag Pada Defek Tulang*; Universitas Airlangga: Surabaya, Indonesia, 2021.
11. Khotib, J.; Lasandara, C.S.; Samirah, S.; Budiatin, A.S. Acceleration of Bone Fracture Healing through the Use of Natural Bovine Hydroxyapatite Implant on Bone Defect Animal Model. *Folia Med. Indones.* **2019**, *55*, 176. [CrossRef]
12. Mandagi, C.A.F.; Bidjuni, H.; Hamel, R.S. Karakteristik Yang Berhubungan Dengan Tingkat Nyeri Pada Pasien Fraktur Di Ruang Bedah Rumah Sakit Umum Gmim Bethesda Tomohon. *J. Keperawatan.* **2017**, *5*, 1–7.
13. Schmidt, K.H.; Wörner, U.M.; Buck, H.J. Examination of New Bone Growth on Aluminium Oxide Implant Contact Surfaces after Oral Administration of Ossein-Hydroxyapatite Compound to Rats. *Curr. Med. Res. Opin.* **1988**, *11*, 107–115. [CrossRef]
14. Straub, D.A. Calcium Supplementation in Clinical Practice: A Review of Forms, Doses, and Indications. *Nutr. Clin. Pract.* **2007**, *22*, 286–296. [CrossRef]
15. Wang, Y.; Von Euw, S.; Fernandes, F.M.; Cassaignon, S.; Selmane, M.; Laurent, G.; Pehau-Arnaudet, G.; Coelho, C.; Bonhomme-Coury, L.; Giraud-Guille, M.M.; et al. Water-Mediated Structuring of Bone Apatite. *Nat. Mater.* **2013**, *12*, 1144–1153. [CrossRef]

16. Olivier, F.; Rochet, N.; Delpeux-Ouldriane, S.; Chancolon, J.; Sarou-Kanian, V.; Fayon, F.; Bonnamy, S. Strontium Incorporation into Biomimetic Carbonated Calcium-Deficient Hydroxyapatite Coated Carbon Cloth: Biocompatibility with Human Primary Osteoblasts. *Mater. Sci. Eng. C* **2020**, *116*, 111192. [CrossRef] [PubMed]
17. Germaini, M.M.; Detsch, R.; Grünewald, A.; Magnaudeix, A.; Lalloue, F.; Boccaccini, A.R.; Champion, E. Osteoblast and Osteoclast Responses to Porous A/B Carbonate-Substituted Hydroxyapatite Ceramics for Bone Regeneration. *Biomed. Mater.* **2017**, *12*, 035008. [CrossRef] [PubMed]
18. Bristow, S.M.; Gamble, G.D.; Stewart, A.; Horne, L.; House, M.E.; Aati, O.; Mihov, B.; Horne, A.M.; Reid, I.R. Acute and 3-Month Effects of Microcrystalline Hydroxyapatite, Calcium Citrate and Calcium Carbonate on Serum Calcium and Markers of Bone Turnover: A Randomised Controlled Trial in Postmenopausal Women. *Br. J. Nutr.* **2014**, *112*, 1611–1620. [CrossRef] [PubMed]
19. Stellon, A.; Davies, A.; Webb, A.; Williams, R. Microcrystalline Hydroxyapatite Compound in Prevention of Bone Loss in Corticosteroid-Treated Patients with Chronic Active Hepatitis. *Postgrad. Med. J.* **1985**, *61*, 791–796. [CrossRef]
20. Adrianto, H.B. Peran Hidroksiapatit Sebagai Bone Graft Dalam Proses Penyembuhan Tulang. *Stomatognatic-J. Kedokt. Gigi* **2011**, *8*, 118–121.
21. Sweetman, S.C. *Martindale*; Pharmaceutical Press: London, UK, 2009.
22. Muliani, I.; Nyoman, M.K.; Ketut, T. Pemberian Kalsium Laktat Dan Berenang Meningkatkan Osteoblast Pada Epiphysis Tulang Radius Mencit Perimenopause. *J. Vet.* **2014**, *15*, 39–45.
23. Wagner, G.; Kindrick, S.; Hertzler, S.; DiSilvestro, R.A. Effects of Various Forms of Calcium on Body Weight and Bone Turnover Markers in Women Participating in a Weight Loss Program. *J. Am. Coll. Nutr.* **2007**, *26*, 456–461. [CrossRef]
24. Szterner, P.; Biernat, M. The Synthesis of Hydroxyapatite by Hydrothermal Process with Calcium Lactate Pentahydrate: The Effect of Reagent Concentrations, PH, Temperature, and Pressure. *Bioinorg. Chem. Appl.* **2022**, *2022*, 3481677. [CrossRef]
25. Gaby, K. Bioavailability and Solubility of Different Calcium-Salts as a Basis for Calcium Enrichment of Beverages. *Food Nutr. Sci.* **2010**, *2010*, 3035.
26. Khotib, J.; Gani, M.A.; Budiatin, A.S.; Lestari, M.L.A.D.; Rahadiansyah, E.; Ardianto, C. Signaling Pathway and Transcriptional Regulation in Osteoblasts during Bone Healing: Direct Involvement of Hydroxyapatite as a Biomaterial. *Pharmaceuticals* **2021**, *14*, 615. [CrossRef] [PubMed]
27. Wölfl, C.; Schweppenhäuser, D.; Gühring, T.; Takur, C.; Höner, B.; Kneser, U.; Grützner, P.A.; Kolios, L. Characteristics of Bone Turnover in the Long Bone Metaphysis Fractured Patients with Normal or Low Bone Mineral Density (BMD). *PLoS ONE* **2014**, *9*, e96058. [CrossRef]
28. Castelo-Branco, C.; Cancelo Hidalgo, M.J.; Palacios, S.; Ciria-Recasens, M.; Fernández-Pareja, A.; Carbonell-Abella, C.; Manasanch, J.; Haya-Palazuelos, J. Efficacy and Safety of Ossein-Hydroxyapatite Complex versus Calcium Carbonate to Prevent Bone Loss. *Climacteric* **2020**, *23*, 252–258. [CrossRef] [PubMed]
29. Castelo-Branco, C.; Ciria-Recasens, M.; Cancelo-Hidalgo, M.J.; Palacios, S.; Haya-Palazuelos, J.; Carbonell-Abelló, J.; Blanch-Rubió, J.; Martinez-Zapata, M.J.; Manasanch, J.; Pérez-Edo, L. Efficacy of Ossein-Hydroxyapatite Complex Compared with Calcium Carbonate to Prevent Bone Loss: A Meta-Analysis. *Menopause* **2009**, *16*, 984–991. [CrossRef] [PubMed]
30. Castelo-Branco, C.; Dávila Guardia, J. Use of Ossein—Hydroxyapatite Complex in the Prevention of Bone Loss: A Review. *Climacteric* **2015**, *18*, 29–37. [CrossRef]
31. Costescu, A.; Pasuk, I.; Ungureanu, F.; Dinischiotu, A.; Costache, M.; Huneau, F.; Galaup, S.; le Coustumer, P.; Predoi, D. Physico-Chemical Properties Of Nano-Sized Hexagonal Hydroxyapatite Powder Synthesized By Sol-Gel. *Dig. J. Nanomater. Biostructures.* **2010**, *5*, 989–1000.
32. Budiatin, A.S.; Samirah; Gani, M.A.; Nilamsari, W.P.; Ardianto, C. The Characterization of Bovine Bone-Derived Hydroxyapatite Isolated Using Novel Non-Hazardous Method. *J. Biomim. Biomater. Biomed. Eng.* **2020**, *45*, 49–56. [CrossRef]
33. Khoo, W.; Nor, F.M.; Ardhyananta, H.; Kurniawan, D. Preparation of Natural Hydroxyapatite from Bovine Femur Bones Using Calcination at Various Temperatures. *Procedia Manuf.* **2015**, *2*, 196–201. [CrossRef]
34. Brahimi, S.; Ressler, A.; Boumchedda, K.; Hamidouche, M.; Kenzour, A.; Djafar, R.; Antunović, M.; Bauer, L.; Hvizdoš, P.; Ivanković, H. Preparation and Characterization of Biocomposites Based on Chitosan and Biomimetic Hydroxyapatite Derived from Natural Phosphate Rocks. *Mater. Chem. Phys.* **2022**, *276*, 125421. [CrossRef]
35. Abdelmagid, S.; Barbe, M.; Hadjiargyrou, M.; Owen, T.; Razmpour, R.; Rehman, S.; Popoff, S.; Safadi, F. Temporal and Spatial Expression of Osteoactivin During Fracture Repair. *J. Cell. Biochem.* **2010**, *111*, 295–309. [CrossRef] [PubMed]
36. Budiatin, A.S.; Gani, M.A.; Samirah; Ardianto, C.; Raharjanti, A.M.; Septiani, I.; Putri, N.P.K.P.; Khotib, J. Bovine Hydroxyapatite-Based Bone Scaffold with Gentamicin Accelerates Vascularization and Remodeling of Bone Defect. *Int. J. Biomater.* **2021**, *2021*, 5560891. [CrossRef]
37. Shi, P.; Liu, M.; Fan, F.; Yu, C.; Lu, W.; Du, M. Characterization of Natural Hydroxyapatite Originated from Fish Bone and Its Biocompatibility with Osteoblasts. *Mater. Sci. Eng. C* **2018**, *90*, 706–712. [CrossRef]
38. He, L.H.; Liu, M.; He, Y.; Xiao, E.; Zhao, L.; Zhang, T.; Yang, H.Q.; Zhang, Y. TRPV1 Deletion Impaired Fracture Healing and Inhibited Osteoclast and Osteoblast Differentiation. *Sci. Rep.* **2017**, *7*, 42385. [CrossRef] [PubMed]
39. Hu, K.; Olsen, B.R. The Roles of Vascular Endothelial Growth Factor in Bone Repair and Regeneration. *Bone* **2016**, *91*, 30–38. [CrossRef] [PubMed]
40. Claes, L.; Recknagel, S.; Ignatius, A. Fracture Healing under Healthy and Inflammatory Conditions. *Nat. Rev. Rheumatol.* **2012**, *8*, 133–143. [CrossRef]

41. Li, B.; Wang, H.; Qiu, G.; Su, X.; Wu, Z. Synergistic Effects of Vascular Endothelial Growth Factor on Bone Morphogenetic Proteins Induced Bone Formation in Vivo: Influencing Factors and Future Research Directions. *BioMed Res. Int.* **2016**, *2016*, 2869572. [CrossRef]
42. Nofikasari, I.; Rufaida, A.; Aqmarina, C.D.; Failasofia, F.; Fauzia, A.R.; Handajani, J. Efek Aplikasi Topikal Gel Ekstrak Pandan Wangi Terhadap Penyembuhan Luka Gingiva. *Maj. Kedokt. Gigi Indones.* **2016**, *2*, 53–59. [CrossRef]
43. Li, J.; Zhang, H.; Yang, C.; Li, Y.; Dai, Z. An Overview of Osteocalcin Progress. *J. Bone Miner. Metab.* **2016**, *34*, 367–379. [CrossRef]
44. Hesaraki, S.; Nazarian, H.; Pourbaghi-Masouleh, M.; Borhan, S. Comparative Study of Mesenchymal Stem Cells Osteogenic Differentiation on Low-Temperature Biomineralized Nanocrystalline Carbonated Hydroxyapatite and Sintered Hydroxyapatite. *J. Biomed. Mater. Res.-Part B Appl. Biomater.* **2014**, *102*, 108–118. [CrossRef]
45. Jafary, F.; Hanachi, P.; Gorjipour, K. Osteoblast Differentiation on Collagen Scaffold with Immobilized Alkaline Phosphatase. *Int. J. Organ. Transplant. Med.* **2017**, *8*, 195. [PubMed]
46. Golub, E.E.; Boesze-Battaglia, K. The Role of Alkaline Phosphatase in Mineralization. *Curr. Opin. Orthop.* **2007**, *18*, 444–448. [CrossRef]
47. Greenblatt, M.B.; Tsai, J.N.; Wein, M.N. Bone Turnover Markers in the Diagnosis and Monitoring of Metabolic Bone Disease. *Clin. Chem.* **2017**, *63*, 464–474. [CrossRef]
48. Lorenzo, J.; Horowitz, M.; Choi, Y.; Takayanagi, H.; Schett, G. *Osteoimmunology: Interactions of the Immune and Skeletal Systems*; Academic Press: Cambridge, MA, USA, 2015.
49. Yudaniayanti, I.S. Aktifitas Alkaline Phosphatase Pada Proses Kesembuhan Patah Tulang Femur Dengan Terapi $CaCO_3$ Dosis Tinggi Pada Tikus Jantan (Sprague Dawley). *Media Kedokt. Hewan* **2005**, *21*, 15–18.
50. Birmingham, E.; Niebur, G.L.; McHugh, P.E. Osteogenic Differentiation of Mesenchymal Stem Cells Is Regulated by Osteocyte and Osteoblast Cells in a Simplified Bone Niche. *Eur. Cells Mater.* **2012**, *23*, 13–27. [CrossRef] [PubMed]
51. Heaney, R.P.; Recker, R.R.; Weaver, C.M. Absorbability of Calcium Sources: The Limited Role of Solubility. *Calcif. Tissue Int.* **1990**, *46*, 300–304. [CrossRef]
52. Ueda, Y.; Taira, Z. Effect of Anions or Foods on Absolute Bioavailability of Calcium from Calcium Salts in Mice by Pharmacokinetics. *J. Exp. Pharmacol.* **2013**, *5*, 65.
53. Buclin, T.; Jacquet, A.F.; Burckhardt, P. Intestinal Absorption of Calcium Gluconate and Oseine-Mineral Complex: An Evaluation by Conventional Analyses. *Schweiz. Med. Wochenschr.* **1986**, *116*, 1780–1783.

Article

In Vitro and In Vivo Cell-Interactions with Electrospun Poly (Lactic-Co-Glycolic Acid) (PLGA): Morphological and Immune Response Analysis

Ana Chor [1,*], Christina Maeda Takiya [2], Marcos Lopes Dias [3], Raquel Pires Gonçalves [3], Tatiana Petithory [4], Jefferson Cypriano [5], Leonardo Rodrigues de Andrade [1,†], Marcos Farina [1] and Karine Anselme [4]

[1] Biomineralization Laboratory, Institute of Biomedical Sciences, Federal University of Rio de Janeiro, Rio de Janeiro 21941-902, Brazil
[2] Immunopathology Laboratory, Institute of Biophysics Carlos Chagas Filho, Federal University of Rio de Janeiro, Rio de Janeiro 20941-902, Brazil
[3] Catalysis Laboratory for Polymerization, Recycling and Biodegradable Polymers, Institute of Macromolecules Professor Eloisa Mano, Federal University of Rio de Janeiro, Rio de Janeiro 21941-598, Brazil
[4] Biomaterials—Biointerfaces Laboratory, Institut de Science des Materiaux de Mulhouse, University of Haute-Alsace, CNRS, IS2M UMR 7361, F-68100 Mulhouse, France
[5] Padrón-Lins Multi-User Microscopy Unit, Institute of Microbiology Paulo Góes, Federal University of Rio de Janeiro, Rio de Janeiro 21941-902, Brazil
* Correspondence: anamedoral@gmail.com
† Present address: Salk Institute for Biological Studies, La Jolla, CA 92037, USA.

Abstract: Random electrospun three-dimensional fiber membranes mimic the extracellular matrix and the interfibrillar spaces promotes the flow of nutrients for cells. Electrospun PLGA membranes were analyzed in vitro and in vivo after being sterilized with gamma radiation and bioactivated with fibronectin or collagen. Madin-Darby Canine Kidney (MDCK) epithelial cells and primary fibroblast-like cells from hamster's cheek paunch proliferated over time on these membranes, evidencing their good biocompatibility. Cell-free irradiated PLGA membranes implanted on the back of hamsters resulted in a chronic granulomatous inflammatory response, observed after 7, 15, 30 and 90 days. Morphological analysis of implanted PLGA using light microscopy revealed epithelioid cells, Langhans type of multinucleate giant cells (LCs) and multinucleated giant cells (MNGCs) with internalized biomaterial. Lymphocytes increased along time due to undegraded polymer fragments, inducing the accumulation of cells of the phagocytic lineage, and decreased after 90 days post implantation. Myeloperoxidase$^+$ cells increased after 15 days and decreased after 90 days. LCs, MNGCs and capillaries decreased after 90 days. Analysis of implanted PLGA after 7, 15, 30 and 90 days using transmission electron microscope (TEM) showed cells exhibiting internalized PLGA fragments and filopodia surrounding PLGA fragments. Over time, TEM analysis showed less PLGA fragments surrounded by cells without fibrous tissue formation. Accordingly, MNGC constituted a granulomatous reaction around the polymer, which resolves with time, probably preventing a fibrous capsule formation. Finally, this study confirms the biocompatibility of electrospun PLGA membranes and their potential to accelerate the healing process of oral ulcerations in hamsters' model in association with autologous cells.

Keywords: PLGA; electrospinning; morphology; immune response; microscopy

Citation: Chor, A.; Takiya, C.M.; Dias, M.L.; Gonçalves, R.P.; Petithory, T.; Cypriano, J.; de Andrade, L.R.; Farina, M.; Anselme, K. In Vitro and In Vivo Cell-Interactions with Electrospun Poly (Lactic-Co-Glycolic Acid) (PLGA): Morphological and Immune Response Analysis. *Polymers* **2022**, *14*, 4460. https://doi.org/10.3390/polym14204460

Academic Editors: Antonia Ressler and Inga Urlic

Received: 29 June 2022
Accepted: 30 September 2022
Published: 21 October 2022

Publisher's Note: MDPI stays neutral with regard to jurisdictional claims in published maps and institutional affiliations.

Copyright: © 2022 by the authors. Licensee MDPI, Basel, Switzerland. This article is an open access article distributed under the terms and conditions of the Creative Commons Attribution (CC BY) license (https:// creativecommons.org/licenses/by/ 4.0/).

1. Introduction

In the history of tissue engineering, degradable materials such as poly (lactic acid—PLA) and poly (glycolic acid—PGA) or their combination poly (lactide-co-glycolide—PLGA), being of smooth or rough surfaces, were capable of inducing tissue or organ regeneration [1]. Recently, degradable and non-degradable FDA-approved polymers [2] with bioactivated surfaces were shown boosting tissue engineering and regenerative medicine applications [3,4]. To this end, medical device design, medical device implantation and immune responses play crucial roles, in combination, to modulate immune responses for tissue regeneration [5,6].

Concerning degradation products in the organism, PLA and PGA by-products have been considered to be nontoxic. After degradation, by-products of lactic acid and glycolic acid are similar to endogenous metabolites. Lactic and glycolic acid enter the tri-carboxylic acid cycle and are eliminated as carbon dioxide and water [7] through respiration, feces and urine [8].

Electrospinning is considered a smart and easy technique to produce 3D synthetic devices with a porous and fibrous architecture similar to biological extracellular matrix [9,10]. Particularly, this technique has been reported to be effective because it allows the mixture of different monomers to improve the degradation time for drug or cell delivery to a specific microenvironment [11]. Such devices are at the forefront in the regenerative medicine area improving cell behavior in a specific milieu. Specifically speaking, translational application of electrospun fibers for oral disease treatment has been used including a variety of oral clinical applications [12]. Regarding oral mucosa regeneration, Wang and colleagues [12] showed tissue engineering strategies using the electrospinning process. Such devices were applied in porcine, rabbits and dog models of oral mucositis induced by anticancer therapies. As there are few studies reporting the use of electrospun fibers for cell therapy to treat oral diseases [12], the present work intends to show how an electrospun PLGA construct, with the addition of fibroblast-like cells from hamster cheek paunch, can serve as a future dressing for chemotherapy-induced oral mucositis ulcerations in hamster model.

It is noteworthy the great potential of this technology to produce nanofibers or microfibers using an electric field. This field can be tunable to control the deposition of aligned or random micro or nanofibers onto a collector, in a 3D pattern, forming a biomimetic scaffold for tissue engineering applications [12]. In general, this system regards advantages, such as the possibility of obtaining a large surface area in relation to the volume, organized interfibrillar spaces, controlled malleability and the facility in generating materials with a variety of sizes, texture and shapes [13]. In addition, 3D electrospun materials keep properties that promote the flow of nutrients from the biological microenvironment, such as oxygenation and vessel penetration throughout the device, accurately as it happens in the human organism [12]. As a result, it runs accelerated cells proliferation [14].

Recently it was shown that, when covered with nanoparticles or bioactive factors, PLGA membranes can turn into a promising bifunctional scaffold [4]. In this way, those scaffolds play a crucial role in cell migration and differentiation, both critical for their future applications in the field of tissue engineering and regenerative medicine. Moreover, medical devices must be sterilized before animals or human applications. Although a variety of techniques can be used to sterilize medical devices, our group previously demonstrated, by physicochemical and structural analysis, that 3D electrospun PLGA membranes treated with gamma radiation showed preserved structures [15].

Over 50 years the term "biocompatibility" has been widely used to qualify the biological properties of a biomaterial. For instance, Liang et al. [16] reported that the selection and fabrications of biomaterials with different structures and forms, such as films, hydrogels, electrospun scaffolds, sponges and foams should be tested to enhance therapeutic efficiency. Accordingly, the present work poses special emphasis on in vitro tests to study cell–biomaterial interactions using MDCK cell-line and primary fibroblast-like cells isolated from hamster cheek paunch with manufactured electrospun PLGA membranes. Moreover, in the present study, tissue-biomaterial immune responses were analyzed after in vivo hamster implantation over time. To our knowledge, this is the first time that 3D electrospun PLGA membranes are studied in a hamster model.

Madin-Darby canine kidney (MDCK) cell-lines are epithelial cells used in experimental research models due to their potential to adhere and form either tubules or flat monolayer sheets in 3D and 2D structures, respectively. This multicellular architecture is formed in virtue of either intercellular interaction between epithelial cells or upon cell–matrix interactions in the appropriate scaffold [17]. On the other hand, primary fibroblast-like cells from hamster' cheek pouch were chosen by our group due to our future tissue engineering challenges, to test this construct in hamster models [18].

Biological responses of cell–material interactions are a function not only of biomaterials design or their intrinsic properties but also of the microenvironment. Specific and long-term reactions can differ as a function of the organ, tissue and or species [4,19–22]. In this milieu, the material's surface properties play an important role in modulating the immune system response, such as the foreign body reactions [23,24]. This immunological response has been analyzed in experimental models as being a natural response of the organism, which will remain as long as polymer fragments are present at tissue–material interfaces [25,26]. Such response recognized as foreign body response (FBR) is classified into two main types of multinucleated giant cells, the Langhans-type giant cells (LCs) and the multinucleated-type giant cells (MNGCs). The LC is present within granulomas of infectious and non-infectious etiology. Morphologically, LCs exhibit a circular or ovoid shape with nuclei arranged in the periphery of the cells, in a limited number (10–20 nuclei per cell), adopting a circular or horse-shoe pattern [27]. On the contrary, MNGCs result from the macrophage response to indigestible substances, exhibiting an irregularly shaped cytoplasm, which may contain hundreds of nuclei per cell [27].

Although PLGA is a widely used polymer in the area of regenerative medicine and tissue engineering, its processing in porous membrane thanks to electrospinning can modify the inflammatory reaction that it will induce and therefore its interaction with tissues [12]. The present study focused on the in vitro morphological analysis of cell–material interactions and on the in vivo immune response after PLGA electrospun membrane implantation. To process the aforementioned analysis over time, our group applied morphological, cytochemical and immunohistochemical approaches associated to light (conventional optical and laser scanning confocal) and electron (scanning and transmission) microscopy. These tools revealed cell proliferation over time onto the PLGA membranes in vitro, and tissue immune responses after membrane implantation, without fibrous tissue formation at the biomaterial interface.

2. Materials and Methods

2.1. PLGA

PLGA 85% L-lactide and 15% glycolide (w/v), (Purac: PLG 8531-lot No.1504000089, Amsterdam, The Netherlands) were produced using the randomly oriented electrospinning process. Polymeric solutions were prepared dissolving 5% PLGA (weight/volume) in the solvent-binary of chloroform (CHCl3-Vetec, Rio de Janeiro, Brazil) and dimethylformamide (DMF-Vetec, Rio de Janeiro, Brazil) systems (80/20 v/v), and magnetically stirred at room temperature.

2.2. Electrospinning Process

The polymeric solution was introduced into a syringe with a needle attached to an ejector pump under the voltage of 15 kV using the Glassman PS/FC 60p02.0-11 (KDS 100 Infusion Syringe Pump) high voltage source. The solution was ejected induced by the potential difference between the needle tip and the screen, with a working distance of 15 cm and a feeding rating of 0.6 mL/h for 6 h. PLGA filaments were deposited onto an aluminum foil and collected as a membrane. The process was standardized for all frameworks produced (dimensions: 10×10 cm^2 and thickness of 6.6 µm, analyzed by scanning electron microscopy). Next, PLGA membranes were sterilized using gamma radiation (Cobalt 60 source—MDS Nordion, model GC 220 E, Ottawa, ON, Canada), with a total dose of 15 KGy [28].

2.3. In Vitro Experiments

2.3.1. MDCK Cells Cultured onto PLGA Membranes

Madin-Darby canine kidney cells (MDCK, ATCC, cat. No. PTA-6503) were plated in 25 mm^2 culture bottles, in complete Cascade Medium 200 (M200-500) culture medium, incubated at 37 °C, in an atmosphere of 5% CO_2 and 95% humidity. After confluence, cells were trypsinized (2 mL TrypLE ™ Express Enzyme (1×), phenol red, cat. No.12605036-Gibco®) for 5 min. Complete culture medium (2 mL) was used to inactivate the process. Next, cells were centrifuged at 700 g for 3 min. PLGA samples (1 × 1 cm^2) were glued on the bottom of the wells of a 6 wells cell culture dish, covered or not with 20µg/mL fibronectin (human plasma fibronectin, Sigma-Aldrich, St. Louis, MI, USA, cat. No. F0895) in culture medium. After 1 h, PLGA samples were washed twice with sterile phosphate saline buffer (PBS) pH 7.4. Next, 1 mL culture medium containing MDCK cells (10^4) was applied only onto the surface of each membrane, and incubated at 37 °C, in an atmosphere of 5% CO_2 and 95% humidity for 1 h. After 1 h, complete culture medium (2 mL) was added in the well to cover each sample, and changed every 48 h. The experiments were carried out in quadruplicate and analyzed after 3, 6 and 12 days. PLGA membranes were fixed using 4% buffered paraformaldehyde solution for 15 min, washed in PBS, pH 7.4, twice; permeabilized with triton X-100 (0,1%) in PBS for 15 min; washed again in PBS, and incubated in PBS containing 0.5% serum albumin (BSA) for 1 h. Samples were treated with Alexa Fluor® 555 phalloidin 1:100 (Invitrogen, cat. No. A34055-Thermo-Fisher Scientific, Waltham, MA, USA), in PBS-0.3% BSA, and stored protected from light for 1.5 h. Next, samples were washed with PBS and treated with Hoechst solution (Bisbenzimide H33342 trihydrochloride solution in PBS, 1:1000, Sigma-Aldrich, cat. No. 14533) for 30 min, washed with PBS (3 times). Confocal microscopy analysis (LSM 700, Zeiss) was performed using the objective lens (W Plan-Apochromatic 63×/1.0 M27). Ten fields per sample were obtained over time, using the X-Z orthogonal axis and captured in the Image J Fiji software [29], to count the cell nuclei. Mean ± sd was used for analysis over time. The difference in cellularity was verified using the GraphPad Prism 5 program. For scanning electron microscopy (SEM) analysis, each sample was fixed in a solution of 1 mL glutaraldehyde 2.5% (Reagen—Quimibras Industrias Químicas S.A., Rio de Janeiro, Brasil); 1 mL of 4% formaldehyde (Merck) and 2 mL of Sorensen phosphate buffer (disodium hydrogen phosphate-Na2HPO4—11.876/L; potassium di-hydrogen phosphate (KH_2PO_4) at 0.2 M, pH 7.2, for 1 h. Then, the samples were washed in the same buffer 3× for 10 min, dehydrated in serial ethanol (Merck) solutions (30, 50, 70 and 90%) for 10 min each, ending with absolute ethanol, twice for 20 min and dried in ambient air. Then, the samples were fixed onto a metallic support with carbon tape, metallized by gold sputtering (20 nm) and observed in a FEI Quanta 400 scanning electron microscope.

2.3.2. Primary Hamsters' Fibroblasts Cultured onto PLGA Membranes

Primary fibroblasts from left hamsters' cheek-pouch were isolated after surgery using intraperitoneal ketamine (80 mg/kg) and xylazine (60 mg/kg). The excised cheek pouch was washed with povidine-iodine for 5 min, next rinsed with a saline solution 1 min, treated with nystatin (100.000 UI/Laboratorio Teuto S/A-Brasil) for 5 min, and finally immersed in sterile saline solution for 5 min. Tissues were sectioned into small pieces under a laminar flow hood (Vertical laminar flow—Pachane LTDA, model PA 320, No.03201. Pirasicaba, Brasil); immersed in collagenase type 1 solution (1 mg/mL-Gibco/Life Technologies) in culture medium (100 mL) inside an incubator at 37 °C under agitation for 3 h and inactivated using fetal bovine serum (10%—Vitrocell/lot 014/18). Finally, the cells were isolated by filtration (cell strainer 100 µm/Sigma-Aldrich) and centrifugation (700 g) of the supernatant. The pellet was suspended in 1 mL low glucose DMEM culture medium, supplemented with 10% fetal bovine serum, 100 IU/mL penicillin, 100 mg/mL streptomycin and 0.01 mg/mL amphotericin (Sigma Aldrich, St. Louis, MO, USA), plated in a 25 cm^2 culture bottle containing 4 mL culture medium, and kept in an incubator at 37 °C in a 5% CO_2 atmosphere. Culture medium was changed every 48 h after 3 washes with PBS, and stored at 37 °C in a

5% CO_2 atmosphere for 4 days. In another assay PLGA membranes (1×1 cm^2) were fixed with silicone glue and treated with 100 µL of a mouse collagen solution (Sigma-Aldrich) 10% (v/v) in serum-free DMEM, incubated for 1 h in the laminar flow and washed using PBS. Next, a suspension of 10^4 or 3×10^5 cells in 100µL of complete DMEM culture medium was added to each membrane and cultured over 1, 3, 6 and 12 days. Over time, the membranes were fixed in 4% paraformaldehyde in PBS for 30 min, washed in PBS for 5 min $3\times$ and kept in a 20 mL Falcon tube with PBS at 4 °C. After 12 days, all the membranes with cells were permeabilized with Triton X-100 (0.1%) in PBS for 10 min, washed $3\times$ in PBS for 5 min, immersed in PBS containing 0.5% BSA for 1 h, washed in PBS for 5 min and incubated with phalloidin-Alexa Fluor 488 (Invitrogen, 1:80) in PBS containing 1% BSA, for 1 h; washed $3\times$ in PBS and finally stained with Hoechst solution (1: 1000 in PBS/Sigma-Aldrich) for 30 min. After washing $3\times$ in PBS for 5 min, membranes were mounted on 3 cm diameter imaging dish with a coverslip. The side of cell seeding was fixed facing the bottom of the coverslips, and covered with a new drop of mounting medium to fix another coverslip. The cell nuclei were counted on the confocal microscope (10 fields per membrane) (LEICA-TCS-SPE, inverted O DMi8, using the ACS-APO $63\times/1.30$ lens). For SEM analysis, membranes were fixed in a solution of 2.5% glutaraldehyde, 4% paraformaldehyde and 0.1 M sodium cacodylate buffer, for 2 h at room temperature, washed with the same buffer for 10 min, post-fixed in 1% buffered osmium tetroxide solution for 40 min, washed with distilled water for 10 min, dehydrated in serial ethanol solutions (30%, 50%, 70% and 90%) for 10 min each and 100% ethanol for 20 min twice, at the end. Membranes were immersed in hexadimethylsilazane in absolute ethanol (v/v) solution for 5 min twice and dried. Next, the membranes were mounted on carbon strips attached to a metal plate and metallized with gold (20 nm) (BAL-TEC SCD 050) for analysis (SEM FEI QUANTA 250, operated at 15 kV).

2.4. In Vivo Experiment

Golden Syrian hamsters were purchased at Oswaldo Cruz Foundation-Rio de Janeiro, Brazil, at six weeks of age and maintained at the Health Science Center bioterium of Federal University of Rio de Janeiro-UFRJ-Brazil, in environmental, nutritional, and health-controlled conditions, according to "The Guide for Care and Use of Laboratory Animals" (DHHS Publication No (NIH) 85-23, Office of Science and Health Reports, Bethesda, MD 20892—available at: http://www.nih.gov). The Ethics Committee on the Use of Animals in Scientific Experimentation at the Health Sciences Center of the Federal University of Rio de Janeiro, registered in the National Council for Animal Experimentation Control (CONCEA-Brazil), certified the use of hamsters in the present study (protocol No. 003/15, approved on 15 April 2015). Animals between 120 and 205 g were included in the present study. The animals were kept in appropriate cages containing 3 animals/cage, lined with sterile shavings, under constant temperature (23 ± 2 °C), in the standard light/dark cycle (12/12 h), with unrestricted access to water and feed. After the experiments animals were euthanized according to the "American Veterinary Medical Association Guidelines on Euthanasia", 2007 (available at http://www.nih.gov).

2.5. PLGA Implantation and Material Collection

PLGA membranes (1×1 cm^2) were implanted in the dorsal region of the hamsters over time (7, 15, 30 and 90 days). The animals were randomly separated in 4 groups and were anesthetized with Ketamine (80 mg/Kg) and Xilasina (60 mg/Kg), in the intraperitoneal region (i.p.) for surgical procedures of PLGA implantation. PLGA was implanted in the upper dorsal region, after asepsis using 70% alcohol and shaving. Two PLGA samples (1×1 cm^2) were implanted in the subcutaneous upper dorsal region, one in each side. Next, the skin edges were sutured (5.0 nylon-Technofio). The animals were kept under observation for 24 h to control post-surgical behavior. Over time (7, 15, 30 and 90 days), PLGA samples were surgically removed with the adjacent skin, and divided in two pieces for morphological analysis. Next, the animals were euthanized, by inhaling CO_2 followed by decapitation through a guillotine.

2.6. Sample Preparation for Morphological Analyzes by Light and Electron Microscopies

The implanted materials were fixed in 4% paraformaldehyde in PBS, pH 7.4 at 4 °C for 2 h. Next, washed in running water, dehydrated in serial solutions of ethanol from 30% to 100% twice for 20 min, clarified in xylene (2 baths of 30 min) and embedded in paraffin. Sections of 5 µm-thick were performed on a rotary microtome (Leica Microsystem RM 2125®, Wetzlar, Germany) and collected on glass slides. Dried histological sections were stained with hematoxylin-eosin (HE) and Masson's trichrome stains. The other samples were immersed in Karnovsky's fixative solution (4% paraformaldehyde solution, 2.5% glutaraldehyde and 0.1 M sodium cacodylate) for 2 h, washed 3 times in the same buffer, cleaved in cubes of about 1 mm^2 and post-fixed with 0.1 M osmium tetroxide solution for 30 min. Next, they were washed with cacodylate buffer 3 times for 10 min, dehydrated in serial ethanol solutions to absolute 2 times for 20 min and embedded in Spurr resin. The selected region was cut into 90 nm thickness and contrasted with 1% uranyl acetate and 2% lead citrate for transmission electron microscopy analysis in a Morgagni 268 (FEI company, Hillsborough, OR, USA) operated at 80 KV.

2.7. Immunohistochemistry

Paraffin sections were collected on silanized histological slides (Sakura Finetek, Staufen, Germany) for immunohistochemistry. After adhesion, histological sections were dewaxed in xylene and hydrated. The samples were washed with 50 mM ammonium chloride solution, in phosphate buffered saline (PBS) pH 8.0, for 15 min, to block free aldehyde residues and washed in PBS; permeabilized with Triton X100 (0.5%) in PBS for 15 min, followed by a bath containing 0.3% hydrogen-peroxide in methanol to inhibit endogenous peroxidase for 15 min, in the dark. After washing with PBS, pH 7.4, sections were submitted to heat-mediated antigen retrieval in either the microwave (potency 800 W) or steamer, according to the antibody used (Table 1). After cooling, the histological sections were incubated with PBS containing 5% bovine serum albumin (BSA), normal 5% goat serum (1 h) in a humid chamber, at room temperature, and then, primary antibodies (Table 1) were incubated. Sections were maintained in a humid chamber, for about 20 h, in the refrigerator. Afterwards, the sections were washed in a PBS solution containing 0.25% Tween-20 (PBS-Tween), followed by incubation with the secondary antibody conjugated to peroxidase (Envision ™ Dual link system-HRP—cat. No. K4601, Dako, CA, USA), for 1 h, followed by washes in PBS-Tween. Peroxidase was developed with the chromogenic substrate diaminobenzidine (Liquid DAB, Dako, cat. No. K3468), followed by washes in PBS-Tween and distilled water and assembly of sections between slide and coverglass with Entellan®. Negative controls were performed by incubating the histological sections with non-immune rabbit or mouse serum or with the antibody diluent in place of the primary antibody.

Table 1. Characteristics of antibodies used and antigenic recovery in the immunohistochemistry assays.

Antibody	Manufacturer	Antigenic Recovery	Dilution
CD3	Dako, polyclonal rabbit, cat.# GA503	Steamer 20 min, Citrate Buffer 0.01 M pH 6.0	1: 400
Myeloperoxidase	AbCam, CA, USA, polyclonal rabbit, cat. # ab9535	Microwave—3 min 3 x, Citrate Buffer 0.01 M pH 6.0	1:100

2.8. Morphometrical and Statistical Analysis

Semi-quantitative analyses of histological sections were performed using an Eclipse E800 light microscope (Nikon, Japan) coupled to a digital camera (Evolution VR Cooled

Color 13 bits (Media Cybernetics, Bethesda, Rockville, MD, USA) using the 2×, 4×, 10×, 20× and 40× objectives lens. The interface capture used was Q-Capture 2.95.0, version 2.0.5 (Silicon GraphicsInc, Milpitas, CA, USA), and the images were transmitted to a color LCD monitor, captured (2048 × 1536 pixel buffer) in TIFF format and digitized. The images were captured after calibration of the appropriate color and contrast parameters and remained constant for each type of staining or immunohistochemistry. The quantification of $CD3^+$ cells (T lymphocytes) or myeloperoxidase + (neutrophils), were performed on the captured images (10 microscopic fields of the slides stained with the respective antibodies, using the 40× objective lens). The results were expressed as number of reactive cells/histological field ± standard error of the mean. The amount of multinucleated giant cells and capillaries (transversal sections) were calculated on 20 photomicrographs randomly obtained from HE stained histological sections (objective lens 20×). Results were expressed as number of multinucleated giant cells or number of capillaries, /histological field ± standard error of the mean.

For statistics, the in vitro analysis of cells was performed by one investigator (AC). Cells nuclei were counted on Image J Fiji software or in Image ProPlus 5.0 (Media Cybernetics, MD, USA). For the in vivo quantification, one investigator (pathologist, CMT) acquired all histological images and performed the quantification, in a blinded manner. The data were analyzed using GraphPad Prism 5.0 software (GraphPad Software, Inc., La Jolla, CA, USA). The differences between the groups were analyzed using one-way ANOVA on ranks (Kruskal–Wallis test), followed by a post hoc test (Dunn's multiple comparison test or Neuman–Keuls). Differences were considered significant at $p < 0.05$, with asterisks indicating the level of significance: * $p < 0.05$, ** $p < 0.01$, *** $p < 0.001$; ns indicates no significant differences.

3. Results

3.1. In Vitro Experiments

3.1.1. MDCK-Cell Line

MDCK cell line is a widely used epithelial cell line to test the monolayer formation. They were used in the present work to test the electrospun PLGA (85:15) membranes produced herein. The cells adhered and proliferated on both fibronectin-coated (FN) and uncoated (no-FN) PLGA membranes, independently, over time. These results showed that electrospun membranes with random fibers, with or without FN promoted cell proliferation, such as shown in Figure 1A–C, with cells forming a monolayer onto the membranes. MDCK cells preferentially adhered and proliferated on the fibronectin (FN) coated PLGA membranes as seen in (Figure 1C). After 3 days in culture, the number of epithelial MDCK cells on FN-coated or no-FN membranes showed no difference (Figure 1C: 3 days FN: 27 ± 2 cells/ mm^2 to specify; 3 days no-FN: 24 ± 1 cells/ mm^2, $n = 4$). An increase in the number of cells was observed after 6 and 12 days (Figure 1C), regardless of treatment with fibronectin. However, after 12 days, there was a greater number of cells on FN-coated membranes in comparison to no-FN membranes (Figure 1C: 6 days FN: 59 ± 8 cells/mm^2; no-FN: 50 ± 8 cells/mm^2; 12 days FN: 131 ± 16 cells/mm^2; no-FN: 102 ± 16 cells/cm^2).

To illustrate, a confocal image is showing MDCK cells through the fibers of no-FN membrane fibers (Figure 1(Aa)) after 3 days in culture. For comparison, image of FN-coated membrane (Figure 1(Af)) is showing cells proliferating through the fibers after 3 days in culture (thick arrow). In an equivalent view of Figure 1(Aa,Af), the cell nuclei (Figure 1(Ab,Ag)) are exhibiting different distribution, showing that cells are following the random design of the fibers. In another equivalent view of A-a (Figure 1(Ac)), confocal image of no-FN membrane (Figure 1(Ac), arrows) after phalloidin staining after 3 days reveals that cells are distributed among the fibers but are not connected since no circumferential actin belt is visible at that time (Figure 1(Ac), circle). For comparison, confocal image of FN-coated membrane (Figure 1(Ah)) exhibits regions where cells appeared to be connected since circumferential actin belt is sometimes visible (Figure 1(Ah), thin arrows) among cells without adhesions (Figure 1(Ah), thick arrows), still not taking the classic cuboidal morphology after 3 days. Image of orthogonal projections (Figure 1(Ad,Ai)) showing cells after 3 days reveals the monolayer formation onto no-FN (Figure 1(Ad)) or FN-coated membranes (Figure 1(Ai)), both with few in-depth cell migration after 3 days.

Image of cells onto no-FN membrane after 12 days (Figure 1(Ba,Bf)) reveals more localized MDCK cells among PLGA fibers (Figure 1(Ba), arrow; double arrow) in comparison to FN-coated membrane showing a dense monolayer of cells among the fibers (Figure 1(Bf), double arrow). An equivalent image of Figure 1(Ba), f reveals the cells' nuclei on no-FN and FN coated membranes (Figure 1(Bb,Bg)), respectively in close contact to each other and exhibiting different directions according to fibers' organization. Moreover, images after phalloidin staining of no-FN or FN-coated membranes (Figure 1(Bc,Bh)), depicts increased cell connections with circumferential actin belt; both taking their classic cuboidal morphology after 12 days (Figure 1(Bc,Bh), arrows). Note that orthogonal projections images after 12 days (Figure 1(Bd,Bi)) show MDCK cells taking decreased size in comparison to cells after 3 days in culture. This can be explained by the proliferation of the cells, causing them to thicken and take the classic cuboidal morphology of epithelial cells at confluence, visible on orthogonal projections. SEM images of MDCK cells reveals the presence of isolated cells after 3 days (Figure 1(Ae,Aj), thin arrows and asterisks), exhibiting less amount of MDCK cells onto no-FN and FN membranes, in comparison with the cluster of cells with extracellular matrix visible after 12 days (Figure 1(Be,Bj), white arrows and asterisks). In addition, SEM image of FN-coated membrane after 12 days (Figure 1(Bj), white arrows) shows cells spreading in close contact with the membrane fibers', by forming cell extensions towards areas without cells following PLGA fibers direction.

Number of cells/mm^2 over time is illustrated in Figure 1C. The results are expressed as mean ± SEM (standard error of the mean). Images of cell nuclei (n = 10) stained with Hoechst on each sample (n = 4) were captured by CLSM (63× objective lens) and counted. Data were submitted to one-way ANOVA followed by Bonferroni's multiple comparison test (ns: not significant, ** p < 0.001, *** p < 0.0001).

3.1.2. In Vitro Experiment Using Primary Fibroblast-like Cells

Primary fibroblasts isolated from hamsters' cheek paunch were seeded onto collagen-coated electrospun PLGA (85:15) membranes. To select the best construct, two cell concentrations (10^4 or 3×10^5 cells) were tested and compared over time (1, 3, 6 and 12 days) using confocal microscopy (CLSM). Both concentrations of cells adhered and proliferated over time onto PLGA-coated membranes. The construct with 3×10^5 cells showed increased cell proliferation after 6 and 12 days in culture; such cells took an elongated morphology and migrated towards the interior of PLGA membrane, posing an ideal dressing for tissue engineering and future tests as cell delivery therapy.

The primary culture of fibroblast-like cells is represented on a phase-contrast image (Figure 2A) (fourth passage) after 4 days, and the random electrospun PLGA fibers covered with collagen is represented on a CLSM image (Figure 2B). After 1 day in culture, cells adhered on the surface of the fibers (Figure 2C, white arrows depict the PLGA fibers), and took an elongated morphology, taking different directions following the fibers (Figure 2D, white arrows). The maximum projection confocal images show cell proliferation (3×10^5) over time and their respective in depth migration (Figure 2(Ee,Ff,Gg,Hh)). After one day, cells were only on the surface of the membrane (Figure 2(Ee)). After 3 days (Figure 2(Ff), white arrows) confocal image shows cells initiating the in-depth migration. After 6 days (Figure 2(Gg), white arrows) more cells were migrating to the interior of the membrane, while a deeper in-depth migration was observed after 12 days (Figure 2(Hh), white arrows), in three different areas of the same membrane. Results of cell proliferation in the assay with 3×10^5 cells after 6 and 12 days, reveal significantly higher proliferation of cells compared to other times (Figure 2I). However, it was possible to observe an increase in the number of cells in each assay with 10^4 cells (1 day: 9 ± 2; 3 days: 30 ± 2; 6 days: 43 ± 2; 12 days: 47 ± 4; or 3×10^5 cells) or 3×10^5 cells (1 day: 35 ± 4; 3 days: 45 ± 10; 6 days: 64 ± 3; 12 days: 76 ± 2) (Figure 2J).

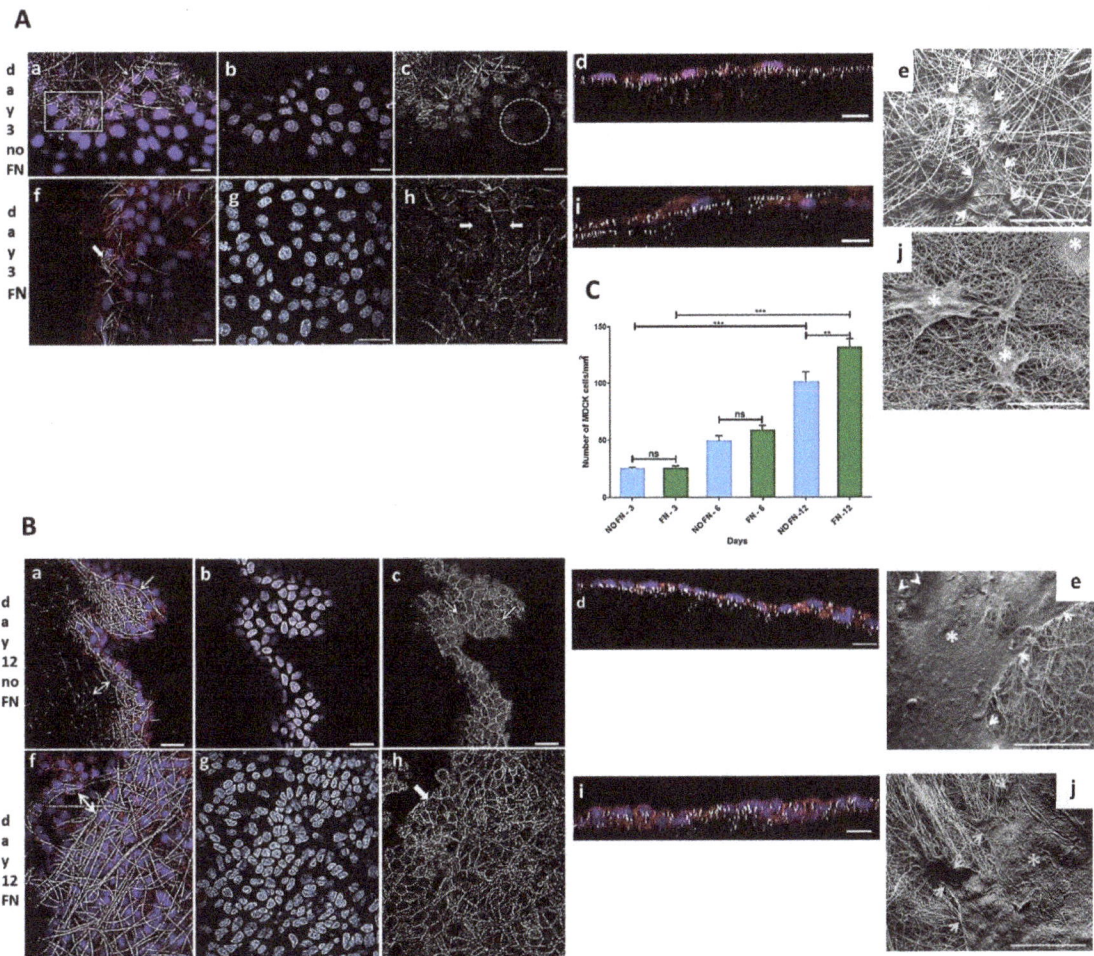

Figure 1. Representative images of MDCK cells cultured onto fibronectin-coated (FN) or uncoated (no-FN) PLGA membranes. (**A**,**B**): Confocal Laser Scanning Microscopy images of cells after 3 and 12 days, stained with Hoechst (blue) and phalloidin conjugated to Alexa 568 (red); (**C**): Graphical analysis. (**Aa**): Merge image of cells among no-FN PLGA fibers (arrow, square) after 3 days; (**Af**): Merge image of FN PLGA fibers and cells (white arrow) after 3 days. (**Ab**,**Ag**): Equivalent fields from (**Aa**,**Af**) processed on Image J software to visualize nuclei. Note the different organization of nuclei following the fibers after 3 days. (**Ac**): confocal images of phalloidin staining of no-FN coated fibers after 3 days (**Ac**, arrows). Note few circumferential actin belt staining revealing adhesions between cells (circle) in comparison with FN membranes (**Ah**) after 3 days showing more adhesions between cells (thin arrows), among cells without adhesions (thick arrows). (**Ad**,**Ai**): cells in a X-Z plane after 3 days forming a monolayer with less cells. Scale bars: (**Aa–Ai**): 20 µm. (**Ae**,**Ai**): SEM images of no-FN or FN membranes exhibiting isolated cells onto the fibers after 3 days (white arrows and asterisks). Scale bars: 100 µm. (**Ba**,**Bf**): Merge image of cells after 12 days. Note fibers and cells among the fibers on no-FN or FN, respectively (white arrows and double arrows). (**Bb**,**Bg**): Equivalent fields from (**Ba–Bf**) showing a greater number of nuclei with a differential organization after 12 days. (**Bc**,**Bh**): Equivalent fields from (**Ba–Bf**) showing the increase in circumferential actin belts demonstrating formation of adhesions between cells exhibiting a cuboidal morphology onto PLGA fibers. Note a

greater number of cells onto FN coated membrane after 12 days. (**Bd,Bi**): cells in an X-Z plane after 12 days showing more proliferation. Scale bars (**Ba–Bi**): 20 µm. (**Be,Bj**): SEM image showing cells covering the majority of the membrane after 12 days (white arrows and asterisks). Scale bars: 200 µm. (**C**): Graphical representation of number of MDCK cells/mm^2 onto FN or no-FN PLGA membranes. ns: not significant, ** $p < 0.001$, *** $p < 0.0001$.

SEM images of collagen-coated membranes showed a collagen network in the interfibrillar spaces (Figure 2K, fiber: arrow; collagen: asterisk). In another SEM image (Figure 2L), a cluster of cells in extracellular matrix spread on the surface of the PLGA membrane in intimate contact with PLGA fibers (black arrows). Interestingly, sometimes cells showed an elongated morphology following PLGA fibers (Figure 2M, arrow heads), while other cells display more spread morphologies in the interfibrillar space when adhering to multiple fibers (Figure 2M, asterisks).

3.2. In Vivo Experiment
3.2.1. Histopathology

The skin of hamsters is constituted by the epidermis, dermis (papillary and reticular layers), hypodermis and a skeletal muscle layer beneath the hypodermis. After PLGA implantation, the inflammatory response induced by PLGA membranes was a well localized chronic inflammatory response in the implantation bed of the membranes, below the muscular layer, remaining during all period of observation. The histological features of the foreign body response induced by the PLGA membrane are represented in Figure 3A–L. The inflammatory response delimited the biomaterial, forming an interface with the adjacent connective tissue without presenting a fibrotic capsule in any timeline (Figure 3A–L), and was constituted by a foreign body reaction consisting mainly by mononuclear cells of the monocytic lineage including macrophages, epithelioid cells and multinucleate giant cells (Figure 3A–L). The membranes were located inside the inflammatory site, such as verified with polarizing microscopy, represented by semi-thin sections after 7 days (Figure 3A,B), and after 15 days (Figure 3D,E) post implantation (PI). Mononuclear cells (Figure 3C) and multinucleate giant cells (MNGCs) (Figure 3C yellow arrows) surrounded the membrane after 7 days PI; and a foreign body response (FBR) beneath the muscle layer was observed after 15 days PI (Figure 3F, arrow). Another histological image after 15 days PI is showing the FBR constituted Langhans-type multinucleate cells (LCs) (Figure 3G arrows), lymphocytes (Figure 3G square) and epithelioid cells (arrow) with an elongated shape in rows surrounding the membrane (Figure 3H arrows). After 30 days histological image shows the FBR formed as a well delimited lesion in the hypodermis (Figure 3I arrow), with bands of connective tissue (Figure 3J, thick arrows) and capillaries (Figure 3J, thin arrow) present inside the lesion. After 90 days PI, histological image shows persistence of the inflammation (Figure 3K, arrow) and presence of both MNGCs (Figure 3L, thin arrow) and LCs (Figure 3L, thick arrow).

The LCs (multiple nuclei disposed at the periphery of the cell) (Figure 4A, black arrows) increased in number from 7 to 30 days PI decreasing successively up to 90 days PI (Figure 4B). Associated to LCs, capillaries sections were present in high numbers at day 7 PI, decreasing along time (Figure 4C,D). A histological image of a region showing the presence of capillaries (Figure 4C, black arrows) along the evolution of the implanted PLGA (Figure 4C, asterisk) represents the PLGA-induced inflammatory reaction. The quantification of the number of capillaries is represented by the number of capillaries over time, showing that the number of capillaries decreased up to 90 PI (Figure 4D). Number of capillaries (transversal sections) were calculated on 20 photomicrographs randomly obtained from HE stained histological sections (objective lens 20×). Results were expressed as number of capillaries, /histological field ± standard error of the mean.

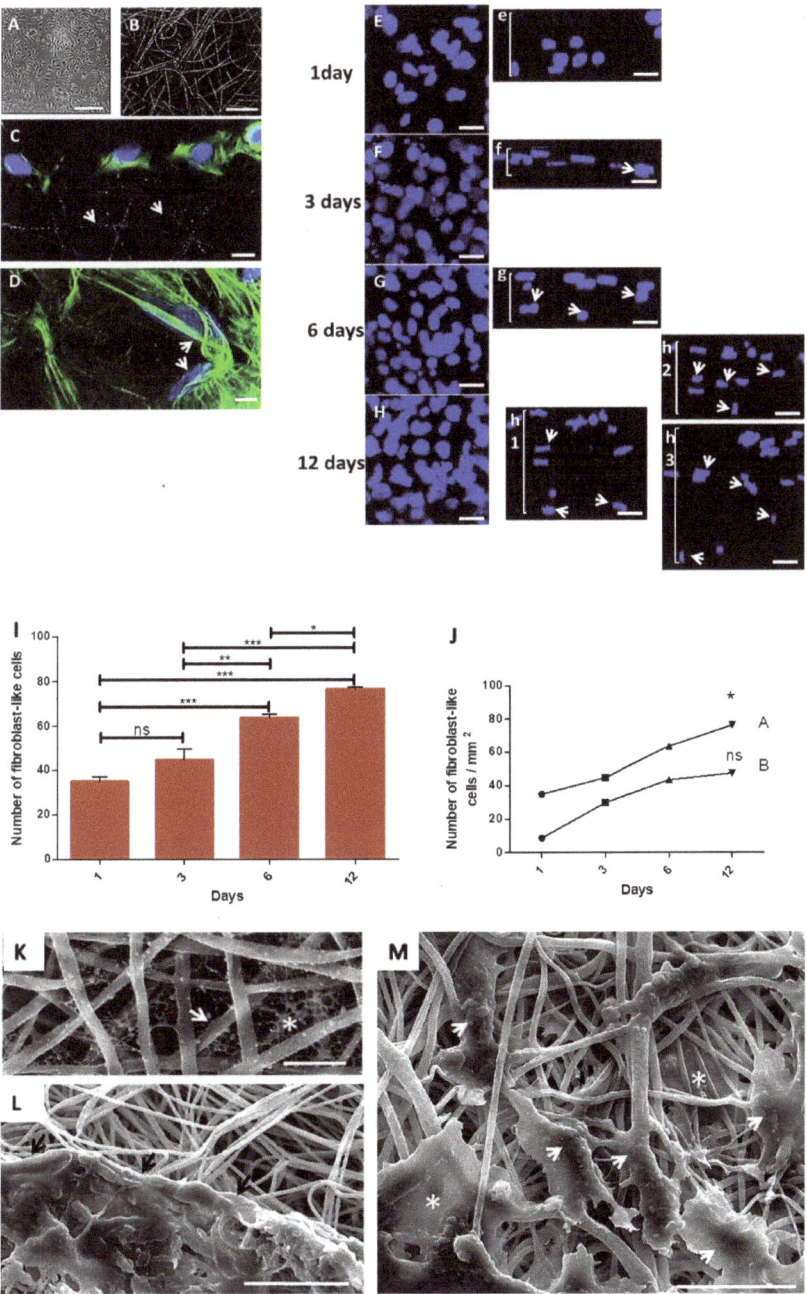

Figure 2. Primary hamster cheek pouch fibroblast-like cells cultured onto PLGA scaffolds coated with collagen I. (**A**): Phase-contrast image of primary fibroblast-like cells after 4 days. Scale bar: 100 μm; (**B**): CLSM image of collagen I-coated PLGA fibers. Scale bar: 20 μm; (**C**): CLSM image of fibroblast-like cells (3×10^5) onto PLGA fibers after 1 day (white arrows). Nuclei in Hoechst stain

(blue) and cytoskeleton in conjugated Alexa 488 phalloidin stain (green) for actin. Scale bar: 30 µm. (**D**): Cells adhered along different directions of PLGA fibers (white arrows). Scale bar: 30 µm. (**E–H**) CLSM images of fibroblast-like cells nuclei exhibiting proliferation after 1, 2, 3, 6 and 12 days on sample surface; (**Ee,Ff,Gg,Hh**): Side view of the nuclei of cells grown onto the scaffold after 1 day (**Ee**), and the in depth migration of cells nuclei to the 3D directions of the scaffolds after 3 days (**Ff**, arrow), 6 days (**Gg**, arrows) and 12 days (**Hh**, arrows)—images of 3 different fields in the same sample exhibiting a great amount of cells through the bulk of the membrane, as depicted in images h1,2,3 (arrow heads); note cells going deeper in image h3; Scale bars: 20 µm. (**I**): Graph of the proliferation of cells onto collagen I-coated membranes seeded with 3×10^5 cells; (**J**): Graph of the expansion of cells in function of the quantity of inoculated cells: 10^4 (A) or 3×10^5 cells (B). *** $p = 0.0001$; ** $p = 0.001$ and * $p = 0.01$; ns: not significant; (**K**): SEM image of PLGA fibers (white arrow) showing the network of collagen (white asterisk) in the interfibrillar spaces (scale bar: 5 µm). (**L**): SEM image of cells in contact with PLGA fibers (black arrows). Scale bar: 30 µm. (**M**): SEM image of elongated cells following PLGA fibers (white arrowheads), and cells in interfibrillar spaces showing different morphologies (white asterisks) (scale bar: 30 µm).

3.2.2. Immunohistochemistry and Morphometry

The inflammatory process was analyzed over time after PLGA membranes implantation. For this goal, the CD3 antigen was used to measure the total number of T cells, and the myeloperoxidase (MPO) reactivity for polymorphonuclear neutrophils (PMNs). Both immunohistochemistry assays quantified these cellular populations, which decreased up to 90 days, showing that the inflammatory process decreased along PLGA fragments degradation.

In the analysis, the total number of T cells (Figure 5A,C,E,G) showed variations over time. The number of T cells increased after 7 and 30 days PI and decreased after 15 and 90 days PI ($p = 0.0002$) (Figure 5A,C,E,G,I), while neutrophils (myeloperoxidase+ cells) levels were significantly elevated at the implantation bed after 15 days PI ($p < 0.05$), decreasing after 90 days PI ($p = 0.004$) (Figure 5B,D,F,H,J). Results are expressed as the number of CD3 or myeloperoxidase+ cells/histological field ± SEM (standard error of the mean).

3.2.3. Transmission Electron Microscopy (TEM)

In this work we chose a high-resolution imaging technique of TEM to show detailed images of PLGA fragments in intimate contact with the cells over time. TEM images showed various populations of macrophages, epithelioid cells and multinucleated giant cells at the granulomatous inflammatory process induced by the PLGA membranes between 7 and 90 days PI (Figure 6A–H). The inflammatory process was observed in the periphery of PLGA, showing mononuclear cells forming a foreign body reaction (FBR) in contact with the extracellular matrix (ECM) 7 days PI (Figure 6A). In the periphery of PLGA, TEM image is showing PLGA fragments (Figure 6B, arrows) in contact with a multinucleate giant cell (Figure 6B(GC)) 7 days PI. Another TEM image shows PLGA fragments in contact with epithelioid cells (EP) (Figure 6C(EP)), presenting different morphologies, with prominent cytoplasmic extensions such as filopodia following the topography of PLGA fragments (Figure 6C, arrows) 7 days PI. PLGA fragments were found inside phagosomes (Figure 6D arrow), in the interior of the cell (Figure 6D, asterisk) or in the extracellular space (Figure 6E, arrow) 15 days PI. However, an interesting and rare light microscopy image shows LCs containing material inside (Figure 6F, arrows) 30 days PI, such as depicted by TEM images showing PLGA fragments in the interior of cells (Figure 6G, arrows) 30 days PI. Finally, TEM image 90 days PI show decreased fragments of PLGA between the cells (Figure 6H, arrows).

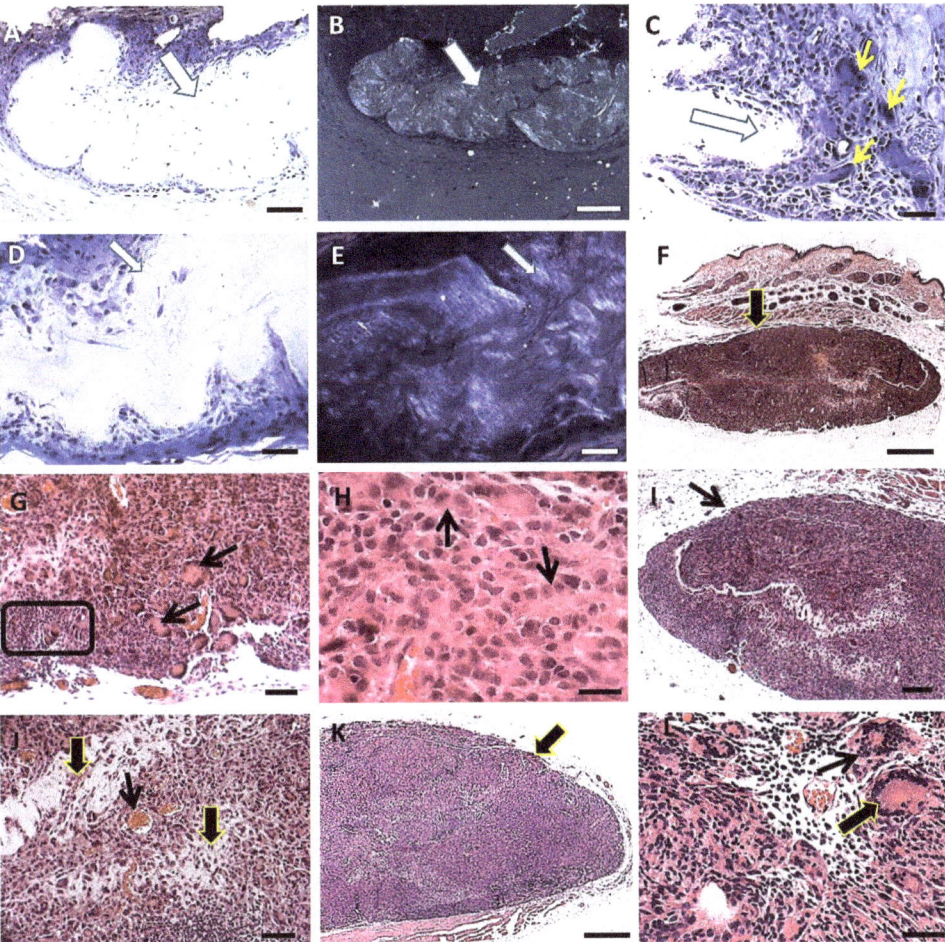

Figure 3. FBR induced by the PLGA membrane: histological features. (**A**): Semi-thin section of a sample in Spurr resin in toluidine blue stain showing the inflammatory response surrounding the membrane (thick arrow) 7 days post implantation (PI). Scale bar: 100 µm; (**B**): Polarization light microscopy of the same image in A. Scale bar: 200 µm; (**C**): Semi-thin section in Spurr resin and toluidine blue stain showing mononuclear cells and MNGCs (thin yellow arrows) surrounding the membrane (thick arrow) 7 days PI. Scale bar: 100 µm; (**D**): Semi-thin section in Spurr resin in toluidine blue stain showing the FBR 15 days PI. Scale bar: 100 µm; (**E**): Polarization light microscopy of the same image in D. Scale bar: 100 µm; (**F**): Section in Masson's trichrome stain showing the FBR in the hypodermis 15 days PI (arrow). Scale bar: 1 mm; (**G**): Section in Masson's trichrome stain showing the FBR 15 days PI; note LCs (arrows) and lymphocytes (square). Scale bar: 200 µm; (**H**): Section in HE stain showing epithelioid cells (arrow) surrounding the membrane. Scale bar: 50 µm; (**I**): Section in HE stain showing the FBR in the hypodermis after 30 days PI (arrow). Scale bar: 500 µm; (**J**): Section in HE stain showing the inflammatory response 30 days PI; connective tissue (thick arrows) and capillaries (thin arrow). Scale bar: 100 µm; (**K**): Section in Masson's trichrome stains showing the FBR in the hypodermis after 90 days PI; note more cells (arrow). Scale bar: 1 mm; (**L**): Section in HE showing MNGC (thin arrow) and LC (thick arrow) in the inflammatory reaction 90 days PI. Scale bar: 50 µm.

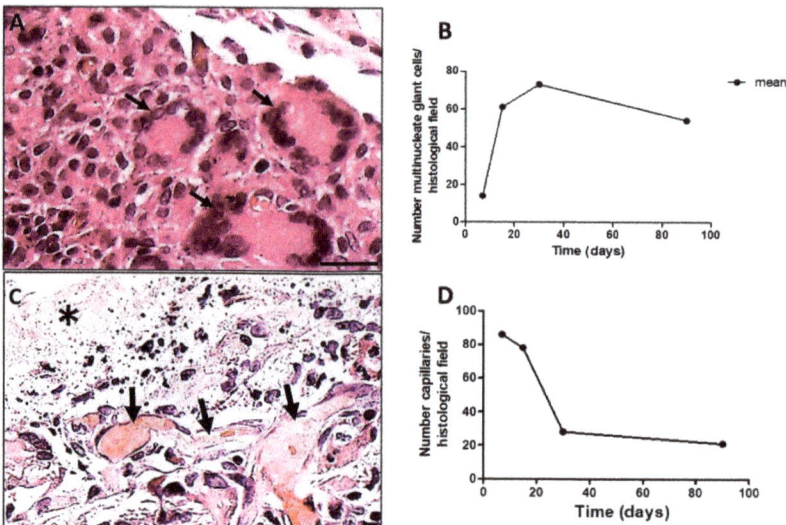

Figure 4. Histological sections of chronic inflammatory response induced by PLGA membranes. (**A**): Section showing the presence of LCs surrounding the chronic inflammation, 30 days PI. Scale bar: 50 μm; (**B**): Graphical representation of the number of LCs per histological field over time, showing decreased numbers up to 90 days PI. Number of MNGCs were calculated on 20 photomicrographs randomly obtained from HE stained sections (objective lens 20×). Results were expressed as number of MNGCs/histological field ± standard error of the mean; (**C**): Section showing the presence of capillaries (black arrows) along the evolution of the implanted PLGA (asterisk). Scale bar: 100 μm; (**D**): Graphical representation of the number of capillaries over time showing decreased number of capillaries up to 90 PI. Number of capillaries were calculated on 20 photomicrographs randomly obtained from HE stained sections (objective lens 20×).

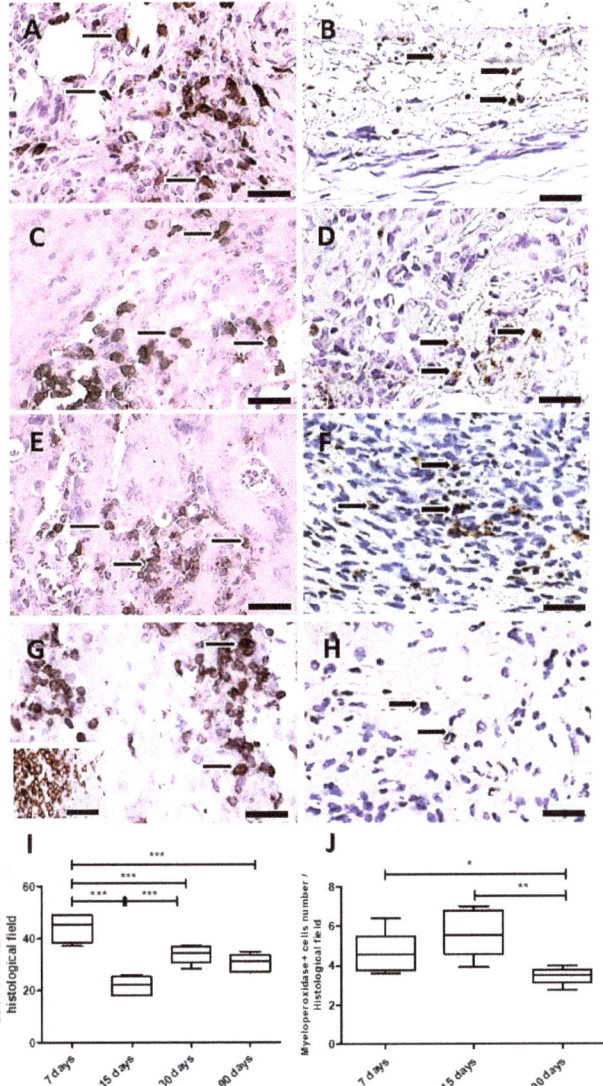

Figure 5. Immunohistochemistry and quantification of CD3 and myeloperoxidase positive cells. Detail of histological sections of PLGA-induced inflammatory response containing CD3$^+$ cells. (**A**): 7 days PI (thin arrows); (**C**): 15 days PI (thin arrows); (**E**): 30 days PI (thin arrows); (**G**): 90 days PI (thin arrows). Detail of histological sections of PLGA-induced inflammatory response containing myeloperoxidase$^+$ cells: (**B**): 7 days PI (thick arrows); (**D**): 15 days PI (thick arrows); (**F**): 30 days PI (thick arrows); (**H**): 90 days PI (thick arrows). In (**G**), inset represents the positive control of the anti-CD3 antibody, depicting CD3$^+$ cells in human tonsil. Scale bars (**A–H**): 50 µm. (**I**): Graphical representation of the CD3$^+$, showing decreased numbers at 15 and 90 days PI. (**J**): Graphical representation of myeloperoxidase$^+$ cells along time showing increased number of myeloperoxidase+ cells a 15 days PI, decreasing 90 days PI. * $p < 0.05$, ** $p < 0.01$, *** $p < 0.001$, ns, indicates significant differences.

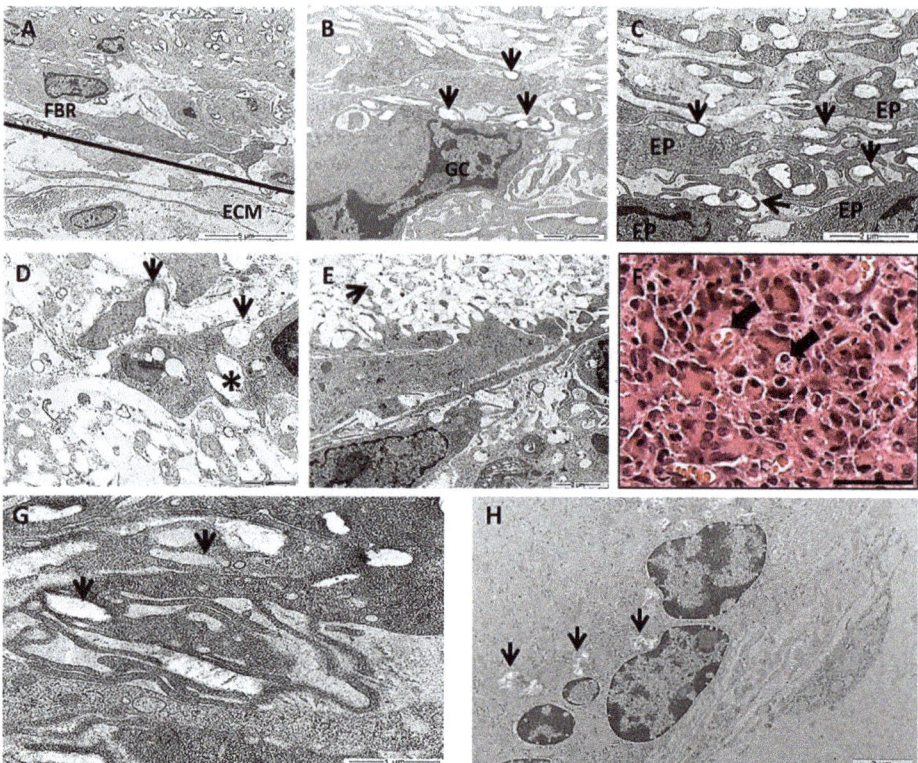

Figure 6. Transmission electron microscopy (TEM) and histological images of the inflammatory response induced by PLGA membranes. (**A**): Image showing the periphery of the PLGA with mononuclear cells forming a FBR in contact with the extracellular matrix (ECM). Scale bar: 5 μm; (**B**): image showing PLGA fragments (arrows) in contact with a MNGC. Scale bar: 2 μm; (**C**): image showing PLGA fragments (arrows) in contact with EP cells taking different morphologies, exhibiting prominent cytoplasmic extensions such as filopodia following the topography of PLGA fragments (arrows). Scale bar: 2 μm; (**D,E**): 15 days PI. D: PLGA fragments inside phagosomes (arrows in **D**) in the interior of the cell (asterisk in **D**) (Scale bar: 1 μm) or in the extracellular space (arrow in **E**). Scale bar: 2 μm; (**F**): Light microscopy image 30 days PI in Masson's trichrome staining showing LC containing material inside (arrows). Scale bar: 100 μm; (**G**): 30 days PI—PLGA fragments were swollen by cells (arrows). Scale bar 1 μm. (**H**). Note less amount of PLGA fragments between the cells 90 days PI (arrows). Scale bar: 2 μm.

4. Discussion

Poly (lactic-co-glycolic acid) (PLGA) is a synthetically manufactured linear copolymer constituted by different proportions of monomeric lactic acid (LA) and glycolic acid (GA). It is used in bioengineering as scaffolds for corneal and skin tissue engineering [30], as well as for bone regeneration [31–33] or for drug delivery [33,34]. PLGA presents a good biocompatibility, excellent mechanical properties, controllable biodegradability and electrospinnability. Herein, we are describing cell-interactions with electrospun PLGA (85:15) membranes in vitro (MDCK cell line or primary culture of hamster oral fibroblasts) or in vivo when implanted in the immunocompetent hamster subcutaneous tissue.

Recently, our group analyzed the physico-chemical properties of the electrospun PLGA membranes used in the present study [15]. The membranes were mostly constituted of fibers less than 1 μm in diameter [15]. In fact, biodegradable electrospun membranes constituted by thin fibers have been shown to promote cell adhesion, migration and proliferation [11,29,35]. In this sense, our team arouses the hypotheses that the fibronectin-coated electrospun PLGA membranes used in the present study should contribute to cell proliferation due to its biomimetic properties, since it has been shown in a previous study that gingival fibroblasts can adhere and proliferate when seeded onto collagen or PLGA substrates [36].

To achieve better results, PLGA hybrid membranes containing collagen or fibronectin have been manufactured [37–40]. Indeed, human dermal fibroblasts cultured onto collagen-coated electrospun PLGA substrates secreted extracellular matrix (ECM) on the substrate forming a fibrillar network in the PLGA interfiber spaces, promoting better cell–material interactions [38]. Our data showed similarities with the study of Sadeghi et al., [38] regarding collagen I-coated PLGA membranes, which promoted fibroblast-like cell proliferation over time and secreted ECM onto the surface of the scaffold. Inanç et al. [41] also showed that periodontal ligament cells seeded onto PLGA membranes exhibiting the same proportions of monomers as herein, produced collagen I and fibronectin which covered the PLGA fibers [40]. Considering the average length of a fibroblast, Lowery et al., [42] demonstrated that membranes possessing interfibrillar spaces above 6 µm and less than 20 µm are permissive to the fibroblast adhesion, locomotion and proliferation inside the scaffold, while this is not the case when the interfibrillar spaces are higher than 20 µm. Although in the present study the interfibrillar spaces were not measured, it can be hypothesized that the net of collagen I formed in the interfibrillar spaces after membrane bioactivation, can favor fibroblast-like cells proliferation over time.

The MDCK cells are a model of canine kidney cell line used in drug screening and biomedical research [43,44]. Moreover, these cells have been widely used in a variety of research studies using synthetic polymers, regardless of polymers fabrication or formats, due to their ability to follow different topographies. In this sense, we appreciated the migration behavior of MDCK cells onto 3D electrospun PLGA scaffolds over time. Indeed, since the 3D electrospun PLGA scaffold used herein possesses randomly oriented fiber topography, conferring interconnected fiber-forming porous, it was considered an ideal architecture to appreciate the monolayer and epithelial barrier formation of these cells. Our in vitro data confirmed that MDCK cells also proliferated onto the PLGA membranes coated or not with fibronectin after three and twelve days, independently. Comparing the two conditions, cell proliferation showed significant better results on the fibronectin-coated PLGA membranes after 12 days. Accordingly, we showed a cluster of cells taking on a classical cuboidal morphology and forming a confluent monolayer with cell polarity after 12 days.

Our morphological study revealed that the implanted membrane promoted a highly cellular foreign body response up to 90 days PI, forming at all studied times a well-delimited structure in the site of implantation. The implantation of biomaterials in vivo frequently elicits a chronic inflammation at the tissue-implant interface followed by wound healing responses and tissue fibrosis [45–47]. Concomitant with the formation of macrophage-derived foreign body multinucleated giant cells, constituting LCs or MNGCs, granulation tissue develops around the biomaterial. Factors secreted by macrophages adhering onto the surface of the biomaterial and the multinucleated giant cells formations attract fibroblasts and transform them into myofibroblasts [48,49]. These cells become aligned around the biomaterial and deposit great amount of collagen and other proteins creating a fibrous capsule around the implanted biomaterial. The fibrous encapsulation is considered as the end stage of the foreign body reaction and healing response to a biomaterial.

In the majority of cases, biomaterials promote the progression of such sterile inflammation to a foreign body response [47]. In the late 1980s and early 1990s, the occurrence of a typical nonspecific foreign-body reaction around internal fracture fixation implants made of pure polyglycolide or polylactide and even of glycolide-lactide copolymer was reported [50–54].

The foreign body response induced by our manufactured electrospun PLGA membranes was extremely cellular with a high amount of LCs derived from fusogenic macrophages. Macrophages are extremely versatile and plastic, being capable of adhering and recognizing foreign materials. At this time, they show typically a classically activated phenotype secreting inflammatory cytokines, ROS, protons release and degradative enzymes. Besides that, they display high phagocytic capacity, being able to phagocytose particles up to a size of 5 mm. In the presence of particles larger than 5 mm, macrophages coalesce to form foreign body multinucleated giant cells [55,56]. At this time, these cells display a reduced phagocytic activity, but an enhanced degradative capacity [56] at the expense of protons secretion, enzymes and ROS [57,58]. Kim [59] reported three steps on the biological pathways of local tissue responses to biomaterials: (i) organization of the inflammatory responses; (ii) monocytes migration to implantation site which differentiate into macrophages and fibrous capsule development and (iii) rapid degradation of the polymer and enhanced formation of fibrous tissues generated in the second step. These biological steps can guide the comprehension and analysis of the immunological responses in different hosts. In this sense, macrophages orchestrate the FBR that remain at the biomaterial/tissue interface when they do not succeed in phagocytizing the material [43,60]. We showed using histological and immunohistochemistry study that the electrospun PLGA membranes induced the recruitment of cells of the monocyte/macrophage lineage with the formation of epithelioid cells, LCs, as well as transitional cells which were similar to the cells seen after the implantation of fragments of MELINEX TM plastic in the subcutaneous tissue of rats [61,62]. Our ultrastructural study using TEM showed the difficulty of identifying these cell types since the transformation of monocytes to macrophages, macrophages into epithelioid cells or MNGCs were gradual, without a clear-cut distinction between the different cell types. Foreign body giant cells were recognized by the numerous nuclei inside cytoplasm, while macrophages, epithelioid cells and even giant cells, showed in the cell surface numerous invaginations and finger-like projections and numerous micropinocytotic vesicles. In addition, some cells mainly in the later stages (after 30 days PI) presented numerous filopodia, which surrounded the PLGA, or phagosomes, some of them containing fragments of the PLGA. The presence of phagosomes containing material similar to that of PLGA fibers could be correlated with the vacuoles seen in light microscopy inside some cells, suggesting that these cells have some capacity for phagocytosis. According to our knowledge, light microscopy images showing cells internalizing PLGA fragments were not found in the literature till now.

Neutrophils are present within the first days (up to day 2) after the biomaterial implantation [25,63]. Host derived chemoattractants released from activated platelets, endothelial cells, mast cells and injured cells direct neutrophils to the site of implantation. Neutrophils trigger a phagocytic response, degranulation and secrete ROS and proteolytic enzymes. The latter can corrode material surface and the degradation remnants could also trigger waves of neutrophils arrival and prolong the inflammatory response (reviewed in [63]). This event could also be occurring in our implanted PLGA membrane since it is susceptible to degradation when implanted in the body or in vitro [15,63,64]. Indeed, neutrophils were present in the foreign body reaction during all timelines of the study. In addition, the persistence of neutrophils in the reaction is a source of chemokines and activations factors for monocytes, macrophages, immature dendritic cells and lymphocytes [65,66]. Effectively a progressive influx of lymphocytes was verified in the late stages. Furthermore, activated lymphocytes at the implantation site could be the source of IL4 and IL13, both known to favor the macrophage fusion on biomaterials [67].

FBR are present within the implantation bed of different biomaterials including collagen-based biomaterials [68–70]. As already described, macrophage fusion is dependent on the presence of the fusogenic molecules on their surface but also on the environmental signals such as the quality and quantity of the adsorbed proteins in the provisional matrix on the biomaterial, on the surface itself and on the topography of the biomaterial surface [48,71]. Few reports elucidate the direct effects of the biomaterial surface on foreign body giant cells formation [72]. It was shown that decrease in monocyte adhesion and foreign body giant cell formation occurs in hydrophilic, anionic and nonionic polyacrylamide/polyacrylic acid surfaces compared to hydrophilic and hydrophobic, cationic surfaces [73]. Besides that, smooth flat surfaces induce considerably more foreign body giant cell formation than rough surfaces [27], while other reports relate that larger PLGA microspheres of 30 mm induce more foreign body giant cell formation than smaller 6 mm microspheres [74]. The implantation of porous materials constituted by high surface-to-volume is prone to show higher ratios of macrophages and foreign body giant cells than smooth-surface implants [75].

Albeit the observed long-standing foreign body response induced by the electrospun PLGA membranes, a fibrous capsule was never seen during all time points studied. Absence of fibrous encapsulation was previously found after the implantation of non-resorbable biomaterials [70] and was also reported in the subcutaneous tissue of rats implanted with PLGA [20,76,77]. Mitragotri and Lahann [75] reported that greater numbers of macrophages and foreign body giant cells on the surface of implants develop more fibrosis and encapsulation of the biomaterials. In addition, Whitaker [78] discussed that high levels of protein adsorption leads to increased cell adhesion and therefore, increased fibrous encapsulation. Considering the aforementioned concepts regarding fibrosis reaction, we can hypothesize that the 3D electrospun porous scaffolds used herein for implantation may have not had high levels of protein adsorption, and may have had decreased cell adhesion, resulting in no fibrous encapsulation. On the other hand, Al-Maawi [79] reported that the accumulation of macrophages and foreign body giant cells on the surface of a biomaterial is correlated with the formation of a highly vascularized granulation tissue and neovessels in the connective tissues. Moreover, Madden [80] reported that, nonporous scaffolds or those with 20 µm pores led to a significant increase in the fibrous capsule thickness, and pores of approximately 30–40 µm reduced the fibrous capsule thickness and the number of M1 macrophages. Several morphological variants of foreign body giant cells are already described but only the LCs and MNGCs participate on the foreign body response to biomaterials. The induction of granulomas around glass implanted in the dorsal subcutaneous tissue of rats allowed to demonstrate that LCs are precursors of MNGCs, and that the epithelioid cells derived from mature macrophages [62]. In view of these arguments, our study suggests that the tissue response induced by the electrospun PLGA membranes could reflect the constant arrival of monocytes to the implantation bed with LCs formation associated to the persistence of fusogenic molecules originated from activated macrophages without fibrous capsule formation.

In the recent years, researchers have become more focused on designing surface-modified PLGA membranes to minimize immune responses. Research studies from Kim et al. [20], Huang et al. [81] and Lee et al. [82], who have implanted PLGA 75:25; 50:50; 75:25, in rats, reported foreign body giant cells formation in intimate contact with PLGA, and decreased inflammatory cells in PLGA-coated membranes in comparison with PLGA alone. In the present study, foreign body giant cells were formed among the inflammatory cells over time. We hypothesized that the degradation products of PLGA 85:15 produced herein induced an inflammatory response in the site of implantation.

While it is difficult to determine the specific role of any individual factor, it is important to highlight that surface topography, surface chemistry and surface energy may influence protein adsorption, platelet activation, cell growth and biocompatibility [83]. In particular, Adabi et al. reported that for some applications, hydrophobic surface could be preferred, since hydrophobic patterns showed more adsorption on intestinal mucus [83]. Since, our future challenges for the use of this membrane as a dressing material aim to accelerate the healing process of oral ulcerations in hamsters' model, we chose a hydrophobic membrane for better adherence on the oral mucus of the animals. Another important issue refers to biocompatibility. Biocompatibility is not only polymer's intrinsic property-dependent, but also biological environment-dependent [4]. For this reason, the intensity and length of specific polymer-tissue interactions can vary greatly in different organs, tissues and species [4,19]. The device used herein was implanted in hamsters to appreciate the immune responses due to our future objective of working with this animal model to develop new future dressing for chemotherapy-induced oral mucositis ulcerations. Indeed, hamster models have been extensively studied to observe pathological effects of radiation exposure or chemotherapy induced oral mucositis and help in the development of effective treatments [84–87]. Furthermore, Syrian golden hamsters seem to be an ideal animal model due to their low cost, small size, easiness to handle and ability to accurately reflect disease progression in humans [88]. Therefore, we used hamsters to verify the biocompatibility of the material. However, there remains a lack of available reagents for studying hamster immune responses which remains an issue to better characterize the type of immune response generated, critical for understanding protection from disease.

It is worthy to mention that, regarding our future challenges of applying the 3D electrospun PLGA (85:15) membrane with the addition of primary fibroblasts as a cell delivery dressing, we performed in vivo tests using a device with a higher surface area to volume which may lead to a higher degradation of the matrix [19].

Moreover, it is also important to mention that the accumulation of PLGA degradative products could cause significant host inflammatory response, a microenvironment favoring tissue fibrosis that is mainly mediated by M1 subtype macrophage [82]. However, considering our in vivo experiments in the present study, the scaffolds became well vascularized and there were no evidence of necrosis or encapsulation, which would have been strong contraindications for future clinical applications [72].

5. Conclusions

The present study appreciated cellular and tissue responses regarding PLGA (85:15) membranes produced by the 3D electrospinning method, for future applications in regenerative medicine. For this goal, both in vitro experiments were performed with primary fibroblast-like cells or MDCK cells that showed proliferation over PLGA electrospun membranes along time. In the in vivo experiment, cell-free irradiated PLGA membranes resulted in a chronic granulomatous inflammatory response in all time points after implantation, mainly constituted of epithelioid cells and LCs. Lymphocytes, myeloperoxidase$^+$ cells, LCs and capillaries decreased after 90 days post implantation. Light microscopy revealed foreign body giant cells showing internalized materials, without fibrous tissue formation. TEM analysis also showed cells exhibiting internalized PLGA fragments decreasing over time. Accordingly, we can conclude that MNGCs constituted a granulomatous reaction around the polymer, which resolved over time, probably preventing a fibrous capsule formation. We expect that the results of this work will boost other studies for translational applications of electrospun fiber membranes as cutaneous or oral dressings.

Author Contributions: Investigation and writing—original draft preparation, A.C.; Investigation and writing—original draft preparation C.M.T.; writing-review and editing M.L.D.; formal analysis, R.P.G.; investigation, T.P.; formal analysis, J.C.; Conceptualization and writing-review and editing, M.F.; writing-review and editing, L.R.d.A.; Conceptualization and writing-review and editing K.A. All authors have read and agreed to the published version of the manuscript.

Funding: The authors thank the Brazilian Agencies Conselho Nacional de Desenvolvimento Científico e Tecnológico (CNPq; Grant No.310917/2014-0, Marcos Lopes Dias.; Grant No.308287/2016-9, Marcos Farina.) PVE fellowship program (Grant No.#406407/2013-4, Karine Anselme.), Fundação de Amparo à Pesquisa do Estado do Rio de Janeiro, FAPERJ (Grant No. E-26/201.304/2014, Marcos Lopes Dias.; Grant No. E-26/203.028/2017, Marcos Farina.) and Coordenaçao de Aperfeiçoamento de Pessoal de Nível Superior (CAPES), for financial support. The authors also thank Centre National de la Recherche Scientifique, CNRS (PICS project Biointerfaces Grant No.#272116, Karine Anselme), for financial support.

Institutional Review Board Statement: The ethical committee for animal use in scientific experiments (Comissão de Ética no Uso de Animais—CEUA) at Health Science Center at Federal University of Rio de Janeiro, Brazil, registered in the National Council of Animal Experimental Control (Conselho Nacional de Controle de Experimentação Animal—CONCEA), process number 01200.001568/2013-

87, certified the use of hamsters in this study (protocol No. 003/15, on 04/15/2015), acquired from Oswaldo Cruz Foundation (Fundação Oswaldo Cruz—Fiocruz).

Data Availability Statement: The raw/processed data required to reproduce these findings can be shared upon request.

Acknowledgments: We would like to thank the electron microscopy and confocal microscopy platforms of Institut de Science des Matériaux de Mulhouse (IS2M), Unidade de Multiusuário Padrón-Lins UNIMICRO/UFRJ for the electron microscopy facilities, the electron microscopy and confocal microscopy platforms of Centro Nacional de Biologia Estrutural e Bioimagem (CENABIO), and INCT-465656/2014-5.

Conflicts of Interest: The authors declare no conflict of interest.

References

1. Gupta, B.; Revagade, N.; Hilborn, J. Poly(Lactic Acid) Fiber: An Overview. *Prog. Polym. Sci.* **2007**, *32*, 455–482. [CrossRef]
2. Zafar, M.; Najeeb, S.; Khurshid, Z.; Vazirzadeh, M.; Zohaib, S.; Najeeb, B.; Sefat, F. Potential of Electrospun Nanofibers for Biomedical and Dental Applications. *Materials* **2016**, *9*, 73. [CrossRef] [PubMed]
3. Hughes, G.A. Nanostructure-Mediated Drug Delivery. In *Nanomedicine in Cancer*; Balogh, L.P., Ed.; Pan Stanford Publishing: Singapore, 2017; pp. 47–72.
4. Elmowafy, E.M.; Tiboni, M.; Soliman, M.E. Biocompatibility, Biodegradation and Biomedical Applications of Poly (Lactic Acid)/Poly (Lactic-Co-Glycolic Acid) Micro and Nanoparticles. *J. Pharm. Investig.* **2019**, *49*, 347–380. [CrossRef]
5. Adusei, K.M.; Ngo, T.B.; Sadtler, K. T Lymphocytes as Critical Mediators in Tissue Regeneration, Fibrosis, and the Foreign Body Response. *Acta Biomater.* **2021**, *133*, 17–33. [CrossRef]
6. Soni, S.S.; Rodell, C.B. Polymeric Materials for Immune Engineering: Molecular Interaction to Biomaterial Design. *Acta Biomater.* **2021**, *133*, 139–152. [CrossRef] [PubMed]
7. Sun, X.; Xu, C.; Wu, G.; Ye, Q.; Wang, C. Poly(Lactic-co-Glycolic Acid): Applications and Future Prospects for Periodontal Tissue Regeneration. *Polymers* **2017**, *9*, 189. [CrossRef] [PubMed]
8. Silva, A.T.C.R.; Cardoso, B.C.O.; e Silva, M.E.S.R.; Freitas, R.F.S.; Sousa, R.G. Synthesis, Characterization, and Study of PLGA Copolymer in Vitro Degradation. *J. Biomater. Nanobiotechnol.* **2015**, *06*, 8–19. [CrossRef]
9. Zafar, M.S.; Khurshid, Z.; Almas, K. Oral Tissue Engineering Progress and Challenges. *Tissue Eng. Regen. Med.* **2015**, *12*, 387–397. [CrossRef]
10. Stratton, S.; Shelke, N.B.; Hoshino, K.; Rudraiah, S.; Kumbar, S.G. Bioactive Polymeric Scaffolds for Tissue Engineering. *Bioact. Mater.* **2016**, *1*, 93–108. [CrossRef]
11. Lobo, A.O.; Afewerki, S.; de Paula, M.M.M.; Ghannadian, P.; Marciano, F.R.; Zhang, Y.S.; Webster, T.J.; Khademhosseini, A. Electrospun Nanofiber Blend with Improved Mechanical and Biological Performance. *Int. J. Nanomed.* **2018**, *13*, 7891–7903. [CrossRef]
12. Wang, Y.; Liu, Y.; Zhang, X.; Liu, N.; Yu, X.; Gao, M.; Wang, W.; Wu, T. Engineering Electrospun Nanofibers for the Treatment of Oral Diseases. *Front. Chem.* **2021**, *9*, 797523. [CrossRef] [PubMed]
13. Dias, M.L.; Dip, R.M.; Souza, D.H.; Nascimento, J.P.; Santos, A.P.; Furtado, C.A. Electrospun Nanofibers of Poly (Lactic Acid)/Graphene Nanocomposites. *J. Nanosci. Nanotechnol.* **2017**, *17*, 2531–2540. [CrossRef] [PubMed]
14. Bye, F.J.; Wang, L.; Bullock, A.; Blackwood, K.A.; Ryan, A.; MacNeil, S. Postproduction Processing of Electrospun Fibres for Tissue Engineering. *J. Vis. Exp.* **2012**, *66*, e4172. [CrossRef] [PubMed]
15. Chor, A.; Gonçalves, R.; Costa, A.; Farina, M.; Ponche, A.; Sirelli, L.; Schrodj, G.; Gree, S.; Andrade, L.; Anselme, K.; et al. In Vitro Degradation of Electrospun Poly(Lactic-Co-Glycolic Acid) (PLGA) for Oral Mucosa Regeneration. *Polymers* **2020**, *12*, 1853. [CrossRef]
16. Liang, Y.; Liang, Y.; Zhang, H.; Guo, B. Antibacterial Biomaterials for Skin Wound Dressing. *Asian J. Pharm. Sci.* **2022**, *17*, 353–384. [CrossRef]
17. Wells, E.K.; Iii, O.Y.; Lifton, R.P.; Cantley, L.G.; Caplan, M.J. Epithelial Morphogenesis of MDCK Cells in Three-Dimensional Collagen Culture Is Modulated by Interleukin-8. *Am. J. Physiol. Physiol.* **2013**, *304*, C966–C975. [CrossRef]
18. Chor, A.; Skeff, M.A.; Takiya, C.; Gonçalves, R.; Dias, M.; Farina, M.; Andrade, L.R.; Coelho, V.D.M. Emerging Approaches of Wound Healing in Experimental Models of High-Grade Oral Mucositis Induced by Anticancer Therapy. *Oncotarget* **2021**, *12*, 2283–2299. [CrossRef]
19. Makadia, H.K.; Siegel, S.J. Poly Lactic-Co-Glycolic Acid (PLGA) As Biodegradable Controlled Drug Delivery Carrier. *Polymers* **2011**, *3*, 1377–1397. [CrossRef]
20. Kim, M.S.; Ahn, H.H.; Na Shin, Y.; Cho, M.H.; Khang, G.; Lee, H.B. An In Vivo Study of the Host Tissue Response to Subcutaneous Implantation of PLGA- and/or Porcine Small Intestinal Submucosa-Based Scaffolds. *Biomaterials* **2007**, *28*, 5137–5143. [CrossRef]
21. Ramot, Y.; Touitou, D.; Levin, G.; Ickowicz, D.E.; Zada, M.H.; Abbas, R.; Domb, A.; Nyska, A. Interspecies Differences in Reaction to a Biodegradable Subcutaneous Tissue Filler: Severe Inflammatory Granulomatous Reaction in the Sinclair Minipig. *Toxicol. Pathol.* **2015**, *43*, 267–271.

22. Ramot, Y.; Haim-Zada, M.; Domb, A.J.; Nyska, A. Biocompatibility and Safety of PLA and Its Copolymers. *Adv. Drug Deliv. Rev.* **2016**, *107*, 153–162. [CrossRef] [PubMed]
23. Zhang, L.; Cao, Z.; Bai, T.; Carr, L.R.; Ella-Menye, J.-R.; Irvin, C.; Ratner, B.D.; Jiang, S. Zwitterionic Hydrogels Implanted in Mice Resist the Foreign-Body Reaction. *Nat. Biotechnol.* **2013**, *31*, 553–556. [CrossRef]
24. Milleret, V.; Simona, B.; Neuenschwander, P.; Hall, H. Tuning Electrospinning Parameters for Production of 3D-Fiber-Fleeces with Increased Porosity for Soft Tissue Engineering Applications. *Eur. Cells Mater.* **2011**, *21*, 286–303. [CrossRef] [PubMed]
25. Anderson, J.M.; Rodriguez, A.; Chang, D.T. Foreign Body Reaction to Biomaterials. *Semin. Immunol.* **2008**, *20*, 86–100. [CrossRef] [PubMed]
26. Anderson, J.M. In Vitro and In Vivo Monocyte, Macrophage, Foreign Body Giant Cell, and Lymphocyte Interactions with Biomaterials. In *Biological Interactions on Materials Surfaces*; Springer: New York, NY, USA, 2009; Volume 1, pp. 225–244.
27. Mcnally, A.K.; Anderson, J.M. Macrophage Fusion and Multinucleated Giant Cells of Inflammation. In *Cell Fusion in Health and Disease*; Dittmar, T., Zänker, K.S., Eds.; Springer: Dordrecht, The Netherlands, 2011; Volume 713.
28. Rediguieri, C.F.; Sassonia, R.C.; Dua, K.; Kikuchi, I.S.; Pinto, T.D.J.A. Impact of Sterilization Methods on Electrospun Scaffolds for Tissue Engineering. *Eur. Polym. J.* **2016**, *82*, 181–195. [CrossRef]
29. Schindelin, J.; Arganda-Carreras, I.; Frise, E.; Kaynig, V.; Longair, M.; Pietzsch, T.; Preibisch, S.; Rueden, C.; Saalfeld, S.; Schmid, B.; et al. Fiji: An open-source platform for biological-image analysis. *Nat. Methods* **2012**, *9*, 676–682. [CrossRef] [PubMed]
30. Kumbar, S.G.; Nukavarapu, S.P.; James, R.; Nair, L.S.; Laurencin, C.T. Electrospun Poly(Lactic Acid-Co-Glycolic Acid) Scaffolds For Skin Tissue Engineering. *Biomaterials* **2008**, *29*, 4100–4107. [CrossRef]
31. Lee, J.H.; Park, J.H.; El-Fiqi, A.; Kim, J.H.; Yun, Y.R.; Jang, J.H.; Han, C.M.; Lee, E.J.; Kim, H.W. Biointerface Control of Electrospun Fiber Scaffolds for Bone Regeneration: Engineered Protein Link to Mineralized Surface. *Acta Biomater.* **2014**, *10*, 2750–2761. [CrossRef]
32. Li, L.; Zhou, G.; Wang, Y.; Yang, G.; Ding, S.; Zhou, S. Controlled Dual Delivery Of BMP-2 and Dexamethasone by Nanoparticle-Embedded Electrospun Nanofibers for the Efficient Repair of Critical-Sized Rat Calvarial Defect. *Biomaterials* **2015**, *37*, 218–229. [CrossRef]
33. Zhang, X.; Shi, X.; Gautrot, J.E.; Peijs, T. Nanoengineered Electrospun Fibers and Their Biomedical Applications: A Review. *Nanocomposites* **2020**, *7*, 1–34. [CrossRef]
34. Chou, S.-F.; Carson, D.; Woodrow, K.A. Current Strategies for Sustaining Drug Release from Electrospun Nanofibers. *J. Control. Release* **2015**, *220*, 584–591. [CrossRef] [PubMed]
35. O'Brien, L.E.; Zegers, M.M.; Mostov, K.E. Building Epithelial Architecture: Insights from Three-Dimensional Culture Models. *Nat. Rev. Mol. Cell Biol.* **2002**, *3*, 531–537. [CrossRef] [PubMed]
36. Yu, P.; Duan, Z.; Liu, S.; Pachon, I.; Ma, J.; Hemstreet, G.P.; Zhang, Y. Drug-Induced Nephrotoxicity Assessment in 3D Cellular Models. *Micromachines* **2021**, *13*, 3. [CrossRef] [PubMed]
37. Bhaskar, P.; Bosworth, L.A.; Wong, R.; O'brien, M.A.; Kriel, H.; Smit, E.; McGrouther, D.A.; Wong, J.K.; Cartmell, S.H. Cell Response to Sterilized Electrospun Poly (ε-Caprolactone) Scaffolds to Aid Tendon Regeneration In Vivo. *J. Biomed. Mater. Res. Part A* **2017**, *105*, 389–397. [CrossRef]
38. Hakki, S.S.; Korkusuz, P.; Purali, N.; Bozkurt, B.; Kuş, M.; Duran, I. Attachment, Proliferation and Collagen Type I Mrna Expression of Human Gingival Fibroblasts on Different Biodegradable Membranes. *Connect. Tissue Res.* **2013**, *54*, 260–266. [CrossRef]
39. Campos, D.M.; Gritsch, K.; Salles, V.; Attik, G.N.; Grosgogeat, B. Surface Entrapment of Fibronectin on Electrospun PLGA Scaffolds for Periodontal Tissue Engineering. *BioRes. Open Access* **2014**, *3*, 117–126. [CrossRef]
40. Sadeghi, A.R.; Nokhasteh, S.; Molavi, A.M.; Khorsand-Ghayeni, M.; Naderi-Meshkin, H.; Mahdizadeh, A. Surface Modification of Electrospun PLGA Scaffold with Collagen for Bioengineered Skin Substitutes. *Mater. Sci. Eng. C* **2016**, *66*, 130–137. [CrossRef]
41. Sadeghi-Avalshahr, A.R.; Khorsand-Ghayeni, M.; Nokhasteh, S.; Molavi, A.M.; Naderi-Meshkin, H. Synthesis and Characterization of PLGA/Collagen Composite Scaffolds as Skin Substitute Produced by Electrospinning Through Two Different Approaches. *J. Mater. Sci. Mater. Med.* **2017**, *28*, 1–10. [CrossRef] [PubMed]
42. Helling, A.L.; Viswanathan, P.; Cheliotis, K.S.; Mobasseri, S.A.; Yang, Y.; El Haj, A.J.; Watt, F.M. Dynamic Culture Substrates That Mimic the Topography of the Epidermal–Dermal Junction. *Tissue Eng. Part A* **2019**, *25*, 214–223. [CrossRef]
43. Inanç, B.; Arslan, Y.E.; Seker, S.; Elçin, A.E.; Elçin, Y.M. Periodontal Ligament Cellular Structures Engineered With Electrospun Poly(DL-Lactide-Co-Glycolide) Nanofibrous Membrane Scaffolds. *J. Biomed. Mater. Res. Part A* **2009**, *90*, 186–195. [CrossRef]
44. Lowery, J.L.; Datta, N.; Rutledge, G.C. Effect of Fiber Diameter, Pore Size and Seeding Method on Growth of Human Dermal Fibroblasts in Electrospun Poly (ε-Caprolactone) Fibrous Mats. *Biomaterials* **2010**, *31*, 491–504. [CrossRef] [PubMed]
45. Anderson, J.M. Biological Responses to Materials. *Annu. Rev. Mater. Sci.* **2001**, *31*, 81–110. [CrossRef]
46. Ratner, B.D. Reducing Capsule Thickness and Enhancing Angiogenesis around Implant Drug Release Systems. *J. Control. Release* **2001**, *78*, 211–218. [CrossRef]
47. Balabiyev, A.; Podolnikova, N.P.; Kilbourne, J.A.; Baluch, D.P.; Lowry, D.; Zare, A.; Ros, R.; Flick, M.J.; Ugarova, T.P. Fibrin Polymer on the Surface of Biomaterial Implants Drives the Foreign Body Reaction. *Biomaterials* **2021**, *277*, 121087. [CrossRef] [PubMed]
48. Ward, W.K. A Review of the Foreign-body Response to Subcutaneously-implanted Devices: The Role of Macrophages and Cytokines in Biofouling and Fibrosis. *J. Diabetes Sci. Technol.* **2008**, *2*, 768–777. [CrossRef]

49. Anderson, J.M.; Cramer, S. Perspectives on the Inflammatory, Healing, and Foreign Body Responses to Biomaterials and Medical Devices. In *Host Response to Biomaterials in Chapter 2*; Academic Press: Cambridge, MA, USA, 2015; pp. 13–36, ISBN 9780128001967. [CrossRef]
50. Böstman, O.; Hirvensalo, E.; Vainionpää, S.; Mäkelä, A.; Vihtonen, K.; Törmälä, P.; Rokkanen, P. Ankle Fractures Treated Using Biodegradable Internal Fixation. *Clin. Orthop. Relat. Res.* **1989**, *138*, 195–203. [CrossRef]
51. Hirvensalo, E. Fracture Fixation with Biodegradable Rods Forty-One Cases of Severe Ankle Fractures. *Acta Orthop. Scand.* **1989**, *60*, 601–606. [CrossRef]
52. Bostman, O.; Hirvensalo, E.; Makinen, J.; Rokkanen, P. Foreign-Body Reactions to Fracture Fixation Implants of Biodegradable Synthetic Polymers. *J. Bone Jt. Surgery. Br. Vol.* **1990**, *72-B*, 592–596. [CrossRef]
53. Poigenfürst, J.; Leixnering, M.; Ben Mokhtar, M. Lokalkomplikationen Nach Implantation von Biorod. *Aktuelle Traumatol.* **1990**, *20*, 157–159.
54. Bostman, O. Osteolytic Changes Accompanying Degradation of Absorbable Fracture Fixation Implants. *J. Bone Jt. Surgery. Br. Vol.* **1991**, *73-B*, 679–682. [CrossRef]
55. Chambers, T.J. Multinucleate Giant Cells. *J. Pathol.* **1978**, *126*, 125–148. [CrossRef] [PubMed]
56. Helming, L.; Gordon, S. The Molecular Basis of Macrophage Fusion. *Immunobiology* **2008**, *212*, 785–793. [CrossRef]
57. Xia, Z.D.; Zhu, T.B.; Du, J.Y.; Zheng, Q.X.; Wang, L.; Li, S.P.; Chang, C.Y.; Fang, S.Y. Macrophages in Degradation of Collagen/Hydroxylapatite (CHA), Beta-Tricalcium Phosphate Ceramics (TCP) Artificial Bone Graft: An In Vivo Study. *Chin. Med. J.* **1994**, *107*, 845–849.
58. Christenson, E.M.; Anderson, J.M.; Hiltner, A. Oxidative Mechanisms of Poly (Carbonate Urethane) and Poly (Ether Urethane) Biodegradation: In Vivo and In Vitro correlations. *J. Biomed. Mater. Res. Part A Off. J. Soc. Biomater. Jpn. Soc. Biomater. Aust. Soc. Biomater. Korean Soc. Biomater.* **2004**, *70*, 245–255. [CrossRef]
59. Santerre, J.; Woodhouse, K.; Laroche, G.; Labow, R. Understanding the Biodegradation of Polyurethanes: From Classical Implants to Tissue Engineering Materials. *Biomaterials* **2005**, *26*, 7457–7470. [CrossRef]
60. Kim, Y.K.; Chen, E.Y.; Liu, W.F. Biomolecular Strategies to Modulate the Macrophage Response to Implanted Materials. *J. Mater. Chem. B* **2015**, *4*, 1600–1609. [CrossRef]
61. McNally, A.K.; Jones, J.A.; MacEwan, S.; Colton, E.; Anderson, J.M. Vitronectin is a Critical Protein Adhesion Substrate For IL-4-Induced Foreign Body Giant Cell Formation. *J. Biomed. Mater. Res. Part A* **2007**, *86*, 535–543. [CrossRef]
62. Van der Rhee, H.J.; Winter, C.P.M.V.D.B.-D.; Daems, W.T. The Differentiation of Monocytes into Macrophages, Epithelioid Cells, And Multinucleated Giant Cells in Subcutaneous Granulomas. *Cell Tissue Res.* **1979**, *197*, 355–378. [CrossRef]
63. Franz, S.; Rammelt, S.; Scharnweber, D.; Simon, J.C. Immune Responses to Implants—A Review of The Implications for the Design of Immunomodulatory Biomaterials. *Biomaterials* **2011**, *32*, 6692–6709. [CrossRef]
64. Gentile, P.; Chiono, V.; Carmagnola, I.; Hatton, P.V. An Overview of Poly(lactic-co-glycolic) Acid (PLGA)-Based Biomaterials for Bone Tissue Engineering. *Int. J. Mol. Sci.* **2014**, *15*, 3640–3659. [CrossRef]
65. Yamashiro, S.; Kamohara, H.; Wang, J.M.; Yang, D.; Gong, W.H.; Yoshimura, T. Phenotypic and Functional Change of Cytokine-Activated Neutrophils: Inflammatory Neutrophils Are Heterogeneous and Enhance Adaptive Immune Responses. *J. Leukoc. Biol.* **2001**, *69*, 698–704. [CrossRef]
66. Gilroy, D.W. The Endogenous Control of Acute Inflammation–from Onset to Resolution. *Drug Discov. Today: Ther. Strateg.* **2004**, *1*, 313–319. [CrossRef]
67. Brodbeck, W.G.; MacEwan, M.; Colton, E.; Meyerson, H.; Anderson, J.M. Lymphocytes and the Foreign Body Response: Lymphocyte Enhancement of Macrophage Adhesion and Fusion. *J. Biomed. Mater. Res. Part A* **2005**, *74A*, 222–229. [CrossRef]
68. Barbeck, M.; Udeabor, S.; Lorenz, J.; Schlee, M.; Holthaus, M.G.; Raetscho, N.; Choukroun, J.; Sader, R.; Kirkpatrick, C.J.; Ghanaati, S. High-Temperature Sintering of Xenogeneic Bone Substitutes Leads to Increased Multinucleated Giant Cell Formation: In Vivo and Preliminary Clinical Results. *J. Oral Implant.* **2015**, *41*, e212–e222. [CrossRef]
69. Ghanaati, S.; Schlee, M.; Webber, M.J.; Willershausen, I.; Barbeck, M.; Balic, E.; Görlach, C.; Stupp, S.I.; Sader, R.A.; Kirkpatrick, C.J. Evaluation of the Tissue Reaction to a New Bilayered Collagen Matrix In Vivo And Its Translation to the Clinic. *Biomed. Mater.* **2011**, *6*, 015010. [CrossRef]
70. Ghanaati, S.; Barbeck, M.; Detsch, R.; Deisinger, U.; Hilbig, U.; Rausch, V.; Sader, R.; Unger, R.E.; Ziegler, G.; Kirkpatrick, C.J. The Chemical Composition of Synthetic Bone Substitutes Influences Tissue Reactions In Vivo: Histological and Histomorphometrical Analysis of the Cellular Inflammatory Response to Hydroxyapatite, Beta-Tricalcium Phosphate and Biphasic Calcium Phosphate Ceramics. *Biomed. Mater.* **2012**, *7*, 015005.
71. Klopfleisch, R.; Jung, F. The Pathology of the Foreign Body Reaction against Biomaterials. *J. Biomed. Mater. Res. Part A* **2016**, *105*, 927–940. [CrossRef]
72. Blackwood, K.A.; McKean, R.; Canton, I.; Freeman, C.O.; Franklin, K.L.; Cole, D.; Brook, I.; Farthing, P.; Rimmer, S.; Haycock, J.W.; et al. Development of Biodegradable Electrospun Scaffolds for Dermal Replacement. *Biomaterials* **2008**, *29*, 3091–3104. [CrossRef]
73. Anderson, J.M. Exploiting the Inflammatory Response on Biomaterials Research and Development. *J. Mater. Sci. Mater. Electron.* **2015**, *26*, 1–2. [CrossRef]
74. Zandstra, J.; Hiemstra, C.; Petersen, A.; Zuidema, J.; van Beuge, M.; Rodriguez, S.; Lathuile, A.; Veldhuis, G.; Steendam, R.; Bank, R.; et al. Microsphere Size Influences the Foreign Body Reaction. *Eur. Cells Mater.* **2014**, *28*, 335–347. [CrossRef]
75. Mitragotri, S.; Lahann, J. Physical Approaches to Biomaterial Design. *Nat. Mater.* **2009**, *8*, 15–23. [CrossRef]

76. Lu, L.; Peter, S.J.; Lyman, M.D.; Lai, H.-L.; Leite, S.M.; Tamada, J.A.; Uyama, S.; Vacanti, J.P.; Langer, R.; Mikos, A.G. In Vitro and In Vivo Degradation of Porous Poly(Dl-Lactic-Co-Glycolic Acid) Foams. *Biomaterials* **2000**, *21*, 1837–1845. [CrossRef]
77. Kaushiva, A.; Turzhitsky, V.M.; Backman, V.; Ameer, G.A. A Biodegradable Vascularizing Membrane: A Feasibility Study. *Acta Biomater.* **2007**, *3*, 631–642. [CrossRef]
78. Whitaker, R.; Hernaez-Estrada, B.; Hernandez, R.M.; Santos-Vizcaino, E.; Spiller, K.L. Immunomodulatory Biomaterials for Tissue Repair. *Chem. Rev.* **2021**, *121*, 11305–11335. [CrossRef]
79. Al-Maawi, S.; Orlowska, A.; Sader, R.; Kirkpatrick, C.J.; Ghanaati, S. In Vivo Cellular Reactions to Different Biomaterials-Physiological and Pathological Aspects and Their Consequences. *Semin. Immunol.* **2017**, *29*, 49–61. [CrossRef]
80. Madden, L.R.; Mortisen, D.J.; Sussman, E.M.; Dupras, S.K.; Fugate, J.A.; Cuy, J.L.; Hauch, K.D.; Laflamme, M.A.; Murry, C.E.; Ratner, B.D. Proangiogenic Scaffolds as Functional Templates for Cardiac Tissue Engineering. *Proc. Natl. Acad. Sci. USA* **2010**, *107*, 15211–15216. [CrossRef]
81. Huang, J.; Zhou, X.; Shen, Y.; Li, H.; Zhou, G.; Zhang, W.; Zhang, Y.; Liu, W. Asiaticoside Loading into Polylactic-Co-Glycolic Acid Electrospun Nanofibers Attenuates Host Inflammatory Response and Promotes M2 Macrophage Polarization. *J. Biomed. Mater. Res. Part A* **2019**, *108*, 69–80. [CrossRef]
82. Lee, Y.; Kwon, J.; Khang, G.; Lee, D. Reduction of Inflammatory Responses and Enhancement of Extracellular Matrix Formation by Vanillin-Incorporated Poly (Lactic-Co-Glycolic Acid) Scaffolds. *Tissue Eng. Part A* **2012**, *18*, 1967–1978. [CrossRef]
83. Adabi, M.; Naghibzadeh, M.; Adabi, M.; Zarrinfard, M.A.; Esnaashari, S.S.; Seifalian, A.M.; Faridi-Majidi, R.; Aiyelabegan, H.T.; Ghanbari, H. Biocompatibility and Nanostructured Materials: Applications in Nanomedicine. *Artif. Cells Nanomed. Biotechnol.* **2017**, *45*, 833–842. [CrossRef]
84. Sonis, S.T.; Tracey, C.; Shklar, G.; Jenson, J.; Florine, D. An Animal Model for Mucositis Induced by Cancer Chemotherapy. *Oral Surg. Oral Med. Oral Pathol.* **1990**, *69*, 437–443. [CrossRef]
85. Tanideh, N.; Tavakoli, P.; Saghiri, M.A.; Garcia-Godoy, F.; Amanat, D.; Tadbir, A.A.; Samani, S.M.; Tamadon, A. Healing Acceleration in Hamsters of Oral Mucositis Induced by 5-Fluorouracil with Topical Calendula Officinalis. *Oral Surg. Oral Med. Oral Pathol. Oral Radiol.* **2013**, *115*, 332–338. [CrossRef]
86. Ribeiro, S.B.; De Araújo, A.A.; Araújo Júnior, R.F.; Brito, G.A.C.; Leitão, R.C.; Barbosa, M.M.; Garcia, V.B.; Medeiros, A.C.; Medeiros, C.A.C.X. Protective Effect of Dexamethasone on 5-FU-Induced Oral Mucositis in Hamsters. *PLoS ONE* **2017**, *12*, E0186511. [CrossRef]
87. Jung, H.; Kim, H.S.; Lee, J.H.; Lee, J.J.; Park, H.S. Wound Healing Promoting Activity of Tonsil-Derived Stem Cells on 5-Fluorouracil-Induced Oral Mucositis Model. *Tissue Eng. Regen. Med.* **2020**, *17*, 105–119. [CrossRef]
88. Warner, B.M.; Safronetz, D.; Kobinger, G.P. Syrian Hamsters as a Small Animal Model for Emerging Infectious Diseases: Advances in Immunologic Methods. In *Emerging and Re-Emerging Viral Infections*; Springer: Cham, Switzerland, 2016; pp. 87–101.

Article

Injectable Cell-Laden Polysaccharide Hydrogels: In Vivo Evaluation of Cartilage Regeneration

Yao Fu [†,‡], Sanne K. Both, Jacqueline R. M. Plass, Pieter J. Dijkstra, Bram Zoetebier and Marcel Karperien *

Department of Developmental BioEngineering, Faculty of Science and Technology, Tech Med Centre, University of Twente, P.O. Box 217, 7500 AE Enschede, The Netherlands
* Correspondence: marcel.karperien@utwente.nl; Tel.: +31-(0)53-4893323
† Present address: Department of Obstetrics, The Second Clinical Medical College, Jinan University (Shenzhen People's Hospital), Shenzhen 518001, China.
‡ Present address: Post-Doctoral Scientific Research Station of Clinical Medicine, Jinan University, Guangzhou 510632, China.

Abstract: Previously, 5% w/v hyaluronic acid-tyramine (HA-TA) and dextran-tyramine (Dex-TA) enzymatically cross-linked hybrid hydrogels were demonstrated to provide a mechanically stable environment, maintain cell viability, and promote cartilaginous-specific matrix deposition in vitro. In this study, 5% w/v hybrid hydrogels were combined with human mesenchymal stem cells (hMSCs), bovine chondrocytes (bCHs), or a combination of both in a 4:1 ratio and subcutaneously implanted in the backs of male and female nude rats to assess the performance of cell-laden hydrogels in tissue formation. Subcutaneous implantation of these biomaterials showed signs of integration of the gels within the host tissue. Histological analysis showed residual fibrotic capsules four weeks after implantation. However, enhanced tissue invasion and some giant cell infiltration were observed in the HA-TA/Dex-TA hydrogels laden with either hMSCs or bCHs but not with the co-culture. Moreover, hMSC-bCH co-cultures showed beneficial interaction with the hydrogels, for instance, in enhanced cell proliferation and matrix deposition. In addition, we provide evidence that host gender has an impact on the performance of bCHs encapsulated in HA-TA/Dex-TA hydrogels. This study revealed that hydrogels laden with different types of cells result in distinct host responses. It can be concluded that 5% w/v hydrogels with a higher concentration of Dex-TA (≥50%) laden with bCH-hMSC co-cultures are adequate for injectable applications and in situ cell delivery in cartilage regeneration approaches.

Keywords: injectable hydrogel; mesenchymal stem cells; chondrocytes; co-cultures; in vivo; subcutaneous implantation; cartilage regeneration

1. Introduction

Articular cartilage injuries may occur as a result of either traumatic mechanical destruction or progressive mechanical degeneration. The combination of lack of blood supply and low mitotic activity of chondrocytes leads to limited ability to repair and regenerate articular cartilage [1–4]. Currently, injectable in-situ-formed hydrogels have emerged as promising cartilage tissue engineering strategies due to the ability to form three-dimensional, highly hydrated scaffolds after injection in aqueous form [5–7].

Injectable materials enable localized and straightforward delivery of cells and biomolecules via minimally invasive procedures without the associated risks of surgical implantation. These materials allow for the ability to fill irregular-shaped defects, avoiding the difficulty of pre-fabricating patient-specific defect shapes [8,9]. Moreover, hydrogels with different physical properties can also be designed and implanted in non-self-healing critical-size defects, temporarily replacing the extracellular matrix and assisting the healing process [10]. Previously, our group developed an injectable hybrid hydrogel composed of tyramine-conjugated hyaluronic acid (HA-TA) and dextran (Dex-TA) [11]. This hydrogel is formed in situ using a biocompatible enzymatic cross-linking reaction and supports survival of chondrocytes (CHs) and mesenchymal stem cells (MSCs) and tissue formation in vitro [12].

Tissue exposure to biomaterials triggers a foreign body response, a non-specific immune response process, which may result in persistent chronic inflammation and fibrotic encapsulation of the material [13–15]. In this inflamed environment, macrophages, lymphocytes, and their granular products contribute to the infiltration of foreign body giant cells (multi-nucleated fused macrophages) into the implantation site and the development of a collagen-rich fibrotic connective tissue layer surrounding the implant [14,16–18]. Degree of host response depends on the extent of homeostasis that is disturbed in the host by the injury, the implantation of the foreign material, and the properties of the material itself. Previously, an implant was considered biocompatible if it was encapsulated by an avascular layer of collagen without affecting its intended performance [18]. However, the impermeable nature of fibrous capsules, in some cases, results in poor mass transport and electrolyte diffusion between cell-laden implants and tissue, which impairs function, safety, and biocompatibility [19–21]. This is particularly relevant when these constructs are used in a tissue regeneration strategy.

In a previous study, we demonstrated that 5% w/v tyramine-conjugated hyaluronic acid and dextran (HA-TA/Dex-TA) hybrid hydrogels laden with bCHs provide a mechanically stable environment, maintain cell viability, and promote a cartilaginous-specific matrix deposition in vitro [12]. In the current study, 5% w/v hybrid hydrogels were combined with human mesenchymal stem cells (hMSCs) and bovine chondrocytes (bCHs) and then subcutaneously implanted in the backs of male and female nude rats for a four-week period. The cell laden hydrogels have a storage modulus (G') of 1.9 kPa for pure Dex-TA, 3.2 kPa for 50/50 HA-TA/Dex-TA, and up to 9.6 kPa for pure HA-TA hydrogels [12]. The main objectives were to assess the response of cell-laden hydrogels with respect to tissue formation, characterize the reaction of neighboring tissues, and investigate the interaction between hybrid hydrogels and co-cultures or mono-cultures. Additionally, we investigated the effect of host gender differences on the outcomes of subcutaneous implantation of HA-TA/Dex-TA hydrogels laden with bCHs and hMSCs in the backs of male and female nude rats.

2. Materials and Methods

2.1. Materials

Dextran (40 kDa, pharmaceutical grade) was purchased from Pharmacosmos, Holbæk, Denmark. Sodium hyaluronate (27 kDa, pharmaceutical grade) was purchased from Contipro Pharma, Dolní Dobrouč, Czech Republic. Tyramine (99%), DMF (anhydrous, 99.8%), LiCl (99.0%), p-nitrophenyl chloroformate (PNC, 96%), pyridine (anhydrous, 99.8%), DMSO-d_6 (99.9%), NaCl ($\geq$99.0%), D_2O (99.9 atom % D), horseradish peroxidase (HRP, 325 U/mg solid), and hydrogen peroxide (30%) were purchased from Sigma-Aldrich, Schnelldorf, Germany. Tyramine·HCl salt (99%) was obtained from Acros Organics, Fair Lawn, NJ, USA. 4-(4,6-Dimethoxy-1,3,5-triazin-2-yl)-4-methyl-morpholinium chloride (DMTMM, 97%) was purchased from Fluorochem Ltd., Hadfield, UK. Ethanol ($\geq$99.9%) and diethyl ether ($\geq$99.7%) were purchased from Merck, Kenilworth, NJ, USA. Milli-Q water was used from the Milli-Q Advantage A10 system (Merck KGaA, Darmstadt, Germany) equipped with a 0.22 µm Millipak® 40 Express filter.

2.2. Cell Culture and Expansion

bCHs were isolated from full-thickness cartilage knee biopsies from ~6-month-old calves according to the previously reported protocol [22]. bCHs were expanded in chondrocyte proliferation medium (Dulbecco's modified Eagle's medium (DMEM; Gibco, Billings, MT, USA) supplemented with 10% fetal bovine serum (FBS; Gibco), 0.2 mM ascorbic acid 2-phosphate (ASAP; Sigma, St. Louis, MO, USA), 0.4 mM proline (Sigma, St. Louis,

MO, USA), 1× non-essential amino acids (Gibco), 100 U/mL penicillin, and 100 µg/mL streptomycin (Invitrogen, Carlsbad, CA, USA)).

Human-bone-marrow–derived MSCs were isolated as previously reported [23] and cultured in MSC proliferation medium (α-MEM (Gibco) supplemented with 10% FBS (Gibco), 1% L-glutamine (Gibco), 0.2 mM ASAP (Sigma), 100 U/mL penicillin, 100 µg/mL streptomycin (Invitrogen, Carlsbad, CA, USA), and 1 ng/mL bFGF)). The use of human material was approved by a local medical ethical committee. The medium was refreshed twice a week, and cells at Passage 3 were used for the experiments.

2.3. Synthesis of Polymers

Dextran-tyramine (Dex-TA) and hyaluronic acid-tyramine (HA-TA) were synthesized as previously reported [12]. Briefly, Dex-TA was synthesized by activation of dextran with PNC and subsequent aminolysis with tyramine adapted from Ramirez et al. [24]. HA-TA was prepared by amidation of the carboxyl groups of HA by tyramine, and the procedure was adapted from Rydergren et al. [25,26]. A detailed description of polymer synthesis can be found in the Supplementary Materials.

2.4. Hydrogel Formation

Hydrogel samples were prepared in a newly designed PTFE mold to produce identical hydrogels of 5 mm in diameter and 2.5 mm in height. After dissolving tyramine-conjugated polymers in sterile phosphate-buffered saline (PBS), polymer solutions were prepared and incubated with horseradish peroxidase (HRP, 40 U/mL) overnight at 4 °C. The mixture was then combined with different types of cells (hMSCs, bCHs, and bCH-hMSC co-cultures) in a concentration of 10 million cells/mL. For co-cultures, hMSCs and bCHs were mixed at a ratio of 80%/20% based on previous observations [27]. Freshly prepared 0.3% hydrogen peroxide (H_2O_2) was added to the mixture and immediately transferred to the mold using a 1 mL pipette after a brief vortex. The final concentrations of the gels were 5 % w/v polymer, 10 million cells/mL, 4 U/mL HRP, and 0.015% H_2O_2. HA-TA and Dex-TA were combined in 3 ratios (100:0, 50:50, and 0:100, represented by group HA, HA/Dex, and Dex, respectively). All conditions are listed in Table 1.

Table 1. Histological scoring of the in vivo sections.

Conditions		Tissue Reaction	Giant Cells	Nuclei Visibility	Neutrophil	Cell Cluster
bCHS	HA	+±	+	++±	±	−
bCHS	HA/Dex	++	+	+	−	+
bCHS	Dex	+	±	++±	−	−
hMSCS	HA/Dex	++	+±	+	±	−
hMSCS	Dex	+	+	++	±	+
bCH/hMSC co-cultures	HA	−	−	+++	−	−
bCH/hMSC co-cultures	HA/Dex	−	±	+	−	−
bCH/hMSC co-cultures	Dex	−	±	++	±	++

Twenty-eight days after subcutaneous implantation in male nude rats (n = 10 rats in each group), sections were assessed via various markers to evaluate tissue response. The presence of the above inflammatory components was scored from absence (−) to profound presence (+++).

2.5. Hydrogel Implantation

After incubation in chondrogenic differentiation medium (DMEM supplemented with 0.2 mM ascorbic acid 2-phosphate (Sigma), 0.4 mM proline (Sigma), 100 U/mL penicillin, 100 µg/mL streptomycin (Invitrogen), 0.1 µM dexamethasone (Sigma), 100 µg/mL sodium pyruvate (Sigma), 50 µg/mL insulin-transferrin-selenite (ITS; Sigma), and 10 ng/mL transforming growth factor β-3 (TGF-β3; R&D Systems)) overnight, the hydrogel samples described above were implanted subcutaneously in the backs of 14-week old nude rats (Crl:NIH-Foxn1rnu) (Figure 1A). Each rat received carprofen (4 mg/kg) as an analgesic before the start of the procedure. Rats were induced with 4% isoflurane and maintained at 1.5–2% during the procedure. Skin was cleaned with 70% ethanol, and 4 subcutaneous pockets were created on each lateral side of the spine on the backs of 10 male rude rats (8 implants in total per rat). In each pocket, 1 hydrogel was inserted. Simultaneously, 2 samples (HA/Dex hydrogels laden with bCHs or hMSCs, respectively) were also implanted subcutaneously in the backs of another 10 female nude rats.

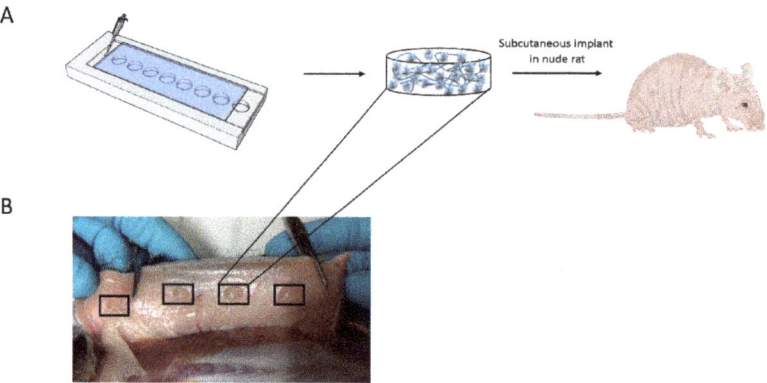

Figure 1. Experimental setup. (**A**) Schematic illustration of the outline of the experiment. Identical cell-laden hydrogels containing different types of cells were formed using a PTFE mold. Next, the prepared hydrogel constructs were implanted subcutaneously in the backs of 10 nude rats for 28 days. (**B**) Macroscopic observation of the different groups 28 days post-implantation. From left to right are representative pictures of groups HA-TA and co-culture, Dex-TA and bCHs, HA-TA/Dex-TA and hMSCs, and HA-TA/Dex-TA and bCHs, respectively. Frames denote the location of the implanted hydrogels.

This experiment was randomized and approved by the local animal experimental committee. After 28 days, implants and respective surrounding tissue were harvested and fixed with 10% formalin and then incubated in cryomatrix (Thermo Scientific) overnight at 4 °C. Samples embedded in cryomatrix were snap-frozen using liquid nitrogen and stored at −80 °C.

2.6. Histology

Cryosections of 10 µm were cut using a cryotome (Thermo Shandon FSE, Thermo Fisher Scientific, Waltham, MA, USA)) and processed for histological evaluation with different staining methods. Hematoxylin-eosin (H&E) staining was performed using a standard protocol. Masson–Goldner Trichrome staining was performed to detect connective tissue and fibrous capsule thickness following the manufacturer's instructions (Merck KGaA, Darmstadt, Germany). Briefly, samples were first placed in Weigert's hematoxylin staining solution for 5 min and washed in tap water for 10 min. Sections were then shortly rinsed in 1% acetic acid, incubated in azophloxine solution for 10 min, then rinsed in 1% acetic acid again, followed by incubation in tungstophosphoric acid orange G solution for 1 min. Sections were again rinsed in 1% acetic acid for 30 s, followed by incubation in light green SF (Merck) for 2 min and another wash in 1% acetic acid. The thickness of the fibrous capsule and the inflammatory cell layer, i.e., a layer of cells mostly consisting of various immune cells, including mast cells, were measured for sections of all conditions. For each section, 4 points on each capsule around the implant were measured (n = 10). Peri-implant fibrotic capsule thickness was defined as the distance between the border of the fibrotic tissue adjacent to the implant and the muscle or fat tissue adjacent to the fibrotic capsule at the other end.

2.7. Naphthol AS-D Chloroacetate Esterase Staining

Naphthol AS-D staining was performed to stain granular cells such as neutrophils and mast cells. Chloroacetate esterase staining was performed following the manufacturer's instructions (Sigma, St. Louis, MO, USA). Briefly, chloroacetate esterase staining solution was first prepared as described by the protocol. Cryosections were then incubated with the staining solution for 20 min at 37 °C in the dark. Sections were subsequently washed in distilled water and mounted with Faramount aqueous mounting medium (Dako, Glostrup, Denmark). All cells containing red granules were regarded as positive.

2.8. Immunohistochemistry Staining

For immunohistochemistry, endogenous peroxidase was blocked by incubating cryosections with 0.3% H_2O_2. After washing with PBS, sections were blocked in 5% bovine serum albumin to prevent non-specific binding. Slides were subsequently incubated overnight at 4 °C with a rabbit polyclonal antibody against COL II (Abcam, Cambridge, UK). Thereafter, sections were incubated with polyclonal goat anti-rabbit HRP-conjugated secondary antibody (Dako) for 30 min at room temperature, followed by development with the DAB Substrate Kit (Abcam). Counterstaining was performed with hematoxylin. Non-immune controls underwent the same procedure without primary antibody incubation.

2.9. Image Analysis and Statistical Analysis

All stained slides were scanned with the NanoZoomer 2.0-RS slide scanner (Hamamatsu, Sendai City, Japan). Stained sections were independently scored by three blinded individuals to assess tissue response. Fibrous capsule thickness and inflammatory cells were evaluated semi-quantitatively using NanoZoomer Digital Pathology software. Data are presented as mean ± standard deviation. Statistical significance was analyzed using one-way analysis of variance (ANOVA) with Tukey's post-hoc analysis. A value of $p < 0.05$ was considered statistically significant.

3. Results

3.1. In Vivo Implantation in Nude Rats

Cell-laden hydrogel constructs were subcutaneously implanted in the backs of nude rats for 28 days to investigate cartilage matrix formation and tissue responses in vivo. The 28-day time point was chosen to focus on the inter-connections between cells and gels as well as co-cultured cells. Macroscopic observation of constructs, in all animals, showed proper integration with host tissue and no signs of edema or toxicity in the tissue surrounding the implants (Figure 1B). All hydrogel samples were clearly visible under the skin and maintained their structural integrity, indicating that they had not yet degraded significantly. In addition, no evident macroscopic inflammation of the tissue at the implantation site was observed.

To investigate the in vivo innate inflammatory response, the explanted samples, as well as the surrounding tissues, were histologically assessed (Figures 2–4). Representative histological images of sections stained with hematoxylin and eosin (H&E) to examine the presence of tissue response are shown in Figure 2. Histological sections were evaluated via light microscopy and scored for tissue reaction, presence of giant cells, nuclei visibility, neutrophils, and cell clusters. The variable degrees of the inflammatory responses are summarized in Table 1. The presence of the above inflammatory responses was scored from absence (−) to profound presence (+++). The stained sections show that the HA hydrogels with encapsulated bCHs do not display a solid gel, but a more porous structure, which could be caused by the degradation of the gel. Mixing in Dex-TA progressively decreased the structure's porosity. Remarkably, it was not present in the HA hydrogels laden with a mixture of hMSCs and bCHs.

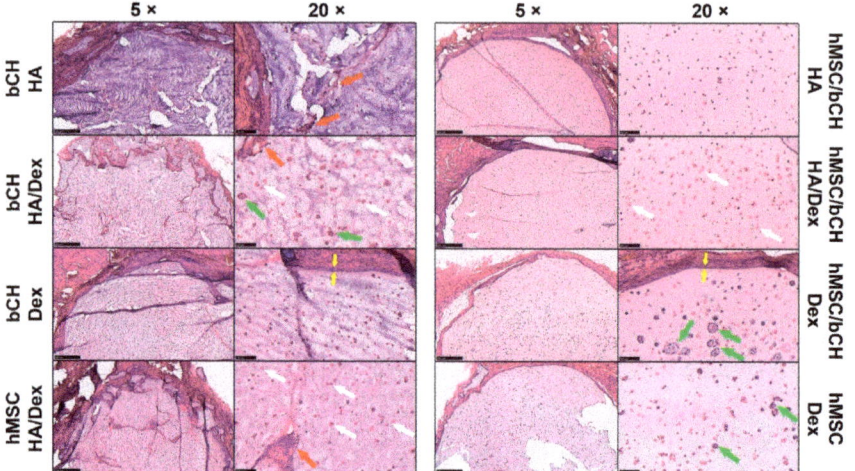

Figure 2. Histological analysis of in vivo response after 28 days of subcutaneous implantation. Representative histological sections of hydrogel samples stained with H&E. The left panel in each condition shows 5× magnification pictures (scale bars represent 500 μm), whereas the right panel shows 20× (scale bars represent 100 μm). The red arrows in the stained sections show giant cells, the green arrows show cell clusters, and the white arrows show the cells without visible nuclei, respectively. Yellow arrows at the material/tissue interface indicate a layer of inflammatory cells.

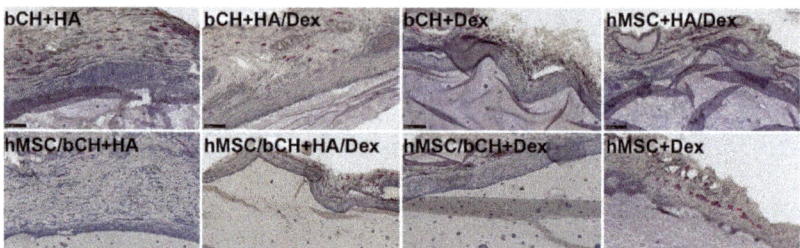

Figure 3. Neutrophil staining after 28 days of subcutaneous implantation. Representative figures of hydrogel samples with cells positively stained with chloroacetate esterase histochemistry. Positively stained cells displayed red granulation. Scale bars represent 100 μm.

A new interface consisting of inflammatory cells (e.g., macrophages) was generated between the hydrogels and host tissues after implantation, but no acute inflammatory reaction was observed in all types of samples. Stained figures show that most of the explanted constructs displayed smooth edges at the material–tissue interface. Hydrogel samples containing only bCHs or hMSCs showed a more abundant chronic tissue reaction, especially in the cell-laden HA-TA/Dex-TA constructs. Enhanced tissue invasion and some giant cell infiltration were observed in the HA-TA/Dex-TA hydrogels either laden with bCHs or hMSCs. However, tissue reaction was barely noticeable in co-culture-laden hydrogels irrespective of the hydrogel composition.

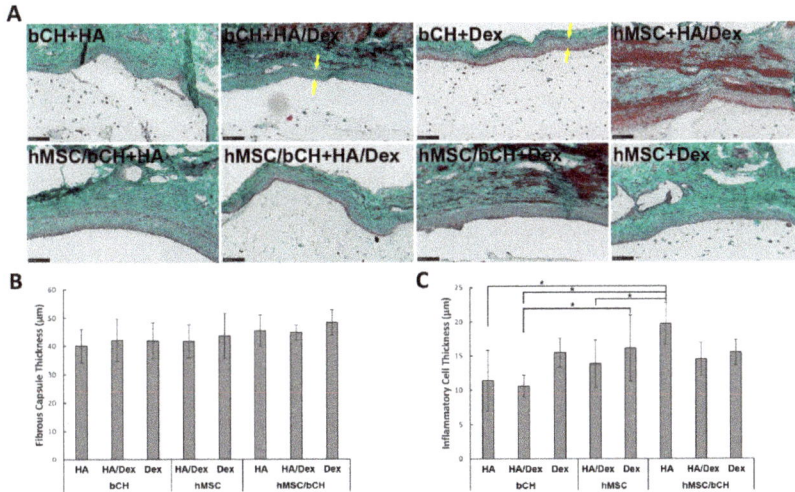

Figure 4. Histological characteristics of implant fibrous capsule after 28 days of subcutaneous implantation. (A) Representative Masson–Goldner trichrome staining of all hydrogels. All images show the interface between the host tissues (top of images) and the implant. As a result of the staining protocol, nuclei will be stained in dark brown to black, cytoplasm and muscles will appear brick red, the connective tissue will appear green, and erythrocytes will be bright orange. Yellow arrows in the stained sections indicate the fibrous capsule at the material/host interface. The thickness of the fibrous capsule (B) and the inflammatory cell layer (C), shown in Figure 2 with yellow arrows, were measured for sections of all conditions. Data are presented as mean with standard deviation as error bars for $n = 10$ biological replicates per hydrogel condition. Scale bars represent 100 μm. * represents $p < 0.05$.

Regarding cell morphology, similar results were found in most of the implanted samples. Encapsulated cells showed a round morphology and homogeneous distribution throughout the constructs. Meanwhile, proliferating cells were found in the conditions of HA/Dex with bCHs, Dex with hMSCs, and Dex with co-cultures, which were present as newly generated cell clusters. In particular, in the Dex-encapsulated co-cultures, enhanced newly formed cell clusters were observed. Of interest, cellularity analysis showed that in HA/Dex construct nuclei visibility significantly decreased compared to other conditions. Most of the cells in the HA/Dex samples were stained pink, so no nuclei were visible, regardless of the cell type encapsulated.

Next, we checked for the presence of neutrophils surrounding the implantation site. Neutrophil granulocytes represent the abundant cell type in peripheral blood and normally arrive as the first immune cells at an implant site and disappear in the course of the following days. Chloroacetate esterase histochemistry is a well-known method to detect granular cells such as neutrophils and mast cells. Positively stained granular cells, which were mostly identified as mast cells, were only sporadically present in the gel-tissue boundary and seldom into the implant. These results demonstrate the absence of significant inflammation processes (Figure 3).

Trichromatic Masson–Goldner staining is most suitable to depict the structure of connective tissues and cells and to assess the fibrous capsule formation by collagen deposition. Representative Masson–Goldner trichrome staining of the cell-laden hydrogel samples at 28 days post-implantation is shown in Figure 4A. This staining revealed that all samples were surrounded by a fibrous capsule. Additionally, a semi-quantitative assessment of capsule thickness and inflammatory cells at the implant surface is shown in Figure 4B,C. The average thickness of the peri-implant capsule in all conditions ranged from 40 μm to 48 μm, and no significant differences were observed between the different hydrogels (Figure 4B). A layer of inflammatory cells was present at the material–tissue interface (Figure 2). The layer thickness was highest around HA hydrogels with co-cultures compared to others (Figure 4C).

3.2. Co-Culture-Laden Hydrogels Present Positively Deposition of Cartilage Matrix

Immunohistochemistry was performed to investigate the expression of proteins that indicate cartilage matrix production. Collagen type II, which is the primary type of collagen present in articular cartilage, was chosen in this study. We rarely observed the expression of COL II in the constructs laden with hMSCs after subcutaneous implantation (Figure 5). We did observe some pericellular staining for COL II within the hydrogels loaded with bCHs. Positively stained protein expression was more apparent in co-culture-laden hydrogels. Notably, intense deposition of COL II was observed in Dex-TA hydrogels laden with co-cultures. Moreover, the histochemical analysis also revealed that the cartilage matrix formation was more dominant at the periphery of the hydrogels. These results indicate that the co-culture system and Dex-TA hydrogel could effectively promote the appropriate interactions or stimulations for chondrogenesis, leading to the facilitated secretion of chondrogenic extracellular matrix and cartilaginous tissue formation during in vivo subcutaneous implantation.

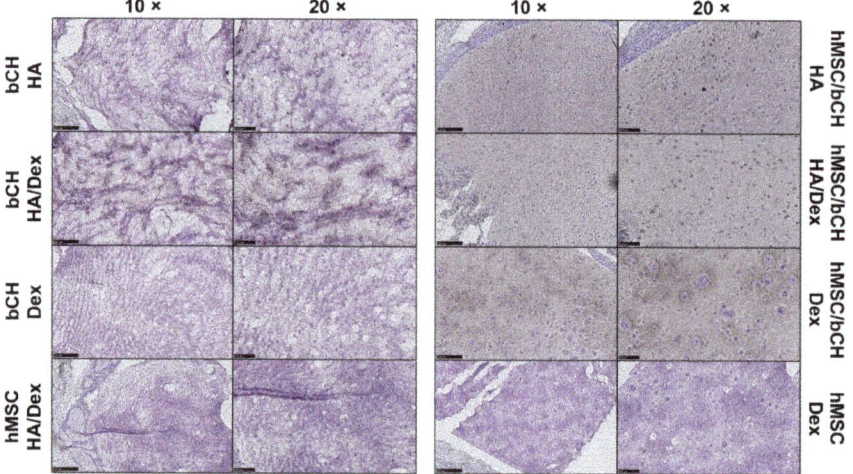

Figure 5. Matrix deposition after 28 days of subcutaneous implantation. Representative immunohistochemical staining of the hydrogel samples for COL II. Positive protein expression stained in dark brown. The left panel in each condition shows 10× magnification pictures (scale bars represent 250 µm), whereas the right panel shows 20× (scale bars represent 100 µm).

3.3. Gender Difference Shows Impact on the Performance of Chondrocyte Laden Hydrogels

To investigate the impact of gender differences on the outcomes, we implanted HA/Dex hydrogels laden with bCHs or hMSCs, respectively, subcutaneously in the backs of another 10 female nude rats. The histological analysis results are summarized in Table 2. Surprisingly, we found gender differences in the performance of implanted hydrogels; however, this impact was mainly observed for hydrogels laden with bCHs but not with hMSCs. Irrespective of gender, all the samples formed a clear thin fibrous capsule surrounding the hydrogels, while no sign of acute inflammatory response was observed. In the hMSC-laden samples, most of the assessed markers were present similarly in samples from both males and females. However, as shown in Figure 6, the outcomes from bCH-laden hydrogels clearly show gender differences. Compared to explant samples from female rats, the in vivo environment in male rats increased the size of encapsulated bCHs, depressed the nuclei visibility of encapsulated bCHs, and promoted the formation of cell clusters.

Table 2. Histological analysis of the in vivo sections from different genders.

Samples	Gender	Tissue Reaction	Giant Cells	Nuclei Visibility	Cell Size (μm)	Collagen Capsule	Capsule Thickness (μm)	Neutrophil	Cell Cluster
bCHS +	M	++	+	+	17.8	+	40.7	−	+
HA/Dex	F	+	+	++±	12.2	+	44.8	−	−
hMSCs +	M	++	+±	+	17.7	+	42.5	±	−
HA/Dex	F	+±	+	+	17.4	+	45.2	±	−

Twenty-eight days after subcutaneous implantation in nude rats of different genders ($n = 10$ rats in each group), sections were assessed via various markers to evaluate the impact of gender difference on the performance of the cell-laden hydrogels. M represents male, and F represents female.

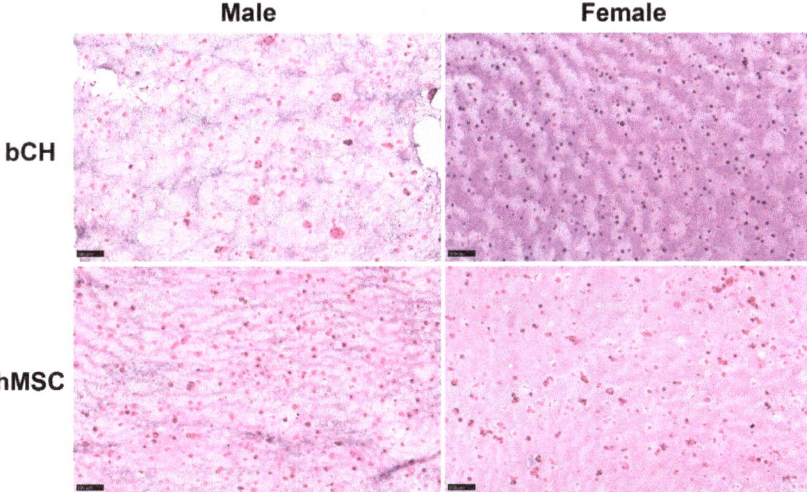

Figure 6. Histological analysis comparing implants laden with bCHs or hMSCs in either male (left) or female (right) nude rats. 5% w/v HA/Dex hydrogels laden with hMSCs or bCHs were implanted subcutaneously into the backs of both male and female rude rats. Representative histological sections of hydrogel samples stained with H&E. Scale bars represent 100 μm.

4. Discussion

In previous research, we developed and characterized an injectable hydrogel system to be applied in cartilage tissue regeneration [12]. In such hydrogels, cells can be homogeneously distributed and in vitro evaluation showed high biocompatibility. We utilized different compositions of hyaluronic acid and dextran hybrid hydrogels and demonstrated that 5% w/v hydrogels showed, after preoperative incubation in chondrogenic differentiation medium, enhanced deposition of cartilage matrix. Cartilage matrix deposition significantly increases mechanical properties. In this work, the safety and biocompatibility of these hybrid hydrogels laden with bCHs and/or hMSCs were tested by subcutaneous implantation in the backs of nude rats for four weeks. This study demonstrated that, as a major result, the hydrogels present limited immune response and formation of a small fibrotic capsule surrounding the material. As a second major aspect, bCHs-hMSCs cocultures show beneficial interaction with the biomaterials, for instance, in enhanced cell proliferation and matrix deposition. Finally, this research revealed that hydrogels with these types of cells resulted in distinct tissue responses, which indicated the possibility of personalized regeneration approaches based on the situation of individual patients.

Of interest, gender appeared to influence the performance of bCH-laden hydrogels after subcutaneous implantation.

The utilization of injectable hydrogels in cartilage regeneration is considered a promising strategy due to the minimally invasive procedure [8,28]. In situ gelation enables the formed hydrogel to quickly set its volume, adapt to the shape of the defect, and establish an efficient integration with the host tissue [9,29]. Moreover, this system can deliver cells and/or bioactive agents of interest in a non-harsh way and keep them at the implantation site [30]. However, implantation of biomaterials triggers a series of host responses at the injury site, including material/tissue interactions, acute and chronic inflammation, granulation tissue development, foreign body reaction, and fibrous capsule development [17,18]. Accordingly, this body response to the foreign material compromises the in vivo functionality and durability of the implanted material [14]. To evaluate the response of a host to hte first contact with this biomaterial, 5% w/v hybrid hydrogels laden with bCHs, hMSCs, or co-cultures thereof were subcutaneously implanted in the backs of nude rats, and histological analysis was conducted to assess the inflammatory processes associated with the implantation, as well as the integration with the host tissue.

After four weeks, all implanted samples showed proper integration with host tissue and no signs of granulation tissue development or toxicity in the tissue surrounding the implants. Histological analysis revealed that all hydrogel conditions elicited no acute inflammatory response. H&E staining of histological sections also revealed a homogeneous distribution of the cells within the matrix, with the cells exhibiting the common round-shaped phenotype characteristic. While no signs of acute inflammation response could be detected, a thin (40–48 μm) fibrotic capsule, as an indication of chronic inflammation, was observed around all the implants. Mostly mast cells, but not neutrophils, were sporadically present in the gel-tissue boundary and seldom into the implant. However, no polymorphonuclear and mono-nuclear cells were visible, and no significant differences were observed between the different hydrogel conditions. These results demonstrate the absence of significant inflammation processes. The in vivo performance of these hydrogels, along with previous data, suggests that HA, Dex, and hybrid hydrogels are suitable for injectable applications in tissue regeneration approaches with good in vivo safety and biocompatibility.

However, it should be noticed that some differences in foreign body reaction to the hydrogel matrix were observed after four weeks. For instance, hydrogels laden with either bCHs or hMSCs induced some tissue reaction, particularly in the hydrogels of HA-TA only. Moreover, the foreign body giant cells, a typical feature of the foreign body reaction, were only observed in somewhat larger numbers in mono-culture cell-laden hydrogels prepared from HA-TA. In this case, the encapsulation of hMSCs did not elicit the formation of foreign body giant cells, as previously reported [31,32].

Nevertheless, Dex-TA hydrogels laden with either bCHs or hMSCs showed moderate tissue reaction in the matrix and less giant cell formation. Moreover, hydrogels with HA-TA showed a porous structure when laden with either bCHs or hMSCs. This can be explained by the degradation of HA in vivo. HA is a major component of the cartilage extracellular matrix and exhibits rapid degradation behaviors in vivo due to its high water-absorbing properties and enzymatic degradation [33]. The space left after HA degradation promotes the surrounding tissue invasion and may explain the infiltration of foreign body giant cells. Moreover, HA-TA/Dex-TA constructs show significantly depressed nuclei visibility compared to other conditions. Most of the cells in HA-TA/Dex-TA gels were stained pink, while nuclei were barely visible, regardless of the cell type encapsulated. We hypothesize that the in vivo environment in the male rats probably increased the size of encapsulated cells in HA/Dex constructs, which impacts the sectioning and in turn affects the staining results.

Interestingly, all these performances were attenuated in the co-culture environment. No tissue reaction and giant cell formation were observed in hydrogels laden with both bCHs and hMSCs. Previous studies demonstrated the beneficial interactions between the cells in bCH-hMSC co-cultures [34]. The in vivo performance indicated that this beneficial interaction also affects the performance of hydrogels compared to hydrogels laden with either hMSCs or bCHs. The underlying mechanism and potential role of this interplay between cells and biomaterials need further investigation.

Moreover, enhanced newly formed cell clusters were observed in Dex-TA-encapsulated bCH-hMSC co-cultures, which indicated that Dex-TA hydrogels provide the environment to support the beneficial interactions in bCH-hMSC co-cultures. Moreover, only hydrogels laden with co-cultures present deposition of cartilage matrix. Notably, coherent with the in vitro study, intense deposition of type II collagen was observed in pure Dex-TA hydrogels with co-cultures. In conclusion, together with results from previous studies, HA-TA/Dex-TA hydrogels with a high concentration of Dex-TA ($\geq 50\%$) provide the opportunity to create optimal biomaterials for cartilage tissue regeneration, while bCH-hMSC co-cultures stimulate interaction with these hydrogels. These data suggest that further in situ study is needed for the development of a fully functional cartilage tissue-engineered construct that can be applied clinically. It should be noted though that the subcutaneous implantation site may have influenced the outcome of cartilage matrix production. It seems feasible that orthotopic implantation may facilitate cartilage matrix production over the subcutaneous implantation site. Additionally, different combinations of hydrogel and cells showed distinct differences in mechanical properties, degradation, and chondro-induction features. These properties are important considerations in the design of precision biomaterials to enable the survival, differentiation, and transplantation of biomaterial-cell-based combination approaches. With the growing interest in personalized therapeutic approaches [35,36], combination therapies have vital potential for their ability to sense and respond to the therapeutic needs of individual patients. The different outcomes of the in vivo performance in this work highlight the potential application of personalized regeneration based on the situation of individual patients.

Animal models are essential to assess the value of current and future tissue engineering therapies, which play a critical role in many domains of study in medicine and biology [37,38]. Multiple factors need to be considered in selecting an appropriate animal model, such as animal size, age, gender, economic cost, ethical concerns, and potential for clinic transition [38,39]. In this work, we investigated the impact of gender differences on our injectable hydrogels for cartilage tissue engineering. The histological assessment indicated that the gender of the host has an effect on the performance of the implanted hydrogels. However, the impact is mainly on the hydrogels laden with bCHs, not with hMSCs. Due to the space limitation in this study, we only chose cell-laden HA/Dex hydrogels. Further studies need to proceed on the animals with normal immune systems and other hybrid hydrogels to check if there is any outcome change.

In this work, the evaluation of the in vivo response upon subcutaneous implantation of hyaluronic acid and dextran hybrid hydrogels was conducted in nude rats, revealing proper integration with the surrounding tissues and the presence of a residual fibrotic capsule. Moreover, the in vivo performance revealed the interaction of bCH-hMSC co-cultures with biomaterials, suggesting their further study towards the development of functional cartilage tissue-engineered constructs for personalized application. Taken together, the results from this work, along with previous data, show that 5% w/v Dex-TA hydrogels laden with bCH-hMSC co-cultures provide an adequate support matrix for chondro-induction between hMSC and bCH co-cultures, stimulating cartilage matrix formation.

Supplementary Materials: The following supporting information can be downloaded at: https://www.mdpi.com/article/10.3390/polymers1933438/s1, Detailed synthetic protocols for Dex-TA and HA-TA.

Author Contributions: Conceptualization, S.K.B., B.Z. and M.K.; methodology, Y.F., J.R.M.P. and BZ; formal analysis, Y.F.; investigation, Y.F., J.R.M.P. and B.Z.; resources, J.R.M.P.; data curation, Y.F. and B.Z.; writing—original draft preparation, Y.F.; writing—review and editing, S.K.B., B.Z. and

M.K.; visualization, Y.F.; supervision, S.K.B.; P.J.D. and B.Z.; project administration, J.R.M.P.; funding acquisition, M.K. All authors have read and agreed to the published version of the manuscript.

Funding: This research was funded by the Netherlands Organisation for Scientific Research (NWO) P15-23 (Project 1) "Activating resident stem cells" and the China Scholarship Council.

Institutional Review Board Statement: The animal study protocol was approved by the Institutional Review Board (or Ethics Committee) of the University of Twente (Protocol 4185-1-1 of Project AVD1100020174185, approved 10 April 2018).

Informed Consent Statement: Not applicable.

Data Availability Statement: Not applicable.

Conflicts of Interest: The authors declare no conflict of interest.

References

1. Hunter, W. Of the structure and disease of articulating cartilages. 1743. *Clin. Orthop. Relat. Res.* **1995**, *317*, 3–6.
2. Mankin, H.J. The response of articular cartilage to mechanical injury. *J. Bone Jt. Surg. Am.* **1982**, *64*, 460–466. [CrossRef]
3. Newman, A.P. Articular cartilage repair. *Am. J. Sports Med.* **1998**, *26*, 309–324. [CrossRef]
4. Lin, W.; Klein, J. Recent Progress in Cartilage Lubrication. *Adv. Mater.* **2021**, *33*, 2005513. [CrossRef] [PubMed]
5. Ekenseair, A.K.; Kasper, F.K.; Mikos, A.G. Perspectives on the interface of drug delivery and tissue engineering. *Adv. Drug Deliv. Rev.* **2013**, *65*, 89–92. [CrossRef] [PubMed]
6. Klouda, L.; Mikos, A.G. Thermoresponsive hydrogels in biomedical applications. *Eur. J. Pharm. Biopharm.* **2008**, *68*, 34–45. [CrossRef] [PubMed]
7. Tan, H.P.; Marra, K.G. Injectable, Biodegradable Hydrogels for Tissue Engineering Applications. *Materials* **2010**, *3*, 1746–1767. [CrossRef]
8. Jeong, B.; Kim, S.W.; Bae, Y.H. Thermosensitive sol-gel reversible hydrogels. *Adv. Drug Deliv. Rev.* **2012**, *64*, 154–162. [CrossRef]
9. Van Tomme, S.R.; Storm, G.; Hennink, W.E. In situ gelling hydrogels for pharmaceutical and biomedical applications. *Int. J. Pharm.* **2008**, *355*, 1–18. [CrossRef]
10. Neffe, A.T.; Pierce, B.F.; Tronci, G.; Ma, N.; Pittermann, E.; Gebauer, T.; Frank, O.; Schossig, M.; Xu, X.; Willie, B.M.; et al. One Step Creation of Multifunctional 3D Architectured Hydrogels Inducing Bone Regeneration. *Adv. Mater.* **2015**, *27*, 1738–1744. [CrossRef]
11. Wennink, J.W.H.; Niederer, K.; Bochynska, A.I.; Moreira Teixeira, L.S.; Karperien, M.; Feijen, J.; Dijkstra, P.J. Injectable Hydrogels by Enzymatic Co-Crosslinking of Dextran and Hyaluronic Acid Tyramine Conjugates. *Adv. Polym. Med.* **2011**, *309–310*, 213–221. [CrossRef]
12. Fu, Y.; Zoetebier, B.; Both, S.; Dijkstra, P.J.; Karperien, M. Engineering of Optimized Hydrogel Formulations for Cartilage Repair. *Polymers* **2021**, *13*, 1526. [CrossRef] [PubMed]
13. Babensee, J.E.; Anderson, J.M.; McIntire, L.V.; Mikos, A.G. Host response to tissue engineered devices. *Adv. Drug Deliv. Rev.* **1998**, *33*, 111–139. [CrossRef]
14. Morais, J.M.; Papadimitrakopoulos, F.; Burgess, D.J. Biomaterials/Tissue Interactions: Possible Solutions to Overcome Foreign Body Response. *AAPS J.* **2010**, *12*, 188–196. [CrossRef]
15. Sheikh, Z.; Brooks, P.J.; Barzilay, O.; Fine, N.; Glogauer, M. Macrophages, Foreign Body Giant Cells and Their Response to Implantable Biomaterials. *Materials* **2015**, *8*, 5671–5701. [CrossRef]
16. Anderson, J.; McNally, A. Biocompatibility of implants: Lymphocyte/macrophage interactions. *Semin. Immunopathol.* **2011**, *33*, 221–233. [CrossRef] [PubMed]
17. Anderson, J.M.; Rodriguez, A.; Chang, D.T. Foreign body reaction to biomaterials. *Semin. Immunol.* **2008**, *20*, 86–100. [CrossRef] [PubMed]
18. Luttikhuizen, D.T.; Harmsen, M.C.; Van Luyn, M.J.A. Cellular and molecular dynamics in the foreign body reaction. *Tissue Eng.* **2006**, *12*, 1955–1970. [CrossRef]
19. Langer, R. Perspectives and Challenges in Tissue Engineering and Regenerative Medicine. *Adv. Mater.* **2009**, *21*, 3235–3236. [CrossRef] [PubMed]
20. Ratner, B.D. Reducing capsular thickness and enhancing angiogenesis around implant drug release systems. *J. Controll. Release* **2002**, *78*, 211–218. [CrossRef]
21. Zhang, L.; Cao, Z.Q.; Bai, T.; Carr, L.; Ella-Menye, J.R.; Irvin, C.; Ratner, B.D.; Jiang, S.Y. Zwitterionic hydrogels implanted in mice resist the foreign-body reaction. *Nat. Biotechnol.* **2013**, *31*, 553–556. [CrossRef] [PubMed]
22. Hendriks, J.; Riesle, J.; Van Blitterswijk, C.A. Effect of stratified culture compared to confluent culture in monolayer on proliferation and differentiation of human articular chondrocytes. *Tissue Eng.* **2006**, *12*, 2397–2405. [CrossRef] [PubMed]
23. Fernandes, H.; Dechering, K.; Van Someren, E.; Steeghs, I.; Apotheker, M.; Leusink, A.; Bank, R.; Janeczek, K.; Van Blitterswijk, C.; de Boer, J. The Role of Collagen Crosslinking in Differentiation of Human Mesenchymal Stem Cells and MC3T3-E1 Cells. *Tissue Eng. Part A* **2009**, *15*, 3857–3867. [CrossRef]

24. Ramirez, J.C.; Sanchezchaves, M.; Arranz, F. Dextran Functionalized by 4-Nitrophenyl Carbonate Groups-Aminolysis Reactions. *Angew. Makromol. Chem.* **1995**, *225*, 123–130. [CrossRef]
25. D'Este, M.; Eglin, D.; Alini, M. A systematic analysis of DMTMM vs EDC/NHS for ligation of amines to Hyaluronan in water. *Carbohydr. Polym.* **2014**, *108*, 239–246. [CrossRef] [PubMed]
26. Rydergren, S. *Chemical Modifications of Hyaluronan Using DMTMM-Activated Amidation, in Synthetical Organic Chemistry*; Uppsala University: Uppsala, Sweden, 2013; Available online: https://www.diva-portal.org/smash/get/diva2:640661/FULLTEXT02.pdf (accessed on 14 April 2022).
27. Wu, L.; Prins, H.-J.; Helder, M.N.; van Blitterswijk, C.A.; Karperien, M. Trophic Effects of Mesenchymal Stem Cells in Chondrocyte Co-Cultures are Independent of Culture Conditions and Cell Sources. *Tissue Eng. Part A* **2012**, *18*, 1542–1551. [CrossRef]
28. Zoetebier, B.; Schmitz, T.C.; Ito, K.; Karperien, M.; Tryfonidou, M.A.; Paez, J.I. Injectable Hydrogels for Articular Cartilage and Nucleus Pulposus Repair: Status Quo and Prospects. *Tissue Eng. Part A* **2022**, *28*, 478–499. [CrossRef]
29. Yu, L.; Ding, J. Injectable Hydrogels as Unique Biomedical Materials. *Chem. Soc. Rev.* **2008**, *37*, 1473–1481. [CrossRef] [PubMed]
30. Hoffman, A.S. Hydrogels for biomedical applications. *Adv. Drug Deliv. Rev.* **2012**, *64*, 18–23. [CrossRef]
31. Aggarwal, S.; Pittenger, M.F. Human mesenchymal stem cells modulate allogeneic immune cell responses. *Blood* **2005**, *105*, 1815–1822. [CrossRef] [PubMed]
32. Bartosh, T.J.; Ylostalo, J.H.; Mohammadipoor, A.; Bazhanov, N.; Coble, K.; Claypool, K.; Lee, R.H.; Choi, H.; Prockop, D.J. Aggregation of human mesenchymal stromal cells (MSCs) into 3D spheroids enhances their antiinflammatory properties. *Proc. Natl. Acad. Sci. USA* **2010**, *107*, 13724–13729. [CrossRef] [PubMed]
33. Burdick, J.A.; Prestwich, G.D. Hyaluronic Acid Hydrogels for Biomedical Applications. *Adv. Mater.* **2011**, *23*, H41–H56. [CrossRef] [PubMed]
34. Fu, Y.; Karbaat, L.; Wu, L.; Leijten, J.; Both, S.K.; Karperien, M. Trophic Effects of Mesenchymal Stem Cells in Tissue Regeneration. *Tissue Eng. Part B-Rev.* **2017**, *23*, 515–528. [CrossRef]
35. Aguado, B.A.; Grim, J.C.; Rosales, A.M.; Watson-Capps, J.J.; Anseth, K.S. Engineering precision biomaterials for personalized medicine. *Sci. Transl. Med.* **2018**, *10*, eaam8645. [CrossRef] [PubMed]
36. Oliva, N.; Unterman, S.; Zhang, Y.; Conde, J.; Song, H.S.; Artzi, N. Personalizing Biomaterials for Precision Nanomedicine Considering the Local Tissue Microenvironment. *Adv. Healthc. Mater.* **2015**, *4*, 1584–1599. [CrossRef]
37. Ahern, B.J.; Parvizi, J.; Boston, R.; Schaer, T.P. Preclinical animal models in single site cartilage defect testing: A systematic review. *Osteoarthr. Cartilage* **2009**, *17*, 705–713. [CrossRef]
38. Moran, C.J.; Ramesh, A.; Brama, P.A.; O'Byrne, J.M.; O'Byrne, F.J.; Levingstone, T.J. The benefits and limitations of animal models for translational research in cartilage repair. *J. Exp. Orthop.* **2016**, *3*, 1. [CrossRef]
39. Mukherjee, P.; Roy, S.; Ghosh, D.; Nandi, S.K. Role of animal models in biomedical research: A review. *Lab. Anim. Res.* **2022**, *38*, 18. [CrossRef] [PubMed]

Article

Bone Healing in Rat Segmental Femur Defects with Graphene-PCL-Coated Borate-Based Bioactive Glass Scaffolds

Ozgur Basal [1,*], Ozlem Ozmen [2] and Aylin M. Deliormanlı [3]

1. Department of Orthopedics and Traumatology, Emsey International Hospital, Pendik, İstanbul 34912, Turkey
2. Department of Pathology, Faculty of Veterinary Medicine, Burdur Mehmet Akif Ersoy University, Istiklal Yerleskesi, Burdur 15030, Turkey
3. Department of Metallurgical and Materials Engineering, Manisa Celal Bayar University, Yunusemre, Manisa 45140, Turkey
* Correspondence: basalozgur@gmail.com; Tel.: +90-54-1304-8338; Fax: +90-262-656-4346

Abstract: Bone is a continually regenerating tissue with the ability to heal after fractures, though healing significant damage requires intensive surgical treatment. In this study, borate-based 13-93B3 bioactive glass scaffolds were prepared though polymer foam replication and coated with a graphene-containing poly (ε-caprolactone) (PCL) layer to support bone repair and regeneration. The effects of graphene concentration (1, 3, 5, 10 wt%) on the healing of rat segmental femur defects were investigated in vivo using male Sprague–Dawley rats. Radiographic imaging, histopathological and immuno-histochemical (bone morphogenetic protein (BMP-2), smooth muscle actin (SMA), and alkaline phosphatase (ALP)) examinations were performed 4 and 8 weeks after implantation. Results showed that after 8 weeks, both cartilage and bone formation were observed in all animal groups. Bone growth was significant starting from the 1 wt% graphene-coated bioactive glass-implanted group, and the highest amount of bone formation was seen in the group containing 10 wt% graphene ($p < 0.001$). Additionally, the presence of graphene nanoplatelets enhanced BMP-2, SMA and ALP levels compared to bare bioactive glass scaffolds. It was concluded that pristine graphene-coated bioactive glass scaffolds improve bone formation in rat femur defects.

Keywords: bioactive glass; graphene; osteogenesis; rat femur defect; in vivo; bone healing

1. Introduction

The treatment of long bone defects resulting from trauma, tumor resections, congenital malformations and osteomyelitis are the most demanding bone pathologies. In cases of failure even after many serial surgeries, the extremity may need to be amputated [1]. To create optimal biological healing conditions in the shortest period, there is a tremendous need for porous biocompatible materials that can provide new bone regeneration and mechanical strength [2–6].

Bioactive glasses (BGs) are emerging materials that can be used to regenerate and heal both bone and skin [3,7]. BGs exhibit osteoconductive, osteoinductive, and osteostimulative characteristics, and are degradable in physiological fluids [8,9]. Some examples of osteoinductive growth factors that promote new bone formation and regeneration are bone morphogenetic proteins (BMP), vascular endothelial growth factor (VEGF), epidermal growth factor (EGF), and transforming growth factor (TGF) [10–14]. Unfortunately, synthesis is costly and growth factors' extensive utility is limited by a low stability in scaffolds. Instead, osteoinductive nanomaterials like graphene nanoplatelets should be incorporated into the architecture of synthetic scaffolds.

One essential element that can enhance osteogenesis is boron, by inducing the osteogenic differentiation-marker gene synthesis during the proliferation and differentiation phase of BMSCs [7,8]. According to an in vitro study conducted by Gu et al. [9], borate-based 13-93B3 bioactive glasses showed better bone-healing performance in rat calvarial defect models compared to silicate-based 13-93 bioactive glass scaffolds [9]. Besides the advantages of boron-based bioactive glass scaffolds in bone regeneration, they have been reported to feature poor mechanical properties compared to silicate-based glasses [10]. Therefore, some biopolymers, such as poly (caprolactone) and poly (L-lactide-co-glycolide), and some inorganic nanomaterials, such as graphene, have been added to the bioactive glass structure to improve its mechanical properties [11–13].

In the past, graphene-containing PCL-coated 13-93B3 bioactive glass scaffolds were prepared by Türk and Deliormanlı [14]. An in vitro cytotoxicity analysis (XTT) showed that the graphene-containing scaffolds were not cytotoxic to pre-osteoblast MC3T3-E1 cells, and cell viability rates were higher compared to a control group after 7 days of incubation [14].

It was previously reported that graphene-containing, grid-like silicate-based bioactive glass-based, and graphene-containing PCL-coated composite scaffolds fabricated by robocasting had no detrimental effect on bone marrow mesenchymal stem cells in vitro [15]. Stem cells implanted onto these composite scaffolds fixed well to the surface and proliferated efficiently. In the absence of transforming growth factors, cells implanted on the scaffolds showed osteogenic differentiation [15].

Wang et al. found in their study that mesoporous bioactive glass-graphene oxide scaffolds had better cytocompatibility and osteogenesis differentiation with rat bone marrow mesenchymal stem cells than bare mesoporous bioactive glass scaffolds [16]. Furthermore, these bioactive scaffolds stimulated vascular ingrowth and promoted bone repair at the lesion site in rat cranial defect models. Similarly, Gao et al. [17] adopted graphene in combination with 58S bioactive glasses for bone tissue engineering using scaffolds fabricated by selective laser sintering. Human MG63 cells adhered to and proliferated on the surface of graphene-containing scaffolds, according to cell culture studies [18].

The potential of graphene derivatives, including some functional groups such as graphene oxide and reduced graphene oxide, for stem cell differentiation into osteogenic, chondrogenic, adipogenic, and neurogenic types has been investigated previously [19]. Functional groups of graphene equivalents are responsible for hydrophobic and electrostatic interactions with proteins which promote osteogenic differentiation. Recently, graphene-containing porous and oriented PCL scaffolds have been used in the regeneration of large osteochondral defects [20], enhancing fibrous, chondroid and osseous tissue regeneration. Additionally, the expressions of bone morphogenetic protein-2, collagen-1, vascular endothelial growth factor and alkaline phosphatase expressions were more prominent in PCL implanted groups in the presence of graphene [20]. However, despite the data available, the contribution of borate-based 13-93B3 scaffolds containing pristine graphene nanoplatelets to bone healing in vivo has not been reported yet.

The hypothesis of this study is that graphene-containing PCL-coated 13-93B3 porous scaffolds are effective bioactive composites in the treatment of segmental bone defects, and are good alternatives to autologous bone grafts. In this study, the effects of borate-based, 13-93B3 scaffolds coated with PCL containing pristine graphene nanopowders at different concentrations on bone healing in segmental bone defect model are investigated.

2. Experimental Studies
2.1. Materials

In the study, melt-derived borate-based 13-93B3 bioactive glass powders (5.5 Na_2O, 11.1 K_2O, 4.6 MgO, 18.5 CaO, 3.7 P_2O_5, 56.6 B_2O_3 wt%) having a median diameter of ~10 μm were used. PCL (Mw 80,000 g/mol) was purchased from Sigma-Aldrich, Darmstadt, Germany. Graphene nanopowders in platelet form were received from Graphene Laboratories Inc. Calverton, NY, USA. The platelets were 60 nanometers thick with particle sizes ranging from 3 to 7 μm, according to the manufacturer.

2.2. Scaffold Preparation

A polymer foam replication technique was used to fabricate porous bioactive glass-based scaffolds [12]. In the presence of 4 wt% ethyl cellulose, a suspension containing borate-based bioactive glass particles (40 vol%) and ethanol was prepared. Scaffolds were then made using cylindirical polyurethane foams (8.5 mm in diameter) as a template. The foams' surfaces were covered using a dip-coating procedure that involved immersion in a bioactive glass suspension. The coated foams were then dried at 25 °C for 48 h before being subjected to a controlled heat treatment at 1 °C/min to degrade the polymeric foam up to 450 °C. They were then sintered for 1 h at 570 °C with a heating rate of 5 °C/min in an air atmosphere [11].

A poly (caprolactone) solution in di-chloromethane was stirred at 25 °C for 4 h at 5 wt%. The graphene nanopowders (at 1, 3, 5, or 10% wt%) were then added to the PCL solution and mixed for 1 h with a magnetic stirrer before homogenization for 15 min with a Bandelin Sonopuls (Bandalin, Berlin, Germany) ultrasonic probe [14]. A graphene-containing polycaprolactone solution was used to coat porous bioactive glass scaffolds using the dip-coating method. Graphene and PCL-coated scaffolds were then dried at room temperature for at least 48 h before characterization. The prepared 0, 1, 3, 5 and 10 wt% graphene nanoparticle-containing PCL-coated composite scaffolds were designated as BG, 1G-P-BG, 3G-P-BG, 5G-P-BG and 10G-P-BG, respectively.

The microstructure of these manufactured scaffolds was examined using an optical and scanning electron microscope (SEM, Zeiss, Gemini 500, Oberkochen, Germany), and a Fourier transform infrared spectrometer with an attenuated total reflectance module (FTIR-ATR, Thermo Scientific, Nicolet, IS20, Waltham, MA, USA) in the range of 525 cm^{-1} to 4000 cm^{-1}. An XRD analysis of the bare borate-based bioactive glass scaffolds was made using a Panalytical Empyrean diffractometer (Malvern Panalytical BV, Brighton, UK) with Cu Kα radiation (λ = 1.5406 Å) in the range 10° to 90° (2θ). The scaffolds' porosity was determined using the Archimedes method. Prior to animal implantation, the scaffolds were soaked in an ethanol bed overnight and then sterilized under UV light for 2 h.

2.3. In Vivo Experiments

All investigations were carried out in line with the Ministry of Health of Turkey, the Declaration of Helsinki, and the National Institutes of Health (NIH) of the United States' Guide for the Care and Use of Laboratory Animals. The Experimental Animal Center and Ethics Committee of Burdur Mehmet Akif Ersoy University (MAKU) approved all experimental procedures in this study (Ethics number: 531, 17 July 2019).

2.3.1. Rat Segmental Defect Model

Male Sprague–Dawley rats, from twenty to twenty-six weeks of age, 260–380 g, were purchased from the Experimental Animal Production and Experimental Research Center of Burdur Mehmet Akif Ersoy University, Turkey. Five groups of 5 specimens each were formed. A right femur segmental defect model was then created for all animals. Rats were fed and watered on a regular basis, and housed in individual cages in controlled rooms with 12-h light/dark cycles at 22 °C and 50% humidity. Prophylactic antibiotic (4 mg/kg gentamicin im.) was administered to all rats approximately 60 min before surgery. 100 mg/kg Alfamine 10% (Alfasan, Woerden, Holland) and 10 mg/kg Xlazinbio 2% (Bioveta, Czech Republic) were injected intraperitoneally to allow for a 30-min anesthetic period. Under anesthesia, the lateral thigh was shaved and a local antiseptic was applied. Animals were positioned on their left side and a 3 cm incision was made using a lateral longitudinal approach from mid-shaft to the distal femur. The incision was extended using a lateral parapatellar arthrotomy approach and a medial dislocation of the patella to gain access to the femoral notch, where a blunt dissection was made to the vastus lateralis muscle. After the right femur shaft was exposed, a 5 mm wide bone cut was performed with a mini saw and a bone segment removed. A prepared scaffold with a hole of 20 Ga in the middle was placed in the femur defect. The graft material was fixed with a 20 Ga- 4 cm retrograde intramedullary pin which was placed from intercondylar notch of the distal femur (Figure 1).

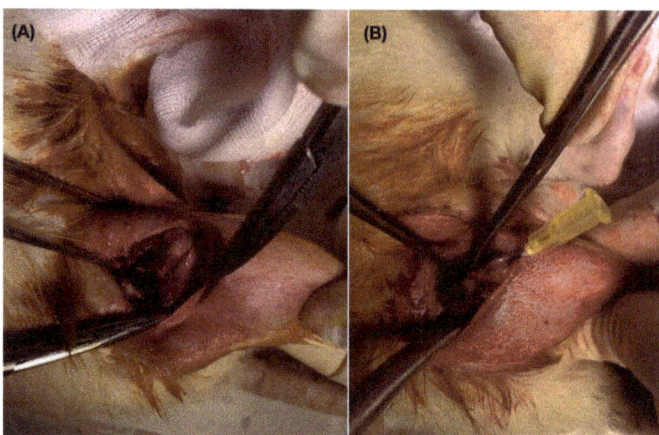

Figure 1. (**A**) 5 mm large femoral defect filled with a porous scaffold. (**B**) Retrograde intramedullary pin fixation.

Rats implanted with bioactive glass-based composite scaffolds with codes BG, 1G-P-BG, 3G-P-BG, 5G-P-BG and 10G-P-BG, were designated as Control Group, Group 1, Group 2, Group 3 and Group 4, respectively. Exclusion criterion is the occurrence of surgical site infection. Two rats from each group were sacrificed at random four weeks after surgery, and three other rats were sacrificed eight weeks after surgery.

2.3.2. Histopathological Method

All rats were anesthetized and euthanized at the end of the study. Bone samples were fixed in 10% neutral-buffering formalin for histological and immuno-histochemical studies during the necropsy. After a 2-day fixation period, bone samples were decalcified for two weeks using a 0.1 M ethylenediaminetetraacetic acid (EDTA) solution, tissues were processed and embedded in paraffin using an automatic tissue processor (Leica ASP300S, Wetzlar, Germany). Using a rotary microtome, five-micron serial sections were obtained from the paraffin blocks (Leica RM 2155; Leica Microsystems, Wetzlar, Germany). A light microscope was used to analyze one slice of each rat stained with hematoxylin and eosin (HE) and one section stained with the Picro Sirius Red method for collagen by a ready to use kit (ab150681, Abcam, Cambridge, UK).

For new bone formation (NFB; mm^2) and the residual material area, histomorphometric variables were calculated (RMA; mm^2). At 400× magnification, osteoblasts and osteoclasts were detected in a 1.23 mm^2 area [21]. The entire faulty area was investigated in two dimensions and calculations were carried out in five distinct parts of each segment. A statistical analysis on the mean values of each group's results was performed. An expert pathologist from a different institution who was unaware of the study design analyzed histopathological changes blindly.

2.3.3. Immunohistochemistry Method

BMP-2 (against BMP-2 antibody (655229.111) (ab6285)), smooth muscle actin (Anti-alpha smooth muscle Actin antibody ((4A4); ab119952)), and ALP (Anti-ALP antibody; (ab67228)) were immuno-stained with the streptavidin biotin method for immuno-histochemical investigation. Abcam provided all primary antibodies and secondary kits (Cambridge, UK). For 60 min, the sections were treated with primary antibodies. A biotinylated secondary antibody and a streptavidin–alkaline phosphatase combination were used for immuno-histochemistry. The secondary antibody was the EXPOSE Mouse and Rabbit Specific HRP/DAB Detection IHC kit (ab80436) Abcam, Cambridge, UK. Antigens were identified using a chromogen diaminobenzidine (DAB). Instead of primary antibodies, an antibody dilution solution was used as a negative control.

As analyses were carried out in a blind approach, the immune-positivity of all slides was evaluated semi-quantitatively by taking staining intensity into account (0, absence of staining; 1, slight; 2, medium and 3, marked). The results of the image analyzer were then subjected to statistical analyses. The Database Manual Cell Sens Life Science Imaging Software System (Olympus Co., Tokyo, Japan) was used to perform morphometric studies.

2.3.4. Statistical Analysis

SPSS version 23.0 was used to conduct statistical analysis (IBM, Armonk, NY, USA), and the Duncan test for independent samples was used to compare groups using a one-way analysis of variance (ANOVA). Statistical significance was defined as p-values of less than 0.001. G power 3.1 (Düsseldorf, Germany) software was used to calculate the sample size of two animals per group; sample size planning revealed an effect size d of 10.5 and an actual power of 0.98.

3. Results

3.1. Bioactive Glass Scaffolds

Figure 2a shows digital images of the bioactive glass-based PCL-coated composite scaffolds containing graphene nanopowders at different concentrations. Accordingly, scaffolds have a porous structure and graphene addition did not alter the morphological features significantly. The diameter and thickness of the fabricated, disc-shape scaffolds were measured at ~7 and ~3 mm, respectively. Optical microscope images shown in Figure 2b reveal the existence of macropores in the range of 300–500 μm. Graphene nanopowders homogeneously distributed within the PCL matrix cover the surface of the bioactive glass samples. Figure 3a,b depict the SEM micrographs of the PCL-coated 13-93B3 bioactive glass scaffolds in both the absence and presence of graphene nanopowders, respectively. The thin PCL layer covering the scaffolds can be clearly observed from the images, and the inclusion of graphene nanopowders created additional sites on the surface of the scaffolds. Figure 3 also reveals that, although the applied coating partially occluded pores at the surface, an interconnected open pore network still existed in the scaffolds.

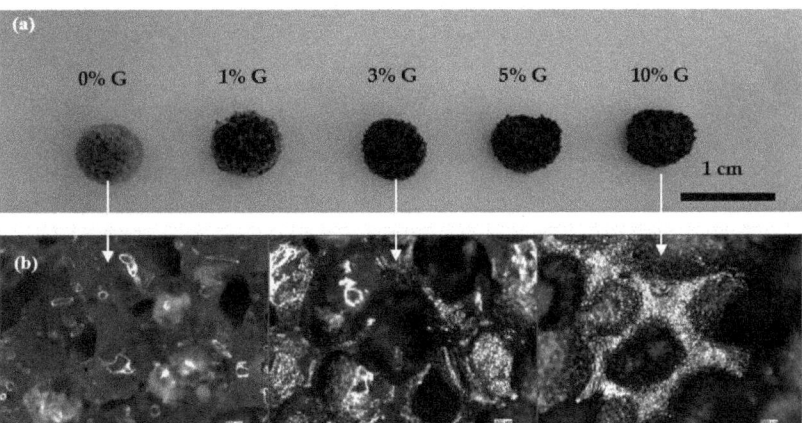

Figure 2. (a) Digital and (b) optical microscope images of the bioactive glass-based composite scaffolds (PCL-coated, containing 0, 1, 3, 5 and 10 wt% graphene) fabricated in the study, scale bar: 500 μm for optical microscope images.

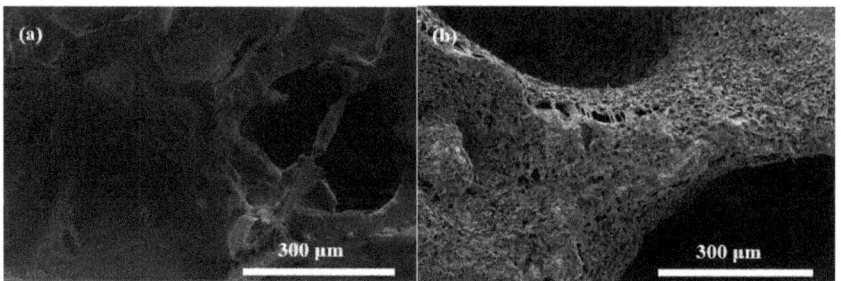

Figure 3. SEM micrographs of the (**a**) bare PCL-coated, (**b**) 10 wt% graphene-containing PCL-coated bioactive glass scaffolds.

Measurements showed that total porosity of the bare bioactive glass scaffolds in absence of surface coating was ~77%. It declined to 69% with the application of a polymeric coating on the surface of the bioactive glass scaffold. The incorporation of graphene nanopowders at the highest concentration further reduced porosity to 65% (Figure 4).

The results of the XRD examination revealed that the sintered bioactive glass scaffolds under investigation were amorphous, with no evidence of crystallization (Figure 5a). The FTIR-ATR spectra of the bare and PCL-coated bioactive glass scaffolds is given in Figure 5b. As a result of the B–O stretching vibrations of BO_4 and triangular BO_3 units, the glasses' FTIR spectra exhibit two broad and notable bands with maxima at about 940 and 1370 cm^{-1}, respectively. The B–O–B bending vibrations of the BO_3 and BO_4 groups are responsible for the medium band at 717 cm^{-1} [22]. The absorbance intensity of these groups was weaker in the spectrum of the PCL-coated bioactive sample compared to the uncoated sample. On the other hand, the hydroxyl and ester groups are represented as a peak around 3500 cm^{-1} and a very strong signal at 1750 cm^{-1}, respectively [23], in the spectrum of PCL (Figure 5c). Functional groups of PCL were not seen in the IR spectrum of the PCL-coated bioactive glass scaffolds due to the application of a thin polymer layer on the surface of the glass. In the case of graphene nanopowders, because of a mismatch in charge states between carbon atoms, they have few absorption signals [24]. This little difference induces a very small electric dipole, resulting in a highly clear IR spectrum (Figure 5d).

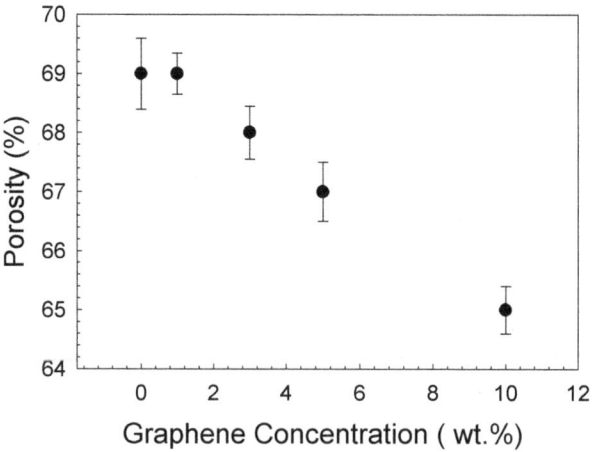

Figure 4. Total porosity of the graphene-containing PCL-coated bioactive glass scaffolds.

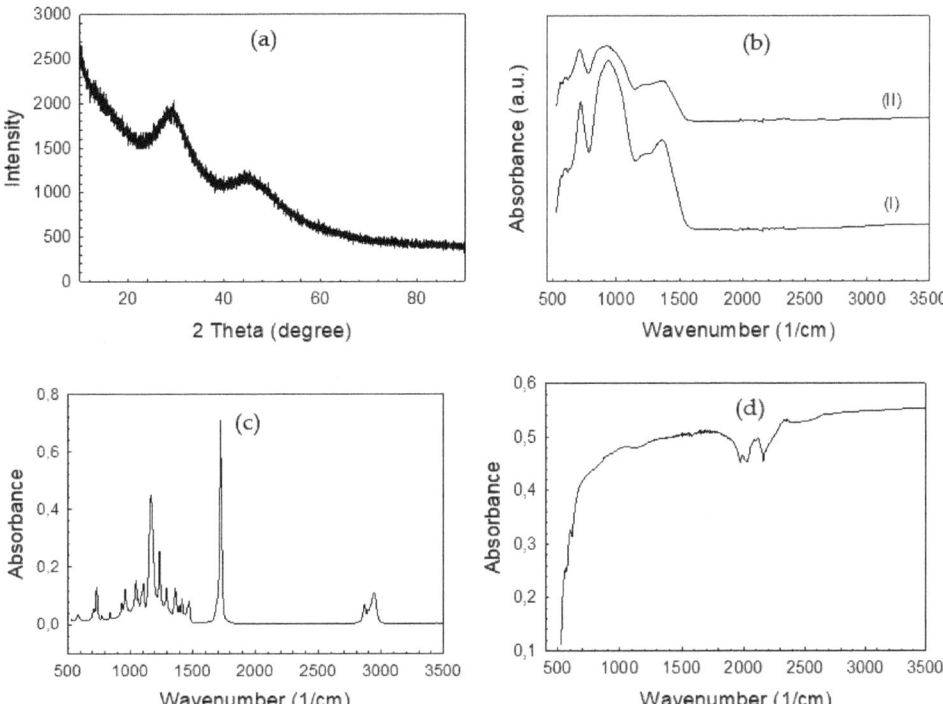

Figure 5. (**a**) XRD pattern of the sintered bare bioactive glass scaffold, (**b**) FTIR-ATR spectra of the bare borate glass (I) and PCL-coated borate glass scaffolds (II), (**c**) FTIR-ATR spectrum of the PCL, (**d**) graphene nanopowders utilized in the study.

3.2. Histopathological Evaluation

Before necropsy, the body weight of all rats was measured at week 0, week 4, and week 8. Week 0's mean body weight was 305 g, week 4's was 262 g, and week 8's was 280 g. Weights did not differ significantly between groups or weeks ($p > 0.05$).

Based on the histopathological examination results of the 4-week groups, the graft material was observed in all animal groups at different ratios. The most significant amount of graft material was detected in group 4. Results also revealed that the absorption rate of the graft material was related to the concentration of graphene nanopowders. As the graphene concentration of the scaffolds decreased, higher absorption behavior was observed. Fibrous tissue proliferation occurred in all groups (Figure 6). Similarly, 8 weeks after implantation, both cartilage and bone formation were observed in all animal groups. Bone formation was significant starting at group 2, with the greatest amount obtained in group 4 (Table 1). Accordingly, new bone area was 41.26 ± 0.71 mm^2 in group 4, whereas it was 5.28 ± 0.61 mm^2 for the control group after 8 weeks. It was also observed that the amount of residual graft material increased for samples containing higher ratio of graphene, which may correlate with a change in the degradation rate of bioactive glass-based scaffolds (Figure 7). The residual material area was calculated to be 12.80 ± 1.92 mm^2 for the 10 wt% graphene-containing scaffold implanted group; however, it was 7.84 ± 0.92 mm^2 for the control group under the same conditions. In the study, osteoblasts and osteoclasts were counted in a 1.23 mm^2 area at 400× magnification. Results showed that after 4 and 8 weeks of implantation, both the osteoblast and osteoclast numbers were significantly higher in group 4 compared to the control group.

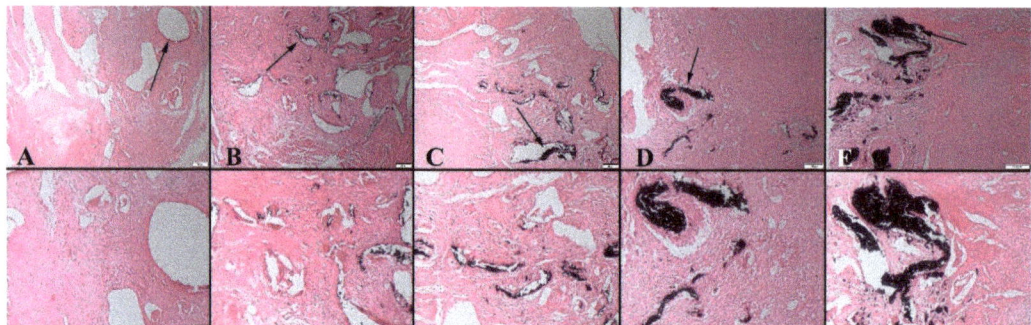

Figure 6. Representative histopathological micrographs of bone fracture areas of(upper row) at 4-week groups, (**A**) Control group, (**B**) Group 1, (**C**) Group 2, (**D**) Group 3 and (**E**) Group 4 lower row higher magnification, graft materials (indicated by arrows), HE, Bars = 200 μm (for upper row) and 100 μm (for lower row).

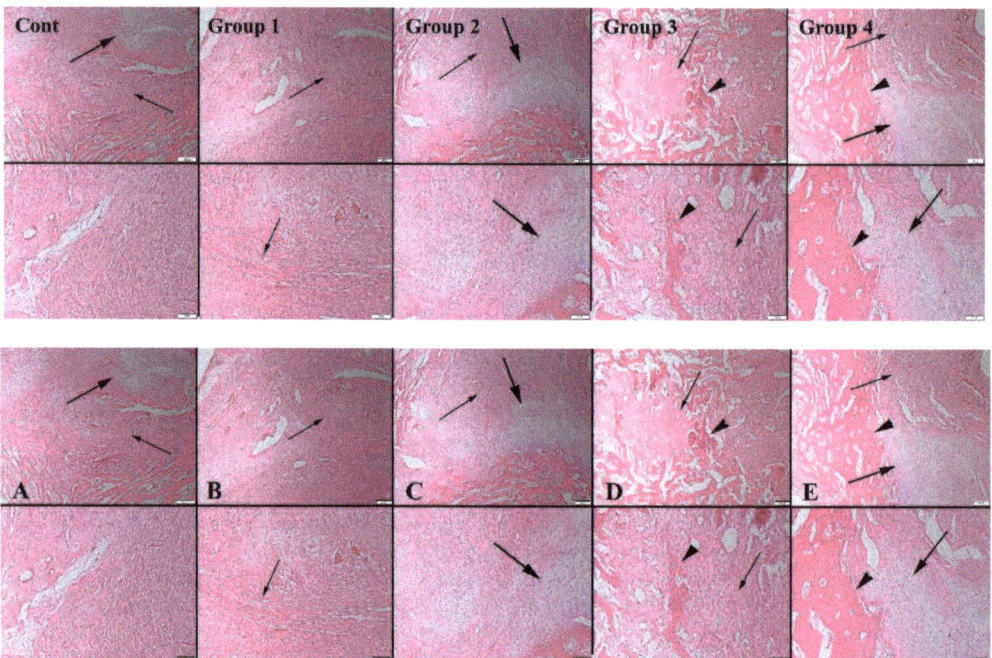

Figure 7. Representative histopathological micrographs of bone fracture areas. (**A**) Control group, (**B**) Group 1, (**C**) Group 2, (**D**) Group 3 and (**E**) Group 4, lower row higher magnification, Fibrous tissue (thin arrows), cartilage (thick arrows) and new bone tissues (arrow heads) (upper row) at 8-week groups, higher magnification (lower row), HE, scale bars = 200 μm (for upper row) and 100 μm (for lower row).

A Picro-Sirius red staining analysis performed to examine fibrous tissue revealed that while marked fibrous tissue was observed in the control group, it decreased in the graphene-coated scaffold implanted groups. After 4 weeks, a decrease in the amount of connective tissue and an increase in bone formation was observed in group 4. After 8 weeks, a decrease in fibrous tissue was recorded in all groups. While cartilage tissue was observed in the control group, bone formation percentages were significant in graphene-coated scaffold implanted groups. The highest amount of bone formation occurred in group 4 (Figure 8).

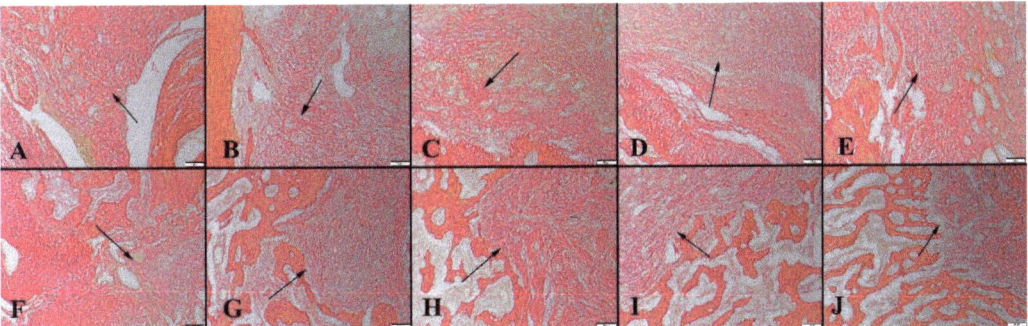

Figure 8. Micrographs showing fibrous tissue formation (arrows) in animal groups obtained via the Picro-Sirius red method. (**A**) Control group, (**B**) Group 1, (**C**) Group 2, (**D**) Group 3 and (**E**) Group 4, for 4-week and (**F**) Control group, (**G**) Group 1, (**H**) Group 2, (**I**) Group 3 and (**J**) Group 4, for 8-week. Decreased fibrous tissue formation is shown in graphene-coated scaffold-implanted groups, scale bar = 100 μm.

Table 1. Statistical analysis of histomorphic data.

		4 Weeks	8 Weeks
	Control	2.66 ± 0.38 [a]	5.28 ± 0.61 [a]
	Group 1	19.40 ± 2.07 [b]	23.98 ± 0.91 [b]
New bone area (mm^2)	Group 2	20.24 ± 1.37 [c]	31.18 ± 1.09 [c]
	Group 3	25.20 ± 1.48 [d]	36.00 ± 3.36 [d]
	Group 4	28.00 ± 1.73 [e]	41.26 ± 0.71 [e]
	Control	15.70 ± 1.78 [a]	7.84 ± 0.92 [a]
	Group 1	19.40 ± 1.14 [b]	8.48 ± 0.91 [a]
Residual material area (mm^2)	Group 2	18.20 ± 2.28 [a]	7.48 ± 1.58 [a]
	Group 3	22.20 ± 1.92 [c]	10.12 ± 1.19 [b]
	Group 4	25.40 ± 2.07 [d]	12.80 ± 1.92 [c]
	Control	10.40 ± 0.54 [a]	14.40 ± 1.14 [a]
	Group 1	13.80 ± 1.48 [b]	18.20 ± 1.64 [b]
Osteoclast number	Group 2	19.00 ± 1.87 [c]	20.00 ± 1.58 [b]
	Group 3	20.00 ± 1.58 [c]	24.00 ± 1.58 [c]
	Group 4	24.00 ± 1.58 [d]	26.60 ± 2.07 [d]
	Control	11.00 ± 1.58 [a]	20.20 ± 2.28 [a]
	Group 1	16.20 ± 1.48 [b]	25.00 ± 1.87 [b]
Osteoblast number	Group 2	17.40 ± 1.14 [b]	27.60 ± 1.14 [c]
	Group 3	19.60 ± 0.89 [c]	28.40 ± 2.96 [c]
	Group 4	23.60 ± 1.34 [d]	35.20 ± 0.83 [d]

Standard deviation of data (SD). A one-way Duncan test was performed for statistical analysis. $p < 0.001$ indicates that differences in the means of groups with different superscript letters in the same column are statistically significant. Differences between groups with the same superscript are statistically insignificant.

3.3. Immunohistochemical Findings

3.3.1. BMP-2 Immunohistochemistry Results

It has been reported that bone morphogenetic protein-2 (BMP-2) is a potent osteoinductive cytokine, which is crucial during bone repair and regeneration. In the current study, a BMP-2 immuno-histochemistry examination revealed the existence of a marked expression in 8-week groups in comparison with 4-week groups. Graphene-coated bioactive glass-implanted groups exhibited more significant expression compared to the bare PCL-coated scaffolds implanted in control group rats (Figure 9).

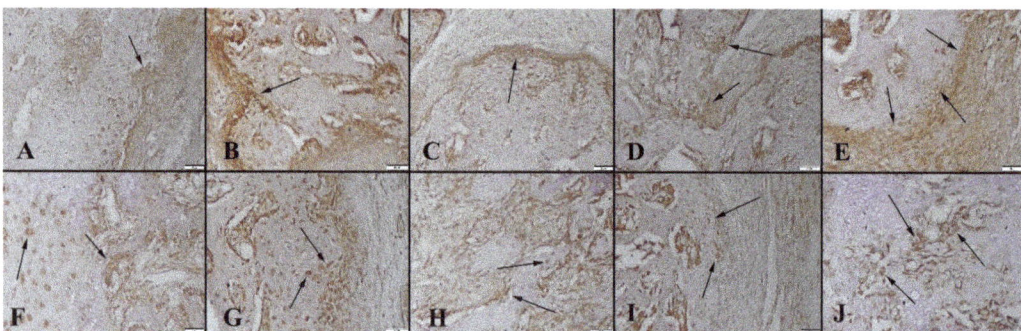

Figure 9. BMP-2 expression between the groups. (**A**) Control group, (**B**) Group 1, (**C**) Group 2, (**D**) Group 3 and (**E**) Group 4, for 4-week and (**F**) Control group, (**G**) Group 1, (**H**) Group 2, (**I**) Group 3 and (**J**) Group 4, for 8-week, arrows indicate immunopositive cells, Streptavidin biotin peroxidase method, scale bars = 50 μm.

3.3.2. Smooth Muscle Actin Immunohistochemistry Results

For evaluation of neo vascularization, sections were immuno-stained with smooth muscle actin. Newly formed vessels expressed the marker and an increased vascularization was observed in the 8-week group compared to the 4-week group. In addition, an increase in new vessel formation was observed in graphene-coated scaffold-implanted groups compared to the control group (Figure 10).

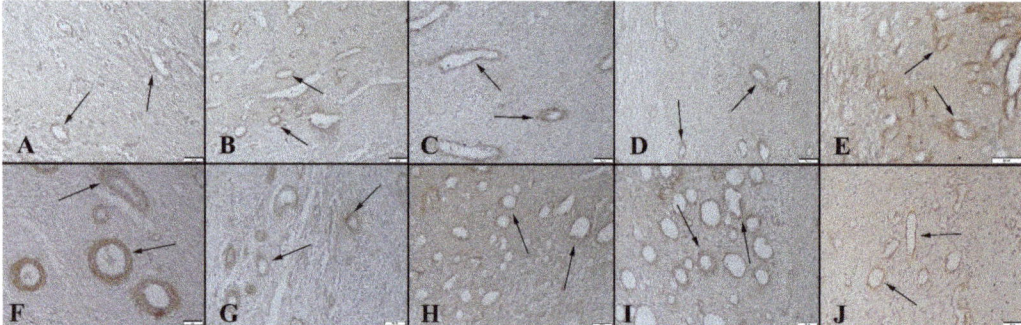

Figure 10. Smooth muscle actin expression among the groups. (**A**) Control group, (**B**) Group 1, (**C**) Group 2, (**D**) Group 3 and (**E**) Group 4, for 4-week and (**F**) Control group, (**G**) Group 1, (**H**) Group 2, (**I**) Group 3 and (**J**) Group 4, for 8-week, arrows indicate immunopositive cells, Streptavidin biotin peroxidase method, scale bars = 50 μm.

3.3.3. ALP Immunohistochemistry Results

ALP immunochemistry results revealed an increased expression in the 8-week groups compared to the 4-week groups. Expressions increased depending on the ratio of graphene (Figure 11).

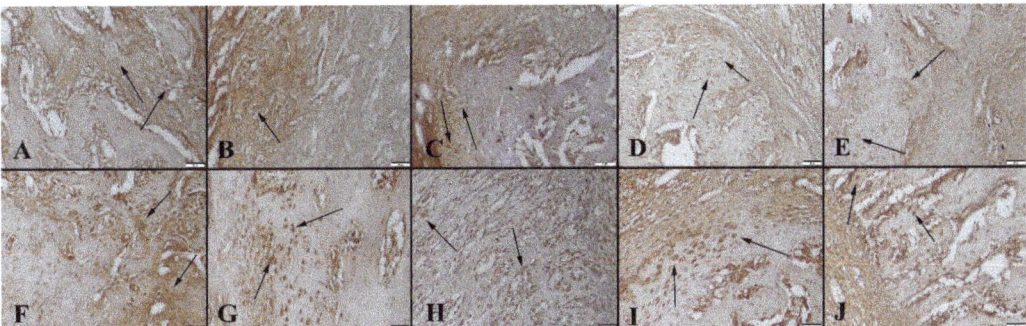

Figure 11. ALP expressions between the groups. (**A**) Control group, (**B**) Group 1, (**C**) Group 2, (**D**) Group 3 and (**E**) Group 4, for 4-week and (**F**) Control group, (**G**) Group 1, (**H**) Group 2, (**I**) Group 3 and (**J**) Group 4, for 8-week, arrows indicate immunopositive cells. Streptavidin biotin peroxidase method, scale bars = 50 µm.

The overall results of the study showed the ameliorative effect of graphene-coated bioactive glass scaffolds on the healing of bone fracture areas. An increase in new bone formation occurred in the graphene-coated graft implanted groups compared to the control group. The healing rate increased proportionally to the increase in graphene concentration.

4. Discussion

In bone tissue engineering and dentistry, bioactive glasses have a wide range of uses [21,25]. Gu et al. examined the ability of porous silicate and borate-based bioactive glass scaffolds to regenerate bone in rat calvarial lesions in vivo in a previous work [9]. At 12 weeks, the volume of new bone formed in the defects implanted with the 13-93 scaffolds was 31%, compared to 20% for the scaffolds with 13-93B3 bioactive glass. The amount of new bone formed in the 13-93 scaffolds was significantly less than in the 13-93B3 scaffolds. Boron liberated from 13-93B3 glass in surrounding defects has been shown to induce bone formation in 13-93 scaffold implanted defect sites [9].

On the other hand, graphene is a carbon-based material with a honeycomb structure one atom thick. It exhibits a distinctive optical, thermal, and mechanical performance and can be manufactured in various structures to provide applications in the field of biomedicine [26–29].

While graphene increases the mechanical strength of bioactive glass, it also makes it electroactive, and so graphene-coated bioactive glass scaffolds maintain excellent electroactivity [6]. In this way, graphene-coated mesoporous scaffolds cause significant increases in the differentiation of MSCs and cell-to-cell communication.

Graphene-containing, PCL-coated borate-based 13-93B3 bioactive glass scaffolds are candidates for effective graft options in filling segmental defects of long bones. In this study, the effects of graphene nanopowders embedded in a PCL matrix coated on bioactive glass scaffolds were investigated for bone healing after 4 and 8 weeks of implantation in rats. Since rigid fixation could not be provided, it was not possible for the bridging callus to form into a hard callus in 8 weeks. The presence of PCL as a coating on the surface of the bioactive glass scaffolds prevented the direct rapid release of graphene nanopowders to the tissue site after implantation. Without using a polymer layer, an increase in local graphene concentration could be detrimental for the bone cells, since graphene has a two-dimensional structure and its toxicity is concentration-dependent.

Furthermore, the presence of a PCL coating may also enhance the mechanical strength of scaffolds, which is crucial for the repair of the bone tissue. The bone remodeling phase begins after formation of a sufficient amount of bone callus (in the first 28 days), and includes the enchondral ossification process (lasting months to years). In other words, in the first 28 days, the bone defect must be knitted with new blood vessels and the bridge callus must be formed [2,30–32]. It is known that an adequate bridge callus cannot be formed in segmental bone defects. For this reason, a bioactive scaffold with a porous structure that imitates cancellous bone is needed [16,33]. The mechanical properties and architectural structure of this scaffold must allow for neo-angiogenesis, ionic interaction and mineralization. Because the vascularization process stagnates after the critical first month, enchondral ossification and mineralization cascades end. At this stage, potent growth factors such as BMP, VEGF, and TGF provide differentiation from MSCs to osteocytes and chondrocytes. Their maintenance and regulation are as important as the initiation of osteoinduction [2,31,34,35].

One of the most important parts of bone defect repair is vascularization, as the amount of new bone required to fill the defect is directly related to the vascularization rate. It has been shown in vitro that borate-based bioactive glass PCL composites can increase VEGF and ANG-1 expression of rBMSCs [7,14]. Moreover, it is known that the incorporation of boron into BGs could enhance osteogenesis in vitro [7]. In the current study, borate-based PCL-coated bioactive glasses constituted the control group. In this way, the stimulating effect of different amounts of graphene nanopowders were clearly observed. According to data obtained with the rat femur defect model, borate-based bare BG scaffolds cannot sufficiently support vascularization and osteogenesis in large bone defects. New vessel formation was examined immunohistochemically (smooth muscle actin) and a significant increase in SMA for the graphene-containing groups observed (Table 2, see also Figure 10).

Table 2. Statistical analysis of immuno-histochemical scores.

	BMP-2	SMA	ALP
4-week			
Control	1.20 ± 0.44 [a]	0.80 ± 0.44 [a]	1.40 ± 0.54 [a]
Group 1	2.60 ± 0.54 [b]	1.40 ± 0.54 [a]	2.20 ± 0.83 [b]
Group 2	2.40 ± 0.54 [b]	1.60 ± 0.89 [a]	2.00 ± 0.70 [b]
Group 3	2.40 ± 0.54 [b]	1.40 ± 0.54 [a]	1.80 ± 0.83 [b]
Group 4	2.60 ± 0.54 [b]	1.80 ± 0.83 [b]	2.00 ± 1.00 [b]
8-week			
Control	1.00 ± 0.00 [a]	1.40 ± 0.54 [a]	1.60 ± 0.54 [a]
Group 1	1.75 ± 0.50 [a]	2.20 ± 0.83 [b]	2.00 ± 0.70 [b]
Group 2	2.00 ± 0.00 [b]	2.20 ± 0.44 [b]	2.20 ± 0.44 [b]
Group 3	2.00 ± 0.81 [b]	2.40 ± 0.54 [b]	2.60 ± 0.54 [b]
Group 4	2.75 ± 0.50 [b]	2.40 ± 0.89 [b]	2.80 ± 0.44 [c]

Standard deviation (SD). A one-way Duncan test was used in the statistical analysis. $p < 0.001$ denotes a statistical significance of differences in the means of groups with different superscript letters in the same column. Differences between groups with the same superscript are statistically insignificant.

The replacement of the scaffold with new bone tissue is directly related to the degradation time of the scaffold. With the onset of the bone remodeling phase, the scaffold should begin to degrade. Early or late biodegradation negatively affects the healing process, especially in long bones.

Bone ALP, OPN, and OCN levels are the biomarkers of the osteogenesis process, and their levels increase during the mineralization process [36,37]. The hormonal response and mechanical factors limit excessive bone formation. Potent growth factors such as

BMP, VEGF, and TGF cannot be synthesized sufficiently in segmental bone defects. For this reason, in the absence of potent growth factors such as BMP, VEGF, and TGF, the remodeling phase is interrupted [2,38,39].

Studies in the literature have reported the effectiveness of bioactive glass scaffolds on bone union with the calvarial bone defect model [40]. However, there are clear differences between long bone healing and flat bone healing [31,40]. Therefore, we think that the defect model preferred in our study will shed more light on orthopedic surgery practices. The major limitation of this study is that only recovery times of 4 and 8 weeks were examined. Further studies on the in vivo response of the bioactive glass-based composite system under investigation can be performed using graphene oxide, using functional groups instead of pristine graphene, and a more hydrophilic polymer coating layer to enhance the cellular interaction.

5. Conclusions

Graphene containing (0, 1, 3, 5 and 10 wt%) PCL-coated porous borate-based 13-93B3 bioactive glass scaffolds were fabricated by polymer foam replication and dip coating methods. The prepared scaffolds were evaluated for their capacity to enhance bone formation in rat segmental defects in vivo. Scaffolds containing graphene showed a better capacity to support new bone formation and alkaline phosphatase activity. The number of osteoclasts and osteoblasts were significantly higher in the 10 wt% graphene-coated bioactive glass scaffold-implanted group compared to the control group. Fibrous tissue formation was observed in the control group, whereas it decreased in the graphene-coated scaffold-implanted groups. Furthermore, the presence of graphene-enhanced BMP-2 and SMA levels was increased compared to bare bioactive glass scaffolds. Pristine graphene-coated bioactive glass scaffolds have a high potential to be used in the repair and regeneration of large segmental bone defects.

Author Contributions: O.B.: research design, animal research, data interpretation, manuscript drafting and approval; O.O.: data acquisition, analysis, interpretation, and drafting; A.M.D.: study design, data collecting and analysis, and data interpretation. The final manuscript has been read and approved by all authors. All authors have read and agreed to the published version of the manuscript.

Funding: This research received no external funding.

Institutional Review Board Statement: The animal study protocol was approved by the Institutional Ethics Committee of Burdur Mehmet Akif Ersoy University (Ethics number: 531, 17 July 2019).

Informed Consent Statement: Not applicable.

Conflicts of Interest: The authors declare no conflict of interest.

References

1. Liu, X.; Rahaman, M.N.; Liu, Y.; Bal, B.S.; Bonewald, L.F. Enhanced Bone Regeneration in Rat Calvarial Defects Implanted with Surface-Modified and BMP-Loaded Bioactive Glass (13–93) Scaffolds. *Acta Biomater.* **2013**, *9*, 7506–7517. [CrossRef] [PubMed]
2. Brydone, A.S.; Meek, D.; MacLaine, S. Bone Grafting, Orthopaedic Biomaterials, and the Clinical Need for Bone Engineering. *Proc. Inst. Mech. Eng. Part H J. Eng. Med.* **2010**, *224*, 1329–1343. [CrossRef] [PubMed]
3. Hench, L.L. The Story of Bioglass®. *J. Mater. Sci. Mater. Med.* **2006**, *17*, 967–978. [CrossRef] [PubMed]
4. Zhang, Y.; Chen, X.; Miron, R.; Zhao, Y.; Zhang, Q.; Wu, C.; Geng, S. In Vivo Experimental Study on Bone Regeneration in Critical Bone Defects Using PIB Nanogels/Boron-Containing Mesoporous Bioactive Glass Composite Scaffold. *Int. J. Nanomed.* **2015**, *10*, 839–846. [CrossRef] [PubMed]
5. Akhavan, O. Graphene Scaffolds in Progressive Nanotechnology/Stem Cell-Based Tissue Engineering of the Nervous System. *J. Mater. Chem. B* **2016**, *4*, 3169–3190. [CrossRef] [PubMed]
6. Yao, Q.; Liu, H.; Lin, X.; Ma, L.; Zheng, X.; Liu, Y.; Huang, P.; Yu, S.; Zhang, W.; Lin, M.; et al. 3D Interpenetrated Graphene Foam/58S Bioactive Glass Scaffolds for Electrical-Stimulation-Assisted Differentiation of Rabbit Mesenchymal Stem Cells to Enhance Bone Regeneration. *J. Biomed. Nanotechnol.* **2019**, *15*, 602–611. [CrossRef] [PubMed]
7. Xia, L.; Ma, W.; Zhou, Y.; Gui, Z.; Yao, A.; Wang, D.; Takemura, A.; Uemura, M.; Lin, K.; Xu, Y. Stimulatory Effects of Boron Containing Bioactive Glass on Osteogenesis and Angiogenesis of Polycaprolactone: In Vitro Study. *BioMed Res. Int.* **2019**, *2019*, 1–12. [CrossRef] [PubMed]
8. Yin, H.; Yang, C.; Gao, Y.; Wang, C.; Li, M.; Guo, H.; Tong, Q. Fabrication and Characterization of Strontium-Doped Borate-Based Bioactive Glass Scaffolds for Bone Tissue Engineering. *J. Alloys Compd.* **2018**, *743*, 564–569. [CrossRef]

9. Gu, Y.; Huang, W.; Rahaman, M.N.; Day, D.E. Bone Regeneration in Rat Calvarial Defects Implanted with Fibrous Scaffolds Composed of a Mixture of Silicate and Borate Bioactive Glasses. *Acta Biomater.* **2013**, *9*, 9126–9136. [CrossRef] [PubMed]
10. Fu, Q.; Rahaman, M.N.; Fu, H.; Liu, X. Silicate, Borosilicate, and Borate Bioactive Glass Scaffolds with Controllable Degradation Rate for Bone Tissue Engineering Applications. I. Preparation and in Vitro Degradation. *J. Biomed. Mater. Res. Part A* **2010**, *95*, 164–171. [CrossRef]
11. Rezwan, K.; Chen, Q.Z.; Blaker, J.J.; Boccaccini, A.R. Biodegradable and Bioactive Porous Polymer/Inorganic Composite Scaffolds for Bone Tissue Engineering. *Biomaterials* **2006**, *27*, 3413–3431. [CrossRef] [PubMed]
12. Türk, M.; Deliormanlı, A.M. Electrically Conductive Borate-Based Bioactive Glass Scaffolds for Bone Tissue Engineering Applications. *J. Biomater. Appl.* **2017**, *32*, 28–39. [CrossRef] [PubMed]
13. Deliormanlı, A.M. Fabrication and Characterization of Poly (ε-Caprolactone) Coated Silicate and Borate based Bioactive Glass Scaffolds. *J. Compos. Mater.* **2016**, *50*, 917–928. [CrossRef]
14. Türk, M.; Deliormanlı, A.M. Graphene-Containing PCL- Coated Porous 13-93B3 Bioactive Glass Scaffolds for Bone Regeneration. *Mater. Res. Express* **2018**, *5*, 45406. [CrossRef]
15. Deliormanlı, A.M.; Türk, M.; Atmaca, H. Response of Mouse Bone Marrow Mesenchymal Stem Cells to Graphene-containing Grid-like Bioactive Glass Scaffolds Produced by Robocasting. *J. Biomater. Appl.* **2018**, *33*, 488–500. [CrossRef]
16. Qi, X.; Wang, H.; Zhang, Y.; Pang, L.; Xiao, W.; Jia, W.; Zhao, S.; Wang, D.; Huang, W.; Wang, Q. Mesoporous Bioactive Glass-Coated 3D Printed Borosilicate Bioactive Glass Scaffolds for Improving Repair of Bone Defects. *Int. J. Biol. Sci.* **2018**, *14*, 471–484. [CrossRef] [PubMed]
17. Gao, C.; Liu, T.; Shuai, C.; Peng, S. Enhancement Mechanisms of Graphene in Nano-58S Bioactive Glass Scaffold: Mechanical and Biological Performance. *Sci. Rep.* **2014**, *4*, 4712. [CrossRef] [PubMed]
18. Deng, Y.; Gao, X.; Shi, X.L.; Lu, S.; Yang, W.; Duan, C.; Chen, Z.G. Graphene Oxide and Adiponectin-Functionalized Sulfonated Poly(Etheretherketone) with Effective Osteogenicity and Remotely Repeatable Photodisinfection. *Chem. Mater.* **2020**, *32*, 2180–2193. [CrossRef]
19. Prasadh, S.; Suresh, S.; Wong, R. Osteogenic Potential of Graphene in Bone Tissue Engineering Scaffolds. *Materials* **2018**, *11*, 1430. [CrossRef] [PubMed]
20. Basal, O.; Ozmen, O.; Deliormanlı, A.M. Effect of Polycaprolactone Scaffolds Containing Different Weights of Graphene on Healing in Large Osteochondral Defect Model. *J. Biomater. Sci. Polym. Ed.* **2022**, *33*, 1123–1139. [CrossRef] [PubMed]
21. Araújo, A.S.; Fernandes, A.B.; Maciel, J.V.; Netto, J.D.N.; Bolognese, A.M. New Methodology for Evaluating Osteoclastic Activity Induced by Orthodontic Load. *J. Appl. Oral Sci.* **2015**, *23*, 19–25. [CrossRef]
22. Ouis, M.A.; Abdelghany, A.M.; ElBatal, H.A. Corrosion Mechanism and Bioactivity of Borate Glasses Analogue to Hench's Bioglass. *Process. Appl. Ceram.* **2012**, *6*, 141–149. [CrossRef]
23. Kim, T.; Lee, S.; Park, S.-Y.; Chung, I. Biodegradable PCL-b-PLA Microspheres with Nanopores Prepared via RAFT Polymerization and UV Photodegradation of Poly (Methyl Vinyl Ketone) Blocks. *Polymers* **2021**, *13*, 3964. [CrossRef]
24. Ruiz, S.; Tamayo, J.A.; Delgado Ospina, J.; Navia Porras, D.P.; Valencia Zapata, M.E.; Mina Hernandez, J.H.; Valencia, C.H.; Zuluaga, F.; Grande Tovar, C.D. Antimicrobial Films Based on Nanocomposites of Chitosan/Poly (Vinyl Alcohol)/Graphene Oxide for Biomedical Applications. *Biomolecules* **2019**, *9*, 109. [CrossRef] [PubMed]
25. Rahaman, M.N.; Liu, X.; Bal, B.S.; Day, D.E.; Bi, L.; Bonewald, L.F. Bioactive Glass in Bone Tissue Engineering. *Ceram. Trans.* **2012**, *237*, 73–82. [CrossRef]
26. Shi, J.; Fang, Y. Biomedical Applications of Graphene. *Graphene Fabr. Charact. Prop. Appl.* **2017**, *2*, 215–232. [CrossRef]
27. Yang, Y.; Asiri, A.M.; Tang, Z.; Du, D.; Lin, Y. Graphene Based Materials for Biomedical Applications. *Mater. Today* **2013**, *16*, 365–373. [CrossRef]
28. Zeimaran, E.; Pourshahrestani, S.; Nam, H.Y.; Razak, N.A.B.A.; Kalantari, K.; Kamarul, T.; Salamatinia, B.; Kadri, N.A. Engineering Stiffness in Highly Porous Biomimetic Gelatin/Tertiary Bioactive Glass Hybrid Scaffolds Using Graphene Nanosheets. *React. Funct. Polym.* **2020**, *154*, 104668. [CrossRef]
29. Fan, Z.; Wang, J.; Liu, F.; Nie, Y.; Ren, L.; Liu, B. A New Composite Scaffold of Bioactive Glass Nanoparticles/Graphene: Synchronous Improvements of Cytocompatibility and Mechanical Property. *Colloids Surf. B Biointerfaces* **2016**, *145*, 438–446. [CrossRef] [PubMed]
30. Bi, L.; Jung, S.; Day, D.; Neidig, K.; Dusevich, V.; Eick, D.; Bonewald, L. Evaluation of Bone Regeneration, Angiogenesis, and Hydroxyapatite Conversion in Critical-Sized Rat Calvarial Defects Implanted with Bioactive Glass Scaffolds. *J. Biomed. Mater. Res. Part A* **2012**, *100*, 3267–3275. [CrossRef] [PubMed]
31. Sun, Y.; Wan, B.; Wang, R.; Zhang, B.; Luo, P.; Wang, D.; Nie, J.-J.; Chen, D.; Wu, X. Mechanical Stimulation on Mesenchymal Stem Cells and Surrounding Microenvironments in Bone Regeneration: Regulations and Applications. *Front. Cell Dev. Biol.* **2022**, *10*. [CrossRef] [PubMed]
32. Chu, C.R.; Szczodry, M.; Bruno, S. Animal Models for Cartilage Regeneration and Repair. *Tissue Eng. Part B Rev.* **2010**, *16*, 105–115. [CrossRef] [PubMed]
33. Deliormanlı, A.M. Direct Write Assembly of Graphene/Poly(ε-Caprolactone) Composite Scaffolds and Evaluation of Their Biological Performance Using Mouse Bone Marrow Mesenchymal Stem Cells. *Appl. Biochem. Biotechnol.* **2019**, *188*, 1117–1133. [CrossRef]

34. Ren, Q.; Cai, M.; Zhang, K.; Ren, W.; Su, Z.; Yang, T.; Sun, T.; Wang, J. Effects of Bone Morphogenetic Protein-2 (BMP-2) and Vascular Endothelial Growth Factor (VEGF) Release from Polylactide-Poly (Ethylene Glycol)-Polylactide (PELA) Microcapsule-Based Scaffolds on Bone. *Braz. J. Med. Biol. Res.* **2017**, *51*, 0809050139. [CrossRef] [PubMed]
35. Yang, J.-J.; Peng, W.-X.; Zhang, M.-B. LncRNA KCNQ1OT1 promotes osteogenic differentiation via miR-205-5p/RICTOR axis. *Exp. Cell Res.* **2022**, *415*, 113119. [CrossRef] [PubMed]
36. Basal, O.; Atay, T.; Ciris, I.M.; Baykal, Y.B. Epidermal Growth Factor (EGF) Promotes Bone Healing in Surgically Induced Osteonecrosis of the Femoral Head (ONFH). *Bosn. J. Basic Med. Sci.* **2018**, *18*, 352–360. [CrossRef]
37. Guan, H.; Kong, N.; Tian, R.; Cao, R.; Liu, G.; Li, Y.; Wei, Q.; Jiao, M.; Lei, Y.; Xing, F.; et al. Melatonin increases bone mass in normal, perimenopausal, and postmenopausal osteoporotic rats via the promotion of osteogenesis. *J. Transl. Med.* **2022**, *20*, 1–15. [CrossRef]
38. Koosha, E.; Eames, B.F. Two Modulators of Skeletal Development: BMPs and Proteoglycans. *J. Dev. Biol.* **2022**, *10*, 15. [CrossRef]
39. Park, K.O.; Lee, J.H.; Park, J.H.; Shin, Y.C.; Huh, J.B.; Bae, J.H.; Kang, S.H.; Hong, S.W.; Kim, B.; Yang, D.J.; et al. Graphene Oxide-Coated Guided Bone Regeneration Membranes with Enhanced Osteogenesis: Spectroscopic Analysis and Animal Study. *Appl. Spectrosc. Rev.* **2016**, *51*, 540–551. [CrossRef]
40. Lim, J.; Lee, J.; Yun, H.S.; Shin, H.I.; Park, E.K. Comparison of Bone Regeneration Rate in Flat and Long Bone Defects: Calvarial and Tibial Bone. *Tissue Eng. Regen. Med.* **2013**, *10*, 336–340. [CrossRef]

Review

Chitosan-Based Biomaterials for Bone Tissue Engineering Applications: A Short Review

Antonia Ressler

Faculty of Chemical Engineering and Technology, University of Zagreb, Marulićev trg 19, HR-10000 Zagreb, Croatia; aressler@fkit.hr; Tel.: +385-01-4597-237

Abstract: Natural bone tissue is composed of calcium-deficient carbonated hydroxyapatite as the inorganic phase and collagen type I as the main organic phase. The biomimetic approach of scaffold development for bone tissue engineering application is focused on mimicking complex bone characteristics. Calcium phosphates are used in numerous studies as bioactive phases to mimic natural bone mineral. In order to mimic the organic phase, synthetic (e.g., poly(ε-caprolactone), polylactic acid, poly(lactide-co-glycolide acid)) and natural (e.g., alginate, chitosan, collagen, gelatin, silk) biodegradable polymers are used. However, as materials obtained from natural sources are accepted better by the human organism, natural polymers have attracted increasing attention. Over the last three decades, chitosan was extensively studied as a natural polymer suitable for biomimetic scaffold development for bone tissue engineering applications. Different types of chitosan-based biomaterials (e.g., molded macroporous, fiber-based, hydrogel, microspheres and 3D-printed) with specific properties for different regenerative applications were developed due to chitosan's unique properties. This review summarizes the state-of-the-art of biomaterials for bone regeneration and relevant studies on chitosan-based materials and composites.

Keywords: biomaterials; bone regeneration; chitosan; composite; polymer; scaffold

Citation: Ressler, A. Chitosan-Based Biomaterials for Bone Tissue Engineering Applications: A Short Review. *Polymers* **2022**, *14*, 3430. https://doi.org/10.3390/polym14163430

Academic Editor: Ángel Serrano-Aroca

Received: 3 July 2022
Accepted: 22 August 2022
Published: 22 August 2022

Publisher's Note: MDPI stays neutral with regard to jurisdictional claims in published maps and institutional affiliations.

Copyright: © 2022 by the author. Licensee MDPI, Basel, Switzerland. This article is an open access article distributed under the terms and conditions of the Creative Commons Attribution (CC BY) license (https://creativecommons.org/licenses/by/4.0/).

1. Introduction

The incidence of bone disorders has increased, as a result of the aging population coupled with increased obesity and poor physical activity, drawing extensive attention to bone repair medicine research [1,2]. When the bone disorder exceeds the critical size defect (>2 cm), the bone tissue cannot heal by itself and clinical treatment is required [2]. Bone grafting is one of the most common methods for bone regeneration, with over two million bone graft procedures conducted worldwide annually. Numerous types of bone grafts have been used in bone tissue engineering in the last few decades; however, increasing attention is directed towards the biomimetic approach in scaffold design, where molecular, structural and biological compatibility with complex native bone tissue is achieved [3]. When fundamental limitations of biomaterials of first and second generations were recognized, studies shifted to the biomimetic approach and biomaterials that stimulate specific cellular responses at the level of molecular biology [4]. For the successful development of biomimetic scaffolds for bone regeneration, and the role of inorganic and organic phases in the bone tissue, a detailed understanding of the bone composition is essential.

Bone is a heterogeneous composite material consisting of a mineral phase, calcium-deficient carbonated hydroxyapatite (CDHAp, $Ca_{10-x}(PO_4)_{6-x}(HPO_4)_x(OH)_{2-x}$), and organic phase, consisting of ~90% collagen type I, ~5% non-collagenous proteins, ~2% lipids and water [5–7]. The various ionic substitutions (e.g., Mg^{2+}, Sr^{2+}, Na^+, K^+, CO_3^{2-}) in the biological CDHAp structure result in a remarkably complex crystal structure with unique biological properties [7]. Proteins in the bone extracellular matrix can be divided into (i) structural proteins (collagen and fibronectin) and (ii) proteins with specialized functions (e.g., regulation of collagen fibril diameter, signaling molecules, growth factors, enzymes) [8]. Although these proteins are present in the bone structure in a relatively small amount of the total protein mass, they modulate a wide variety of bone key role functions such as regulation of mineralization, cell adhesion and bone resorption/remodeling [6]. Cortical bone is a dense outer surface of bone that forms a protective layer around the inner part, spongy or trabecular bone, in which the main metabolism functions occur [5]. The building blocks of bone tissue are mineralized collagen fiber, composite biomaterial of collagen type I and nano-sized CDHAp. The CDHAp crystals are deposited in parallel with the collagen fibers, and they are later formed by self-assembly of the collagen triple helix [7,9]. Type I collagen is a right-handed helix composed of three left-handed helix polypeptide chains with nonhelical ends, with molecular dimensions of ~300 nm in length and ~1.5 nm in diameter. The collagen triple helix is stabilized via direct inter-chain hydrogen bonds and inter- and intra-chain water-mediated hydrogen bonds [10]. Along with type I collagen, osteocalcin is the next most abundant protein within the bone organic matrix and plays a major role as a structure-directing molecule. It is assumed that osteocalcin mediates the nucleation and growth of platelet-shaped (~50 × 25 × 2 nm) CDHAp crystals [11]. Figure 1 shows a hierarchical structure of typical bone at various length scales. On the macro-length scale, the structure of cortical or compact bone consists of circles in cross-section (Haversian systems) with osteonic canals, while the trabecular part of the bone has a highly porous structure. On the nano-length scale, the structure framework is collagen fibers composed of bundles (triple helix) of mineralized collagen fibers [9].

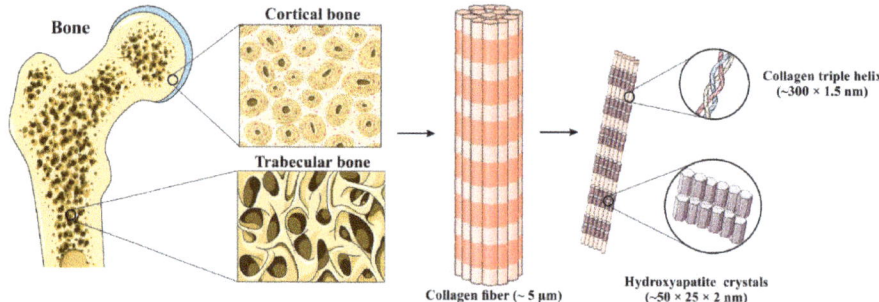

Figure 1. The hierarchical structure of bone at various length scales. Adapted from [12] with permission from Elsevier.

As biological apatites are characterized by various ionic substitutions that are crucial for bone metabolism, numerous studies have focused on the synthesis and characterization of biomimetic ionic-substituted hydroxyapatite, which is used as a bioactive phase in biomaterials for bone regeneration [12]. As a lot of efforts are put into mimicking the inorganic phase of the bone tissue, the same efforts are directed towards mimicking the organic phase. Combination of these mimicking biomaterials leads to composite material with a complex structure similar to natural bone tissue. Naturally derived polymers (e.g., collagen, gelatin, chitosan, glycosaminoglycans, silk fibrin) have been widely used in a variety of tissue engineering applications, as they can mimic a natural extracellular matrix. As natural polymers are building components of biological tissues, they demonstrate excellent biocompatibility in vivo and present a range of ligands and peptides that facilitate cell adhesion and osteogenic differentiation [3]. One of the most widely studied biopolymers is chitosan, a natural aminopolysaccharide with a unique structure and multidimensional properties suitable for a wide range of applications in biomedicine [13]. Along with bone tissue engineering applications, chitosan has been widely applied in drug delivery and gene therapy because of its excellent biocompatibility and biodegradability under physiological conditions [14]. In addition, the chitosan structure allows chemical and mechanical modifications in order to obtain novel properties, functions and applications [13]. Prior to the development of scaffolds with appropriate regenerative properties, the in vivo regenerative process steps need to be taken into account. After scaffold implantation (1), proteins are absorbed in the scaffold interface (2), followed by infiltration of immune cells (3), the release of chemical signals by immune cells to recruit stem cells (4), microenvironment remodeling (5) and vascularization (6), as schematically shown in Figure 2 [14,15].

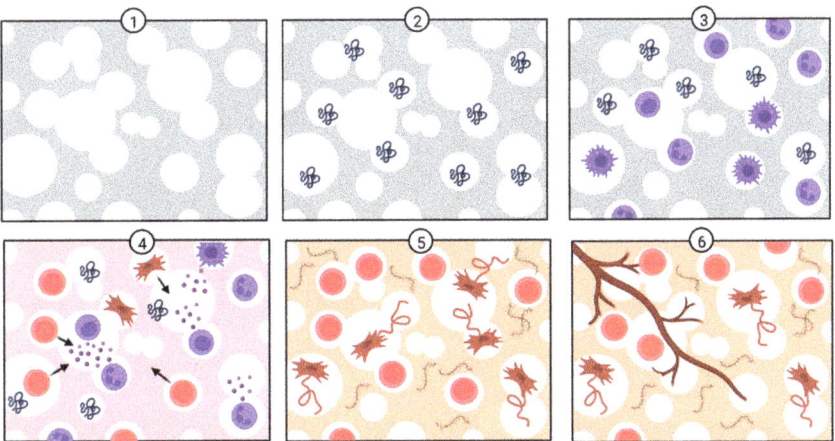

Figure 2. In vivo bone regeneration process after scaffold implantation. Created with BioRender.com (accessed on 30 June 2022).

Constant progress in bone tissue engineering is ensuring the development of novel functional biomaterials that can solve current challenges in the field. However, more efforts are required to ensure the reproducibility of developed biomaterials and standardization of characterization methods, which could increase the ability to compare biomaterials properties conducted in different studies. Joined efforts and frequent analysis of available literature and design requirements could increase the development of scaffolds with appropriate characteristics. In this short review paper, which is organized into several sections, the author first outlines general knowledge about natural bone tissue and natural polymers, followed by chitosan structure and characteristics. The third section provides a summary of requirements for bone scaffold development (biocompatibility, porosity and pore size distribution, mechanical strength, biodegradation) with a focus on chitosan-based materials. The fourth section provides a summary of different methods for the preparation of chitosan-based biomaterials (molded macroporous, fiber-based, hydrogel, microspheres and 3D-printed scaffolds). The last section provides relevant and recent viewpoints from the literature on the composite scaffolds based on chitosan and calcium phosphates, calcium silicate and bioactive glass. Current trends in the design of chitosan-based scaffolds are highlighted and future perspectives are discussed.

2. Chitosan Structure and Characteristics

Chitosan is a partially deacetylated derivate of chitin, one of the most abundant polymers in nature found in the shells of crustaceans and walls of fungi. It is composed of randomly distributed β-(1-4)-linked D-glucosamine (glucosamine) and N-acetyl-D-glucosamine (N-acetylglucosamine) structure units, structurally similar to glycosaminoglycan, a key component of the bone matrix and cell surface which modulates the bioavailability and activity of various osteoclastic and osteogenic factors [5,16–18]. Deacetylation of chitin is almost never complete and the chitosan chain still contains amide groups to some extent [16]. The degree of deacetylation (DD, %) is defined as the molar fraction of glucosamine in the chitosan composed of N-acetylglucosamine and glucosamine structure units [19]. The DD of chitosan is defined as low (55–70%), middle (70–85%), high (85–95%) or ultrahigh (95–100%), where ultrahigh is difficult to achieve [20]. In Figure 3, the structures of chitin, chitosan and protonated chitosan as a water-soluble poly-electrolyte are shown.

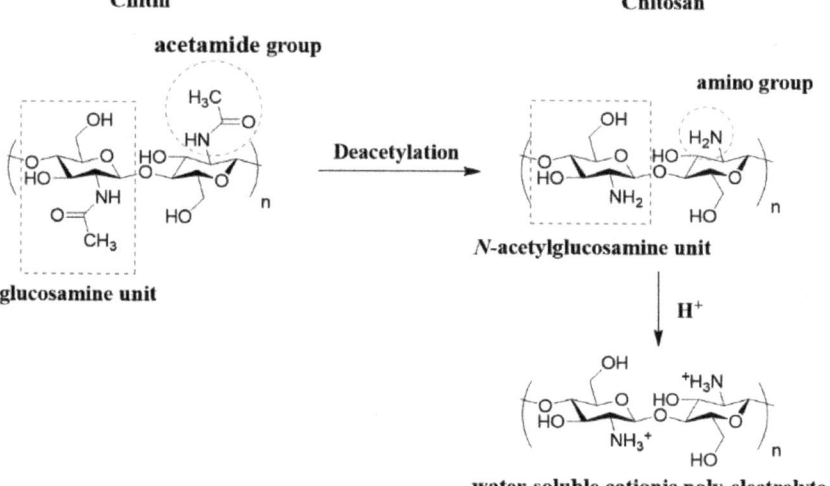

Figure 3. Structure of chitin, chitosan and protonated chitosan (water-soluble poly-electrolyte). The structures were obtained in the ChemDraw 7.0 (PerkinElmer, Waltham, MA, USA) software.

Chitosan has poor solubility in physiological solvents (e.g., water) due to its strong intermolecular hydrogen bonding and it is considered a strong base due to primary amino groups with a pKa value of 6.3 [13,16]. Chitosan solution can be obtained in acidic aqueous (pH < 6) media, which protonate chitosan amino groups, rendering the polymer positively charged and thereby overcoming associative forces between chains [13,16,21]. If the pH of chitosan solution increases above 6, chitosan amino groups become deprotonated and the polymer chain loses its charge, which leads to insolubility. The solubility is highly dependent on the degree of the deacetylation, the used deacetylation method and molecular weight. The solubility of chitosan can be increased by chemical modifications possible at two hydroxyl functional groups in the polymer chain [13]. The detailed review paper by Upadhyaya et al. [22] provides an overview of the water-soluble carboxymethyl chitosan as a modification of the non-soluble chitosan. The highly desired properties of biomaterials designed for applications in the human organism are antibacterial properties without harmful effects on healthy cells. The polycationic nature of the chitosan chain is essential for antibacterial activity. The most probable pathway of chitosan antibacterial activity is by binding to the negatively charged bacterial cell wall (disruption of the cell membrane), followed by attachment to DNA (inhibition of DNA replication) and subsequently cell death. Electrostatic interaction between the polycationic structure and the predominantly anionic components of the microorganisms' surface, such as Gram-negative lipopolysaccharide and cell surface proteins, plays a key role in antibacterial activity [23].

As previously mentioned, protein adsorption is the first step to take place upon implantation. Protein adsorption occurs within a few minutes or even seconds after scaffold implantation and the cells that reach the biomaterial surface no longer attach directly to the biomaterial but to the adsorbed protein layer. Through cell membrane-bound receptors or ligands, cells identify bioactive binding sites on the protein layer and behave according to the stimuli received [15]. As a natural positive-charged polysaccharide, protonable amino groups on the chitosan backbone electrostatically interact with the various negatively charged proteins [24]. Electrostatic interactions between biomaterial and proteins depend on the biomaterials' surface and protein charges, which are a function of pH and the solution ionic content. Usually, at low pH, proteins are positively charged, whereas at high pH they are negatively charged [15]. Bovine serum albumin protein is often used as a model protein for biomaterial characterization regarding protein adsorption capacity, because of its high stability, availability at high purity and water solubility [25]. Interactions between BSA protein and chitosan chain depend on the pH and the interaction mechanism is highly complex. BSA protein is negatively charged at neutral pH and the electrostatic interaction of BSA with chitosan is governed by the following two factors: (i) the interaction between protonated chitosan amino groups and the dissociated carboxyl groups of BSA and (ii) the repulsion of the protonated amino groups of chitosan and BSA, as explained by Kim et al. [26]. The protein adsorption capacity of scaffolds needs to be determined, as protein adsorption is the first and crucial step after biomaterial implantation. However, to develop and design a suitable scaffold for bone tissue regeneration, numerous requirements need to be addressed.

3. Requirements for Bone Scaffold Development

Scaffolds for bone tissue regeneration must be biocompatible, non-toxic and biodegradable, with an ability to mold into various geometries and forms suitable for cell seeding, migration, growth and differentiation. The structure should mimic the porous and phase structure of the natural bone while maintaining suitable mechanical properties [27,28].

3.1. Biocompatibility

Biocompatibility is one of the essential requirements for materials used in tissue engineering applications. Biocompatible materials do not produce a toxic or immunological response in the human body [5]. In almost all published papers, chitosan is described as a non-toxic and biocompatible biopolymer safe for use in the human organism as a scaffold or drug carrier. However, the biocompatibility must be confirmed by biological evaluation for each chitosan-based biomaterial, as they might have different physicochemical characteristics due to different biogenic sources, chitosan type, molecular weight, DD of chitosan and different phases incorporated into the chitosan to obtain composite biomaterials with multifunctional characteristics. In addition, non-cytotoxicity is commonly assessed for 3 or 7 days of cell culture; however, the extended time period of evaluation should be considered. Along with the required extended cell culture time, the appropriate cell lines for bone applications should be used.

3.2. Porosity and Pore Size Distribution

Porosity, pore size distribution and pore diameter are some of the most important factors for efficient cell attachment, migration, vascularization and tissue regeneration [29]. Cortical bone has a low porosity of 5–10%, whereas trabecular bone has a porosity of 50–90% [5]. During bone regeneration, interconnected pores in the scaffold are essential for the efficient diffusion of nutrient, oxygen and metabolic waste [30]. In order to design a functional scaffold, along with the porosity in the range of 50–90%, micro- (<20 μm) and macroporosity (>100–400 μm) need to be considered [5]. Microporosity is crucial for cell seeding and retention, capillaries growth, vascularization and cell-matrix interactions. Macroporosity promotes osteogenesis by enhancing cell migration, cell–cell network formation, vascularization, nutrient supply and metabolic waste diffusion [3,30]. Oh et al. [31] systematic study on pore size gradient scaffolds has shown that 380–405 μm pore size has better cell growth for chondrocytes and osteoblasts, whereas the scaffolds with 186–200 μm pore size were better for fibroblasts' growth. In addition, scaffolds with 290–310 μm pore size showed faster new bone formation than those of other pore sizes. Zhou et al. [32] obtained chitosan-based scaffolds with different bioactive phases, hydroxyapatite and whitlockite, and pore size of ~105 μm. In vivo studies have shown new bone formation within the scaffolds, meaning that pores of ~105 μm meet the requirements for efficient cell seeding and bone ingrowth. An innovative approach to obtain a multi-layered chitosan scaffold with a gradient of pore size (160–275 μm) for osteochondral defect repair was developed by Pitrolino et al. [33]. Osteogenic and chondrogenic differentiation of human mesenchymal stem cells (MSCs) preferentially occurred in selected layers of the scaffold in vitro, driven by the distinct pore gradient and material composition. In the study by Ressler et al. [34], a multi-substituted (Sr^{2+}, Mg^{2+}, Zn^{2+} and SeO_3^{2-}) calcium phosphate/chitosan composite scaffold with a pore size in the range of 20–350 μm and a porosity of ~75% was prepared by the freeze-gelation method. The requirements for micro- and macroporosity were successfully achieved by adjusting the polymer concentration in the starting solution. Different pore size distributions in these studies indicate that, by using different preparation methods and chitosan concentration of the starting solution, the pore size distribution and porosity can be adjusted and controlled. The pore size distribution and porosity should be some of the main scaffold characteristics considered prior to scaffold development. If the pores are mainly micropores, seeded cells can clog the pores on the scaffold surface and disable diffusion, tissue ingrowth and regeneration. If the pores are mainly macropores, seeding of the cell would not be efficient and that might lead to parts of the scaffold where tissue regeneration is not possible.

3.3. Mechanical Strength

The mechanical strength is a critical feature in bone regeneration and it is primarily controlled by pore volume and characteristics of used materials [22]. Optimum balance between porosity, pore size distribution and mechanical properties requirements is still a major challenge in the development of the scaffold. The compressive strength of a trabecular bone is 2–12 MPa, whereas for the cortical bone it is 100–230 MPa [35–37]. The mechanical properties of scaffolds for load-bearing applications should be such to successfully replace hard bone tissue [30]. The mechanical characteristics of chitosan scaffolds are significantly lower than the compressive strength and modulus of natural bone tissue. Reported compressive modulus and strength differ depending on the scaffold characteristics, but fall in the ranges of 0.0038–2.56 MPa [38]. Due to poor mechanical properties, chitosan-based scaffolds can be used for non-load-bearing applications, mainly as support for osteoblast cells to adhere, proliferate and differentiate into mature bone cells, producing mineralized extracellular matrix or as drug carriers [39]. Poor mechanical properties limit chitosan-based scaffolds to small bone loss in non-load-bearing implantation areas and improvement of such biomaterials is needed if they would be used for load-bearing applications [39,40]. An innovative approach to improve the mechanical properties of hydroxyapatite/chitosan scaffolds was reported by Rogina et al. [40]. A 3D-printed poly (lactic acid) (PLA) construct was used as a mechanical support, where large pores of 960 ± 50 μm allowed enough space to form a porous composite hydrogel by freeze-gelation technique. PLA and PLA/chitosan scaffolds show similar linear region behavior under loading with a modulus of 32.3 ± 5.4 and 27.3 ± 3.2 MPa, respectively, whereas composite scaffolds based on PLA and hydroxyapatite/chitosan hydrogel possessed lower stiffness with the modulus of 16.4 ± 2.5 MPa [40]. Depending on the application of chitosan-based scaffolds, mechanical properties are one of the main characteristics that should be considered during scaffold design. The design of a scaffold with appropriate porosity, pore diameter and mechanical properties is still a challenge, as these parameters are correlated and their compensation is required.

3.4. Biodegradation

The ideal scaffold for bone regeneration should degrade at the same rate as the new tissue formation. If the rate of degradation is higher than the regeneration rate, the scaffold cannot provide support for the host tissue and the regeneration would not be efficient. At physiological conditions, chitosan undergoes physical (e.g., swelling, cracking, dissolution) and chemical (e.g., depolymerization, oxidation, non-enzymatic and enzymatic hydrolysis) degradation. Hydrolytic degradation of the glycosidic bonds between polysaccharide units occurs at a higher rate, making non-enzymatic hydrolytic mechanisms a minor part of chitosan degradation [41]. Chitosan can be enzymatically degraded in vivo by lysozyme, a polycationic protein present in the extracellular matrix in human bone tissue [30]. Lysozyme breaks the chitosan chain by cleaving the glycosidic bonds between polysaccharide units in the polymer. As a result, the molecular weight of the polymer is reduced until eventual solubility and removal of degradation products occur. The degradation products are non-toxic, mainly composed of glucosamine and saccharide, which can then be easily extracted from the body without interference with organs. The degradation rate by each mechanism is related to the degree of crystallinity, which is controlled by the DD, where higher DD results in a lower degradation rate due to closer chain packing and hydrogen bonding [5]. The lysozyme concentration in the extracellular matrix of human tissues can increase up to 1000-fold the amount usually found in serum (0.95–2.45 µm) [42–44]. Therefore, it is important to determine the degradation rate of chitosan scaffolds at lower and higher concentrations of lysozyme to examine scaffold stability under physiological conditions.

4. Chitosan Three-Dimensional Scaffolds

Due to chitosan's physical and chemical properties, various types of scaffolds (molded macroporous, fiber-based, hydrogel, microspheres and 3D-printed) can be obtained (Figure 4) for specific treatments that require unique properties. Increasing attention has been gained by 3D-printed chitosan-based scaffolds in recent years, as this technique enables the biofabrication of patient-personalized scaffolds with highly complex geometries. In recent years, a few high-quality review papers on the 3D printing of chitosan, including bioprinting, were published by Rajabi et al. [45], Taghizadeh et al. [46] and Yadav et al. [47].

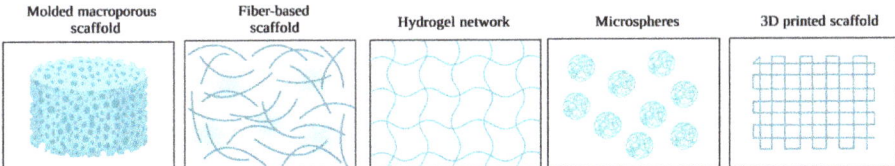

Figure 4. Different designs of chitosan-based three-dimensional scaffolds. Created with BioRender.com (accessed on 30 June 2022).

4.1. Molded Macroporous Scaffolds

Compared to the fibers, hydrogel, microspheres and 3D-printed scaffolds, molded chitosan scaffolds are the most studied. The most commonly used method is phase separation and lyophilization, where molded chitosan solution is frozen to allow phase separation [48]. As acetic acid is most commonly used for dissolving chitosan, after the lyophilization the neutralization of chitosan acetate salt is required to prevent scaffold dissolution in the aqueous media. The freeze-gelation method is similar to a previously explained method where after phase separation due to freezing, scaffolds are placed in the gelation solution of sodium hydroxide and ethanol below the chitosan freezing temperature. Following the gelation, scaffolds are washed with ethanol and lyophilized (dried). A combination of the described methods can be used [34,40]. To obtain the desired pore dimension and shape, the polymer concentration, freezing speed and freezing temperature need to be adjusted. In addition, prior to the step phase separation/lyophilization method, porogens can be added to the chitosan solution. The porogens are later leached from the scaffold, leading to additional porosity. When porogens are used without combination with the separation/lyophilization method, the drawback is that this method leads to a lack of control over the interconnectivity of pores inside the scaffold structure. Further, the gas foaming technique can be used alone or in combination with porogens to obtain an open porosity of scaffolds. The high-pressure carbon dioxide (CO_2) is allowed to saturate the polymeric solution, which causes clusters in the solution and induces porosity [30].

4.2. Fiber-Based Scaffolds

Electrospinning is a process that utilizes an electric field to control the deposition of polymer fibers onto target substrates [49]. Compared to synthetic polymers, natural polymers are less spinnable because of limited solubility in most organic solvents, high molecular weight, a polycationic character in solution and three-dimensional networks of strong hydrogen bonds [50]. Fiber-based chitosan scaffolds obtained by electrospinning were highly studied in the 2000s. Homayoni et al. [51] resolved the problem of chitosan high viscosity, which limits its spinnability, through the application of an alkali treatment that hydrolyzes chitosan chains and decreases its molecular weight. Solutions of the treated chitosan in aqueous 70–90% acetic acid produce nanofibers with appropriate quality and processing stability. Optimum nanofibers are achieved with chitosan that is hydrolyzed for 48 h, with a nanofiber diameter of 140 nm. The fiber diameter is strongly affected by the electrospinning conditions and solvent concentration. Recent reviews of the literature for the electrospinning of chitosan-based solutions for tissue engineering and regenerative medicine applications were provided by Qasim et al. [52] and Anisiei et al. [53].

4.3. Hydrogel

Conventional methods for applications in tissue engineering (pre-formed hydrogels and scaffolds) face the problem of surgical implantation, increasing the risk of infections and improper scaffold shape and size [54]. In the last decade, smart injectable hydrogels have gained increasing attention because they can be used in minimally invasive treatments [55]. Detailed review papers, regarding mechanisms of injectable hydrogel formation and application in adipose, bone, cartilage, intervertebral discs and muscle tissue engineering were published by Sivashanmugam et al. [54] and Gasperini et al. [56]. Smart injectable gelling systems are liquid at room temperature, and then form gels when injected into the fractured location, filling the complex shape of the defect [55]. Such hydrogels should shorten the surgical operation time, minimize the damage effects of large muscle retraction, reduce the size of scars and lessen post-operative pain, allowing patients to achieve rapid recovery in a cost-effective manner [57]. Hydrogels can be used in non-load-bearing applications to carry and protect cells, proteins, growth factors or drugs, and ensure adequate permeability for the transport of cells' nutrients and metabolites [58]. A highly important characteristic of injectable hydrogels is gelation time, as slow gelation can cause delocalized gel formation due to the gel precursor diffusion [59]. Hydrogels derived from naturally occurring polysaccharides mimic many features of the extracellular matrix. Therefore, they can direct the migration, growth and organization of encapsulated cells during tissue regeneration [16]. As previously mentioned, chitosan is a pH-responsive polymer, as in mild acids it is soluble and upon neutralization it forms a hydrogel. This occurs due to the removal of repulsive electrostatic interactions during the neutralization process, thereby allowing the amino groups to interact via intermolecular hydrogen bonding [46]. The anionic nature of most human tissues, due to the presence of glycosaminoglycans in the extracellular matrix and the cationic character of chitosan, allows adherence of these hydrogels to tissue sites [19]. In addition, as the cells are negatively charged, positively charged scaffolds are expected to provide a more suitable environment for attachment due to ionic or electrostatic interactions [54]. Due to the polycationic nature of chitosan, pH and thermally induced physical cross-linked hydrogels are highly interesting, as they can be obtained without using cross-linking agents that might be toxic to the human organism. Glycerophosphate salts are widely used for obtaining pH and thermo-sensitive injectable chitosan hydrogels; however, sodium bicarbonate ($NaHCO_3$) can also be used as a gelling agent [60,61]. In situ synthesized hydroxyapatite in a chitosan solution (10 °C) was used to obtain pH-responsive hydrogel at 37 °C. A slightly acidic environment of prepared composite solution favors $NaHCO_3$ dissociation that releases HCO_3^- ions responsible for carbon dioxide production and pH increases (Figure 5). Although the sol-gel transitions in chitosan solutions with $NaHCO_3$ as a gelling agent appeared to be thermally sensitive upon the temperature increase, these systems performed the pH-induced gelation process. The decrease in the apparent charge density of chitosan molecules allows the formation of the three-dimensional chitosan network due to physical junctions of hydrogen bonds [61].

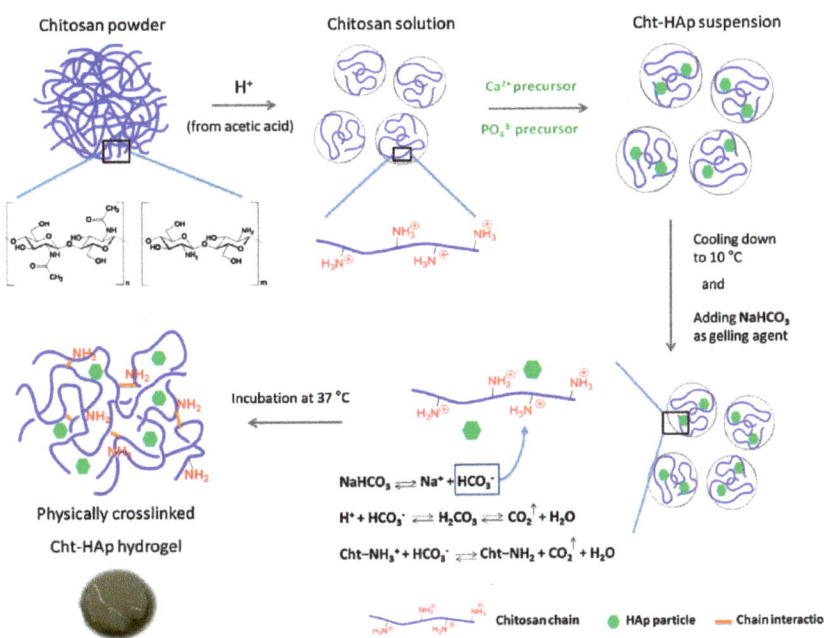

Figure 5. Preparation of physically cross-linked chitosan-hydroxyapatite (Cht-HA) hydrogel, with permission from Elsevier [61].

4.4. Microspheres

Due to biocompatibility and biodegradability, chitosan microsphere systems have been proposed for use as injectable bone-filling (non-load-bearing) biomaterial and/or drug delivery matrices [62]. Chitosan microspheres for drug delivery and preparation methods were summarized in the review paper by Mitra and Dey [63]. Cell microcarriers in the form of injectable scaffolds offer advantages similar to ones characteristic of injectable hydrogels that repair complex-shaped tissue defects with minimal surgical intervention [64]. Wang et al. [65] fabricated collagen/chitosan-based microspheres (diameter of 200 μm) via the emulsification method by using glutaraldehyde as a cross-linking agent. Obtained microspheres have shown stability for at least 90 days and good biological properties by supporting attachment and proliferation of the cells. As previously described, chitosan degradation products are not toxic for cells and the human organism; however, to obtain stable chitosan microspheres, chemical cross-linking is required to cross-link amino groups in the chitosan chain. As suggested by Fang et al. [64], residual cross-linking agents in microspheres might have a toxic effect towards cells, surrounding tissue and the human organism. Complete removal of unreacted cross-linking agents from obtained scaffolds remains a challenge. To overcome these drawbacks of chemically cross-linked chitosan, the introduction of a bioactive polyanionic biopolymer to interact electrostatically with the amino groups of chitosan to form polyelectrolyte complexes (PEC) has been proposed. Fang et al. [64] obtained poly(L-glutamic acid)/chitosan PEC porous microspheres by electrostatic interactions. It was determined that the pore size distribution, porosity, structure and stability of microspheres are dependent on freezing temperature and polymer concentration (Figure 6). An additional study with an approach free of toxic cross-linking agents was conducted by Huang et al. [66]. Highly porous chitosan microspheres were prepared through an emulsion-based thermally induced phase method with an average diameter of microspheres of ~150 μm and with interconnected pores in the range of 20–50 μm. Obtained microspheres showed excellent biocompatibility with multidirectional cell–cell interactions. Another approach to avoid cross-linking agents to produce stable chitosan-based microspheres is through physical cross-linking via chelation interactions of copper and zinc with chitosan, as recently reported by Lončarević et al. [67] and Rogina et al. [68]. The studies highlight the alternative approach to produce stable chitosan-based microspheres by using simple complexation reactions through transition metal ions.

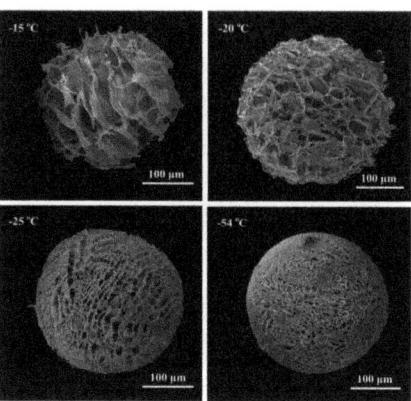

Figure 6. Scanning electron micrographs of porous chitosan microspheres fabricated from chitosan at a concentration of 2% at different temperatures. Reprinted from [64] with permission from Elsevier.

For usage as a bone-filling biomaterial, the drawbacks of pure chitosan microspheres are lack of bone-binding ability and burst release problems. To overcome limitations, Ding et al. [62] proposed that obtaining composite microspheres based on chitosan and hydroxyapatite can lead to an increase in bone-binding ability. In vivo studies on apatite-coated chitosan microsphere conducted by Xu et al. [69] showed bone formation after 7 days. Further, hydroxyapatite/sodium alginate/chitosan composite microspheres, reported by Bi et al. [70], were prepared by an emulsion cross-link technique where calcium ions were used as a cross-linking agent. However, the microspheres as microcarriers often only enabled cell attachment and growth on the surface due to low or closed porosity [62,66]. Although multiple chitosan-based microspheres have been developed, minority studies report highly porous microspheres with open porosity that enable cell migration and cell–cell interactions in the 3D environment. However, even with low or closed porosity, the advantage of microspheres for use in biomedical applications over a granular approach for bone repair is a larger specific surface area, which can improve cell adhesion and proliferation [70]. In addition, chitosan microspheres can be used as a filler component in molded scaffolds based on other polymers [71], or microspheres can be molded to obtain highly porous scaffolds [72].

5. Chitosan Composite Scaffolds

The extracellular matrix in natural bone tissue supports cell attachment, proliferation and differentiation. Scaffolds for bone regeneration should mimic the natural ECM as much as possible. In particular, integration of multiple stimuli in scaffolds including physical (e.g., porosity, pore size distribution, topography, stiffness) and biochemical (e.g., growth factors, key role elements, genes, proteins) factors similar to natural bone tissue will improve scaffold efficacy [3,12]. MSCs are used in regenerative medicine because of their potential for self-renewal and multipotency. Multipotent stem cells can differentiate into multiple lineages (e.g., myocyte, adipocyte, osteoblast, chondrocyte, neuron), as schematically shown in Figure 7 [73,74]. Biomaterials can direct MSC attachment, proliferation and differentiation into different cell types and this can be controlled by the optimization of material characteristics such as composition, geometry, pore size, porosity, topography, stiffness, etc. [75]. Chitosan and its polymer-based composites are often used for the development of materials due to the similarity of the polysaccharide structure to the glycosaminoglycans of cartilage and as it can direct MSCs differentiation towards chondrocyte cell type (chondrogenesis) [76–78]. However, in order for chitosan-based scaffolds to mimic natural bone tissue and stimulate differentiation into osteoblast cell types (osteogenesis), chitosan is often combined with inorganic phases. Calcium phosphates, calcium silicates and bioactive glass are among the most studied bioactive components within the chitosan matrix to mimic naturally occurring mineral phases and stimulate the osteogenic differentiation of cells. An innovative approach was reported by Pitrolino et al. [33] and Erickson et al. [79], who obtained a multi-layered chitosan scaffold for osteochondral defect repair with the incorporation of the highly porous layer based on chitosan and hydroxyapatite for bone regeneration.

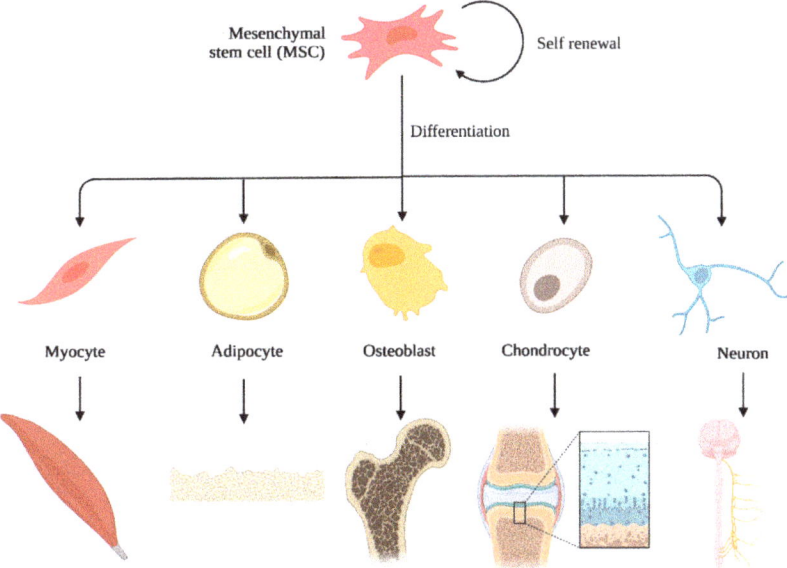

Figure 7. Mesenchymal stem cells differentiation. Created with BioRender.com (accessed on 2 July 2022).

5.1. Calcium Phosphates

Calcium phosphates (e.g., hydroxyapatite, α-tricalcium phosphate, β-tricalcium phosphate) are among the most studied bioactive phases combined with the chitosan matrix. Inspired by natural bone tissue, numerous studies are focused on the scaffolds, where chitosan provides an organic matrix mimicking naturally occurring collagen, while calcium phosphate crystals mimic naturally occurring minerals (apatite). Numerous studies have confirmed the osteogenic properties of chitosan/calcium phosphate-based scaffolds. Rogina et al. [80,81] confirmed that compressive strength and swelling capacity measured in physiological conditions have shown that the critical hydroxyapatite portion, which improves chitosan properties, does not exceed 30 wt%, while further studies in perfusion conditions confirmed the best influence of hydroxyapatite on hMSC proliferation and osteoinduction on composite scaffolds with 30% of hydroxyapatite. A higher apatite fraction indicated poor mineralization of hMSCs extracellular matrix. Siddiqui et al. [82] prepared composite scaffolds based on chitosan and β-tricalcium phosphate cross-linked with genipin, confirming osteogenic differentiation during 21 days of cell culture. In recent years, to mimic the chemical composition of natural apatite, calcium phosphates were substituted with various key role ions and combined with chitosan. Ran et al. [83] obtained Mg-, Zn-, Sr- and Si-doped hydroxyapatite/chitosan hydrogels and confirmed that the Sr-chitosan/hydroxyapatite hydrogel exhibited the highest proliferation potential among the cultured cells compared to the samples with other ions. To further enhance osteogenic properties and improve mechanical properties, Zhang et al. [84] obtained composite scaffolds based on silk fibroin, carboxymethyl chitosan, strontium substituted hydroxyapatite and cellulose. Mansour et al. [85] prepared a chitosan-based scaffold loaded with Ag/Mg-co-substituted hydroxyapatite. Ghorbani et al. [86] prepared an electrospun scaffold (fiber diameter 210 nm) based on PCL, chitosan and zinc-doped hydroxyapatite, showing a positive effect on cell attachment and proliferation. Further, the synergic effect of Sr^{2+}, Mg^{2+}, Zn^{2+} and SeO_3^{2-} ions was confirmed in perfusion conditions. It has been determined that ions have a significant influence on the expression of characteristic bone genes (alkaline phosphatase, bone sialoprotein, collagen type I and dentin matrix protein I), phosphate deposits and newly formed bone tissue. Recent studies have shown better osteogenic properties of substituted calcium phosphate/chitosan scaffolds compared to the non-substituted scaffolds [34]. Therefore, further studies can be focused on using substituted calcium phosphates with chitosan and other polymers, such as collagen, which can further increase osteogenic properties.

5.2. Calcium Silicate

Along with calcium phosphates and silicates, calcium silicates are promising biocompatible ceramic materials that can provide a microenvironment suitable for bone tissue regeneration. Along with Ca^{2+} ions, silicate ions have a key role in the bone regeneration process, as they can regulate MEK and PKC pathways [87]. In a recent study by Zhou et al. [88], it has been demonstrated that calcium silicate had significantly greater osteoinductive capacity both in vitro and in vivo compared with the traditional clinically used β-tricalcium phosphate bioceramics. Further, ionic substitutions are not only studied for calcium phosphates. In recent years, increasing attention has been directed towards substituted calcium silicates, especially for substitutions with Sr^{2+} ions [89–91]. Due to the positive effect of calcium silicates on bone regeneration, significant efforts have been directed toward obtaining scaffolds based on chitosan and calcium silicates. Peng et al. developed a lanthanum- [87] and gadolinium [92] -doped mesoporous calcium silicate/chitosan scaffold, which supports the adhesion, proliferation and differentiation of MSCs. Lin et al. [93] enhanced calcium silicate properties by obtaining composite scaffolds based on chitosan to ensure the antibacterial properties of the scaffolds. Mukherjee et al. [94] reported improved osteoblast function (viability, adhesion and proliferation) on titanium implant surfaces coated with a nanocomposite based on apatite, wollastonite ($CaSiO_3$) and chitosan. A significant increase in the expression of osteocalcin and mineralization, compared to a non-treated substrate, confirmed the biocompatibility of the composite coating and its ability to initiate early osseointegration. Further, Genasan et al. [95] confirmed that the addition of calcium silicate (40% w/w) into gellan-chitosan scaffolds induces osteogenic differentiation of mesenchymal stromal cells where significant depositions of minerals, along with the expression of osteogenic genes, including BMP2, Run2, osteocalcin and osteonectin, were detected. Even though the calcium silicate-based scaffolds have shown desired properties for bone regeneration, more studies are focused on scaffold development and characterization based on calcium phosphates. Future studies should be focused on the development of scaffolds based on calcium silicates and substituted calcium silicates within the chitosan matrix and should be compared to similar scaffolds based on calcium phosphates and chitosan.

5.3. Bioactive Glass

Along with calcium phosphates, bioactive glasses are used as bioactive fillers in chitosan-based scaffolds to increase cell response and osteogenic properties. Bioactive glasses are widely used for bone tissue regeneration due to their chemical interactions in vivo, where osseointegration is promoted by the formation of a calcium phosphate layer [96]. When included in the chitosan matrix, bioactive glasses enhance the metabolic activity of cells and mineralization [97]. Saatchi et al. [98] reported chitosan/polyethylene oxide nanofibrous scaffolds containing different amounts of cerium-doped bioactive glass. It has been determined that increasing the content of cerium-doped bioactive glass, cell adhesion and spreading have been enhanced. Further, fibroblast cells spread across the composite scaffold and took a 3D shape; however, there was no sign of cell expansion on the polymer scaffold without cerium-doped bioactive glass. In addition, composites based on chitosan and bioactive glasses are used as coating materials on AZ91 Mg alloy [99], 316 L stainless steel [100], WE43 Mg alloy [101,102] and Ti-6Al-4V [103] to improve the biocompatibility and bioactivity of metallic substrates for biomedical applications. To prevent the formation of biofilm on orthopedic implants, coatings based on chitosan and bioactive glasses can be combined with different drugs (e.g., vancomycin) to prevent the adhesion and proliferation of bacteria, as reported by Zarghami et al. [104]. Further, Sergi et al. [105] prepared a composite based on commercial passive gauzes, chitosan and bioactive glass doped with Sr^{2+}, Mg^{2+} and Zn^{2+} ions for wound healing. It was determined that wound dressings with obtained composite material showed higher bioactivity compared to wound dressings with pure chitosan. The release of Sr^{2+}, Mg^{2+} and Zn^{2+} ions enhanced cell proliferation and wound healing rate. A composite system based on chitosan and doped bioactive glass could be further examined for bone tissue engineering applications. In addition to a research paper, a detailed review paper by Sergi et al. [106] provided an overview of studies on bioactive glasses and natural polymer composites for medical devices for both soft and hard tissues. Due to different ions present in the bioactive glasses, further studies should provide additional comparative studies and a better understanding of the significance of each ionic component in bioactive glasses and its influence on the osteogenic properties of chitosan scaffolds.

6. Conclusions and Future Perspective

Various chitosan-based materials have been designed and reported in the literature. However, more efforts are required to address current challenges to bring developed biomaterials to clinical use and application. Prior to the scaffold design, the researcher should consider all requirements for in vivo studies, clinical trials and mass productions. *ISO 10993 Biological evaluation of medical devices—Part 1: Evaluation and testing within a risk management process* should be considered prior to biological evaluation. Further, the characterization of chitosan-based materials for bone regeneration is not standardized. Even if there is a large number of papers on chitosan-based scaffolds as potential materials for bone regeneration, different methods for characterization are used. Therefore, the results of different studies cannot be properly compared and a final conclusion on material potential cannot be conducted. In recent years, the standardization of protocols and regulation of biomaterials has become highly required in order to improve technology transfer and increase the amount of commercially available products [107–109]. An additional challenge characteristic for naturally derived polymers is that the different properties depend on the source and preparation method. This further disables comparison between obtained scaffolds from different studies. With joint efforts from researchers by following requirements for biomaterials design and characterization, more developed biomaterials could be translated to clinical trials and be approved for commercial use.

Funding: This research received no external funding.

Acknowledgments: Smart Servier Medical Art (http://smart.servier.com/, accessed on 30 June 2022) and BioRender.com are gratefully acknowledged.

Conflicts of Interest: The author declares no conflict of interest.

References

1. Amini, A.R.; Laurencin, C.T.; Nukavarapu, S.P. Bone Tissue Engineering: Recent Advances and Challenges. *Crit. Rev. Biomed. Eng.* **2012**, *40*, 363–408. [CrossRef] [PubMed]
2. Zhu, G.; Zhang, T.; Chen, M.; Yao, K.; Huang, X.; Zhang, B.; Li, Y.; Liu, J.; Wang, Y.; Zhao, Z. Bone physiological microenviroment and healing mechanism: Basis for future bone-tissue engineering scaffolds. *Bioact. Mater.* **2021**, *6*, 4110–4140. [CrossRef] [PubMed]

3. Fernandez-Yague, M.A.; Abbah, S.A.; McNamara, L.; Zeugolis, D.I.; Pandit, A.; Biggs, M.J. Biomimetic approaches in bone tissue engineering: Integrating biological and physicomechanical strategies. *Adv. Drug Deliv. Rev.* **2014**, *84*, 1–29. [CrossRef] [PubMed]
4. Hench, L.L.; Thompson, I. Twenty-first century challenges for biomaterials. *J. R. Soc. Interface* **2010**, *7*, S379–S391. [CrossRef]
5. Cordonnier, T.; Sohier, J.; Rosset, P.; Layrolle, P. Biomimetic Materials for Bone Tissue Engineering—State of the Art and Future Trends. *Adv. Eng. Mater.* **2011**, *13*, B135–B150. [CrossRef]
6. Gordon, J.A.R.; Tye, C.E.; Sampaio, A.V.; Underhill, T.M.; Hunter, G.K.; Goldberg, H.A. Bone sialoprotein expression enhances osteoblast differentiation and matrix mineralization in vitro. *Bone* **2007**, *41*, 462–473. [CrossRef]
7. Dorozhkin, S.V.; Epple, M. Biological and medical significance of calcium phosphates. *Angew. Chem. Int. Engl.* **2002**, *41*, 3130–3146. [CrossRef]
8. Boskey, A.L. Bone composition: Relationship to bone fragility and antiosteoporotic drug effects. *Bonkey Rep.* **2013**, *2*, 447. [CrossRef]
9. Sadat-Shojai, M.; Khorasani, M.T.; Dinpanah-Khoshdargi, E.; Jamshidi, A. Synthesis methods for nanosized hydroxyapatite with diverse structures. *Acta Biomater.* **2013**, *9*, 7591–7621. [CrossRef]
10. Colaco, E.; Brouri, D.; Aissaoui, N.; Cornette, P.; Dupres, V.; Domingos, R.F.; Lambert, J.F.; Maisonhaute, E.; El Kirat, K.; Landoulsi, J. Hierarchical Collagen–Hydroxyapatite Nanostructures Designed through Layer-by-Layer Assembly of Crystal-Decorated Fibrils. *Biomacromolecules* **2019**, *20*, 4522–4534. [CrossRef]
11. Simon, P.; Grüner, D.; Worch HPompe, W.; Lichte, H.; El Khassawna, T.; Heiss, C.; Wenisch, S.; Kniep, E. First evidence of octacalcium phosphate osteocalcin nanocomplex as skeletal bone component directing collagen triple–helix nanofibril mineralization. *Sci. Rep.* **2018**, *8*, 13696. [CrossRef] [PubMed]
12. Ressler, A.; Žužić, A.; Ivanišević, I.; Kamboj, N.; Ivanković, H. Ionic substituted hydroxyapatite for bone regeneration applications: A review. *Open Ceram.* **2021**, *6*, 100122. [CrossRef]
13. Pillai, C.K.S.; Paul, W.; Sharma, S.P. Chitin and chitosan polymers: Chemistry, solubility and fiber formation. *Prog. Polym. Sci.* **2009**, *34*, 641–678. [CrossRef]
14. Xia, B.; Deng, Y.; Lv, Y.; Chen, G. Stem cell recruitment based on scaffold features for bone tissue engineering. *Biomater. Sci.* **2021**, *9*, 1189–1203. [CrossRef] [PubMed]
15. Felgueiras, H.P.; Antunes, J.C.; Martins, M.C.L.; Barbosa, M.A. 1—Fundamentals of protein and cell interactions in biomaterials. In *Martins, Peptides and Proteins as Biomaterials for Tissue Regeneration and Repair*; Mário, A., Barbosa, M., Cristina, L., Eds.; Woodhead Publishing: Sawston, UK, 2018; pp. 1–27.
16. Tan, H.; Chu, C.R.; Payne, K.A.; Marra, K.G. Injectable in situ forming biodegradable chitosan-hyaluronic acid based hydrogels for cartilage tissue engineering. *Biomaterials* **2009**, *30*, 2499–2506. [CrossRef] [PubMed]
17. Mansouri, R.; Jouan, Y.; Hay, E.; Blin-Wakkach, C.; Ostertag, A.; Le Henaff, C.; Marty, C.; Geoffroy, V.; Marie, P.J.; Cohen-Solal, M.; et al. Osteoblastic heparan sulfate glycosaminoglycans control bone remodeling by regulating Wnt signaling and the crosstalk between bone surface and marrow cells. *Cell Death Dis.* **2017**, *8*, e2902. [CrossRef]
18. Islam, S.; Bhuiyan, M.A.R.; Islam, M.N. Chitin and Chitosan: Structure, Properties and Applications in Biomedical Engineering. *J. Polym. Environ.* **2017**, *25*, 854–866. [CrossRef]
19. Jiang, Y.; Fu, C.; Wu, S.; Liu, G.; Guo, J.; Su, Z. Determination of the Deacetylation Degree of Chitooligosaccharides. *Mar. Drugs* **2017**, *15*, 332. [CrossRef]
20. Lv, S.H. 7—High-performance superplasticizer based on chitosan. In *Biopolymers and Biotech Admixtures for Eco-Efficient Construction Materials*; Fernando, P.-T., Volodymyr, I., Niranjan, K., Henk, J., Eds.; Woodhead Publishing: Sawston, UK, 2016; pp. 131–150.
21. Chenite, A.; Buschmann, M.; Wang, D.; Chaput, C.; Kandani, N. Rheological characterization of thermogelling chitosan/glycerol-phosphate solutions. *Carbohydr. Polym.* **2001**, *46*, 39–47. [CrossRef]
22. Upadhyaya, L.; Singh, J.; Agarwal, V.; Tewari, R.P. Biomedical applications of carboxymethyl chitosans. *Carbohydr. Polym.* **2013**, *91*, 452–466. [CrossRef]
23. Yilmaz Atay, H. Antibacterial Activity of Chitosan-Based Systems. In *Functional Chitosan*; Jana, S., Jana, S., Eds.; Springer: Singapore, 2019.
24. Aguilar, A.; Zein, N.; Harmouch, E.; Hafdi, B.; Bornert, F.; Offner, D.; Clauss, F.; Fioretti, F.; Huck, O.; Benkirane-Jessel, N.; et al. Application of Chitosan in Bone and Dental Engineering. *Molecules* **2019**, *24*, 3009. [CrossRef] [PubMed]
25. Phan, H.T.; Bartelt-Hunt, S.; Rodenhausen, K.B.; Schubert, M.; Bartz, J.C. Investigation of Bovine Serum Albumin (BSA) Attachment onto Self-Assembled Monolayers (SAMs) Using Combinatorial Quartz Crystal Microbalance with Dissipation (QCM-D) and Spectroscopic Ellipsometry (SE). *PLoS ONE* **2015**, *100*, e0141282. [CrossRef] [PubMed]
26. Kim, U.J.; Lee, Y.R.; Kang, T.H.; Choi, J.W.; Kimura, S.; Wada, M. Protein adsorption of dialdehyde cellulose-crosslinked chitosan with high amino group contents. *Carbohydr. Polym.* **2017**, *163*, 34–42. [CrossRef] [PubMed]
27. Roseti, L.; Parisi, V.; Petretta, M.; Cavallo, C.; Desando, G.; Bartolotti, I.; Grigolo, B. Scaffolds for Bone Tissue Engineering: State of the art and new perspectives. *Mater. Sci. Eng. C* **2017**, *78*, 1246–1262. [CrossRef]
28. Venkatesan, J.; Kim, S.K. Chitosan composites for bone tissue engineering-an overview. *Mar. Drugs* **2010**, *8*, 2252–2266. [CrossRef]
29. Bružauskaitė, I.; Bironaitė, D.; Bagdonas, E.; Bernotienė, E. Scaffolds and cells for tissue regeneration: Different scaffold pore sizes-different cell effects. *Cytotechnology* **2016**, *68*, 355–369. [CrossRef]

30. Deb, P.; Deoghare, A.B.; Borah, A.; Baruna, E.; Lala, S.D. Scaffold Development Using Biomaterials: A review. *Mater. Today Proc.* **2018**, *5*, 1209–12919. [CrossRef]
31. Oh, S.H.; Park, I.K.; Kim, J.M.; Lee, J.H. In vitro and in vivo characteristics of PCL scaffolds with pore size gradient fabricated by a centrifugation method. *Biomaterials* **2007**, *9*, 1664–1671. [CrossRef]
32. Zhou, D.; Qi, C.; Chen, Y.; Zhu, Y.J.; Sun, T.; Chen, F.; Zhang, C. Comparative study of porous hydroxyapatite/chitosan and whitlockite/chitosan scaffolds for bone regeneration in calvarial defects. *Int. J. Nanomed.* **2017**, *12*, 2673–2687. [CrossRef]
33. Pitrolino, K.A.; Felfel, R.M.; Pellizzeri, L.M.; McLaren, J.; Popov, A.A.; Sottile, V.; Scotchford, C.A.; Scammell, B.E.; Roberts, G.A.F.; Grant, D.M. Development and in vitro assessment of a bi-layered chitosan-nano-hydroxyapatite osteochondral scaffold. *Carbohydr. Polym.* **2022**, *282*, 119126. [CrossRef]
34. Ressler, A.; Anutnović, M.; Teruel-Biosca, L.; Gallego Ferrer, G.; Babić, S.; Urlić, I.; Ivanković, M.; Ivanković, H. Osteogenic differentiation of human mesenchymal stem cells on substituted calcium phosphate/chitosan composite scaffold. *Carbohydr. Polym.* **2022**, *282*, 119126. [CrossRef] [PubMed]
35. Turnbull, G.; Clarke, J.; Picard, F.; Riches, P.; Jia, L.; Han, F.; Li, B.; Shu, W. 3D bioactive composite scaffolds for bone tissue engineering. *Bioact. Mater.* **2018**, *3*, 278–314.
36. Jacob, J.; More, N.; Kalia, K.; Kapusetti, G. Piezoelectric smart biomaterials for bone and cartilage tissue engineering. *Inflamm. Regen.* **2018**, *38*, 2. [CrossRef] [PubMed]
37. Henkel, J.; Woodruff, M.A.; Epari, D.R.; Steck, R.; Glatt, V.; Dickinson, I.C.; Choong, P.F.M.; Schuetz, M.A.; Hutmacher, D.W. Bone regeneration based on tissue engineering conceptions—A 21st century perspective. *Bone Res.* **2014**, *1*, 216–248. [CrossRef]
38. Jana, S.; Florczyk, S.J.; Leung, M.; Zhang, M. High-strength pristine porous chitosan scaffolds for tissue engineering. *J. Mater. Chem.* **2012**, *22*, 6291–6299. [CrossRef]
39. Przejora, A.; Palka, K.; Ginalska, G. Biomedical potential of chitosan/HA and chitosan/β-1,3-glucan/HA biomaterials as scaffolds for bone regeneration—A comparative study. *Mater. Sci. Eng. C* **2016**, *58*, 891–899. [CrossRef]
40. Rogina, A.; Pubolšan, L.; Hanžek, A.; Gomez-Estrada, L.; Gallego Ferrer, G.; Marijanović, I.; Ivanković, M.; Ivanković, H. Macroporous poly(lactic acid) construct supporting the osteoinductive porous chitosan-based hydrogel for bone tissue engineering. *Polymer* **2016**, *98*, 172–181. [CrossRef]
41. Jennings, J.A. 7—Controlling chitosan degradation properties in vitro and in vivo. In *Bumgardner, Chitosan Based Biomaterials Volume 1*; Jennings, J.A., Bumgardner, J.D., Eds.; Woodhead Publishing: Sawston, UK, 2017; pp. 159–182.
42. Hu, J.; Hou, Y.; Park, H.; Choi, B.; Hou, S.; Chung, A.; Lee, M. Visible light crosslinkable chitosan hydrogels for tissue engineering. *Acta Biomater.* **2012**, *8*, 1730–1738. [CrossRef]
43. Jin, R.; Moreira Teixeira, L.S.; Dijkstra, P.J.; Karperien, M.; van Blitterswijk, C.A.; Zhong, Z.Y.; Feijen, J. Injectable chitosan-based hydrogels for cartilage tissue engineering. *Biomaterials* **2009**, *30*, 2544–2551. [CrossRef]
44. Park, H.; Choi, B.; Hu, J.; Lee, M. Injectable chitosan hyaluronic acid hydrogels for cartilage tissue engineering. *Acta Biomater.* **2013**, *9*, 4779–4786. [CrossRef]
45. Rajabi, M.; McConnell, M.; Cabral, J.; Ali, M.A. Chitosan hydrogels in 3D printing for biomedical applications. *Carbohydr. Polym.* **2021**, *260*, 117768. [CrossRef] [PubMed]
46. Taghizadeh, M.; Taghizadeh, A.; Yazdi, M.K.; Zarrintaj, P.; Stadler, F.J.; Ramsey, J.D.; Habibzadeh, S.; Rad, S.H.; Naderi, G.; Saeb, M.R.; et al. Chitosan-based inks for 3D printing and bioprinting. *Green Chem.* **2022**, *24*, 62–101. [CrossRef]
47. Yadav, L.R.; Chandran, S.V.; Lavanya, K.; Selvamurugan, N. Chitosan-based 3D-printed scaffolds for bone tissue engineering. *Int. J. Biol. Macromol.* **2021**, *183*, 1925–1938. [CrossRef]
48. Capuana, E.; Lopresti, F.; Carfi Pavia, F.; Brucato, V.; La Carrubba, V. Solution-Based Processing for Scaffold Fabrication in Tissue Engineering Applications: A Brief Review. *Polymers* **2014**, *13*, 2041. [CrossRef]
49. Matthews, J.A.; Wnek, G.E.; Simpson, D.G.; Bowlin, G.L. Electrospinning of collagen nanofibers. *Biomacromolecules* **2002**, *3*, 232–238. [CrossRef] [PubMed]
50. Lan Levengood, S.K.; Zhang, M. Chitosan-based scaffolds for bone tissue engineering. *J. Mater. Chem. B* **2014**, *2*, 3461. [CrossRef] [PubMed]
51. Homayoni, H.; Hosseini Ravandi, S.A.; Valizadeh, M. Electrospinning of chitosan nanofibers: Processing optimization. *Carbohydr. Polym.* **2009**, *77*, 656–661. [CrossRef]
52. Qasim, S.B.; Zafar, M.S.; Najeeb, S.; Khurshid, Z.; Shah, A.H.; Husain, S.; Rehman, I.U. Electrospinning of Chitosan-Based Solutions for Tissue Engineering and Regenerative Medicine. *Int. J. Mol. Sci.* **2018**, *19*, 407. [CrossRef]
53. Anisiei, A.; Oancea, F.; Marin, L. Electrospinning of chitosan-based nanofibers: From design to prospective applications. *Rev. Chem. Eng.* **2021**. [CrossRef]
54. Sivashanmugam, A.; Arun Kumar, R.; Vishnu Priya, M.; Nair, S.V.; Jayakumar, R. An overview of injectable polymeric hydrogels for tissue engineering. *Eur. Polym.* **2015**, *72*, 543–565. [CrossRef]
55. Mishra, D.; Bhunia, B.; Banerjee, I.; Datta, P.; Dhara, S.; Maiti, T.K. Enzymatically crosslinked carboxymethyl–chitosan/gelatin/nano-hydroxyapatite injectable gels for in situ bone tissue engineering application. *Mater. Sci. Eng. C* **2011**, *31*, 1295–1304. [CrossRef]
56. Gasperini, L.; Mano, J.F.; Reis, R.L. Natural polymers for the microencapsulation of cells. *J. R. Soc. Interface* **2014**, *11*, 20140817. [CrossRef] [PubMed]
57. Liu, H.; Li, H.; Cheng, W.; Yang, Y.; Zhu, M.; Zhou, C. Novel injectable calcium phosphate/chitosan composites for bone substitute materials. *Acta Biomater.* **2006**, *2*, 557–565. [CrossRef]

58. Couto, D.S.; Hong, Z.; Mani, J.F. Development of bioactive and biodegradable chitosan-based injectable systems containing bioactive glass nanoparticles. *Acta Biomater.* **2009**, *5*, 115–123. [CrossRef] [PubMed]
59. Wu, J.; Zhou, T.; Liu, J.; Wan, Y. Injectable chitosan/dextran-polylactide/glycerophosphate hydrogels and their biodegradation. *Polym. Degrad. Stab.* **2015**, *120*, 273–282. [CrossRef]
60. Liskova, J.; Bačakova, L.; Skwarczynska, A.L.; Musial, O.; Bliznuk, V.; Schamphelaere, K.D.; Modrzejewska, Z.; Douglas, T.E.L. Development of Thermosensitive Hydrogels of Chitosan, Sodium and Magnesium Glycerophosphate for Bone Regeneration Applications. *J. Funct. Biomater.* **2015**, *6*, 192–203. [CrossRef] [PubMed]
61. Rogina, A.; Ressler, A.; Matić, I.; Gallego Ferre, G.; Marijanović, I.; Ivanković, M.; Ivanković, H. Cellular hydrogels based on pH-responsive chitosan-hydroxyapatite system. *Carbohydr. Polym.* **2017**, *166*, 173–182. [CrossRef] [PubMed]
62. Ding, C.C.; Teng, S.H.; Pan, H. In-situ generation of chitosan/hydroxyapatite composite microspheres for biomedical application. *Mater. Lett.* **2012**, *79*, 72–74. [CrossRef]
63. Mitra, A.; Dey, B. Chitosan microspheres in novel drug delivery systems. *Indian J. Pharm. Sci.* **2011**, *73*, 355–366. [PubMed]
64. Fang, J.; Zhang, Y.; Yan, S.; Liu, Z.; He, S.; Cui, L.; Yin, J. Poly(L-glutamic acid)/chitosan polyelectrolyte complex porous microspheres as cell microcarriers for cartilage regeneration. *Acta Biomater.* **2014**, *10*, 276–288. [CrossRef]
65. Wang, D.; Wang, M.; Wang, A.; Li, J.; Li, X.; Jian, H.; Bai, S.; Yin, J. Preparation of collagen/chitosan microspheres for 3D macrophage proliferation in vitro. *Colloids Surf. A Physicochem. Eng. Asp.* **2019**, *572*, 266–273. [CrossRef]
66. Huang, L.; Xiao, L.; Poudel, A.J.; Li, J.; Zhou, P.; Gauthier, M.; Liu, H.; Wu, Z.; Yang, G. Porous chitosan microspheres as microcarriers for 3D cell culture. *Carbohydr. Polym.* **2018**, *202*, 611–620. [CrossRef] [PubMed]
67. Lončarević, A.; Ivanković, M.; Rogina, A. Electrosprayed Chitosan-Copper Complex Microspheres with Uniform Size. *Materials* **2021**, *14*, 5630. [CrossRef] [PubMed]
68. Rogina, A.; Vidović, D.; Antunović, M.; Ivanković, M.; Ivanković, H. Metal ion-assisted formation of porous chitosan-based microspheres for biomedical applications. *Int. J. Polym. Mater. Polym.* **2021**, *7*, 1027–1035. [CrossRef]
69. Xu, F.; Wu, Y.; Zhang, Y.; Yin, P.; Fang, C.; Wang, J. Influence of in vitro differentiation status on the in vivo bone regeneration of cell/chitosan microspheres using a rat cranial defect model. *J. Biomater. Sci. Polym. Ed.* **2019**, *30*, 1008–1025. [CrossRef]
70. Bi, Y.G.; Lin, Z.T.; Deng, S.T. Fabrication and characterization of hydroxyapatite/sodium alginate/chitosan composite microspheres for drug delivery and bone tissue engineering. *Mater. Sci. Eng. C Mater. Biol. Appl.* **2019**, *100*, 576–583. [CrossRef]
71. Guo, Z.; Bo, D.; He, Y.; Luo, X.; Li, H. Degradation properties of chitosan microspheres/poly(L-lactic acid) composite in vitro and in vivo. *Carbohydr. Polym.* **2018**, *193*, 1–8. [CrossRef]
72. Hu, X.; He, J.; Yong, X.; Lu, J.; Xiao, J.; Liao, Y.; Li, Q.; Xiong, C. Biodegradable poly (lactic acid-co-trimethylene carbonate)/chitosan microsphere scaffold with shape-memory effect for bone tissue engineering. *Colloids Surf. B* **2020**, *195*, 111218. [CrossRef]
73. Zakrzewski, W.; Dobrzyński, M.; Szymonowicz, M.; Rybak, Z. Stem cells: Past, present, and future. *Stem Cell. Res. Ther.* **2019**, *10*, 68. [CrossRef]
74. Yang, Y.H.K.; Ogando, C.R.; Wang See, C.W.; Chang, T.Y.; Barabino, G.A. Changes in phenotype and differentiation potential of human mesenchymal stem cells aging in vitro. *Stem Cell. Res. Ther.* **2018**, *9*, 131. [CrossRef]
75. Leach, J.K.; Whitehead, J. Materials-Directed Differentiation of Mesenchymal Stem Cells for Tissue Engineering and Regeneration. *ACS Biomater. Sci. Eng.* **2018**, *4*, 1115–1127. [CrossRef] [PubMed]
76. Oryan, A.; Sahvieh, S. Effectiveness of chitosan scaffold in skin, bone and cartilage healing. *Int. J. Biol. Macromol.* **2017**, *104*, 1003–1011. [CrossRef] [PubMed]
77. Gossla, E.; Bernhardt, A.; Tonndorf, R.; Aibibu, D.; Cherif, C.; Gelinsky, M. Anisotropic Chitosan Scaffolds Generated by Electrostatic Flocking Combined with Alginate Hydrogel Support Chondrogenic Differentiation. *Int. J. Mol. Sci.* **2021**, *22*, 9341. [CrossRef] [PubMed]
78. Sadeghianmaryan, A.; Naghieh, S.; Sardroud, H.A.; Yazdanpanah, Z.; Soltani, Y.A.; Sernaglia, J.; Chen, X. Extrusion-based printing of chitosan scaffolds and their in vitro characterization for cartilage tissue engineering. *Int. J. Biol. Macromol.* **2020**, *164*, 3179–3192. [CrossRef]
79. Erickson, A.E.; Sun, J.; Lan Levengood, S.K.; Swanson, S.; Chang, F.C.; Tsao, C.T.; Zhang, M. Chitosan-based composite bilayer scaffold as an in vitro osteochondral defect regeneration model. *Biomed. Microdevices* **2019**, *21*, 34. [CrossRef] [PubMed]
80. Rogina, A.; Rico, P.; Gallego Ferrer, G.; Ivanković, M.; Ivanković, H. In Situ Hydroxyapatite Content Affects the Cell Differentiation on Porous Chitosan/Hydroxyapatite Scaffolds. *Ann. Biomed. Eng.* **2016**, *44*, 1107–1119. [CrossRef]
81. Rogina, A.; Antunović, M.; Pribolšan, L.; Caput Mihalić, K.; Vukasović, A.; Ivković, A.; Marijanović, I.; Gallego Ferrer, G.; Ivanković, M.; Ivanković, H. Human Mesenchymal Stem Cells Differentiation Regulated by Hydroxyapatite Content within Chitosan-Based Scaffol under Perfusion Conditions. *Polymers* **2017**, *9*, 387. [CrossRef]
82. Siddiqui, N.; Pramanik, K.; Jabbari, E. Osteogenic differentiation of human mesenchymal stem cells in freeze-gelled chitosan/nano β-tricalcium phosphate porous scaffolds crosslinked with genipin. *Mater. Sci. Eng. C* **2015**, *54*, 76–83. [CrossRef]
83. Ran, J.; Jiang, P.; Sun, G.; Ma, Z.; Hu, J.; Shen, X.; Tong, H. Comparisons among Mg, Zn, Sr, and Si doped nano-hydroxyapatite/chitosan composites for load-bearing bone tissue engineering applications. *Mater. Chem. Front.* **2017**, *1*, 900–910. [CrossRef]

84. Zhang, X.Y.; Chen, J.P.; Han, J.; Mo, J.; Dong, P.F.; Zhuo, J.H.; Feng, Y. Biocompatiable silk fibroin/carboxymethyl chitosan/strontium substituted hydroxyapatite/cellulose nanocrystal composite scaffolds for bone tissue engineering. *Int. J. Biol. Macromol.* **2019**, *136*, 1247–1257. [CrossRef]
85. Mansour, S.F.; El-dek, S.I.; Dorozhkin, S.V.; Ahmed, M.K. Physico-mechanical properties of Mg and Ag doped hydroxyapatite/chitosan biocomposite. *New J. Chem.* **2017**, *43*, 13773. [CrossRef]
86. Ghorbani, F.M.; Kaffashi, B.; Shokrollahi, P.; Seyedjafari, E.; Ardeshirylajimi, A. PCL/chitosan/Zn-doped nHA electrospun nanocomposite scaffold promotes adipose derived stem cells adhesion and proliferation. *Carbohydr. Polym.* **2015**, *118*, 133–142. [CrossRef] [PubMed]
87. Peng, X.Y.; Hu, M.; Liao, F.; Yang, F.; Ke, Q.K.; Guo, Y.P.; Zhu, Z.H. La-Doped mesoporous calcium silicate/chitosan scaffolds for bone tissue engineering. *Biomater. Sci.* **2019**, *7*, 1565–1573. [CrossRef]
88. Zhou, P.; Xia, D.; Ni ZOu, T.; Wang, Y.; Zhang, H.; Mao, L.; Lin k Xu, S.; Liu, J. Calcium silicate bioactive ceramics induce osteogenesis through oncostatin M. *Bioact. Mater.* **2021**, *6*, 810–822. [CrossRef] [PubMed]
89. Chiu, Y.C.; Shie, M.Y.; Lin, Y.H.; Lee, A.K.; Chen, Y.W. Effect of Strontium Substitution on the Physicochemical Properties and Bone Regeneration Potential of 3D Printed Calcium Silicate Scaffolds. *Int. J Mol. Sci.* **2019**, *20*, 2729. [CrossRef]
90. Liu, L.; Yu, F.; Li, L.; Zhou, L.; Zhou, T.; Xu, Y.; Lin, K.; Fang, B.; Xia, L. Bone marrow stromal cells stimulated by strontium-substituted calcium silicate ceramics: Release of exosomal miR-146a regulates osteogenesis and angiogenesis. *Acta Biomater.* **2021**, *119*, 444–457. [CrossRef]
91. Huang, T.H.; Kao, C.T.; Shen, Y.F.; Lin, Y.T.; Liu, Y.T.; Yen, S.Y.; Ho, C.C. Substitutions of strontium in bioactive calcium silicate bone cements stimulate osteogenic differentiation in human mesenchymal stem cells. *J. Mater. Sci. Mater. Med.* **2019**, *30*, 68. [CrossRef]
92. Liao, F.; Peng, X.Y.; Yang, F.; Ke, Q.F.; Zhu, Z.H.; Guo, Y.P. Gadolinium-doped mesoporous calcium silicate/chitosan scaffolds enhanced bone regeneration ability. *Mater. Sci. Eng. C Mater. Biol. Appl.* **2019**, *104*, 109999. [CrossRef]
93. Lin, M.C.; Chen, C.C.; Wu, I.T.; Ding, S.J. Enhanced antibacterial activity of calcium silicate-based hybrid cements for bone repair. *Mater. Sci. Eng. C Mater. Biol. Appl.* **2020**, *110*, 110727. [CrossRef]
94. Mukherjee, S.; Sharma, S.; Soni, V.; Joshi, A.; Gaikwas, A.; Bellare, J.; Kode, J. Improved osteoblast function on titanium implant surfaces coated with nanocomposite Apatite-Wollastonite-Chitosan- an experimental in-vitro study. *J. Mater. Sci. Mater. Med.* **2022**, *33*, 25. [CrossRef]
95. Genasan, K.; Mehrali, M.; Veerappan, T.; Talebian, S.; Malliga Raman, M.; Singh, S.; Swamiappan, S.; Mehrali, M.; Kamarul, T.; Balaji Raghavendran, H.R. Calcium-Silicate-Incorporated Gellan-Chitosan Induced Osteogenic Differentiation in Mesenchymal Stromal Cells. *Polymers* **2021**, *13*, 3211. [CrossRef] [PubMed]
96. Luz, G.M.; Mano, J.F. Chitosan/bioactive glass nanoparticles composites for biomedical applications. *Biomed. Mater.* **2012**, *7*, 054104. [CrossRef] [PubMed]
97. Mota, J.; Yu, N.; Caridade, G.; Luz, G.M.; Gomes, M.E.; Reis, R.L.; Jansen, J.A.; Walboomers, X.F.; Mano, J.F. Chitosan/bioactive glass nanoparticle composite membranes for periodontal regeneration. *Acta Biomater.* **2012**, *8*, 4173–4180. [CrossRef]
98. Saatchi, A.; Arani, A.R.; Moghanian, A.; Mozafari, M. Synthesis and characterization of electrospun cerium-doped bioactive glass/chitosan/polyethylene oxide composite scaffolds for tissue engineering applications. *Ceram. Int.* **2021**, *47*, 260–271. [CrossRef]
99. Alaei, M.; Atapour, M.; Labbaf, S. Electrophoretic deposition of chitosan-bioactive glass nanocomposite coatings on AZ91 Mg alloy for biomedical applications. *Prog. Org. Coat.* **2020**, *147*, 105803. [CrossRef]
100. Vafa, E.; Bazargan-Lari, R.; Bahroloom, E.B. Electrophoretic deposition of polyvinyl alcohol/natural chitosan/bioactive glass composite coatings on 316L stainless steel for biomedical application. *Prog. Org. Coat.* **2021**, *151*, 106059. [CrossRef]
101. Heise, S.; Wirth, T.; Höhlinger, M.; Torres Hernández, Y.; Rodriquez Ortiz, J.A.; Wagener, V.; Virtanen, S.; Boccaccini, A.R. Electrophoretic deposition of chitosan/bioactive glass/silica coatings on stainless steel and WE43 Mg alloy substrates. *Surf. Coat. Technol.* **2018**, *344*, 553–563. [CrossRef]
102. Höhlinger, M.; Christa, D.; Zimmermann, V.; Heise, S.; Boccaccini, A.R.; Virtanen, S. Influence of proteins on the corrosion behavior of a chitosan-bioactive glass coated magnesium alloy. *Mater. Sci. Eng. C* **2019**, *100*, 706–714. [CrossRef]
103. Mahlooji, E.; Atapour, M.; Labbaf, S. Electrophoretic deposition of Bioactive glass—Chitosan nanocomposite coatings on Ti-6Al-4V for orthopedic applications. *Carbohydr. Polym.* **2019**, *226*, 115299. [CrossRef]
104. Zarghami, V.V.; Ghorbani, M.; Bagheri, K.P.; Shokrgozar, M.A. Prevention the formation of biofilm on orthopedic implants by melittin thin layer on chitosan/bioactive glass/vancomycin coatings. *J. Mater. Sci. Mater. Med.* **2021**, *32*, 75. [CrossRef]
105. Sergi, R.; Bellucci, D.; Salvatori, R.; Cannillo, V. Chitosan-Based Bioactive Glass Gauze: Microstructural Properties, In Vitro Bioactivity, and Biological Tests. *Materials* **2020**, *13*, 2819. [CrossRef] [PubMed]
106. Sergi, R.; Bellucci, D.; Cannillo, V. A Review of Bioactive Glass/Natural Polymer Composites: State of the Art. *Materials* **2020**, *13*, 5560. [CrossRef] [PubMed]
107. Ebrahimi, M. Chapter 12—Standardization and regulation of biomaterials. In *Woodhead Publishing Series in Biomaterials, Handbook of Biomaterials Biocompatibility*; Mozafari, M., Ed.; Woodhead Publishing: Sawston, UK, 2020; pp. 251–265.

108. González Vázquez, A.G.; Blokpoel Ferreras, L.A.; Bennett, K.E.; Casey, S.M.; Brama, P.A.; O'Brien, F.J. Systematic Comparison of Biomaterials-Based Strategies for Osteochondrial and Chondral Repair in Large Animal Models. *Adv. Healthc. Mater.* **2021**, *10*, 2100878. [CrossRef] [PubMed]
109. Noordin Kahar, N.N.F.N.M.; Jaafar, M.; Ahmad, N.; Sulaiman, A.R.; Yahaya, B.H.; Abdul Hamid, Z.A. Biomechanical evaluation of biomaterial implants in large animal model: A review. *Mater. Today Proc.* 2022; *in press*.

Article

Sodium Alginate/Chitosan Scaffolds for Cardiac Tissue Engineering: The Influence of Its Three-Dimensional Material Preparation and the Use of Gold Nanoparticles

Nohra E. Beltran-Vargas [1,*], Eduardo Peña-Mercado [1], Concepción Sánchez-Gómez [2,†], Mario Garcia-Lorenzana [3], Juan-Carlos Ruiz [1], Izlia Arroyo-Maya [1], Sara Huerta-Yepez [4] and José Campos-Terán [1,*]

1. Process and Technology Department, Division of Natural Science and Engineering, Universidad Autonoma Metropolitana-Cuajimalpa, Ciudad de Mexico C.P. 05300, Mexico
2. Research Laboratory of Developmental Biology and Experimental Teratogenesis, Children's Hospital of Mexico Federico Gomez, Ciudad de Mexico C.P. 06720, Mexico
3. Department of Reproduction Biology, Division of Biological and Health Sciences, Universidad Autonoma Metropolitana-Iztapalapa, Ciudad de Mexico C.P. 09340, Mexico
4. Research Laboratory of Hematooncology, Children's Hospital of Mexico Federico Gomez, Ciudad de Mexico C.P. 06720, Mexico

* Correspondence: nbeltran@cua.uam.mx (N.E.B.-V.); jcampos@cua.uam.mx (J.C.-T.); Tel.: +52-55-5814-6500 (ext. 3807) (N.E.B.-V.); +52-55-5814-6500 (ext. 6530) (J.C.-T.)
† Deceased author.

Citation: Beltran-Vargas, N.E.; Peña-Mercado, E.; Sánchez-Gómez, C.; Garcia-Lorenzana, M.; Ruiz, J.-C.; Arroyo-Maya, I.; Huerta-Yepez, S.; Campos-Terán, J. Sodium Alginate/Chitosan Scaffolds for Cardiac Tissue Engineering: The Influence of Its Three-Dimensional Material Preparation and the Use of Gold Nanoparticles. *Polymers* 2022, 14, 3233. https://doi.org/10.3390/polym14163233

Academic Editors: Antonia Ressler and Inga Urlic

Received: 16 June 2022
Accepted: 12 July 2022
Published: 9 August 2022

Publisher's Note: MDPI stays neutral with regard to jurisdictional claims in published maps and institutional affiliations.

Copyright: © 2022 by the authors. Licensee MDPI, Basel, Switzerland. This article is an open access article distributed under the terms and conditions of the Creative Commons Attribution (CC BY) license (https://creativecommons.org/licenses/by/4.0/).

Abstract: Natural biopolymer scaffolds and conductive nanomaterials have been widely used in cardiac tissue engineering; however, there are still challenges in the scaffold fabrication, which include enhancing nutrient delivery, biocompatibility and properties that favor the growth, maturation and functionality of the generated tissue for therapeutic application. In the present work, different scaffolds prepared with sodium alginate and chitosan (alginate/chitosan) were fabricated with and without the addition of metal nanoparticles and how their fabrication affects cardiomyocyte growth was evaluated. The scaffolds (hydrogels) were dried by freeze drying using calcium gluconate as a crosslinking agent, and two types of metal nanoparticles were incorporated, gold (AuNp) and gold plus sodium alginate (AuNp+Alg). A physicochemical characterization of the scaffolds was carried out by swelling, degradation, permeability and infrared spectroscopy studies. The results show that the scaffolds obtained were highly porous (>90%) and hydrophilic, with swelling percentages of around 3000% and permeability of the order of 1×10^{-8} m^2. In addition, the scaffolds proposed favored adhesion and spheroid formation, with cardiac markers expression such as tropomyosin, troponin I and cardiac myosin. The incorporation of AuNp+Alg increased cardiac protein expression and cell proliferation, thus demonstrating their potential use in cardiac tissue engineering.

Keywords: alginate; chitosan; scaffolds; nanoparticles; cardiac tissue engineering

1. Introduction

Acute myocardial infarction (AMI) remains the primary cause of death worldwide. AMI occurs when blood flow to the coronary arteries is blocked, leading to necrosis, tissue remodeling and fibrosis that can cause progressive cardiac damage and heart failure [1]. A major problem in the recovery of AMI patients is the low proliferation percentage for heart cell regeneration [2]. For effective AMI treatment, it is necessary to prevent tissue remodeling, attenuate scar formation and promote cell proliferation and regeneration to replace damaged tissue. For this reason, tissue engineering emerges as a therapeutic alternative for the development of functional tissue that can be used for the regeneration of affected tissues. In cardiac tissue engineering, attempts are made to combine cells with biocompatible materials to generate a three-dimensional construct that can restore the damaged myocardium [3]. The ideal scaffolds for cardiac repair must have high porosity and biocompatibility as well as be permeable to nutrients and metabolic waste. In addition, they must have an adjustable degradation time to minimize the formation of fibrous capsules and promote incorporation into the host tissue to avoid a chronic inflammatory response [4]. Likewise, they must recreate the microenvironment, structure and three-dimensional organization of the myocardium; improve cell survival and promote cell adhesion, differentiation and maturation. In addition, the scaffolds must allow for vascularization to ensure the flow of oxygen and nutrients to the cells and favor the transmission of electrical and mechanical impulses for proper host-tissue coupling [5].

Natural polysaccharides such as alginate and chitosan have been widely used for tissue engineering because of their biocompatibility, biodegradability and structural similarity to the extracellular matrix components [6–10]. Chitosan, a polycationic polymer (the presence of positively charged amine groups), promotes the cell adhesion, proliferation and differentiation of different cell types [7,11]. Alginate, a polyanionic polymer (presence of negatively charged carboxyl groups), promotes regeneration and favors the vascularization and restoration of electrical conductivity and cell growth [6]. The chemical natures of alginate and chitosan polymers make them sensitive to pH (pKa = 3.4–3.7 and pKa = 6.3, respectively) of aqueous media, swelling in opposite directions [5].

Previous reports have shown that alginate/chitosan scaffolds improve mechanical and biological properties [8,9,12], in addition to promoting growth and maintaining cardiac cell viability [10,13–16]. They have a gradual degradation and favor cell retention, survival and migration to the affected area, allowing the formation of blood vessels, with reduction in fibrosis and hypertrophy area [14,17].

The addition or functionalization of biomaterials with metallic nanomaterials can improve the physical and electrical properties of the scaffolds [18–21]. Recently, the need and importance of designing and developing new cardiac patches based on conductive biomaterials for possible therapeutic application has been reported [22–24]. The properties of gold nanoparticles (AuNp) promote cardiac cell growth and contractility [25–28]. Biomaterials functionalized with AuNp favor cardiomyocyte elongation and alignment, with an increased expression of cardiac proteins and improved cell contraction [29–33]; however, the use of AuNp has not been studied in alginate/chitosan scaffolds.

Although there have been good results in the use of alginate/chitosan scaffolds for cardiomyocyte growth, it is interesting to test the use of star-type AuNp and tubular AuNp+Alg in this type of scaffold to favor cell adhesion and growth and increase cardiac protein expression, identifying the best conditions for scaffold fabrication for therapeutic purposes. Thus, the aim of this work was to compare different methods of fabrication of alginate/chitosan scaffolds to improve swelling percentages, permeability, porosity and degradation rate and to evaluate the effect of the functionalization of the proposed scaffolds with AuNp and AuNp+Alg on cardiomyocyte growth.

2. Materials and Methods

2.1. Preparation of Alginate–Chitosan Scaffolds

Sodium alginate (Sigma Aldrich, Mannheim, Germany, # 9005-38-3) and chitosan (medium molecular weight, Sigma Aldrich, Mannheim, Germany, # 448877) (0.75–1.25% w/v) powders were mixed and dissolved in ultrapure water (Milli-Q system, 18.2 M-cm) and acetic acid (1% w/v, Sigma Aldrich, Mannheim, Germany, #1005706). The pH was adjusted to be between 5 and 6 to favor the interaction between the biomaterials. The resulting mixture/solution was ready for undergoing the corresponding experimental method (1 to 4), see Figure 1. In all cases, the scaffold solutions placed within the 24-well box were frozen at $-20\ °C$ for 12 h and subsequently freeze-dried in a lyophilizer (Labconco Corporation, Kansas City, MO, USA) for 8 h at $-49\ °C$ under vacuum with a pressure of 0.100 mBar.

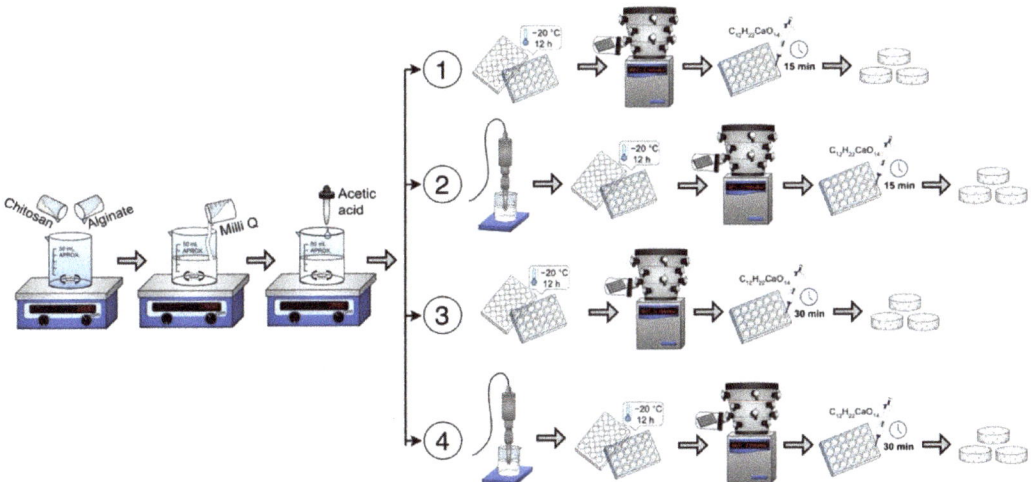

Figure 1. Four scaffold preparation methods: Method 1 (without sonication), Method 2 (with sonication), Method 3 (longer crosslinking time) and Method 4 (with sonication and longer crosslinking time).

Method 1 (without sonication). An alginate–chitosan solution of 0.5 mL was deposited in each well of a 24-well box. Then it was frozen, lyophilized and crosslinked with 1 mL of 1% w/v calcium gluconate in water for 15 min. Washes were performed with ultrapure water and again frozen at −20 °C for 12 h. Finally, they were lyophilized again for 8 h.

Method 2 (with sonication). The alginate–chitosan solution was sonicated (Sonics Vibra Cell VCX 750, Newtown, CT, USA) for 5 min at 20 kHz and 750 W. Subsequently, the alginate–chitosan solution was deposited in each well of a 24-well box. After freezing and lyophilization, crosslinking was performed with 1% calcium gluconate for 15 min. Subsequently, they were washed with ultrapure water, frozen and lyophilized for 8 h.

Method 3 (longer crosslinking time). The alginate–chitosan solution was deposited in each well of a 24-well box. After freezing and lyophilization, crosslinking was performed with 1% calcium gluconate for 30 min. Subsequently, washes were performed with ultrapure water, frozen and freeze-dried for 8 h.

Method 4 (with sonication and longer crosslinking time). The alginate–chitosan solution was sonicated at 20 kHz and 750 W. The solution was then deposited in the 24-well box, frozen and lyophilized. Crosslinking was performed for 30 min with 1% calcium gluconate. Subsequently, washes were performed with ultrapure water, frozen and freeze-dried for 8 h.

2.2. Synthesis of Metallic Nanoparticles

For the functionalization of alginate–chitosan scaffolds, metallic nanoparticles (Np) were prepared using a novel methodology with modifiable topography. This synthesis method is based on the preparation of citrate-stabilized gold nanoparticles [34], which are attached to the surface of a polymeric core consisting of Poly(D,L-lactide-co-glycolide) acid (PLGA) and stabilized with the copolymer Pluronic F-127. After the preparation of these metallic nanoprecursors, the growth of a gold shell is promoted on their surfaces, finally generating gold nanoparticles (AuNp).

The AuNp synthesis process is described below: first, the synthesis of PLGA (Sigma Aldrich, 102229183, Mannheim, Germany) cores was carried out. For this, a 10% w/v solution of PLGA in acetone (C_3H_6O) was prepared (Sigma Aldrich, #67641, Mannheim, Germany), which was drip-added to 38 mL of a 1% w/v aqueous solution of the triblock copolymer Pluronic F-127 (C_3H_6O-C_2H_4O)x (Sigma Aldrich, #9003116, Mannheim, Germany), under stirring (250 rpm) at a constant temperature of 10 °C. Subsequently, it was homogenized for 10 min in an ice bath with a sonic tip of 750 W, 20 kHz frequency and 40% amplitude and left in agitation for 4 h. At this stage of the AuNp synthesis, a modification to the PLGA nuclei preparation method was also carried out, which consisted of coating some of them with a sodium alginate solution. Briefly, 20 mL of PLGA nuclei was mixed with 1 mL of 1% (w/v) sodium alginate and the mixture was stirred (250 rpm) for 4 h at room temperature. PLGA nuclei were purified through three cycles of centrifugation (9000 rpm, 18 °C) and resuspension. The supernatant was discarded, and the pellets were dispersed in 30 mL of ultrapurified water. Subsequent to the PLGA cores preparation, gold nanoseeds were synthesized. These were obtained by mixing the following solutions: 0.125 mL of 0.01 M chloroauric acid ($HAuCl_4$) (Sigma Aldrich, 1001642619, Mannheim, Germany), 10 mL of 0.0256 M trisodium citrate ($Na_3C_6H_5O_7$) (Sigma Aldrich, 1001851140, Mannheim, Germany) and 0.3 mL of 0.1 M sodium borohydride ($NaBH_4$, Sigma Aldrich, 1002918750, Mannheim, Germany). The latter was added at a temperature of 4 °C. The formation of gold nanoprecursors was carried out by mixing both the PLGA cores and alginate-modified PLGA cores (PLGA+Alg) with the gold nanoseeds in a 1:1 ratio, under constant stirring at 250 rpm for 24 h, followed by centrifugation (7000 rpm, 18 °C) for 20 min. The supernatant was discarded and the pellets were dispersed in 30 mL of ultrapurified water. The gold nanoprecursors were sonicated for 10 min to avoid particle aggregation. Once the PLGA cores were coated with the gold nanoseeds, the metal shell was assembled on their surface by mixing 2.025 mL of AuNp nanoprecursors with 45 mL of a solution containing 3.69 mM potassium carbonate (K_2CO_3, Sigma Aldrich, 1002055627, Mannheim, Germany) and 0.025 M gold (III) chloride trihydrate ($HAuCl_4 \cdot 3H_2O$) (Sigma Aldrich, 1001642619, Mannheim, Germany). The latter was carried out under stirring (250 rpm) at room temperature. After 5 min, 225 µL of fresh 0.5 M ascorbic acid was added to the mixture.

The above was performed for both PLGA cores and sodium-alginate-modified cores (PLGA+Alg). The functionalization of the alginate–chitosan scaffolds was carried out through the addition of AuNp in a calculated concentration range between 1×10^{-12} and 3×10^{-9} mg/mL. The AuNp concentration was calculated using the Lambert–Beer equation. For the above, the determination of the molar extinction coefficient was calculated based on the following equation:

$$\varepsilon = \frac{N_a \cdot \sigma}{2303} \tag{1}$$

where ε is the molar extinction coefficient, N_a is Avogadro's number, σ is the effective cross section in cm^2 and 2303 is (ln 10) × 1000.

2.3. Characterization of Alginate/Chitosan Scaffolds

The alginate/chitosan scaffolds synthesized by the four methods and their subsequent modification with AuNp or AuNp+Alg were characterized by means of swelling, permeability, porosity and degradation tests in different aqueous media, recording their weight and thickness prior to their use. Fourier transform infrared spectroscopy (FTIR) was used as a chemical characterization technique.

2.4. Swelling Degree Studies

For swelling measurements over time in a typical experiment, the scaffold (approximately 1.3 cm in diameter) is placed in contact with 2 mL of the aqueous medium at a constant temperature of 20 °C at different times, after which the sample is removed from the medium, weighed and placed back in the medium. The degree of swelling, S, in %, was calculated gravimetrically using Equation (2), where W_s and W_0 are the weights of the swollen scaffold and dry scaffold (initial), respectively. The following aqueous media were used: ultrapure water (pH 7), PBS (phosphate buffer saline, pH 7.4) and buffers prepared with mixtures of Na_2HPO_4/citric acid solutions to obtain pHs of 3, 5, 8 and 9 (measured with the Conductronic potentiometer model PC45 (Puebla, Mexico).

$$\text{Swelling, } S, (\%) = \left(\frac{W_S - W_0}{W_0}\right) \times 100 \quad (2)$$

2.5. Permeability Value

The intrinsic permeability coefficient (k) was calculated according to Darcy's law:

$$k = K\frac{\mu}{\rho g} \quad (3)$$

$$K = \frac{a}{A}\frac{L}{t}\ln\frac{H_1}{H_2} \quad (4)$$

where (μ) is the viscosity of the medium, (ρ) is the density of the medium, (g) is the gravity acceleration, (a) is the tube area, (A) is the cross-sectional area at the sample flow, (L) is the sample thickness (in this case of the scaffold) and (H_1) and (H_2) are the initial and final heights of the tube through which the medium passes. We used 27.5 cm for H_2. A detailed description of the custom device is presented in the Supplementary Materials (Figure S1), which is similar to the other reported systems [35,36].

2.6. Porosity

The scaffolds' porosity was determined by the liquid displacement method and ethanol was used as the penetrating medium because it does not induce shrinkage or swelling, is not a solvent for polymers and is able to easily penetrate the pores. Each scaffold was placed in a cylinder with a known volume of ethanol, in which it was left for 48 h, the scaffold was removed, and the final volume was recorded. Finally, the following equation was used:

$$\text{Porosity } (\%) = \left(\frac{W_S - W_0}{\rho V}\right) \times 100 \quad (5)$$

where (W_S) is the weight of the saturated scaffold, (W_0) is the initial weight of the scaffold, (ρ) is the ethanol density and (V) is the volume of liquid displaced.

2.7. Degradation

Degradation studies were divided into two main studies. The first study was performed at conditions similar to cell cultures in which the scaffolds (5 mm in diameter approximately) were immersed in M199 medium (11150067, Gibco, Thermo Fisher Scientific, Waltham, MA, USA) supplemented with fetal bovine serum (A4766801, Gibco, Thermo Fisher Scientific, Waltham, MA, USA) to a pH of 7.4 and were placed in an incubator at 37 °C for 17 days to evaluate their degradation degrees at the following times: days 1, 2, 3, 6, 10 and 17. At all times, as much water as possible was removed from the container with the scaffold and weighed, and new medium was placed before returning to the incubator. For the second study, the material degradation (approximately 1.3 cm in diameter) exposed to different aqueous media for 7 days at laboratory conditions (20 °C) was calculated. In a typical experiment, for the scaffold immersed in the medium, 2 mL was used with each wash and proceeded with the following times and number of washes: twice for 15 min, once for 16 h and twice for 15 min. Finally, as much water as possible was removed from the container with the scaffold, frozen at −20 °C for at least 12 h and freeze-dried in a lyophilizer at −49 °C and a pressure of 0.09 mBar for 6 h (followed gravimetrically until there was no weight change). The degradation degree, D, in %, was calculated according to Equation (5), where W_F and W_0 are the final weight of the dry scaffold and exposed to either degradation condition and the weight of the initial dry scaffold (before being exposed to either degradation condition), respectively.

$$\text{Degradation degree} = n, D, (\%) = \left(\frac{W_0 - W_F}{W_0}\right) \times 100 \tag{6}$$

2.8. Infrared

Fourier transform infrared spectra with attenuated total reflectance (FTIR-ATR) were taken from 650 cm^{-1} to 4000 cm^{-1} using a Perkin–Elmer model 100 spectrometer (Waltham, MA, USA) equipped with a diamond tip.

2.9. Characterization of the Metallic Nanoparticles

The particle sizes, their distributions and the zeta potentials of AuNp and AuNp+Alg were analyzed by dynamic light scattering using a Nanosizer Nano ZS (Malvern Instruments Ltd., Malvern, UK). Samples were diluted 1:100 in ultrapure water and analyzed at 25 °C, with a scattering angle of 90.

The Np morphology was determined by scanning electron microscopy, using a TM3030PLUS scanning electron microscope (Hitachi, Germany) with an operating voltage of 15 kV; and moreover, the AuNp morphology was examined using a JEM-1010 transmission electron microscope (Jeol, Tokyo, Japan) with a voltage of 60 kV. The AuNp aqueous dispersion was diluted 1:100 and 15 µL was deposited on a copper grid (200 mesh). The grid was allowed to dry at room temperature before analysis. UV-Vis spectroscopy analysis was conducted in order to visualize the plasmon of the gold nanoparticles.

2.10. Primary Culture of Chicken Embryonic Cardiomyocytes

Cardiomyocytes were obtained from chicken embryos after 7 days of incubation. Detailed information on cardiomyocytes isolation and culture is presented in the Supplementary Materials. For each scaffold, 1×10^6 cells were cultured and incubated for 7 days at 37 °C in supplemented medium 199. Animal use protocols and study procedures were based on the Official Mexican Standard (NOM-062-ZOO-1999). The project was approved by the research, ethics and biosafety committees of the Hospital Infantil de México Federico Gómez (HIM-2020-059).

2.11. Indirect Cytotoxicity Assay

To demonstrate that the different elaborated scaffolds did not generate cytotoxic particles, an indirect cytotoxicity assay was performed using the MTT method. The viability of cells cultured with medium that had contact with scaffolds functionalized with AuNp and AuNp+Alg was also compared. Indirect cytotoxicity assays were performed using a monolayer of cardiomyocytes by triplicate.

2.12. Scanning Electron Microscopy

Scaffolds and constructs were fixed with glutaraldehyde (4%) and dehydrated through a series of graded ethanol concentrations (50° to absolute). They were critical-point-dried (Samdri 789A, Tousimins Research Co., Rockville, MD, USA) and coated with gold film (Denton Vacuum Desk 1A, Cherry Hill Industrial Center, Moorestown, NJ, USA). The samples were observed under a JEOL JSM 5300 (Tokyo, Japan) scanning electron microscope, and the accelerating voltage was 15 kV.

2.13. Histological Analysis

Constructs were fixed with paraformaldehyde (4%) and processed according to standard histological technique. Transverse sections of 3 µm thickness were made and stained with hematoxylin–eosin (H&E) and scanned and digitized with Aperio CS2 equipment (Leica Biosystems, Deer Park, IL, USA).

2.14. Immunohistochemical Analysis

The sections were deparaffinized and subjected to antigenic recovery in sodium citrate buffer (pH 6). Endogenous peroxidase blocking was performed with hydrogen peroxide (3%) for 30 min. A nonspecific binding blockade was performed for 3 h. The sections were incubated with the primary antibodies anti-tropomyosin (SC:74480, 1:500) and anti-PCNA (AB-2426, 1:1000) at room temperature overnight. They were then incubated with horseradish peroxidase (HRP)-conjugated secondary antibody for 30 min. The antigen–antibody complex was revealed with an immunodetection kit (Vector Laboratories, Inc. Cat # 30026, Berlingame, CA, USA) and counterstaining was performed with hematoxylin. Quantitative analysis (Intensity (Int)) was performed by digital pathology.

2.15. Western Blot

Total protein extraction from the constructs was performed with lysis buffer (T-PER, Thermo Fisher Scientific, Waltham, MA, USA) and added to protease inhibitor (Sigma Aldrich, Mannheim, Germany). Protein concentration was quantified using the direct microdrop method (NanoDrop lite, Thermo Fisher Scientific, Waltham, MA, USA). A total of 30 µg of total protein was subjected to 10% sodium dodecylsulfate polyacrylamide gel electrophoresis (SDS-PAGE) and transferred to a nitrocellulose membrane (Bio-Rad). Antibodies were used at the following dilutions: primary antibodies troponin I (SC-365446 1:1000), MYH (SC-376157, 1:1000) and GAPDH (SC-48167), followed by incubation with HRP-coupled secondary antibodies in blocking solution for 1 h at room temperature, anti-mouse (SC-516102, 1:10,000) and anti-goat (SC-2020, 1:20,000). Finally, immunodetection by chemiluminescence (Super Signal® West Femto, Thermo Scientific, Waltham, MA, USA) was performed with autoradiographic plates. Densitometry analysis was performed with Image J software, version 1.45 (National Institute of Health, Bethesda, MD, USA).

2.16. Statistical analysis

GraphPad Prism 9 software (San Diego, CA, USA) was used to perform the statistical analysis, using a significance value of $p < 0.001$. Data are presented as the mean ± standard error (SE). An ANOVA was performed to evaluate physical changes in scaffold fabrication and changes in percent swelling and permeability between scaffold fabrication methods and to analyze cell viability. To compare the four scaffold fabrication methods with respect to functionalization (without Np, with AuNp and with AuNp+Alg) a multivariate analysis was performed, with Tukey's post hoc test for multiple comparisons.

3. Results

3.1. Characterization of Alginate–Chitosan Scaffolds

Method 1 (without sonication) had the highest weight with respect to the other three fabrication methods ($p < 0.001$). Regarding thickness, the scaffolds generated by Method 2 (with sonication) were the thinnest, presenting significant changes with respect to Methods 1 and 3 ($p < 0.001$), see Supplementary Materials, Figure S2.

3.2. Swelling Degree

In Figure 2A, the swelling percentages of the scaffolds according to the processing method can be observed. The swelling percentage of the scaffolds was higher than 2500% in three of the methods. The scaffolds elaborated with Method 3 (30 min crosslinking) were the ones with the highest swelling percentage. Figure 2B shows that the swelling percentage increases when functionalizing the scaffolds with AuNp and AuNp+Alg, particularly in Method 3, which had the highest values.

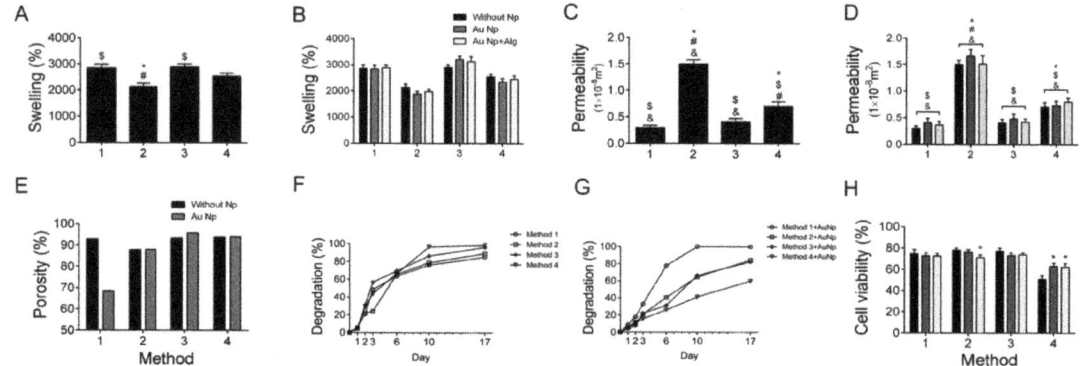

Figure 2. (**A**) Swelling percentages (%) for the four scaffold preparation methods. (**B**) Swelling percentages for the four preparation methods functionalized with AuNp and AuNp+Alg. (**C**) Permeability (m^2) for the four scaffold preparation methods. (**D**) Permeability for the four preparation methods functionalized with AuNp and AuNp+Alg. (*) $p < 0.001$ vs. Method 1, ($) $p < 0.001$ vs. Method 2, (#) $p < 0.001$ vs. Method 3, (&) $p < 0.001$ vs. Method 4. (**E**) Porosity percentage (%) for the four scaffold preparation methods and the scaffolds functionalized with AuNp. (**F**) Degradation degree for the four scaffold preparation methods and (**G**) the functionalization of the four methods with AuNp. (**H**) Graphical representation of cell viability analysis for the four preparation methods functionalized with AuNp and AuNp+Alg. (*) $p < 0.001$ vs. without Np. Data are presented as the mean ± SE.

The scaffolds obtained with Method 3 were studied in their swelling in different aqueous media (pH dependence) over 7 days (Figure 3). Similar behavior can be observed in each medium over time, in which the swelling increases, reaches a maximum and does not decrease significantly at the total time studied. The Table 1 shows the percentage reached with respect to the maximum swelling (S_{max}) of the scaffolds. At 5 min, the hydrogels reached between 53 and 94% of the S_{max}, and at 60 min, most of the scaffolds reached between 90 and 100% of the S_{max}. The swelling dependence of the alginate/chitosan scaffolds on pH showed a decrease in the swelling when going from acidic to basic pHs, from 2945 to 1880% (Figure 3C).

Table 1. Maximum swelling (S_{max}) as a function of pH.

	Swelling Referred to S_{max} (%)		
Medium	**5 min**	**40 min**	**60 min**
pH 3	75	92	95
pH 5	76	97	100
Milli-Q	71	88	90
PBS	81	93	93
pH 8	53	57	58
pH 9	94	99	95

3.3. Permeability

Method 2 (sonication) has the highest permeability coefficient, being more than twice the one obtained with the other scaffold processing methods (Figure 2C). When functionalizing the scaffolds with AuNp and AuNp+Alg, a slight increase in permeability is observed in the four proposed scaffold elaboration methods (Figure 2D).

3.4. Porosity

It is observed that the scaffolds of Method 3 (30 min crosslinking) are the most porous, and this property increases to more than 95% when functionalized with AuNp (Figure 2E).

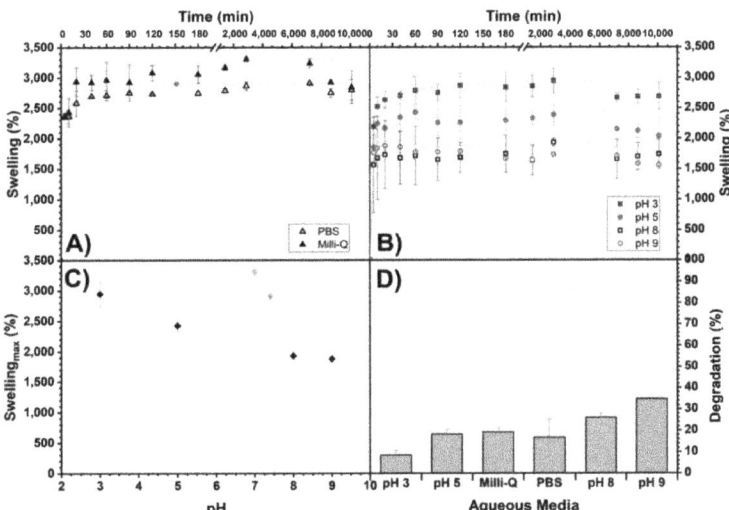

Figure 3. Swelling of alginate/chitosan scaffolds (Method 3) as a function of time in different aqueous media: (**A**) Milli-Q and PBS and (**B**) buffers prepared with Na_2HPO_4/citric acid for pH 3, 5, 8 and 9. In addition, their maximum swelling (S_{max}) as a function of pH (**C**) and the percentages reached with respect to S_{max} at 5, 40 and 60 min are shown (see Table 1). Maximum swelling trend lines are shown as an eye guide. (**D**) Percentage of degradation (%) of the scaffolds of Method 3 (more crosslinking time) as a function of the aqueous medium exposed for 7 days.

3.5. Degradation

Degradation analysis under cell culture conditions was performed on scaffolds made with the four proposed methods (Figure 2F) and on scaffolds of the four methods functionalized with AuNp (Figure 2G). It is observed that the scaffolds of Method 1 (without sonication) are the ones that degrade slower without AuNp, but when functionalized with AuNp, they degrade faster. In the case of Method 4, the opposite happens; they are the scaffolds that degrade faster without AuNp and with AuNp they degrade slower. In general, they degrade slightly faster with AuNp, with the exception of Method 3 (longer crosslinking time) where functionalization with AuNp causes them to degrade slightly slower, although the changes are not statistically significant.

The scaffolds' degradation by Method 3 (more crosslinking time) with respect to time and subjected to different aqueous media for 7 days are shown in Figure 3D. For all samples, there is a weight decrease, corroborating a degradation between 8 and 35%, and, as observed, it increases with increasing pH.

3.6. FTIR-ATR

Figure 4 shows the chemical characterization (IR spectra) of the alginate/chitosan scaffolds before and after crosslinking by Method 3 (30 min crosslinking), followed by the incorporation of AuNp and AuNp+Alg, in addition to the initial sodium alginate and chitosan spectra. For sodium alginate (Figure 4(A1)), the presence of the characteristic groups is confirmed [37]. Figure 4(A2) shows the characteristic peaks of chitosan [38]. In the case of the scaffold with alginate/chitosan before crosslinking (Figure 4(A3)), peaks that overlap due to the presence of both components can be observed (follow the dotted lines). The spectrum of the scaffold crosslinked with calcium gluconate by Method 3 (Figure 4(A4)) is similar to the spectrum of the scaffold before crosslinking, but an incorporation of OH and COO− due to the gluconate is present, and an increase in the intensity of the −CH peak is observed. The IR spectra of the scaffolds crosslinked and with the incorporation of AuNp and AuNp+Alg are presented in Figure 4(A5,A6), respectively; they are similar to the scaffolds before doping with AuNp, see Supplementary Materials.

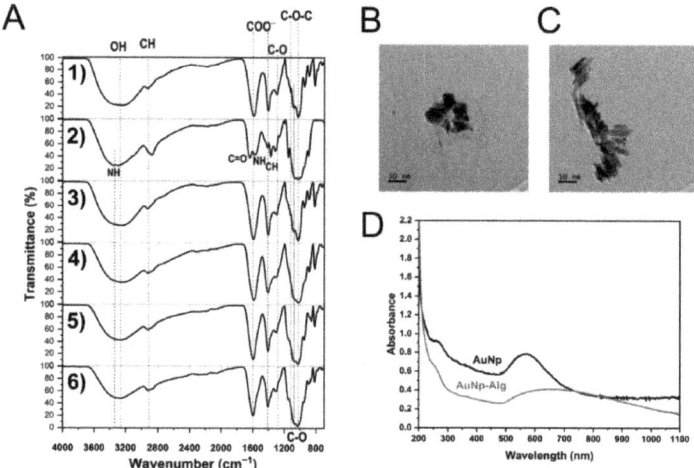

Figure 4. (**A**) FTIR-ATR spectra of: sodium alginate (1), chitosan (2) and alginate/chitosan scaffolds before (3) and after crosslinking (4, Method 3) and with incorporation of gold (5) or gold plus sodium alginate (6) nanoparticles. Vertical dotted and continuous lines are presented as eye guides. Representative electromicrographs of (**B**) AuNp and (**C**) AuNp+Alg obtained by TEM. (**D**) Gold nanoparticles plasmon.

3.7. Characterization of AuNp

The obtained metallic nanoparticles presented an average hydrodynamic diameter of 74.5 ± 1 nm and 91.1 ± 1 nm for PLGA and PLGA+Alg core-shell nanoparticles, respectively. As for the surface charge value, it was determined to be -25.4 ± 0.8 mV for PLGA-cored samples and -36.8 ± 1 mV for those modified with sodium alginate (see Supplementary Materials).

Particle morphology was determined by SEM and transmission electron microscopy (TEM). According to these techniques, AuNp presented a spheroidal structure (Figure 4B), while AuNp+Alg showed cylindrical particle characteristics (Figure 4C). The UV-Vis spectra of AuNp and AuNp+Alg solutions (before dilution) were taken (Figure 4D), presenting both a plasmon starting at ~500 nm and reaching a maximum wavelength at 571 nm and 639 nm, respectively, confirming the sizes observed by TEM. As a reference, Peña et al. showed a maximum plasmon at a wavelength of 528 nm for homogeneous spherical AuNp ~32 nm of diameter [30].

3.8. Cytotoxicity

The presence of Np does not affect cell viability. The scaffolds elaborated by Methods 1, 2 and 3 were the ones that allowed the maintenance of up to 80% of viable cells (Figure 2H). The results demonstrate that the alginate/chitosan scaffolds elaborated by the different proposed methods without Np and functionalized with AuNp and with AuNp+Alg do not release toxic components that could negatively impact the viability of the cells cultured on them.

3.9. Scanning Electron Microscopy (SEM)

The electromicrographs of the scaffolds show a porous appearance (Figure 5). The presence of spheroids of different diameters can be observed (Figure 5B,G–I,L). Functionalization with Np allowed the increase in the number, size and distribution of spheroids, especially for the scaffolds of Method 3 (Figure 5H,I).

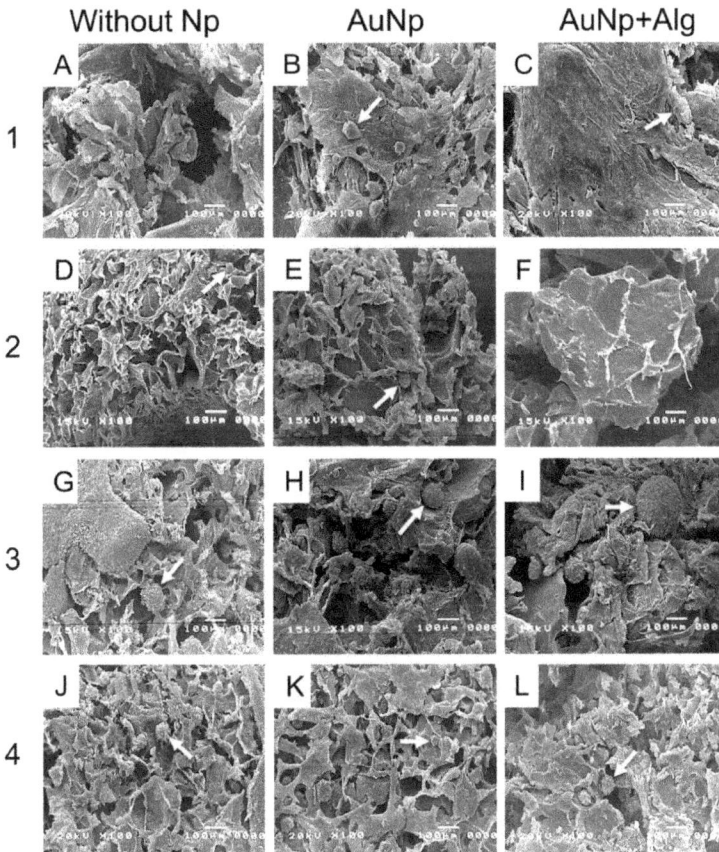

Figure 5. Representative electromicrographs of cross sections of the constructs. Method 1: (**A**) without Np, (**B**) AuNp and (**C**) AuNp+Alg. Method 2: (**D**) without Np, (**E**) AuNp and (**F**) AuNp+Alg. Method 3: (**G**) without Np, (**H**) AuNp and (**I**) AuNp+Alg. Method 4: (**J**) without Np, (**K**) AuNp and (**L**) AuNp+Alg. Identifier: arrow (spheroids).

3.10. Histological Analysis

Histological analysis shows that there was high cell migration and penetration through the pores of the scaffolds. Consistent with the SEM, the scaffolds of Method 3 were the ones that presented the largest spheroids, reaching 200 μm in diameter, in addition to being present both in the periphery and inside the scaffold (Figure 6). All the scaffolds with Np presented a greater number of spheroids compared to the scaffolds without Np (Figure 6B,E,H,K). However, spheroids were observed in all four proposed methods.

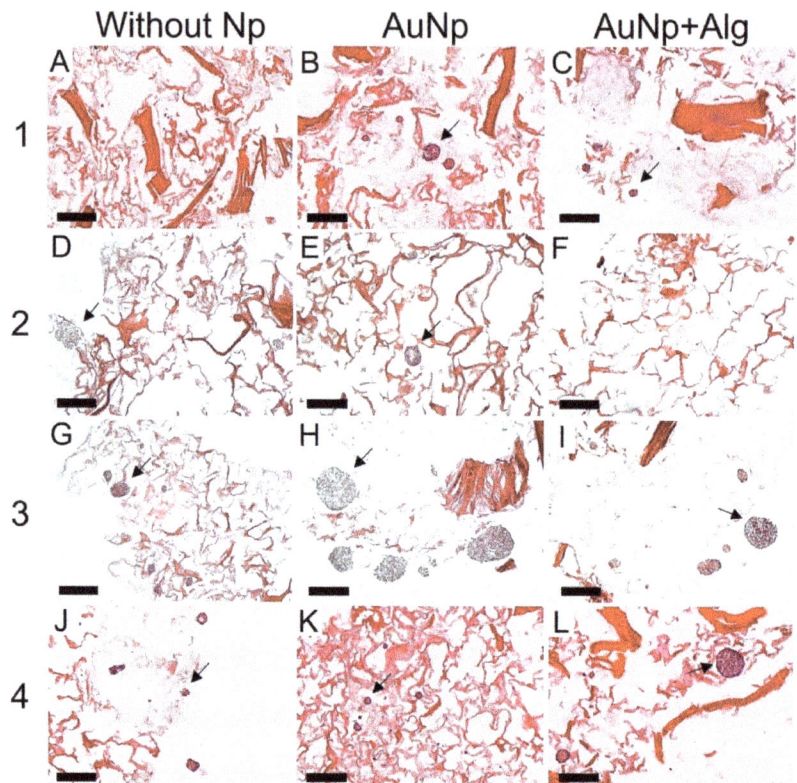

Figure 6. Representative photomicrographs of cross sections of the constructs. Method 1: (**A**) without Np, (**B**) AuNp and (**C**) AuNp+Alg. Method 2: (**D**) without Np, (**E**) AuNp, and (**F**) AuNp+Alg. Method 3: (**G**) without Np, (**H**) AuNp and (**I**) AuNp+Alg. Method 4: (**J**) without Np, (**K**) AuNp and (**L**) AuNp+Alg. H&E. Identifier: arrow (spheroids). Scale bar: 200 µm, original magnification: ×100.

3.11. Immunodetection of Proliferation and Cardiomyocyte Markers

Immunohistochemical analysis was performed to measure the levels of PCNA (proliferation marker, Figure 7) and tropomyosin (Figure 8). There was increased proliferation in the scaffolds prepared by Methods 3 and 4, and in particular in Method 3, functionalization with AuNp+Alg significantly increased cell proliferation compared to scaffolds without Np (Figure 7M). For the case of tropomyosin, an increase in this protein was observed mainly in the periphery of the spheroid in the scaffolds elaborated by Methods 1, 3 and 4. The scaffolds elaborated by Methods 3 and 4 showed the highest tropomyosin expression, particularly when functionalization with AuNp+Alg was present. Figure 8C shows the interaction between the scaffold and the cardiomyocytes (amplification is shown in the Supplementary Materials).

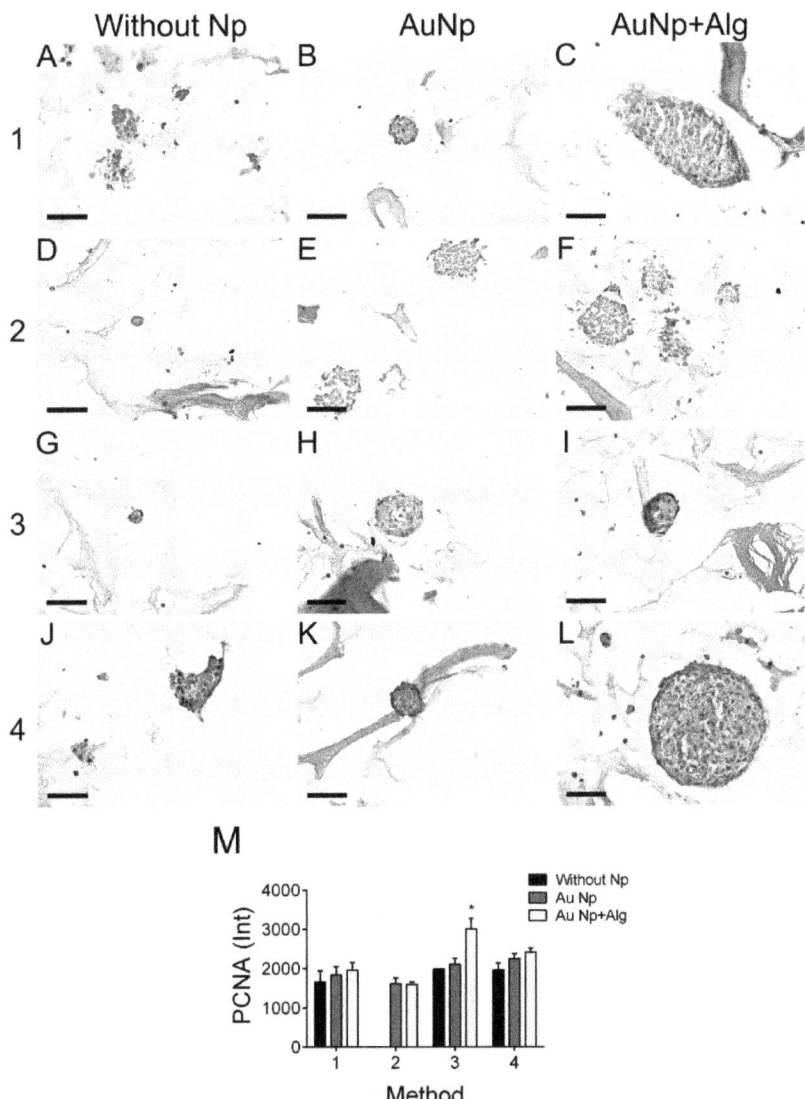

Figure 7. Representative photomicrographs of PCNA expression in different constructs. Method 1: (**A**) without Np, (**B**) AuNp and (**C**) AuNp+Alg. Method 2: (**D**) without Np, (**E**) AuNp and (**F**) AuNp+Alg. Method 3: (**G**) without Np, (**H**) AuNp and (**I**) AuNp+Alg. Method 4: (**J**) without Np, (**K**) AuNp and (**L**) AuNp+Alg. Scale bar: 50 µm, original magnification: ×400. (**M**) Graphical representation of the quantitative analysis of PCNA expression (Int) in the different constructs. (*) $p < 0.001$. Data are presented as the mean ± SE.

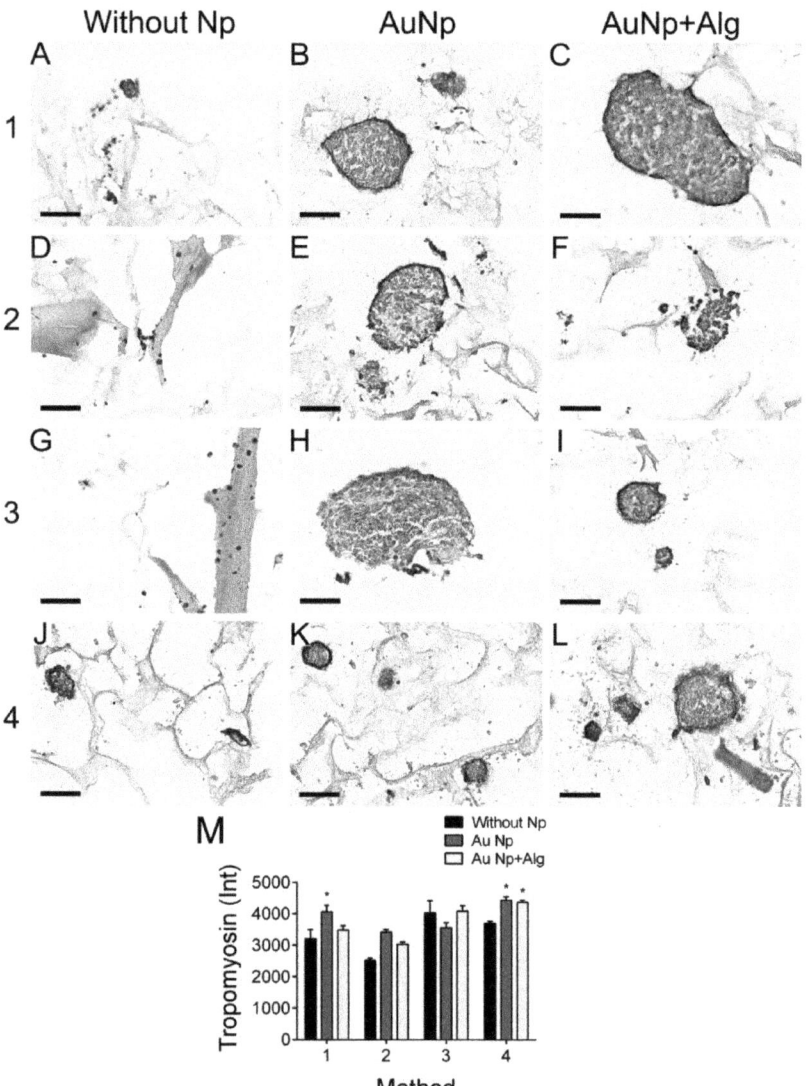

Figure 8. Representative photomicrographs of tropomyosin expression in different constructs. Method 1: (**A**) without Np, (**B**) AuNp and (**C**) AuNp+Alg. Method 2: (**D**) without Np, (**E**) AuNp and (**F**) AuNp+Alg. Method 3: (**G**) without Np, (**H**) AuNp and (**I**) AuNp+Alg. Method 4: (**J**) without Np, (**K**) AuNp and (**L**) AuNp+Alg. Scale bar: 50 μm, original magnification: ×400. (**M**) Graphical representation of the quantitative analysis of tropomyosin expression (Int) in the different constructs. (*) $p < 0.001$. Data are presented as the mean ± SE.

Immunodetection by western blot revealed troponin I and cardiac myosin expression in all scaffold elaboration methods (Figure 9A). In addition, densitometric analysis was performed to demonstrate that for myosin, in the case of the scaffolds of Methods 2 and 4, functionalization with both AuNp and AuNp+Alg caused an increase in expression. In the case of troponin I, it can be seen that in the scaffolds prepared by Methods 2 and 3, functionalization with AuNp+Alg also promoted increased expression (Figure 9B).

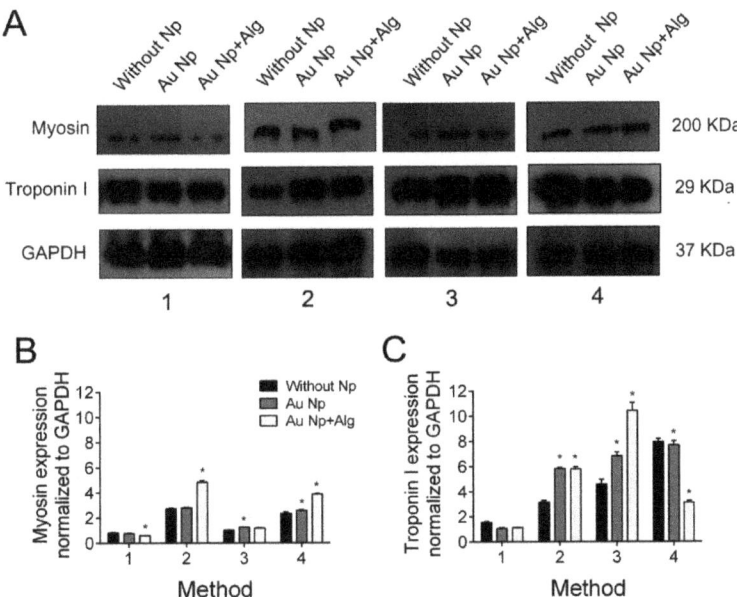

Figure 9. (**A**) Representative Western blot figure of myosin and troponin I in the different constructs: without AuNp and with AuNp and AuNp+Alg. Graphical representation of the quantitative analysis of (**B**) myosin and (**C**) troponin I expression. (*) $p < 0.001$. Data are presented as mean ± SE.

4. Discussion

AMI generates a high incidence of deaths worldwide; thus, it is necessary to develop new therapeutic strategies to regenerate cardiac tissue that often loses its function [39].

In this work, four different methods for the elaboration of sodium alginate–chitosan scaffolds are proposed. Crosslinking between polymers is important because it provides the scaffolds with increased stability, higher mechanical strength and hydrolysis resistance [40]. In addition, crosslinking agents promote chemical interactions between exposed functional groups. Although various crosslinking agents have been proposed to improve the biological and mechanical properties of these hydrogels [13,41], in this work the use of calcium gluconate as a crosslinking agent was proposed, favoring cell adhesion and cardiomyocyte growth. In addition, it was found that the fabrication method has an important impact on the structure and properties of the scaffolds and that in general the proposed methods favor the cardiomyocytes growth and contractility.

Scaffolds that have low electrical conductivity can be modified with gold, carbon, selenium or silver Np to increase their conductivity and improve their biocompatibility [29,42]. In particular, AuNp considerably increases scaffold conductivity, favors cell organization, enhances cell contractile activity and promotes cardiomyocyte maturation, proliferation and migration [21,27,31,33,43].

Nanoparticles tend to be more stable and interact better with biological systems when they are coated with materials of biological origin such as alginate [44,45] or albumin [46,47], among others. In this work, we proposed to use sodium alginate as a stabilizing agent and we evaluated the physicochemical and biological properties of the scaffolds and the metallic nanoparticles that were developed.

It has been reported that a high swelling percentage is favorable for cell growth, adhesion and vascularization [48]. Our results show that the scaffolds that were prepared had swelling percentages above 3000%, a percentage higher or similar to what is reported in other alginate/chitosan scaffolds, in addition to corroborating their pH dependence [14,49]. Baei et al. [27] reported that the addition of AuNp in chitosan scaffolds decreases the hydrophilic functional groups exposed, which translates into lower swelling percentages. Contrary to what they report, in our alginate/chitosan scaffolds, when functionalized with AuNp, the swelling percentage increased, probably due to the fabrication and crosslinking method that was proposed.

Permeability is a property that is directly related to the pore interconnectivity degree, as it controls the nutrient flow through the scaffold to the cells migrating inside it, thus achieving efficient cell growth [50]. Tresoldi et al. [51] reported permeabilities of 9×10^{-15} m^2 in alginate–gelatin scaffolds, while in studies performed by Rai et al. [52], permeabilities of 2×10^{-15} m^2 were obtained for fibrous poly(glycerol sebacate)-poly(ε-caprolactone) scaffolds. In the case of this work, a 1×10^{-8} m^2 permeability was obtained in all the proposed fabrication methods, which favors the flow of nutrients and oxygen to the interior of the scaffolds, for greater cell proliferation.

It was demonstrated that our scaffolds were highly porous, which is a feature that directly influences cell penetration and nutrient and oxygen transport [53], with adequate pore size to favor cell infiltration and colonization. The scaffolds elaborated by Method 3 (longer crosslinking time) were the most porous, and functionalizing them with AuNp was found to increase their porosity, to increase their swelling percentage and to decrease their degradation time, which favors the formation of cardiac tissue.

The degradation rate depends on the regeneration rate of the tissue to be replaced. In this work, the degradation percentage was analyzed by placing the scaffolds in culture medium supplemented with fetal bovine serum, which had not been previously reported, to know their behavior under culture conditions. Similar to what has been reported in other works, functionalization with AuNp decreased the degradation rate [54], which in our case favored cardiomyocyte cell growth.

The physicochemical characteristics of proposed gold Np (size and zeta potential) were similar to those reported with potential applications in the biomedical field, specifically in tissue engineering [18–21]. Stable Np do not agglomerate and help improve the electrical conductivity of the scaffold. Our AuNp+Alg had a Z potential (-36 mV) similar to the AuNp+Alg (-30 mV) elaborated by Shen, K, et al. [45], which were stable and helped to generate percentages of cell viability around 95%.

The difference in morphological characteristics between AuNp and AuNp+Alg could be explained by the presence of sodium alginate surrounding the PLGA nuclei, which acts as a structural growth director in a favored specific direction, as has also been reported by Pal et al. [55].

Nanoparticles used in tissue engineering have various sizes. Small Np (20 nm) [56] and large Np (156 nm) [19] have been reported to increase cell viability by up to 60%, suggesting that the size of the Np does not influence their biocompatibility. In our case, the AuNp+Alg (91 nm) were the ones that increased cardiac cell proliferation (Figure 6).

This work demonstrated that natural biomaterials such as alginate and chitosan are not cytotoxic and allow for cardiomyocyte growth as reported in other study groups [57]. The final architecture of our constructs allowed for the adhesion, growth and maintenance of cardiac cell integrity; the cultures remained healthy for 7 days; and the cells were able to enter the scaffolds, grow between the pores and form spheroids. In addition, histological and molecular analysis allowed us to observe the expression of the characteristic proteins of cardiac tissue, which increased with the incorporation of the AuNp+Alg that was proposed (Figures 7 and 8), which is in agreement with the work of Dvir et al., where functionalized scaffolds were shown to increase the expression of cardiac markers [58]. In our case, it was probably due to the biological properties reported for alginate [6].

In addition, an increase in cardiac cell differentiation has been reported when aggregated into spheroids, particularly if these aggregates can bind to a substrate [59], a situation that occurred in this study system.

Similar proliferation was observed in all scaffolds, but the proposed new functionalization with AuNp+Alg considerably increased cell proliferation.

It is interesting to note that in addition to functionalization, the proposed processing methods also have an impact on the expression of cardiac markers, confirming that scaffold topology plays a fundamental role in cell distribution and development. Several factors influence bioengineered tissue generation: on the one hand, the scaffolds and their structure, and on the other hand, the integrity of the cells to be cultured and the appropriate culture conditions. Taken together, the results suggest that these cardiac tissue constructs may be a novel implant material to study cardiac patch–cell interactions and can be used to study their biocompatibility, degradation and vascularization in an in vivo model.

5. Conclusions

The data as a whole demonstrate that the proposed alginate/chitosan scaffolds elaborated by the working group favor the growth of cardiomyocytes and spheroid formation and that functionalization with AuNp and AuNp+Alg not only increases cell proliferation but also promotes the increase in cardiac proteins such as troponin I, myosin and tropomyosin.

The presence of a carbohydrate-based biopolymer such as sodium alginate in AuNp+Alg proved to be an effective means to modify the structural and physicochemical properties of AuNp. Furthermore, it was demonstrated that AuNp+Alg scaffolds can be used as a therapeutic alternative in cardiac tissue engineering.

The proposed scaffolds were found to be useful for cardiomyocyte growth, which presents the possibility of applying them to the generation of a patch that can be used as a therapeutic alternative in AMI.

6. Patents

Soporte de hidrogel de alginato y quitosano para crecimiento de tejidos. Beltran NE, Francisco E., Vaquero D, Arroyo I. Instituto Mexicano de la propiedad industrial, MX/a/2020/012621. 24/11/2020.

Supplementary Materials: The following supporting information can be downloaded at https://www.mdpi.com/article/10.3390/polym14163233/s1. Figure S1: Custom device constructed to measure scaffold's permeability. Figure S2: Weight and thickness for the four scaffold preparation methods; File S1: S_database.

Author Contributions: Conceptualization, N.E.B.-V. and E.P.-M.; methodology, N.E.B.-V., E.P.-M., C.S.-G., M.G.-L., J.-C.R., I.A.-M., S.H.-Y. and J.C.-T.; validation, N.E.B.-V., E.P.-M., M.G.-L., J.-C.R., I.A.-M. and J.C.-T.; formal analysis, N.E.B.-V., E.P.-M., J.-C.R., I.A.-M. and J.C.-T.; investigation, N.E.B.-V., E.P.-M. and J.-C.R.; resources, N.E.B.-V. and C.S.-G.; data curation, N.E.B.-V., E.P.-M., C.S.-G., M.G.-L., J.-C.R., I.A.-M., S.H.-Y. and J.C.-T.; writing—original draft preparation, N.E.B.-V., E.P.-M., J.-C.R and I.A.-M.; writing—review and editing, N.E.B.-V., E.P.-M., C.S.-G., M.G.-L., J.-C.R., I.A.-M., S.H.-Y. and J.C.-T.; visualization, N.E.B.-V., E.P.-M., C.S.-G., M.G.-L., J.-C.R., I.A.-M., S.H.-Y. and J.C.-T.; supervision, N.E.B.-V., C.S.-G., M.G.-L. and J.C.-T.; project administration, N.E.B.-V.; funding acquisition, N.E.B.-V., C.S.-G. and M.G.-L. All authors have read and agreed to the published version of the manuscript.

Funding: This research received support from Secretaria de Educación, Ciencia, Tecnología e Innovación (SECTEI), project SECTEI/211/10357c19; and also from the Hospital Infantil de México Federico Gómez, protocol HIM/2020/059 SSA.1688.

Institutional Review Board Statement: Not applicable.

Informed Consent Statement: Not applicable.

Data Availability Statement: The data presented in this study are contained within the article or are available as Supplementary Materials.

Acknowledgments: The authors thank Daniela Ángeles, Cinthya González and Yenifer Alba for their support in the elaboration and characterization of the scaffolds, Daniela Angeles for Figure 1 design and implementation, Daniela Vaquero for her support in developing the nanoparticles, Anahis Cruz-Ledesma and Mayra Montecillo-Aguado for processing the tissues and Carlos C. Patiño for cell culture and sample processing. We are also grateful for the support received from the Hospital Infantil de México Federico Gómez, protocol HIM/2020/059 SSA.1688.

Conflicts of Interest: The authors declare no conflict of interest.

References

1. Saleh, M.; Ambrose, J.A. Understanding myocardial infarction. *F1000Research* **2018**, *7*, 1378. [CrossRef] [PubMed]
2. Bar, A.; Cohen, S. Inducing Endogenous Cardiac Regeneration: Can Biomaterials Connect the Dots? *Front. Bioeng. Biotechnol.* **2020**, *8*, 126. [CrossRef] [PubMed]
3. Chaudhuri, R.; Ramachandran, M.; Moharil, P.; Harumalani, M.; Jaiswal, A.K. Biomaterials and cells for cardiac tissue engineering: Current choices. *Mater. Sci. Eng. C Mater. Biol. Appl.* **2017**, *79*, 950–957. [CrossRef]
4. Huyer, L.D.; Montgomery, M.; Zhao, Y.; Xiao, Y.; Conant, G.; Korolj, A.; Radisic, M. Biomaterial based cardiac tissue engineering and its applications. *Biomed. Mater.* **2015**, *10*, 034004. [CrossRef] [PubMed]
5. Vunjak-Novakovic, G.; Lui, K.O.; Tandon, N.; Chien, K.R. Bioengineering Heart Muscle: A Paradigm for Regenerative Medicine. *Annu. Rev. Biomed. Eng.* **2011**, *13*, 245–267. [CrossRef] [PubMed]
6. Cattelan, G.; Gerbolés, A.G.; Foresti, R.; Pramstaller, P.P.; Rossini, A.; Miragoli, M.; Malvezzi, C.C. Alginate Formulations: Current Developments in the Race for Hydrogel-Based Cardiac Regeneration. *Front. Bioeng. Biotechnol.* **2020**, *8*, 414. [CrossRef]
7. Sultankulov, B.; Berillo, D.; Sultankulova, K.; Tokay, T.; Saparov, A. Progress in the Development of Chitosan-Based Biomaterials for Tissue Engineering and Regenerative Medicine. *Biomolecules* **2019**, *9*, 470. [CrossRef]
8. Reed, S.; Wu, B.M. Biological and mechanical characterization of chitosan-alginate scaffolds for growth factor delivery and chondrogenesis. *J. Biomed. Mater. Res. Part B Appl. Biomater.* **2017**, *105*, 272–282. [CrossRef]
9. Kumbhar, S.G.; Pawar, S.H. Synthesis and characterization of chitosan-alginate scaffolds for seeding human umbilical cord derived mesenchymal stem cells. *Biomed. Mater. Eng.* **2016**, *27*, 561–575. [CrossRef]
10. Zhu, T.; Jiang, J.; Zhao, J.; Chen, S.; Yan, X. Regulating Preparation of Functional Alginate-Chitosan Three-Dimensional Scaffold for Skin Tissue Engineering. *Int. J. Nanomed.* **2019**, *14*, 8891–8903. [CrossRef]
11. Rodríguez-Vázquez, M.; Vega-Ruiz, B.; Ramos-Zúñiga, R.; Saldaña-Koppel, D.A.; Quiñones-Olvera, L.F. Chitosan and Its Potential Use as a Scaffold for Tissue Engineering in Regenerative Medicine. *BioMed Res. Int.* **2015**, *2015*, 821279. [CrossRef] [PubMed]
12. Li, Z.; Ramay, H.R.; Hauch, K.D.; Xiao, D.; Zhang, M. Chitosan–alginate hybrid scaffolds for bone tissue engineering. *Biomaterials* **2005**, *26*, 3919–3928. [CrossRef] [PubMed]
13. Baysal, K.; Aroguz, A.Z.; Adiguzel, Z.; Baysal, B.M. Chitosan/alginate crosslinked hydrogels: Preparation, characterization and application for cell growth purposes. *Int. J. Biol. Macromol.* **2013**, *59*, 342–348. [CrossRef] [PubMed]
14. Bushkalova, R.; Farno, M.; Tenailleau, C.; Duployer, B.; Cussac, D.; Parini, A.; Sallerin, B.; Fullana, S.G. Alginate-chitosan PEC scaffolds: A useful tool for soft tissues cell therapy. *Int. J. Pharm.* **2019**, *571*, 118692. [CrossRef] [PubMed]
15. Dai, M.; Zheng, X.; Xu, X.; Kong, X.; Li, X.; Guo, G.; Luo, F.; Zhao, X.; Wei, Y.Q.; Qian, Z. Chitosan-Alginate Sponge: Preparation and Application in Curcumin Delivery for Dermal Wound Healing in Rat. *J. Biomed. Biotechnol.* **2009**, *2009*, 595126. [CrossRef]
16. Deng, B.; Shen, L.; Wu, Y.; Shen, Y.; Ding, X.; Lu, S.; Jia, J.; Qian, J.; Ge, J. Delivery of alginate-chitosan hydrogel promotes endogenous repair and preserves cardiac function in rats with myocardial infarction. *J. Biomed. Mater. Res. Part A* **2015**, *103*, 907–918. [CrossRef]
17. Zhao, S.; Xu, Z.; Wang, H.; Reese, B.E.; Gushchina, L.V.; Jiang, M.; Agarwal, P.; Xu, J.; Zhang, M.; Shen, R.; et al. Bioengineering of injectable encapsulated aggregates of pluripotent stem cells for therapy of myocardial infarction. *Nat. Commun.* **2016**, *7*, 13306. [CrossRef]
18. Navaei, A.; Saini, H.; Christenson, W.; Sullivan, R.T.; Ros, R.; Nikkhah, M. Gold nanorod-incorporated gelatin-based conductive hydrogels for engineering cardiac tissue constructs. *Acta Biomater.* **2016**, *41*, 133–146. [CrossRef]
19. Nezhad-Mokhtari, P.; Akrami-Hasan-Kohal, M.; Ghorbani, M. An injectable chitosan-based hydrogel scaffold containing gold nanoparticles for tissue engineering applications. *Int. J. Biol. Macromol.* **2020**, *154*, 198–205. [CrossRef]
20. Lee, J.; Manoharan, V.; Cheung, L.; Lee, S.; Cha, B.-H.; Newman, P.; Farzad, R.; Mehrotra, S.; Zhang, K.; Khan, F.; et al. Nanoparticle-Based Hybrid Scaffolds for Deciphering the Role of Multimodal Cues in Cardiac Tissue Engineering. *ACS Nano* **2019**, *13*, 12525–12539. [CrossRef]
21. Kalishwaralal, K.; Jeyabharathi, S.; Sundar, K.; Selvamani, S.; Prasanna, M.; Muthukumaran, A. A novel biocompatible chitosan–Selenium nanoparticles (SeNPs) film with electrical conductivity for cardiac tissue engineering application. *Mater. Sci. Eng. C Mater. Biol. Appl.* **2018**, *92*, 151–160. [CrossRef] [PubMed]
22. Morsink, M.; Severino, P.; Luna-Ceron, E.; Hussain, M.A.; Sobahi, N.; Shin, S.R. Effects of electrically conductive nano-biomaterials on regulating cardiomyocyte behavior for cardiac repair and regeneration. *Acta Biomater.* **2022**, *139*, 141–156. [CrossRef] [PubMed]
23. Li, Y.; Wei, L.; Lan, L.; Gao, Y.; Zhang, Q.; Dawit, H.; Mao, J.; Guo, L.; Shen, L.; Wang, L. Conductive biomaterials for cardiac repair: A review. *Acta Biomater.* **2022**, *139*, 157–178. [CrossRef] [PubMed]
24. Esmaeili, H.; Patino-Guerrero, A.; Hasany, M.; Ansari, M.O.; Memic, A.; Dolatshahi-Pirouz, A.; Nikkhah, M. Electroconductive biomaterials for cardiac tissue engineering. *Acta Biomater.* **2022**, *139*, 118–140. [CrossRef] [PubMed]
25. Gentemann, L.; Kalies, S.; Coffee, M.; Meyer, H.; Ripken, T.; Heisterkamp, A.; Zweigerdt, R.; Heinemann, D. Modulation of cardiomyocyte activity using pulsed laser irradiated gold nanoparticles. *Biomed. Opt. Express* **2017**, *8*, 177–192. [CrossRef]

26. Fleischer, S.; Shevach, M.; Feiner, R.; Dvir, T. Coiled fiber scaffolds embedded with gold nanoparticles improve the performance of engineered cardiac tissues. *Nanoscale* **2014**, *6*, 9410–9414. [CrossRef]
27. Baei, P.; Jalili-Firoozinezhad, S.; Rajabi-Zeleti, S.; Tafazzoli-Shadpour, M.; Baharvand, H.; Aghdami, N. Electrically conductive gold nanoparticle-chitosan thermosensitive hydrogels for cardiac tissue engineering. *Mater. Sci. Eng. C Mater. Biol. Appl.* **2016**, *63*, 131–141. [CrossRef]
28. Nair, R.S.; Ameer, J.M.; Alison, M.R.; Anilkumar, T.V. A gold nanoparticle coated porcine cholecyst-derived bioscaffold for cardiac tissue engineering. *Colloids Surf. B Biointerfaces* **2017**, *157*, 130–137. [CrossRef]
29. Ashtari, K.; Nazari, H.; Ko, H.; Tebon, P.; Akhshik, M.; Akbari, M.; Alhosseini, S.N.; Mozafari, M.; Mehravi, B.; Soleimani, M.; et al. Electrically conductive nanomaterials for cardiac tissue engineering. *Adv. Drug Deliv. Rev.* **2019**, *144*, 162–179. [CrossRef]
30. Peña, B.; Maldonado, M.; Bonham, A.J.; Aguado, B.A.; Dominguez-Alfaro, A.; Laughter, M.; Rowland, T.J.; Bardill, J.; Farnsworth, N.L.; Ramon, N.A.; et al. Gold Nanoparticle-Functionalized Reverse Thermal Gel for Tissue Engineering Applications. *ACS Appl. Mater. Interfaces* **2019**, *11*, 18671–18680. [CrossRef]
31. Shevach, M.; Fleischer, S.; Shapira, A.; Dvir, T. Gold Nanoparticle-Decellularized Matrix Hybrids for Cardiac Tissue Engineering. *Nano Lett.* **2014**, *14*, 5792–5796. [CrossRef] [PubMed]
32. Shevach, M.; Maoz, B.M.; Feiner, R.; Shapira, A.; Dvir, T. Nanoengineering gold particle composite fibers for cardiac tissue engineering. *J. Mater. Chem. B* **2013**, *1*, 5210–5217. [CrossRef] [PubMed]
33. Sridhar, S.; Venugopal, J.R.; Sridhar, R.; Ramakrishna, S. Cardiogenic differentiation of mesenchymal stem cells with gold nanoparticle loaded functionalized nanofibers. *Colloids Surf. B Biointerfaces* **2015**, *134*, 346–354. [CrossRef]
34. Topete, A.; Alatorre-Meda, M.; Villar-Álvarez, E.M.; Cambón, A.; Barbosa, S.; Taboada, P.; Mosquera, V. Simple Control of Surface Topography of Gold Nanoshells by a Surfactant-less Seeded-Growth Method. *ACS Appl. Mater. Interfaces* **2014**, *6*, 11142–11157. [CrossRef]
35. Pennella, F.; Cerino, G.; Massai, D.; Gallo, D.; D'Urso Labate, G.F.; Schiavi, A.; Deriu, M.A.; Audenino, A.L.; Morbiducci, U. A Survey of Methods for the Evaluation of Tissue Engineering Scaffold Permeability. *Ann. Biomed. Eng.* **2013**, *41*, 2027–2041. [CrossRef] [PubMed]
36. Wang, Y.; Meng, H.; Yuan, X.; Peng, J.; Guo, Q.; Lu, S.; Wang, A. Fabrication and in vitro evaluation of an articular cartilage extracellular matrix-hydroxyapatite bilayered scaffold with low permeability for interface tissue engineering. *Biomed. Eng. Online* **2014**, *13*, 80. [CrossRef]
37. Badita, C.R.; Aranghel, D.; Burducea, C.; Mereuta, P. Characterization of sodium alginate based films. *Rom. J. Phys.* **2020**, *65*, 602.
38. Sosnik, A.; Imperiale, J.C.; Vázquez-González, B.; Raskin, M.M.; Muñoz-Muñoz, F.; Burillo, G.; Cedillo, G.; Bucio, E. Mucoadhesive thermo-responsive chitosan-g-poly(N-isopropylacrylamide) polymeric micelles via a one-pot gamma-radiation-assisted pathway. *Colloids Surf. B Biointerfaces* **2015**, *136*, 900–907. [CrossRef]
39. Birnbach, B.; Höpner, J.; Mikolajczyk, R. Cardiac symptom attribution and knowledge of the symptoms of acute myocardial infarction: A systematic review. *BMC Cardiovasc. Disord.* **2020**, *20*, 445. [CrossRef]
40. Fang, Y.; Zhang, T.; Song, Y.; Sun, W. Assessment of various crosslinking agents on collagen/chitosan scaffolds for myocardial tissue engineering. *Biomed. Mater.* **2020**, *15*, 045003. [CrossRef]
41. Rosellini, E.; Madeddu, D.; Barbani, N.; Frati, C.; Graiani, G.; Falco, A.; Lagrasta, C.; Quaini, F.; Cascone, M.G. Development of Biomimetic Alginate/Gelatin/Elastin Sponges with Recognition Properties toward Bioactive Peptides for Cardiac Tissue Engineering. *Biomimetics* **2020**, *5*, 67. [CrossRef] [PubMed]
42. de Almeida, D.A.; Sabino, R.M.; Souza, P.R.; Bonafé, E.G.; Venter, S.A.S.; Popat, K.C.; Martins, A.F.; Monteiro, J.P. Pectin-capped gold nanoparticles synthesis in-situ for producing durable, cytocompatible, and superabsorbent hydrogel composites with chitosan. *Int. J. Biol. Macromol.* **2020**, *147*, 138–149. [CrossRef] [PubMed]
43. Dong, Y.; Hong, M.; Dai, R.; Wu, H.; Zhu, P. Engineered bioactive nanoparticles incorporated biofunctionalized ECM/silk proteins based cardiac patches combined with MSCs for the repair of myocardial infarction: In vitro and in vivo evaluations. *Sci. Total Environ.* **2020**, *707*, 135976. [CrossRef] [PubMed]
44. Sood, A.; Arora, V.; Shah, J.; Kotnala, R.K.; Jain, T.K. Multifunctional gold coated iron oxide core-shell nanoparticles stabilized using thiolated sodium alginate for biomedical applications. *Mater. Sci. Eng. C Mater. Biol. Appl.* **2017**, *80*, 274–281. [CrossRef]
45. Shen, K.; Huang, Y.; Li, Q.; Chen, M.; Wu, L. Self-Assembled Polysaccharide–Diphenylalanine/Au Nanospheres for Photothermal Therapy and Photoacoustic Imaging. *ACS Omega* **2019**, *4*, 18118–18125. [CrossRef]
46. Prabha, G.; Raj, V. Sodium alginate–polyvinyl alcohol–bovin serum albumin coated Fe_3O_4 nanoparticles as anticancer drug delivery vehicle: Doxorubicin loading and in vitro release study and cytotoxicity to HepG2 and L02 cells. *Mater. Sci. Eng. C Mater. Biol. Appl.* **2017**, *79*, 410–422. [CrossRef]
47. Bolaños, K.; Kogan, M.J.; Araya, E. Capping gold nanoparticles with albumin to improve their biomedical properties. *Int. J. Nanomed.* **2019**, *14*, 6387–6406. [CrossRef]
48. Annabi, N.; Nichol, J.W.; Zhong, X.; Ji, C.; Koshy, S.; Khademhosseini, A.; Dehghani, F. Controlling the Porosity and Microarchitecture of Hydrogels for Tissue Engineering. *Tissue Eng. Part B Rev.* **2010**, *16*, 371–383. [CrossRef]
49. Tamimi, M.; Rajabi, S.; Pezeshki-Modaress, M. Cardiac ECM/chitosan/alginate ternary scaffolds for cardiac tissue engineering application. *Int. J. Biol. Macromol.* **2020**, *164*, 389–402. [CrossRef]

50. Macías-Andrés, V.I.; Li, W.; Aguilar-Reyes, E.A.; Ding, Y.; Roether, J.A.; Harhaus, L.; Patiño, C.A.L.; Boccaccini, A.R. Preparation and characterization of 45S5 bioactive glass-based scaffolds loaded with PHBV microspheres with daidzein release function. *J. Biomed. Mater. Res. Part A* **2017**, *105*, 1765–1774. [CrossRef]
51. Tresoldi, C.; Pacheco, D.P.; Formenti, E.; Pellegata, A.F.; Mantero, S.; Petrini, P. Shear-resistant hydrogels to control permeability of porous tubular scaffolds in vascular tissue engineering. *Mater. Sci. Eng. C Mater. Biol. Appl.* **2019**, *105*, 110035. [CrossRef] [PubMed]
52. Rai, R.; Tallawi, M.; Frati, C.; Falco, A.; Gervasi, A.; Quaini, F.; Roether, J.A.; Hochburger, T.; Schubert, D.W.; Seik, L.; et al. Bioactive Electrospun Fibers of Poly(glycerol sebacate) and Poly(ε-caprolactone) for Cardiac Patch Application. *Adv. Healthc. Mater.* **2015**, *4*, 2012–2025. [CrossRef] [PubMed]
53. Croisier, F.; Jérôme, C. Chitosan-based biomaterials for tissue engineering. *Eur. Polym. J.* **2013**, *49*, 780–792. [CrossRef]
54. Yadid, M.; Feiner, R.; Dvir, T. Gold Nanoparticle-Integrated Scaffolds for Tissue Engineering and Regenerative Medicine. *Nano Lett.* **2019**, *19*, 2198–2206. [CrossRef]
55. Pal, A.; Esumi, K.; Pal, T. Preparation of nanosized gold particles in a biopolymer using UV photoactivation. *J. Colloid Interface Sci.* **2005**, *288*, 396–401. [CrossRef] [PubMed]
56. Tentor, F.R.; de Oliveira, J.H.; Scariot, D.B.; Lazarin-Bidóia, D.; Bonafé, E.G.; Nakamura, C.V.; Venter, S.A.; Monteiro, J.P.; Muniz, E.C.; Martins, A.F. Scaffolds based on chitosan/pectin thermosensitive hydrogels containing gold nanoparticles. *Int. J. Biol. Macromol.* **2017**, *102*, 1186–1194. [CrossRef] [PubMed]
57. Jiang, L.; Chen, D.; Wang, Z.; Zhang, Z.; Xia, Y.; Xue, H.; Liu, Y. Preparation of an Electrically Conductive Graphene Oxide/Chitosan Scaffold for Cardiac Tissue Engineering. *Appl. Biochem. Biotechnol.* **2019**, *188*, 952–964. [CrossRef] [PubMed]
58. Dvir, T.; Timko, B.P.; Brigham, M.D.; Naik, S.R.; Karajanagi, S.S.; Levy, O.; Jin, H.; Parker, K.K.; Langer, R.; Kohane, D.S. Nanowired three-dimensional cardiac patches. *Nat. Nanotechnol.* **2011**, *6*, 720–725. [CrossRef]
59. Ashur, C.; Frishman, W.H. Cardiosphere-Derived Cells and Ischemic Heart Failure. *Cardiol. Rev.* **2018**, *26*, 8–21. [CrossRef]

Article

Characterization and Evaluation of Composite Biomaterial Bioactive Glass–Polylactic Acid for Bone Tissue Engineering Applications

Georgina Carbajal-De la Torre [1,*], Nancy N. Zurita-Méndez [1], María de Lourdes Ballesteros-Almanza [2], Javier Ortiz-Ortiz [1], Miriam Estévez [3] and Marco A. Espinosa-Medina [1,*]

1. Facultad de Ingeniería Mecánica, Universidad Michoacana de San Nicolás de Hidalgo, Morelia C.P. 58000, Mexico; nancy.zurita@umich.mx (N.N.Z.-M.); javier.ortiz@umich.mx (J.O.-O.)
2. Facultad de Biología, Universidad Michoacana de San Nicolás de Hidalgo, Morelia C.P. 58000, Mexico; balmanza@umich.mx
3. Centro de Física Aplicada y Tecnología Avanzada, Universidad Nacional Autónoma de México, Querétaro C.P. 76230, Mexico; miries@fata.unam.mx
* Correspondence: georginacar@gmail.com (G.C.-D.); marco.espinosa@umich.mx (M.A.E.-M.)

Abstract: The limitations associated with the clinical use of autographs and allografts are driving efforts to develop relevant and applicable biomaterial substitutes. In this research, 3D porous scaffolds composed of bioactive glass (BG) obtained through the sol-gel technique and polylactic acid (PLA) synthesized via lactic acid (LA) ring-opening polymerization were prepared by the gel-pressing technique. Two different weight compositions were evaluated, namely, BG70-PLA30 and BG30-PLA70. The structure and morphology of the resulting scaffolds were analysed by FTIR, XRD, SEM, and under ASTM F1635 standard characterizations. The results confirmed that BG promotes the formation of a hydroxy-carbonated apatite (HAp) layer on composites when immersed in simulated body fluid (SBF). Biodegradability evaluations were carried out according to the ISO 10993-13:2010 standard. In addition, electrochemical evaluations were performed in both Hank's and SBF solutions at 37 °C in order to analyse the degradation of the material. This evaluation allowed us to observe that both samples showed an activation mechanism in the early stages followed by pseudo-passivation due to physical bioactive glass characteristics, suggesting an improvement in the formation of the HAp nucleation. The described composites showed excellent resistance to degradation and outstanding suitability for bone tissue engineering applications.

Keywords: bioactive glass; polylactic acid; scaffolds; electrochemical evaluations

1. Introduction

Materials that substitute bone tissues are of great interest to the scientific community, as traumatic injuries and pathologies in which the skeletal structure is damaged are extremely common [1–3]. Several years ago, it was thought that human tissues or organs were only replaceable by transplants or metallic and polymeric devices. However, many of these materials can cause an undesirable immune response, leading to inflammation and rejection. Biomaterials based on the SiO_2–CaO–$Na_2O$$P_2O_5$ system, commonly called bioactive glass (BG), have the ability to form bonds with bone and connective tissues; this ability is attributed to the formation of a silica layer with a high surface area and the formation of polycrystalline hydroxyapatite layers on the bioactive glass surface [4,5]. BGs have been studied in soft-tissue engineering applications such as peripheral nerve regeneration and chronic pain treatment as well [6].

BG is frequently obtained by two methods: 1. melt-derived glass, in which the oxides are silica, calcium, phosphate, and sodium precursors, which then undergo further solidification; and 2. sol–gel synthesis, which employs low processing temperatures for an economical method in which the properties can easily be controlled. Overall, the bioglasses obtained by this method exhibited high surface areas and suitable porosity, providing osteogenic potential [7–9]. Furthermore, lactic acid (LA) production in its L(+) isomerism is promoted by the intense physical activity of the muscles; although LA is unassimilable by the organism, its produced polymer (polylactic acid, or PLA) has a high biodegradability rate and is bio-compatible, immunologically inert, non-toxic, and absorbable [10]. Consequently, this polymer could be used for the elaboration of a composite biomaterial for bioengineering applications such as controlled drug release systems, bioabsorbable fixation devices, and bone regeneration implants. Through the method of direct polycondensation, it is possible to obtain low molecular weight products, which are important in biomedical applications [11]. As is known, the surface reactions of materials with their biological environment occur a few seconds after they are implanted in the body, interacting with proteins present in the physiological environment; hence, it is important to evaluate the in vitro biological behaviour of the biomaterials. Simulated body fluid (SBF) and Hank's saline solution at 37 °C are the aqueous media that allow for understanding of the corrosion mechanism of composite biomaterials, as their ionic compositions are close to those of human plasma [8–12].

This research synthesized BG using the sol–gel method, using tetraethyl orthosilicate (TEOS) as an initial precursor. The polymeric material (PLA) was synthesized by the ring-opening polymerization of lactic acid, and they were subsequently mixed by employing a solvent in two different weight compositions (BG70-PLA30 and BG30-PLA70), then deposited by dip coating on 316L stainless steel sheets of about 0.2 mm in thickness. Assessment of the corrosive behaviour in Hank's solution and simulated body fluid (SBF) was performed using electrochemical techniques. Their bioactivity in PBS was evaluated by the ASTM F1635 standard test method for in vitro degradation testing of poly (L-lactic Acid) resin and fabricated as scaffolds. In contrast, SBF bioactivity was evaluated using the methodology of Kokubo et al. [13]. This project discusses the results of the methods and measurements of the properties of these scaffolds, describes a desirable resistance to degradability and bioactivity in simulated body solutions, and uses extensive electrochemical analysis to evaluate the degradation conditions of the composite biomaterial.

2. Materials and Methods

The sol–gel technique was used to synthesize the bioactive glass, while polylactic acid was synthesized via lactic acid (LA) polycondensation. Furthermore, BG, PLA, and BG-PLA composite samples were prepared and characterized by Fourier transform infrared spectroscopy (FTIR), X-Ray diffraction (XRD) analysis, and scanning electron microscopy (SEM). The corrosion behaviour of the composites in Hank's balanced salt solution and simulated body fluid (SBF) at 37 °C were performed using electrochemical techniques. The biomaterial bioactivity in SBF and phosphate-buffered saline (PBS) was measured each week for 28 days.

2.1. PLA Synthesis

In order to reproduce a more efficient PLA production process, the ring-opening polymerization (ROP) method was carried out. The initial lactic acid of reactive grade (Meyer ®) was put inside using a rotary evaporator (Hahnshin Scientific Co., model: HS-2000NS, Michoacán, México), applying a 35-rpm rotation, heating temperature, and vacuum at −200 mmHg. Tin(II) 2-ethyl hexanoate (Sigma-Aldrich ® ~95%, Michoacán, México) was added, and the temperature was raised to 175 °C. The reaction takes 6 h under these conditions. The obtained PLA was dissolved in propanone (Meyer ®, Michoacán, México) and precipitated with distilled water. The white PLA powder was then washed, filtered, and dried.

2.2. BG-PLA Composite Synthesis

All precursors were reactive grade and used without further purification. The method was performed in two steps. In the first step, tetraethyl silicate (98% Sigma-Aldrich ®, Michoacán, México) and 0.1 M nitric acid (JT Baker ®, Michoacán, México) were mixed at room temperature. Then, triethyl phosphate (TEP: 99.8%, Sigma-Aldrich ®, Michoacán, México) and calcium nitrate tetrahydrate (99%, Sigma-Aldrich ®, Michoacán, México) were added at intervals. The reaction lasted for an additional hour after the last compound was added. The bioactive glass was obtained in this step. The general synthesis reactions can be observed in Equation (1). In the second step, the material obtained as a gel form was mixed with the PLA obtained in two weight percentages (wt.%), namely, BG30-PLA70 and BG70-PLA30, employing dissolvent propanone (Meyer ®, Michoacán, México). The composite biomaterial was first kept in a sealed glass jar at room temperature for ten days and later at 70 °C for three days. Thermal treatment was performed at 120 °C for two days.

$$SiC_8H_{20}O_4 \xrightarrow{HNO_3\ 0.1\ M} Si(OH)_4 \xrightarrow{TEP,\ Ca(NO_3)_2 \cdot 4H_2O} (SiO_2)_x (CaO)_y (P_2O_5)_z \qquad (1)$$

2.3. SBF and PBS Preparation

The SBF was prepared according to the Kokubo protocol [13]. The reaction was controlled at a pH of 7.45 ± 0.01. The obtained solution was cooled at 20 °C and kept under refrigeration at 7 °C. At the same time, the PBS solution was obtained by dissolving a phosphate-buffered saline tablet (Sigma®, Michoacán, México) in 200 mL of deionized water to obtain a 0.01 M phosphate buffer with 0.0027 M KCl and 0.137 M NaCl contents, with a pH of 7.4 at 25 °C.

2.4. BG-PLA Coatings

A 316L stainless steel sheet approximately 0.2 mm thick was used as the substrate material. These sheets were polished using sandpapers of 230, 300, 500, 600, and 1000 grades. The 316L substrate surface was chemically treated by immersion in NaOH (Sigma-Aldrich ®, Michoacán, México) 6M solution for 24 h and cleaned with deionized water and acetone. The composite material for the coating application was obtained by dissolving 5 g of BG-PLA powders in 20 mL of acetone (reagent grade, Meyer ®, Michoacán, México). The coatings were applied by immersion using the dip-coating method with a 176 mm/min speed rate and residence time of 30 s. The obtained layers were dried at 120 °C for 24 h.

2.5. BG-PLA Scaffolds Design

BG-PLA scaffolds were obtained by the gel-pressing technique, in which 10 g of each composite material (BG70-PLA30 and BG30-PLA70) was dissolved in 40 mL of chloroform ($CHCl_3$, Meyer ® 99.8%, Michoacán, México). Subsequently, porosity was achieved by particle leaching using NaCl crystals a maximum of 500 µm of diameter in a proportion of 60 wt.% of the weight of the total components. The homogeneous phase was pressed into containers with 0.635 cm diameter and 1.2 cm height. The solvent was first evaporated at room temperature for two days, then heated in an oven at 50 °C for 24 h. The scaffold samples were immersed in distilled water to eliminate salt, and the porosity formation was dried again in an oven and stored in sterile Petri dishes.

2.6. BG-PLA Bioactivity in SBF

The main characteristic of bioactivity in SBF is the formation of hydroxyapatite (HAp) on the material surface. For this evaluation, the BG-PLA scaffolds were immersed in triplicate in a polyethylene bottle (three scaffolds per bottle) with SBF solution in a 100 mL/g ratio at a controlled temperature of 37 °C and pH 7.4. Bioactivity measurements were obtained at 7, 14, 21, and 28 days. The nomenclature identification for the samples is shown in Table 1. At the end of each test, the scaffolds were removed from the SBF solution, gently rinsed with deionized water, and allowed to dry for 4 to 5 days in an incubator at 37 °C. The pH of the solution was monitored, and the SBF solution was replaced every week due to the cation concentration decreasing during the experiments. Bioactivity results were complemented by XRD, FTIR, and SEM characterization.

Table 1. Identification of samples evaluated in SBF.

Sample BG70-PLA30	SBF Evaluation (Days)	Sample BG30-PLA70	SBF Evaluation (Days)
BG70-PLA30-SBF-7	7	BG30-PLA70-SBF-7	7
BG70-PLA30-SBF-14	14	BG30-PLA70-SBF-14	14
BG70-PLA30-SBF-21	21	BG30-PLA70-SBF-21	21
BG70-PLA30-SBF-28	28	BG30-PLA70-SBF-28	28

2.7. BG-PLA Degradation in PBS

Degradation monitoring was carried out by measuring the change in weight during the sample's immersion in PBS. Similarly, the BG-PLA pieces were immersed at 37 °C for 1, 2, 3, and 4 weeks, with the weight change measured after each period. The PBS solutions were replaced every seven days. This study was performed according to the standard ISO of 10993-13:2010 [14]. The percentage of weight loss was calculated from Equation (1). The nomenclature identification for the samples is shown in Table 2.

$$\text{Weightloss}(\%) = 100 \left(\frac{W_1 - W_2}{W_1} \right) \qquad (2)$$

where W_1 and W_2 are the weight of the dry composite before and after immersion, respectively.

Table 2. Identification of samples evaluated in PBS.

Sample BG70-PLA30	PBS Evaluation (Days)	Sample BG30-PLA70	PBS Evaluation (Days)
BG70-PLA30-PBS-7	7	BG30-PLA70-PBS-7	7
BG70-PLA30-PBS-14	14	BG30-PLA70-PBS-14	14
BG70-PLA30-PBS-21	21	BG30-PLA70-PBS-21	21
BG70-PLA30-PBS-28	28	BG30-PLA70-PBS-28	28

2.8. FTIR and XRD Characterizations

Fourier transform infrared spectroscopy analysis was performed with a Bruker spectrometer model Tensor 27. The applied measurement range was 4000 to 400 cm^{-1}, with a 4 cm^{-1} resolution and sample and background scan times of 32 scans. The samples were obtained by mixing 0.0020 g of the powders and 0.20 g of KBr, then compressed by applying 9.9 tons of pressure for 1 min with a PIKE Technologies CrushIR hydraulic press machine. Then, the compacted sample was characterized with FTIR equipment. XRD measurements were conducted using a D8 Advanced Da-Vinci equipment X-Ray diffractometer. Scans were taken with a 2θ step size of 0.04° from 20° to 90° and a counting time of 0.3 s using Cu Kα radiation. The phases were identified by matching the observed patterns to the entries in the indexing software.

2.9. Electrochemical Tests

Electrochemical tests were performed using a potentiostat/galvanostat Gill-AC (ACM Instruments) controlled by a computer. A three-electrode cell arrangement was used with an Ag/AgCl saturated reference electrode (SSCE-RE), platinum wire as an auxiliary electrode (AE), and the coating samples (WE). Hank's balanced salt solution (Sigma-Aldrich ®, Michoacán, México) modified with sodium bicarbonate (without phenol red, calcium chloride, or magnesium sulphate, sterile-filtered, and suitable for cell culture) and simulated body fluid (SBF) at 37 ± 1 °C was the electrolyte used to emulate human body temperature, which was controlled by an electric heating band.

A polarization potential scan obtained potentiodynamic polarization curves (TF) from −500 mV to +1500 mV vs. open circuit potential (OCP) at a scan rate of 1 mV/s. Corrosion current density values, i_{corr}, and other parameters were calculated using the Tafel extrapolation method between an extrapolation range of ±100 mV around the OCP. Before running the experiments, a 10 min delay time was set until the OCP reached the steady-state condition. The LPR measurements were obtained in a range of ±15 mV vs. the OCP with a scan rate of 1 mV/s every 15 min for 48 h. Polarization resistance (R_p) and current density kinetics were obtained by Ohm's law and the Stern and Geary equations [15]. The electrochemical impedance spectroscopy (EIS) measurements were carried out at OCP using a voltage signal with an amplitude of 30 mV and a frequency interval between 23,000 and 0.01 Hz.

3. Results

3.1. X-ray Diffraction Analysis

The diffraction patterns obtained for the biomaterials are shown in Figure 1. As can be observed, the obtained BG presents a ceramic formulation system composed of SiO_2–Na_2O–CaO–P_2O_5. The presence of P_2O_5 allows formation of a network, promoting the glass crystallization process [16], and induces the formation of a calcium phosphate layer that crystallizes into biomimetic hydroxyapatite due to the incorporation of hydroxide and carbonate ions from the biological fluid [17]. The X-ray diffraction results for the orthorhombic lattice PLA were compared with the crystallographic PDF data 00-064-1624, presenting diffractions in 2θ angles positioned at 12.42°, 16.63°, 19.08°, and 22.3°, correlated to the Miller's indices of the planes (103), (200), (203), and (211) of the polymeric material. As expected, the composition of the BG-PLA composite at both proportions (BG70-PLA30 and BG30-PLA70) agrees with the presence of each phase.

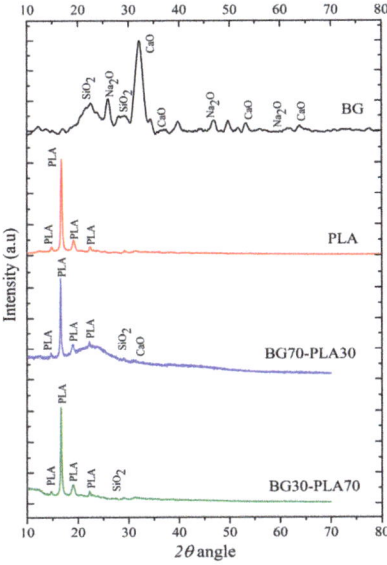

Figure 1. X-ray diffraction for the obtained materials.

3.2. FTIR Characterization

Fourier transform infrared spectroscopy with the KBr technique has been used recently to study the structure–composition relationship in various glasses and glass ceramics [18]. The FTIR results for the PLA and BG samples are shown in Figure 2. According to the analysis of the BG spectrum (Figure 2a), absorption peaks could be observed at 1386, 838, and 461 cm^{-1}, representing the bending and stretching vibrations of Si−O−Si bonds. The vibrational band with low intensity at 566 cm^{-1} corresponds to the bending vibrations of the phosphate (PO$_4^{3-}$) groups [19], suggesting that the phosphate can be considered as a network former [20]. The broad band at 3427 cm^{-1} could be ascribed to the vibration of different OH^{-} groups, and represents the surface silanol groups related to different hydroxyl groups. This indicates the superposition of stretching modes of non-hydrogen-bonded silanols (isolated silanol groups) and hydrogen-bonded-silanol (vicinal silanol groups) [21]. As can be observed in the polylactic acid FTIR spectra in Figure 2b, the stretching vibrations of C-C bonds are found at 865 cm^{-1}, the asymmetric and symmetric C−O−C stretching peaks are related to 1132 and 1211 cm^{-1}, respectively, the C−H symmetric bending can be located at 1375 cm^{-1}, −CH$_3$ asymmetric bending can be seen at 1457 cm^{-1}, C=O stretching bonds are represented at 1755 cm^{-1}, and C−H symmetric and asymmetric stretching at the 2944 and 3000 cm^{-1} peaks, respectively [22].

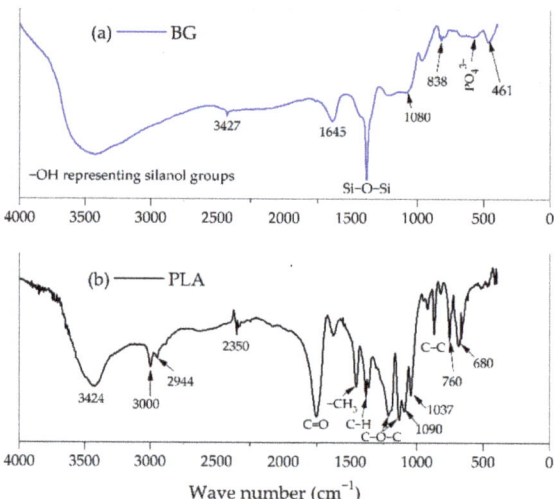

Figure 2. FTIR spectra for the synthesized species (a) BG and (b) PLA.

The analysis of the bioactivity of the composite scaffolds BG70-PLA30 in SBF by FTIR is shown in Figure 3. The results show vibrational bands related to the silanol groups, C−H, C=O, C−O−C, and P-O bonds; the broad band at 3000–3600 cm^{-1} is present due to the silanol groups on the composite surface. The Hap formation on the surface of the composite immersed in SBF is associated with the presence of the bands around 560–600 cm^{-1}, which correspond to the bending vibrations of P−O bonds that are visible in the Si−Na−P system [21]. The FTIR spectrum of the BG70-PLA30 scaffolds shows a broad spectrum, reflecting the Si−O−Si symmetric stretching vibrations.

The silanol groups at 3500 cm^{-1} are present in the FTIR spectrum of the BG30-PLA70 sample in SBF (Figure 4). Due to the higher composition of PLA in the composite, the presence of stretching vibrations of C-C, C-O-C, and C=O bonds are observed, and are associated with the crystallinity of the PLA phase. The bending and stretching vibrations of Si-O-Si bonds correspond to the BG phase. At 21 and 28 days of immersion, the formation of the Hap phase was observed in the vibrational bands with low intensity at 573 cm^{-1} and 610 cm^{-1}, which are related to the bending vibrations of the phosphate (PO$_4^{3-}$) groups. Chen et al. [23] observed the vibrational bands at 608 and 561 cm^{-1} to be associated with the strengthened intermolecular interaction of the molecules in the crystal lattice in highly-ordered arrangements. Thus, it is possible that the phosphate formation acts as a molecular link. The BG dissolution mechanism in the biological fluids was associated with ions leaching from BG into PBS, followed by decomposition of silica–oxygen bonds of the

BG network and redeposition of the calcium and phosphorus ions onto the biomaterial surface [24].

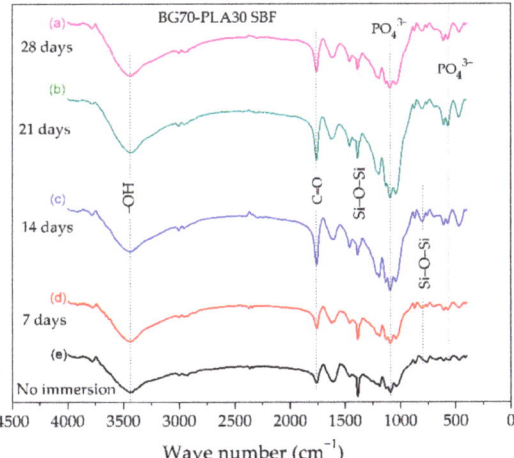

Figure 3. FTIR spectra of the BG70-PLA30 composite after different soaking times in SBF.

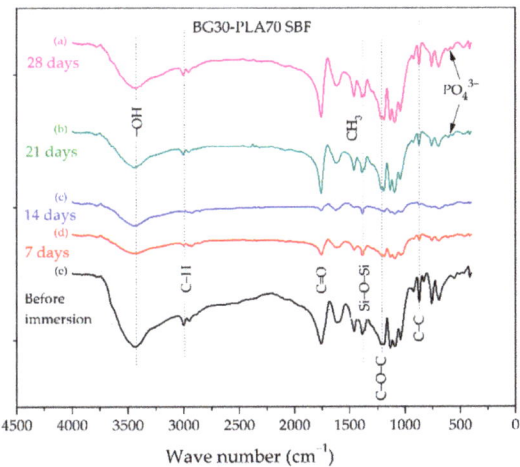

Figure 4. FTIR spectra of the BG30-PLA70 composite after different soaking times in SBF.

Figures 5 and 6 show the FTIR results of the BG-PLA composites in PBS. The FTIR results of the BG70-PLA30 scaffolds in PBS immersed for 7, 14, 21, and 28 days (Figure 5) present consistent degradation due to the presence of the phosphate groups' bending vibrations, which are more defined with longer immersion times. The mineralization process was associated with the intensity increase of the 1037 cm^{-1} peak due to P-O stretching vibration. The FTIR results for the biomaterial BG30-PLA70 (Figure 6) present the formation of a vibrational band with low intensity at 566 cm^{-1} after 21 days of immersion, corresponding to the deposition of phosphorous ions on the surface. The peak intensity decrease in the vibrational band at 1385 cm^{-1}, which corresponds to Si−O−Si bending and stretching vibrations at 28 days of immersion, indicates the process decomposition of the BG phase.

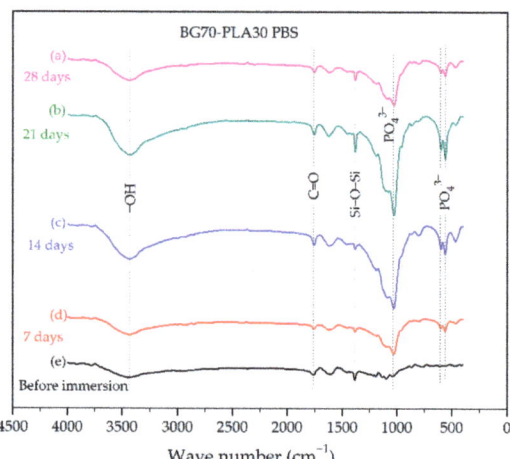

Figure 5. FTIR spectra of the BG70-PLA30 composite after different soaking times in PBS.

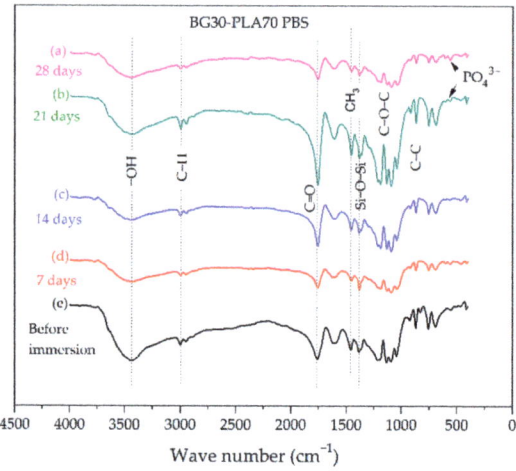

Figure 6. FTIR spectra of the BG30-PLA70 composite after different soaking times in PBS.

3.3. SEM Characterization

Figure 7a shows that the morphology of the BG presents an irregular morphology and particles with a size between 2.31 to 15.47 μm. Figure 7c shows the chemical composition by EDS of the BG sample, indicating the presence of Ca, O, Si, P, and C, which are constituents of the bioactive bioglass and were observed in the FTIR and XRD characterization results as well. Furthermore, the BG morphology here is similar to that reported by Sharifianjazi et al. [25] and Xia et al. [26]. The PLA morphology is shown in Figure 7b, while its components are indicated in EDS spectrum of Figure 7d.

The BG-PLA scaffolds were characterized by SEM as well. The BG70-PLA30 composite (Figure 8a) presented a dense morphology with a homogeneous phase, with no differentiation between the BG and PLA phases. On the other hand, the BG30-PLA70 scaffolds (Figure 8b) showed a cracked surface morphology, which is associated with the presence of tension stress at the grain interfaces of the polymer and BG phase during sintering due to the higher wt.% quantity of the PLA in the composite. The elemental chemical analysis by EDS of both composite samples is shown in Table 3. As expected, the chemical composition for the samples is in agreement with the quantity of the polymeric and vitreous phases, denoting a major percentage of C when the PLA phase was higher in the scaffolds (BG30-PLA70).

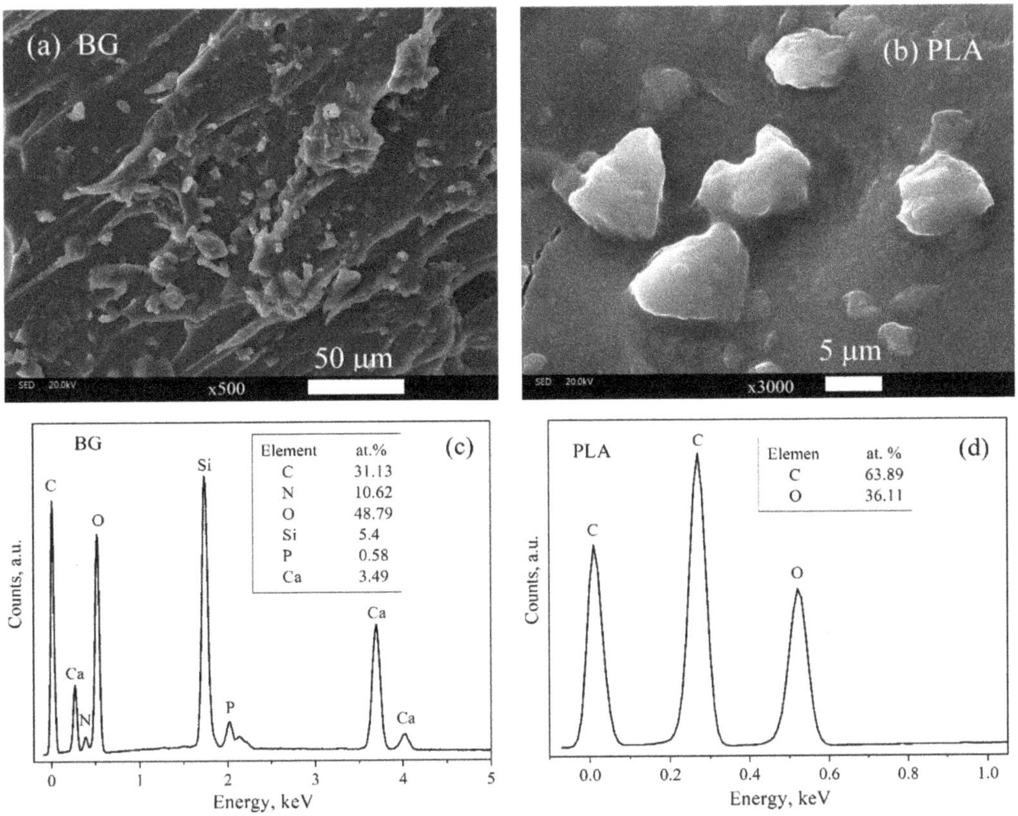

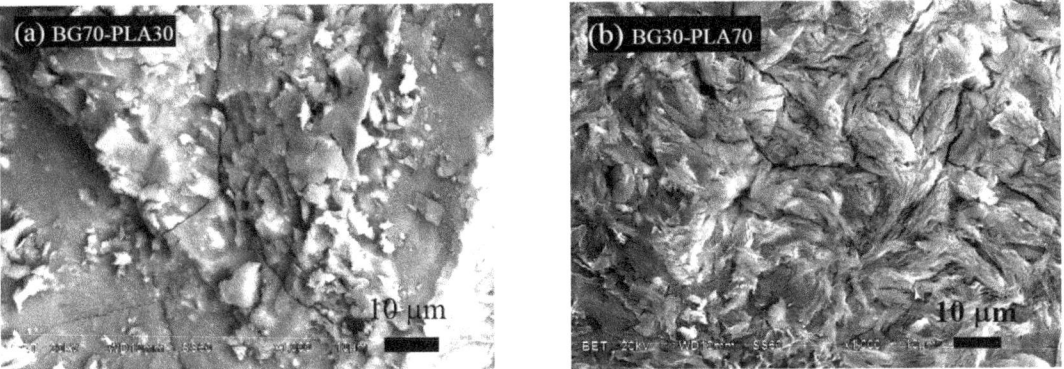

Figure 7. Morphology by SEM of (**a**) BG and (**b**) PLA; EDS chemical analyses of (**c**) BG and (**d**) PLA.

Figure 8. Scaffold SEM analysis for (**a**) BG70-PLA30 and (**b**) BG30-PLA70.

Table 3. Energy-dispersive X-ray spectroscopy for both compositions.

EDS Element	BG70-PLA30 at.%	BG30-PLA70 at.%
C	38.485	50.103
O	48.356	46.811
Si	9.704	0.588
Ca	3.454	2.498

3.4. Evaluation of BG-PLA Bioactivity in SBF

The evaluation of the bioactivity of the BG-PLA composites was achieved as described in Section 2.3, and is supported by the FTIR (shown in Figures 3 and 4) and SEM characterization. In accordance with Equation (2), the average weight loss of the BG70-PLA30 remained fairly constant over the different time periods of immersion. The average weight loss values ($\overline{X}$) across the four time periods did not show any significant variation, with measured values between 37.85 and 38.5 wt.%, as seen in Table 4. A similar profile was presented by the BG30-PLA70 composite, with an average weight loss of between 39.6 and 38.6 wt.%. Nevertheless, the differences in the weight loss values for both composites during the four time periods are associated with the evolution of HAp formation, which is integrated into the measurement. After immersion, the scaffold composites presented degradation indications due to water interaction in the ion exchange mechanism between the BG/PLA phases and the solution. The water molecules disassociate the Si-O bonds in the BG network forming Si-OH groups, which attracts the Ca^{2+}, H_2PO^{4-}, HPO_4^{2-} and PO_4^{3-} ions present in the SBF solution. This favourably promoted HAp nucleation site formation on the sample's surface [23]. The degradation behaviour of the BG (30 wt.%) composite showed an initial increase as a result of water uptake, then a subsequent decrease due to mass loss attributed to the polymer achieving a critical molecular weight sufficiently small to allow diffusion out of the matrix [27], and finally a small mass increase due to room humidity. These three steps are key to sample degradation.

Table 4. Average weight loss ($\overline{X}$) and standard deviation (S_X) for the scaffolds immersed in SBF and PBS over time periods of 7, 14, 21, and 28 days.

Time, (Days)	SBF				PBS			
	BG30 PLA70		BG70 PLA30		BG30 PLA70		BG70 PLA30	
	$\overline{x}$, (%)	S_X	$\overline{x}$, (%)	S_X	$\overline{x}$, (%)	S_X	$\overline{x}$, (%)	S_X
7	39.60	1.34	37.86	1.15	27.80	1.86	24.12	0.03
14	38.88	0.81	38.95	0.27	28.42	1.32	22.42	0.22
21	39.72	1.47	39.33	0.04	28.05	1.06	23.00	0.57
28	38.60	0.92	38.53	0.23	27.91	0.32	24.93	1.06

Figure 9 shows the SEM morphology of the BG70-PLA30 scaffolds after 14 days and 28 days of immersion in SBF. The morphology surfaces show the evolution of HAp formation, with greater presence at 28 days of immersion. The EDS chemical composition related to Figure 9 is presented in Table 5 and confirms the HAp growth associated with the quantity of calcium adhesion to the surface (spectrum 2), which is substantial after 28 days of immersion of the biomaterial in SBF. Similarly, Figure 10 shows the SEM morphology of the BG30-PLA70 scaffolds after 14 days and 28 days of immersion in SBF. After 14 days of immersion, the sample morphology shows the presence of cracks on the surface which represent the early stages of degradation, promoting the first interstitial condition for HAp phase nucleation. The EDS results of the chemical analysis of the composite at 14 and 28 days is shown in Table 5. The chemical composition is quite similar in the elements and atomic percentages between both time periods, indicating that the BG concentration in the biomaterial scaffolds is important for optimal bioactivity.

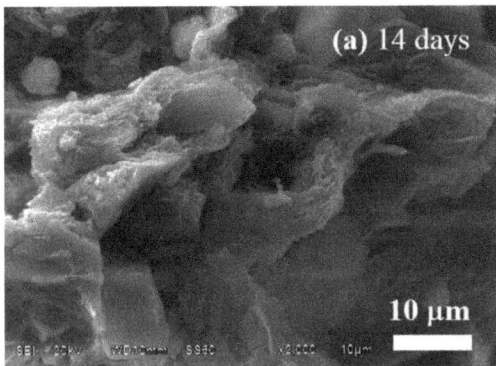 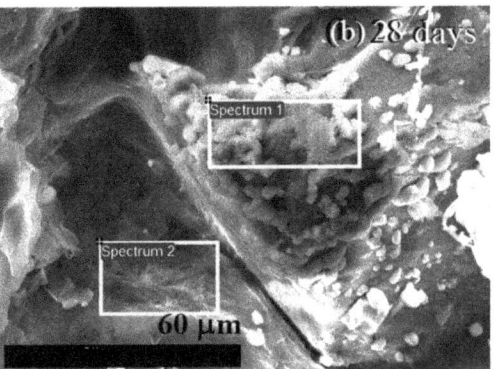

Figure 9. EDS analysis for BG70-PLA30 after (**a**) 14 and (**b**) 28 days of immersion in SBF.

Table 5. EDS chemical analysis for BG-PLA composites after immersion in SBF.

Element at.%	BG70-PLA30			BG30-PLA70	
	14 Days	28 Days		14 Days	28 Days
		S. 1	S. 2		
C	41.25			58.97	51.34
O	45.59	37.9	58.7	34.58	43.6
Na	0.88	25.1	17.3	2.35	1.26
Si	4.82	13.5	8.7	1.05	2.59
P	2.91			0.24	
Cl	0.41	19.1	9.15	2.02	0.85
Ca	4.13	4.41	6.15	0.8	0.36

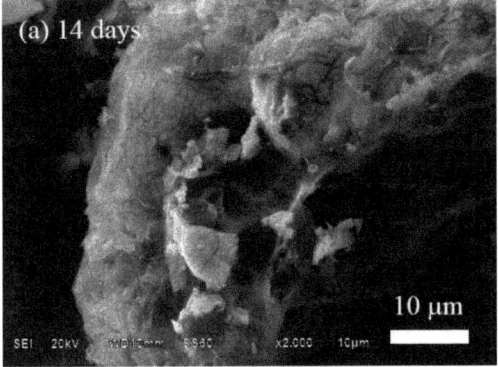 

Figure 10. EDS analysis for BG30-PLA70 after (**a**) 14 and (**b**) 28 days of immersion in SBF.

3.5. Evaluation of BG-PLA Degradation in PBS

Because biodegradability is an essential property when designing scaffolds, the evaluation of this property was realized and supported by FTIR (Figures 5 and 6) and SEM characterizations. The weight loss of the samples after immersion in phosphate-buffered saline solution (PBS) is shown in Table 4. The degradation behaviour of the BG70-PLA30 sample (blue line) shows the lower percentage change in mass in PBS. This profile can be divided into two regions: an initial increase due to the water uptake from the amorphous areas with the presence of terminal groups, folds, and chains with free rotation, and subsequent mass loss represented by a final decrease related to the degradation rate due to attack on the crystalline areas [28]. Meanwhile, the red line represents the average weight loss profile for the BG30-PLA70 composite in Table 4. The weight loss of the sample was approximately 28% over the four time periods, showing similar behaviour in this solution.

The morphology of the BG70-PLA30 composite after 14 and 28 days immersed in PBS is shown in Figure 11. After 28 days of immersion the surface sample showed more dissolution than at 14 days. The highest mass dissolution occurred at 28 days, diminishing the formation of reaction products deposited on the composite surface (Figure 11b). The sample's surface did not show higher product growth than the sample immersed for 14 days (Figure 11a). Additionally, the results of EDS analysis in both time periods confirm the degradation behaviour when comparing the elements presented in the samples after 14 and 28 days of soaking in PBS (Table 6). The carbon, sodium, and chloride elements in the initial sample were degraded into the solution after 28 days of immersion, as noted in Table 6. Similarly, the morphology of the degradation of the BG30-PLA70 scaffolds can be observed in Figure 12 after 14 and 28 days of immersion. The formation of spherical growths formed by the HAp phase and the presence of calcium in a high concentration confirms this. For this composite, the results of the EDS chemical analyses shown in Table 5 after 14 and 28 days of immersion present the peak bioactivity of the prepared scaffolds. The increase in Ca content is more evident in the samples with longer immersion times.

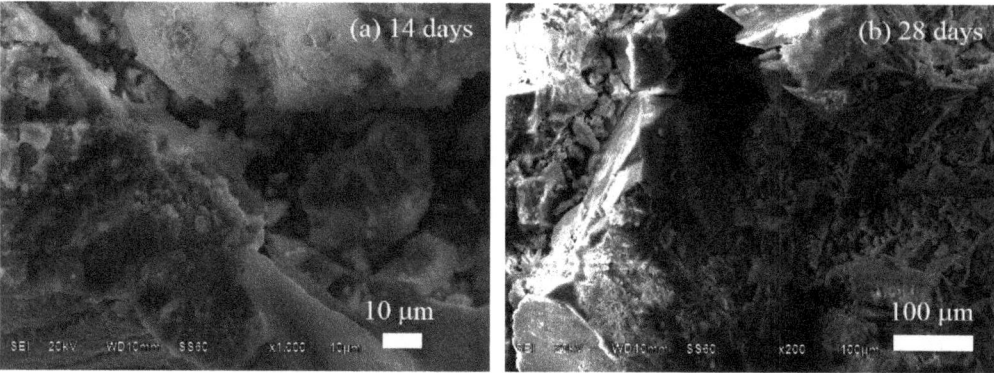

Figure 11. Scaffold SEM analysis for BG70-PLA30 after (**a**) 14 days and (**b**) 28 days of immersion in PBS.

Table 6. EDS chemical analysis for BG-PLA composites after immersion in PBS.

Element at.%	BG70-PLA30		BG30-PLA70		
	14 Days	28 Days	14 Days	28 Days	
				S.1	S.2
C	18.96		35.73	34.87	16.31
O	62.28	72.89	51.23	44.62	57.6
Na	1.07		0.85	1.44	0.38
Si	7.28	22.54	4.82	0.89	1.42
P	0.7	1.37	3.08	6.85	3.7
Cl	0.82		0.39		
Ca	8.88	3.2	3.9	11.33	20.6

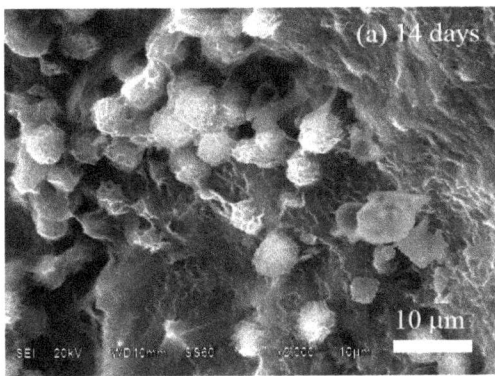 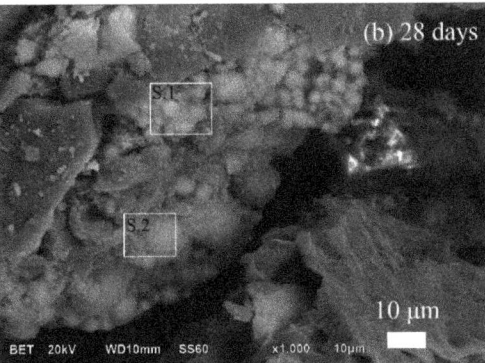

Figure 12. Scaffold SEM analysis for BG30-PLA70 after (**a**) 14 days and (**b**) 28 days of immersion in PBS.

3.6. Electrochemical Evaluation

This section shows the in vitro results for the biomaterials in Hank's and SBF solutions; electrochemical techniques were used to identify the mass transport mechanism through the developed biomaterial applied as a coating on the 316L SS substrate.

3.6.1. Potentiodynamic Tests

Figure 13 shows the corrosion behaviour of the BG-PLA biomaterial samples (BG, BG70-PLA30, and BG30-PLA70) in both Hank's (Figure 13b) and SBF (Figure 13b) saline solutions at 37 °C. This condition of saline solutions is representative of the behaviour of the biomaterials in corporeal applications. All samples showed an activation mechanism in the early stages, followed by a pseudo-passivation or current limited behaviour associated with the inhibited corrosion due to the physical barrier formed by the coatings. The behaviour presented after activation was associated with the physical bioglass characteristics; as a semiconductor material, charge transfer is limited by this property. After that, a wide over-potential range inhibiting corrosion (as passivation behaviour) up to breakdown over-potential was observed in the coated samples. Table 7 shows the potentiodynamic parameter obtained from polarization plots (Figure 13). In general, i_{corr} showed low current density values between 0.1 to 0.3 µA/cm^2, which is lower than the i_{corr} presented by the 316L SS (around 0.733 µA/cm^2) under similar conditions (Figure 13a, curve 4). Furthermore, the i_{pass} values were observed in the same order of magnitude. The coated samples showed a corrosion potential E_{corr} more positive than the 316L SS as a correlation of minor electrochemical activity, as indicated in Table 7. The electrochemical behaviour of the bioglass materials could be utilized in biomedical applications as biomaterial supports.

Table 7. Comparison of potential of corrosion between the samples evaluated in Hank's and SBF solutions.

Solution	Sample	i_{corr} µA/cm^2	E_{corr} mV	β_a mV	β_c mV	E_{transp} mV	i_{pass} µA/cm^2
Hank	BG	0.108	−211.7	123	67	545	0.593
	BG70-PLA30	0.241	−257.5	111	62	394	0.817
	BG30-PLA70	0.236	−181.1	92	79	448	1.540
	316L SS	0.733	−301.8	75	67	396	6.310
SBF	BG	0.145	−166.9	134	103	440	1.490
	BG70-PLA30	0.228	−210.0	135	76	428	0.149
	BG30-PLA70	0.190	−168.0	117	122	289	1.040

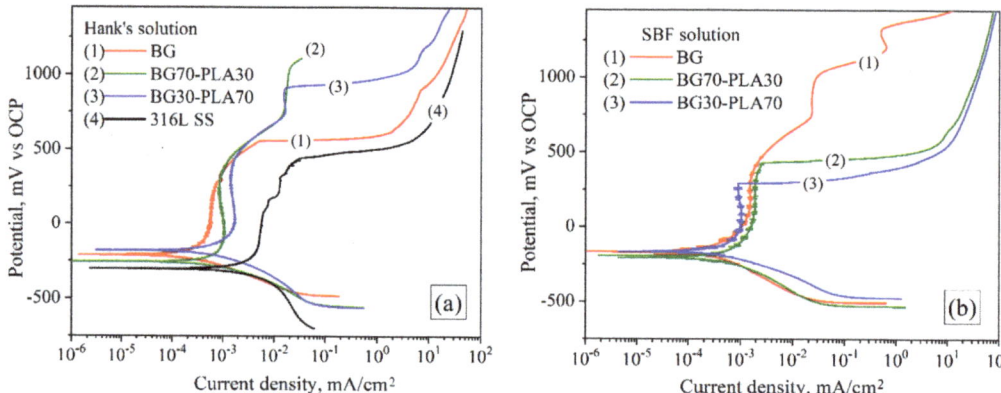

Figure 13. Potentiodynamic behaviour of both composite scaffolds in (**a**) Hank's saline solution and (**b**) SBF.

3.6.2. LPR Measurements

To observe the behaviour of the coatings as a function of time, linear polarization resistance (LPR) measurements were made. The polarization resistance (R_p) and E_{corr} kinetics obtained by the LPR measurements of the BG, BG70-PLA30, and BG30-PLA70 coatings in both saline solutions are shown in Figures 14 and 15 (the substrate was measured in Hank's solution only). In Hank's solution, the substrate alloy showed the highest R_p values during the first 10 h of immersion (about the 2.5 M Ohm·cm^2), although it showed a decrease of around 1.2 M Ohm·cm^2. This was associated with localized anodic dissolution and the breaking of the passive film present from the beginning of immersion due to the activity of chlorine in the solution. However, the BG coating displayed stability during complete immersion, with R_p values around 2 M Ohm·cm^2 (Figure 14, curve 1), which was associated with the homogeneous and continuous covering of the coating on the substrate. On the other hand, BG coating in the SBF solution showed the lowest R_p kinetic values, between 0.5 to 1 M Ohm·cm^2.

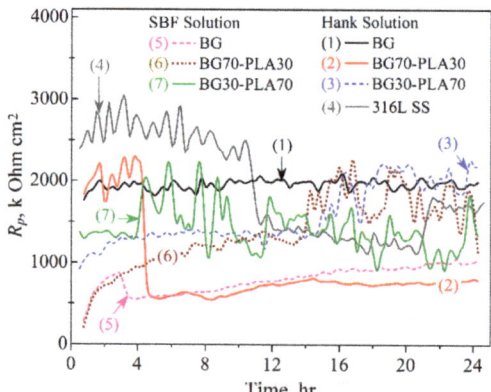

Figure 14. R_p kinetics of the BG-PLA coatings in Hank's and SBF solutions at 37 °C.

The addition of PLA to the BG phase caused variability in the Hank's and SBF solutions In particular, the hybrid coatings in the SBF solution presented increased corrosion resistance in the second half of immersion, as did the BG30-PLA70 in Hank's solution (Figure 14, curve 3). Although the BG70-PLA30 coating in Hank's solution did not show

R_p kinetics as the others did, the coated samples generally showed improved corrosion resistance in a stable range between 1 to 2 M Ohm·cm². The R_p fluctuations were correlated with the porous characteristics of hybrid coatings that promote a finite diffusion corrosion mechanism, as described below in the EIS results. Likewise, the potential kinetics present the evolution of the activity of the coatings, as can be seen in Figure 15; as a result, the BG70-PLA30 coating in Hank's solution showed more negative potentials, as did the BG in SBF; thus, the R_p values were the lowest. This behaviour was associated with the characteristics of the coating microstructures. However, the E_{corr} kinetics of the other coating samples remained stable during the immersion time; the BG30-PLA70 hybrid coating kept the potentials in both solutions higher, in accordance with those shown by the metallic substrate. In addition, the BG70-PLA30 coating in the SBF solution developed exponential growth of the E_{corr} kinetic to the more positive potentials; thus, the corrosion resistance increased.

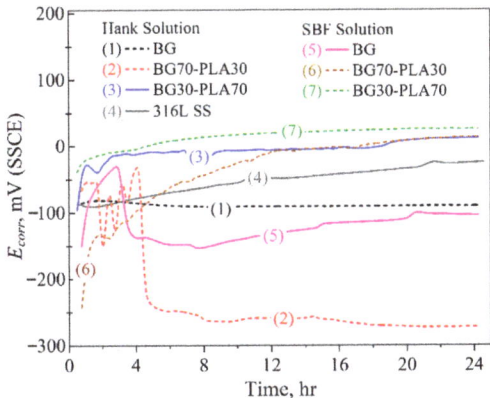

Figure 15. E_{corr} kinetics of the BG-PLA coatings in Hank's and SBF solutions at 37 °C.

The kinetic current density (i_{corr}) showed the opposite behaviour in terms of R_p kinetics because of the indirect correlation of the current density with the resistance, as described by Ohm's law; these were calculated using the Stern and Geary function [15]. Figure 16 shows the corrosion current behaviour of the coatings in both solutions. According to the R_p results, in Hank's solution the lowest i_{corr} values were observed with the BG and the BG30-PLA70 coatings and the 316L SS substrate, as well as the BG30-PLA70 hybrid coating in the SBF solution, with i_{corr} values around 0.1 µA/cm² in the second half of the immersion time. However, the BG70-PLA30 hybrid coating showed i_{corr} kinetic instability in the SBF solution. According to the R_p results, the highest current densities were displayed by the BG coating in the SBF solution, followed by the BG70-PLA30 in Hank's solution (Figure 16 curves 5 and 2, respectively).

The estimation of the corrosion rate (CR), as described in the ASTM G102 [15] using the i_{corr} kinetics (Figure 16 right scale), is valid for the data obtained by the substrate, presenting a CR between 0.06 to 0.12 µm/year. These lower CR values are a consequence of the chromium oxide protective film formed previously on the self-protected 316L SS alloy. The application of the hybrid coatings did not lead to an increase in corrosion resistance. However, this was not the main purpose of coating the metallic substrate with the BG-PLA biomaterials; rather, it was to improve their functionality due to their high bioactivity, osteoconductivity, and biodegradability for potential applications and physiological functionality as implantable devices. Therefore, the kinetics of the hybrid coatings showed CR values as low as the substrate and in the same order of magnitude. In general, the kinetics showed stability of current density and CR at around 1 to 2 µm/year during the immersion time, with the exception of the BG coating in SBF solution and the BG70-PLA30 in Hank's solution.

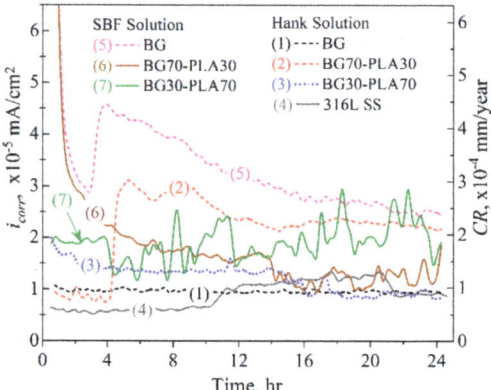

Figure 16. i_{corr} kinetics for the BG-PLA coatings in Hank's and SBF solutions at 37 °C.

3.6.3. EIS Analysis

Electrochemical impedance spectroscopy (EIS) was used to identify the probable corrosion mechanisms present at the solution/coating and coating/substrate interfaces and through the thickness scale. Additionally, the substrate alloy was evaluated to establish a baseline or reference curve. The 316L SS was evaluated in Hank's solution, considering that similar results could be obtained in SBF. Two EIS measurements were obtained for the substrate and the coated samples to identify the corrosion mechanisms both at the beginning and after approximately 24 h of immersion. Figure 17 shows the EIS results of the coatings and substrate at the beginning of immersion in the Hank's and SBF solutions, represented in Nyquist and Bode diagrams (Figure 17a,b, respectively). Likewise, the EIS measurements obtained after 24 h of immersion are presented in Figure 18.

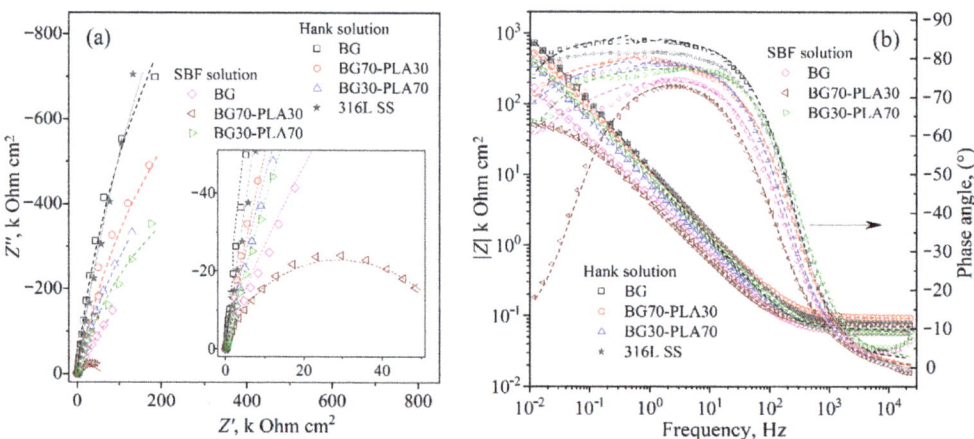

Figure 17. EIS results for BG-PLA coatings obtained at the beginning of immersion in Hank's and SBF solutions: (**a**) Nyquist and (**b**) Bode plots. Scatter and lines indicate the experimental data and the fitting results, respectively.

The Nyquist plots obtained at both immersion times (beginning and 24 h.) show the representative form of high resistance corrosion mechanisms for both coated and uncoated samples. The impedance module ($|Z|$) for all materials showed high values above 100 k Ohm·cm^2, as shown in the Bode plots in Figures 17b and 18b. The substrate as the Nyquist curves of the hybrid coatings presented characteristic capacitive and resistive elements mixed with finite diffusion through a physical barrier composed by the Cr_2O_3 film (for uncoated substrate) and Cr_2O_3 film/BG-PLA mixed thickness (for the coated samples).

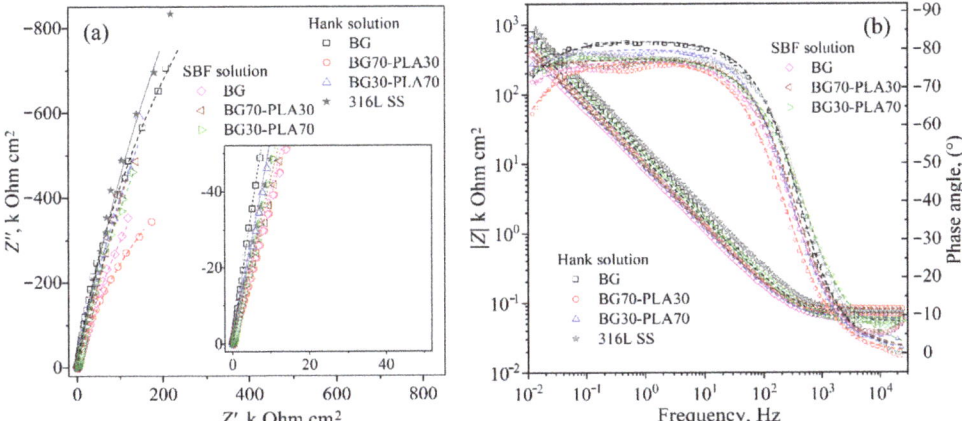

Figure 18. EIS results for BG-PLA coatings obtained after 24 h of immersion in Hank's and SBF solutions: (**a**) Nyquist and (**b**) Bode plots. Scatter and lines indicate the experimental data and the fitting results, respectively.

The physical barrier had a high effect on the current density, increasing the time for species diffusion through the scale thickness, retarding the activation mechanism at the metallic interface [29,30]. The phase angle Bode plots show a zone within a wide range of frequencies (from 500 0.5 Hz, approximately) with phase angle values above to 70°, which according to the R_p results described above correspond to the capacitive and resistive behaviour. This could be of interest in biomedical applications as a scaffold in tissue engineering, allowing the controlled transport of mass through the porous microstructure. Similar to the observed LPR results, the BG coating in the SBF solution and the BG70-PLA30 showed lower initial impedance values (Figure 17a). At 24 h of immersion, the total impedance had increased to values within the same order of magnitude as the other coatings, as shown in Figure 18b.

The proposed corrosion mechanisms were associated with the effect of the microstructural morphology on the coating behaviour composed of the mixed BG and PLA phases at different ratios, as an electrode with a microstructure with a superimposed porous layer [31] acts as a barrier against electron and ion diffusion, reducing the surface area for electrochemical reactions at the metallic interface [29]. Nevertheless, the micro-galvanic cell formation at the coating/substrate could be increased. However, the corrosion behaviour of the substrate is associated with the activation mechanism and with diffusion through the Cr_2O_3 protective film. Figure 19 shows the electric circuit models (ECM) that were used as an analogy to better explain the governing corrosion mechanism at the active surfaces and the coating thickness. For the substrate, the analogue ECM correspond to Model 1 (Figure 19), which is composed of the electrolyte resistance (R_s), set up in series with a parallel arrangement of a constant phase element (CPE_1, as capacitive behaviour at the double layer) and the polarization resistance (R_{L1}) of the inner layer (composed of Cr_2O_3), which represents the activation mechanism at the Cr_2O_3 film/electrolyte interface. The arrangement of CPE_2 in parallel with the R_{ct} represents the diffusive element presented by the Cr_2O_3 protective layer inherent in stainless-steel alloys. When the hybrid coating is applied, the ECM incorporates an element of Warburg diffusion impedance (Z_D) in a serial arrangement before the charge transference resistance (R_{ct}) at the coating/substrate interface (Model 2 of Figure 19), describing the diffusion through the coating thickness with enough roughness and porosity and allowing the fluid permeation and/or the ion diffusion. Consequently, the electrochemical mechanism governed by limited mass transport was observed.

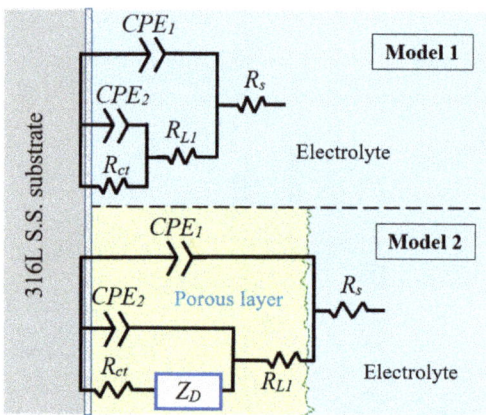

Figure 19. Equivalent circuit models of the corrosion mechanisms observed at the active interfaces: Model 1 for uncoated substrate, Model 2 for BG-PLA coatings.

In the ECMs used here, the Z_D element (finite-length Warburg) represents the short Warburg (W_s) element. In general, the ECM described here represents the analogue equivalent electrical circuit of the impedance for a coated electrode by a hybrid porous layer [31]. The elements of ECM are defined using the following equations:

$$Z(CPE_i) = \frac{1}{T_{CPE_i}(j\omega)^\alpha} \quad (3)$$

$$Z(R_i) = R_S, R_{L1}, R_{ct} \quad (4)$$

$$Z_{W_s} = \sigma \frac{\tanh(jT_D\omega)^P}{(jT_D\omega)^P} \quad (5)$$

where R_s, R_{L1}, and R_{ct} are the electrolyte resistance, inner layer, and charge transference resistance, respectively, T_{CPEi} is the i constant phase capacitance, and α is a dimensionless potential number ($0 < \alpha \leq 1$, while $\alpha = 1$ assumes that CPE is a perfect capacitance C_{dl}). Angular frequency is $\omega = 2\pi f$ with f = linear frequency, complex number $j = \sqrt{(-1)}$, and Z_f is the Faradaic impedance at the metal/scale interface. Hence, the term T_D represents the ratio of scale thickness L and the effective diffusion coefficient D_{eff} of that scale, $T_D = L^2 D_{eff}^{-1}$ power is between $0 < P < 1$, and σ is the constant of diffusion or the modulus of the Warburg resistance. Here, the CPE_i elements were applied instead of the perfect capacitance for better fitting. Tables 8 and 9 show the fitting values obtained for each equivalent electric element of the ECM used in the fitting analysis of the experimental data. The lines in Figures 17 and 18 correspond to the fitting results of the experimental data using the proposed ECM.

Table 8. EIS Parameters obtained by substrate experimental data fitting (Model 1, Figure 19).

316L SS	Time Immersion	
Model 1	Beginning	24 h
R_s (Ω cm^2)	78	75
T_{CPE1} (μF cm^2)	11.88	10.28
α_1 (*)	0.924	0.896
R_{L1} (Ω cm^2)	167	80,654
T_{CPE1} (μF cm^2)	2.911	1.032
α_2 (*)	0.892	0.875
R_{ct} (kΩ cm^2)	8250.5	8.693

Table 9. EIS Parameters obtained by fitting the experimental data for the BG-PLA coatings using Model 2, Figure 19.

	Hank's Solution			SBF Solution		
Model 2	Measurement at the Beginning of Immersion					
	BG	BG70-PLA30	BG30-PLA70	BG	BG70-PLA30	BG30P-LA70
R_s (Ω cm^2)	62	94	71	61	71	57
T_{CPE1} (μF cm^2)	2.316	7.215	1.693	8.954	3.376	15.82
α_1 (*)	0.982	0.884	0.415	0.447	0.867	0.811
R_{L1} (Ω cm^2)	6.41	1.88	1.24	0.70	52.95	12.52
T_{CPE1} (μF cm^2)	12.77	10.23	25.07	38.09	15.52	6.073
α_2 (*)	0.939	0.883	0.896	0.867	0.822	0.969
σ (kΩ cm^2/s)	4842.1	3122.1	4561.9	2847.2	17.76	1025.9
T_D (s)	1.019	0.872	0.967	0.982	7.8×10^{-16}	1.7×10^{-16}
P (*)	0.942	0.935	0.971	0.976	0.937	0.941
R_{ct} (kΩ cm^2)	15.84	22.76	5.07	3.77	40.35	476.26
	Measurement at the 24 h					
	BG	BG70-PLA30	BG30-PLA70	BG	BG70-PLA30	BG30-PLA70
R_s (Ω cm^2)	60	81	64	58	63	53
T_{CPE1} (μF cm^2)	10.80	22.34	15.80	19.18	17.49	8.05
α_1 (*)	0.900	0.848	0.874	0.833	0.840	0.880
R_{L1} (Ω cm^2)	12.94	3.08	20.36	14.83	20.88	20.63
T_{CPE1} (μF cm^2)	2.285	1.426	2.456	5.729	2.471	8.751
α_2 (*)	0.979	0.992	0.992	0.934	0.992	0.852
σ (kΩ cm^2/s)	2070.7	825.02	5595.5	3390.4	10,397	6207.1
T_D (s)	2.5×10^{-16}	3.3×10^{-11}	1.9×10^{-17}	5.6×10^{-18}	6.5×10^{-16}	6.17
P (*)	0.811	0.830	0.754	0.772	0.581	0.847
R_{ct} (kΩ cm^2)	2612.7	828.52	653.23	83.42	10,252	5.07

Note: (*) Dimensionless Warburg element.

In the experimental data fitting, the CPE_i elements were applied instead of the perfect capacitance for better fitting. These CPE elements were associated with the heterogeneous morphology of the metallic surface and the surface of the coating. For the purposes of EIS analysis, a homogeneously distributed porous microstructure of the BG-PLA coatings was considered, which was formed during the drying and sintering process after applying them to the substrate. Although the coating surface SEM images are not presented here, the fitting Model 2 matches the experimental data associated with a porous microstructure. Additionally, the surface roughness of the substrate and the coatings promotes a depression of the semicircle (Nyquist plot) of the activation process (for the typical *Randles* circuit), and the α parameter values (Equation (3)) are lower than unity, as shown in Tables 8 and 9. Thus, the Nyquist plots present a depression, and the phase angles have lower values (Figures 17 and 18). However, the values of parameter P (Equation (5)) were higher than 0.5, which is associated with the Warburg diffusion mechanism and a 45° angle of the phase at low frequencies. Thus, p values close to 1 (Table 9) represent a mechanism associated with capacitive behaviour associated with the high resistance characteristic of the coatings. The Bode plots show a combination of the effect of the time delay of the mass transfer mechanism due to the finite diffusion of species through the coating thickness and the resistive characteristic of the Cr_2O_3 inner layer, causing a wide range in the loop, with about 70–80° of the phase angle formed from the middle to the low frequencies. Similar results have been previously reported [32], and were associated with high resistance and capacitive behaviour.

4. Discussion

The electrochemical results with the porous hybrid coatings allow for mass transport and fluid permeation, which was observed in this work, suggesting potential applications in the tissue engineering area, results which are of interest for future study. Although study of the coatings' bioactivity for bone regeneration was not within the scope of this work, it is proposed for further study. Accordingly, the formation of particles with Ca and P contents during the bioactivity testing of BG-PLA in SBF and PBS solutions (described in Sections 3.4 and 3.5 above) due to the interaction of the H_2O molecules with the Si-O bonds in the BG microstructure promoted the formation of Si-OH groups, which attract the ions of Ca^{2+}, $H_2PO_4^-$, HPO_4^{2-}, and PO_4^{3-} in the SBF solution. This favours the HAp nucleation sites, and they precipitate after a period of immersion time; similarly, observation has been made for a hybrid composite with a PCL matrix [29,33,34]. The amorphous inorganic formation of the nuclei crystallizes into the apatite phase [34], and the addition of nanopowders to the polymer matrix can improves apatite nucleation [29]. In accordance with these results, the BG phased improves of the formation of the HAp phase, as has previously been suggested.

Based on the R_p kinetics behaviour, the proposed mechanisms associated with the EIS fitting results, and the formation of HAp particles, the physical characteristics of the microstructure of the coatings allowed redox reactions to take place at the porous surface. Therefore, Ca^{2+}, $H_2PO_4^-$ and HPO_4^{2-} concentrations were increased at those sites and the associated current density was added to the total measured at the metallic surface. Because the application of the coatings formed a physical barrier at the metallic surface, the charge transfer consequently decreases and the corrosion resistance should increase. However, the presence of intermediate electrochemical reactions through the coating thickness maintains R_p kinetics values of the coated samples slightly lower than those presented by the substrate. Thus, the electrochemical results support the potential application of the BG-PLA composite in biomedical applications. Consequently, further studies to determine the in vitro degradation behaviour and adhesion performance of the hybrid coatings will be undertaken.

5. Conclusions

In summary, BG-PLA composite scaffolds with two different compositions synthetized by the sol–gel technique were evaluated and characterised. The morphology of the BG70-PLA30 composite structure was dense, with a well-distributed phase. The surface morphology of the BG30-PLA70 composite presented crack formation associated with tension stress concentrations in the polymeric phase during the drying process. Their potential as bone tissue engineering scaffolds was assessed by in vitro testing using Hank's and SBF solutions, confirming the bioactivity of the composites by their ability to form HAp on the surfaces and their adequate biodegradation when immersed in PBS after 21 days of immersion. Both properties were confirmed by SEM and FTIR characterization.

The electrochemical evaluation of the scaffolds in Hank's saline solution and SBF as a coating in a 316L SS substrate allowed us to observe that both samples showed activation mechanisms at the early stages, followed by pseudo-passivation or current-limited behaviour due to the physical characteristics of the bioactive glass, which suggests that improvements in the formation of HAp nucleation consequently allow redox reactions at the surface of the coating.

Author Contributions: Conceptualization, G.C.-D.l.T.; methodology, N.N.Z.-M.; investigation, M.A.E.-M. and G.C.-D.l.T. writing—original draft preparation, N.N.Z.-M.; resources, J.O.-O.; writing—review and editing, G.C.-D.l.T. and M.E.; formal analysis, M.A.E.-M. and N.N.Z.-M.; supervision, G.C.-D.l.T., M.E. and M.A.E.-M.; project administration, M.d.L.B.-A.; funding acquisition, G.C.-D.l.T. All authors have read and agreed to the published version of the manuscript.

Funding: This research did not receive any specific grant from funding agencies in the public, commercial, or not-for-profit sectors.

Institutional Review Board Statement: Not applicable.

Informed Consent Statement: Not applicable.

Data Availability Statement: The datasets generated during the current study are available from the corresponding author on reasonable request.

Acknowledgments: The authors acknowledge Hugo I. Medina-Vargas from the Biochemical Engineering Faculty, Technological Institute of Morelia, and the support of the Material Degradation Laboratory of the Mechanical Engineering Faculty (UMSNH) for research development. The present research was supported by research project number 243236 of CB-2014-02 from the I0017 fund of the CONACYT.

Conflicts of Interest: The authors declare no conflict of interest.

References

1. Fernandez, G.; Keller, L. Bone substitutes: A review of their characteristics, clinical use, and perspectives for large bone defects management. *J. Tissue Eng.* **2018**, *9*, 1–18. [CrossRef]
2. Hasan, A.; Byambaa, B. Advances in osteobiologic materials for bone substitutes. *J. Tissue Eng. Regen. Med.* **2018**, *12*, 1448–1468. [CrossRef]
3. Kashirina, A.; Yao, Y. Biopolymers for bone substitutes: A review. *Biomater. Sci.* **2019**, *7*, 3961–3983. [CrossRef]
4. Hasan, M.; Kim, B. In vitro and in vivo evaluation of bioglass microspheres incorporated brushite cement for bone regeneration. *Mater. Sci. Eng. C* **2019**, *103*, 1–12. [CrossRef]
5. Paiva, A.; Risso, J. Bone substitutes and photobiomodulation in bone regeneration: A systematic review in animal experimental studies. *J. Biomed. Mater. Res. A* **2021**, *109*, 1765–1775.
6. Mazzoni, E.; Laquinta, M. Bioactive materials for soft tissue repair. *Front. Bioeng. Biotechnol.* **2021**, *9*, 1–17. [CrossRef]
7. Wajda, A.; Sitarz, M. Structural and microstructural comparison of bioactive melt-derived and gel-derived glasses from CaO-SiO$_2$ binary system. *Ceram. Int.* **2018**, *44*, 8727–10020. [CrossRef]
8. Baino, F.; Fiume, E. Bioactive sol-gel glasses: Processing, properties, and applications. *Int. J. Appl. Ceram. Technol.* **2018**, *15*, 841–860. [CrossRef]
9. Khurshid, Z.; Husain, S. Novel techniques of scaffold fabrication for bioactive glasses. In *Biomedical, Therapeutic and Clinical Applications of Bioactive Glasses*; Kaur, G., Ed.; Woodhead Publishing: Cambridge, UK, 2019; pp. 497–519.
10. Fan-Long, J.; Rong-Rong, H. Improvement of thermal behaviour of biodegradable poly(lactic acid) polymer: A review. *Compos. B Eng.* **2018**, *164*, 287–296.
11. Ramírez-Herrera, C.A.; Flores-Vela, A.I. PLA degradation pathway obtained from direct polycondensation of 2-hydroxypropanoic acid using different chain extenders. *J. Mater. Sci.* **2018**, *53*, 10846–10871. [CrossRef]
12. Waizy, H.; Weizbauer, A. In vitro corrosion of ZEK100 plates in Hank's Balanced Salt Solution. *Biomed. Eng. OnLine* **2012**, *11*, 1–14. [CrossRef]
13. Kokubo, T.; Takadama, H. How useful is SBF in predicting in vivo bone bioactivity? *Biomaterials* **2006**, *27*, 2907–2915. [CrossRef]
14. *ISO-10993-13*; Biological Evaluation of Medical Devices—Part 13: Identification and Quantification of Degradation Products from Polymeric Medical Devices. 2nd ed. ISO: Geneva, Switzerland, 2010.
15. *ASTM G 102*; Standard Practice for Calculation of Corrosion Rates and Related Information from Electrochemical Measurements. ASTM: West Conshohocken, PA, USA, 1999.
16. Cacciotti, I.; Lombardi, M. Sol-gel derived 45S5 bioglass: Synthesis, microstructural evolution and thermal behaviour. *J. Mater. Sci. Mater. Med.* **2012**, *23*, 1849–1866. [CrossRef]
17. Vichery, C.; Nedelec, J.M. Bioactive glass nanoparticles: From synthesis to materials design for biomedical applications. *Materials* **2016**, *9*, 288. [CrossRef]
18. El-Badry, H.; Moustaffa, F. Infrared absorption spectroscopy of some bio-glasses before and after immersion in various solutions. *Indian J. Pure App. Phys.* **2000**, *38*, 741–761.

19. Chen, W.; Long, T. Hydrothermal synthesis of hydroxyapatite coatings with oriented nanorod arrays. *RSC Adv.* **2014**, *4*, 185–191. [CrossRef]
20. Radev, L.; Vladov, D. In vitro Bioactivity of Polycaprolactone/Bioglass Composites. *Int. J. Mater. Chem.* **2013**, *5*, 91–98.
21. Vaid, C.; Murugavel, S. Alkali oxide containing mesoporous bioactive glasses: Synthesis, characterization and in vitro bioactivity. *Mater. Sci. Eng. C* **2013**, *33*, 959–968. [CrossRef]
22. Doganay, D.; Coskun, S. Electrical, mechanical and thermal properties of aligned silver nanowire/polylactide nanocomposite films. *Compos. Part B Eng.* **2016**, *99*, 288–296. [CrossRef]
23. Chen, J.; Zeng, L. Preparation and characterization of bioactive glass scaffolds and evaluation of bioactivity and cytotoxicity in vitro. *Bioact. Mater.* **2018**, *3*, 315–321. [CrossRef]
24. El-Meliegy, E.; Mabrouk, M. Novel Fe_2O_3-doped glass/chitosan scaffolds for bone tissue replacement. *Ceram. Int.* **2018**, *44*, 8727–10020. [CrossRef]
25. Sharifianjazi, F.; Parvin, N. Synthesis and characteristics of sol-gel bioactive SiO_2-P_2O_5-CaO-Ag_2O glasses. *J. Non-Cryst. Solids.* **2017**, *476*, 108–113. [CrossRef]
26. Xia, W.; Chang, J. Well-ordered mesoporous bioactive glasses (MBG): A promising bioactive drug delivery system. *J. Control. Release.* **2006**, *110*, 522–530. [CrossRef]
27. Felfel, R.; Zakir, K. Accelerated in vitro degradation properties of polylactic acid/phosphate glass fibre composites. *J. Mater. Sci.* **2015**, *50*, 3942–3955. [CrossRef]
28. Zuluaga, F. Algunas aplicaciones del ácido Poli-L-Láctico. *Rev. Acad. Colomb. Cienc.* **2013**, *37*, 125–142.
29. Jokar, M.; Darvishi, S.; Torkaman, R.; Kharaziha, M.; Karbasi, M. Corrosion and bioactivity evaluation of nanocomposite PCL-forsterite coating applied on 316L stainless steel. *Surf. Coat. Technol.* **2016**, *307*, 324–331. [CrossRef]
30. Zurita-Mendez, N.N.; Carbajal-De la Torre, G.; Estevez, M.; Ballesteros-Almanza, M.L.; Cadenas, E.; Espinosa-Medina, M.A. Evaluation of the electrochemical behavior of TiO_2/Al_2O_3/PCL composite coatings in Hank's solution. *Mater. Chem. Phys.* **2019**, *235*, 1–10. [CrossRef]
31. Orazem, M.E.; Tribollet, B. *Electrochemical Impedance Spectroscopy*; John Wiley & Sons, Inc.: Hoboken, NJ, USA, 2008.
32. González-Reyna, M.A.; Espinosa-Medina, M.A.; Esparza, R.; Hernández-Martinez, A.R.; Maya-Cornejo, J.; Estevez, M. Anticorrosive Effect of the Size of Silica Nanoparticles on PMMA-Based Hybrid Coatings. *J. Mater. Eng. Perform.* **2021**, *30*, 1054–1065. [CrossRef]
33. Abdal-hay, A.; Amna, T.; Lim, J.K. Biocorrosion and osteoconductivity of PCL/nHAp composite porous film-based coating of magnesium alloy. *Solid State. Sci.* **2013**, *18*, 131–140. [CrossRef]
34. Abdal-hay, A.; Hwang, M.G.; Lim, J.K. In vitro bioactivity of titanium implants coated with bicomponent hybrid biodegradable polymers. *J. Sol-Gel Sci. Technol.* **2012**, *64*, 756–764. [CrossRef]

Article

Impact of Three Different Processing Techniques on the Strength and Structure of Juvenile Ovine Pulmonary Homografts

Johannes J van den Heever [1,*], Christiaan J Jordaan [1], Angélique Lewies [1], Jacqueline Goedhals [2], Dreyer Bester [1], Lezelle Botes [3], Pascal M Dohmen [1,4] and Francis E Smit [1]

1. Department of Cardiothoracic Surgery, Faculty of Health Sciences, University of the Free State (UFS), P.O. Box 339 (Internal Box G32), Bloemfontein 9300, South Africa; jordaancj@ufs.ac.za (C.J.J.); lewiesa@ufs.ac.za (A.L.); besterd@ufs.ac.za (D.B.); pascal.dohmen@med.uni-rostock.de (P.M.D.); smitfe@ufs.ac.za (F.E.S.)
2. Department of Anatomical Pathology, Faculty of Health Sciences, University of the Free State (UFS), P.O. Box 339 (Internal Box G32), Bloemfontein 9300, South Africa; goedhalsj@ufs.ac.za
3. Department of Health Sciences, Central University of Technology, Free State (CUT), Private Bag X20539, P.O. Box 339 (Internal Box G32), Bloemfontein 9300, South Africa; botesl@cut.ac.za
4. Klinikdirektor (k), Klinik und Poliklinik für Herzchirurgie, Universitätsmedizin Rostock, Schillingallee 35, 18057 Rostock, Germany
* Correspondence: vdheeverjj@ufs.ac.za; Tel.: +27-834614052

Abstract: Homografts are routinely stored by cryopreservation; however, donor cells and remnants contribute to immunogenicity. Although decellularization strategies can address immunogenicity, additional fixation might be required to maintain strength. This study investigated the effect of cryopreservation, decellularization, and decellularization with additional glutaraldhyde fixation on the strength and structure of ovine pulmonary homografts harvested 48 h post-mortem. Cells and cellular remnants were present for the cryopreserved group, while the decellularized groups were acellular. The decellularized group had large interfibrillar spaces in the extracellular matrix with uniform collagen distribution, while the additional fixation led to the collagen network becoming dense and compacted. The collagen of the cryopreserved group was collapsed and appeared disrupted and fractured. There were no significant differences in strength and elasticity between the groups. Compared to cryopreservation, decellularization without fixation can be considered an alternative processing technique to maintain a well-organized collagen matrix and tissue strength of homografts.

Keywords: homografts; ischaemic harvesting; decellularization; cryopreservation; glutaraldehyde-fixation

Citation: van den Heever, J.J.; Jordaan, C.J.; Lewies, A.; Goedhals, J.; Bester, D.; Botes, L.; Dohmen, P.M.; Smit, F.E. Impact of Three Different Processing Techniques on the Strength and Structure of Juvenile Ovine Pulmonary Homografts. Polymers 2022, 14, 3036. https://doi.org/10.3390/polym14153036

Academic Editors: Antonia Ressler and Inga Urlic

Received: 19 May 2022
Accepted: 20 June 2022
Published: 27 July 2022

Publisher's Note: MDPI stays neutral with regard to jurisdictional claims in published maps and institutional affiliations.

Copyright: © 2022 by the authors. Licensee MDPI, Basel, Switzerland. This article is an open access article distributed under the terms and conditions of the Creative Commons Attribution (CC BY) license (https://creativecommons.org/licenses/by/4.0/).

1. Introduction

End-stage heart valve disease mandates the repair or replacement of a patient's diseased heart valve/s with either mechanical or biological valve prostheses. Mechanical prostheses demonstrate superior durability and longevity in patients; however, recipients require lifelong anticoagulation therapy. In contrast, bioprosthetic valves (including glutaraldehyde (GA)-fixed porcine valves or bovine pericardium mounted onto a frame, free xenograft valves, or donor homograft valves) do not require continuous anticoagulation therapy but have limited durability and require more frequent reoperation [1]. Currently, cryopreserved pulmonary homografts remain the valve of choice for the replacement of the native pulmonary valve in the Ross procedure [2], as well as for the reconstruction of the right ventricle outflow tract (RVOT) in children with congenital abnormalities [3]. Unfortunately, the early degeneration of these homografts occurs in younger patients [4], and there is a lack of availability, especially for smaller-sized conduits suitable for neonates [5].

According to the internationally accepted guidelines, homografts from either beating or non-beating heart donors should be harvested and processed within 24 h after death to retain maximum cell viability. This guideline restricts the available post-mortem donor pool significantly. Various efforts have been made to address the shortages in homograft availability, and our research group has proven that the post-mortem ischaemic time can be extended safely to around 48 h without affecting valve performance [6]. Although cryopreservation is currently the most frequently used and probably the best method for long-term storage of homografts [7], it does damage the collagen scaffold, irrespective of ischaemic harvesting time [8]. The effect of cryopreservation on the collagen scaffold might be of greater importance in determining the long-term survival of the homograft than the impact of extending the post-mortem ischaemic time prior to harvesting. The presence of donor cell antigens in cryopreserved homografts is also associated with an adverse immunological response from the recipient, resulting in valve calcification and degeneration [9–11].

Decellularization processes remove interstitial and endothelial cellular components from a homograft valve and can potentially lead to the creation of a valve with significantly reduced immunogenicity and reduced calcification [12]. There are concerns about maintaining the strength of the conduit once the cellular components are removed [13]. Furthermore, using multi-detergent, enzyme-based decellularization methods might also decrease flexural stiffness, disrupt the extracellular matrix (ECM) structure [14] and reduce the glycosaminoglycan (GAG) content [15]. These GAGs fill most of the extracellular space and provide mechanical support to the tissue [16]. Therefore, the additional fixation and stabilization of the collagen scaffold with GA might be required [17]. However, the free aldehyde groups of GA are associated with cellular toxicity [18], while conjugated GA may contribute to graft calcification [19,20]. Washing and detoxifying with solutions that can bind the free aldehyde groups prior to implantation can improve durability and biocompatibility [21]. EnCap technology describes the fixation of biological tissue with GA and the subsequent treatment of the GA-fixated tissue with a high concentration liquid polyol like propylene glycol (PG). The PG binds to the free aldehyde groups of the GA, blocking the free aldehyde groups as a potential binding site for calcium and other minerals and thereby mitigating the calcification of the tissue [22]. From a study evaluating the performance of commercially stent-mounted GA-preserved aortic homografts (not decellularized), it was concluded that better long-term performance of the homograft tissue compared to GA-fixed heterologous biological valves could be expected [23]; however, other studies using GA-fixed decellularized homograft valves are lacking.

The aim of this study was to compare the morphology and mechanical properties of standard cryopreserved pulmonary homografts to decellularized pulmonary homografts and decellularized pulmonary homografts treated with EnCap™ AC technology, evaluating the impact of the processing method on ovine homografts harvested after 48 h cold ischaemia. Our proprietary decellularization protocol with proven synergy was used [24].

2. Materials and Methods

2.1. Study Design

Pulmonary homografts harvested from juvenile Dorper sheep ($n = 15$), with a post-mortem cold ischaemic harvesting time of 48 h, were divided into three groups: Group 1, cryopreserved ($n = 5$); Group 2, decellularized ($n = 5$); and Group 3, decellularized plus EnCap™-treated ($n = 5$). After processing, the structural integrity, strength, and elasticity of the leaflet and wall tissue were evaluated and compared. Strength and elasticity analysis included tensile strength (TS) and Young's modulus (YM), and morphological evaluation included DAPI staining to confirm decellularization, hematoxylin and eosin staining (H&E), von Kossa staining, modified Verhoeff's van Gieson staining, scanning electron microscopy (SEM), and transmission electron microscopy (TEM). A schematic representation of the study design is given in Figure 1. The interfaculty Animal Ethics Committee of the University of the Free State (UFS-AED2016/0101) approved the study.

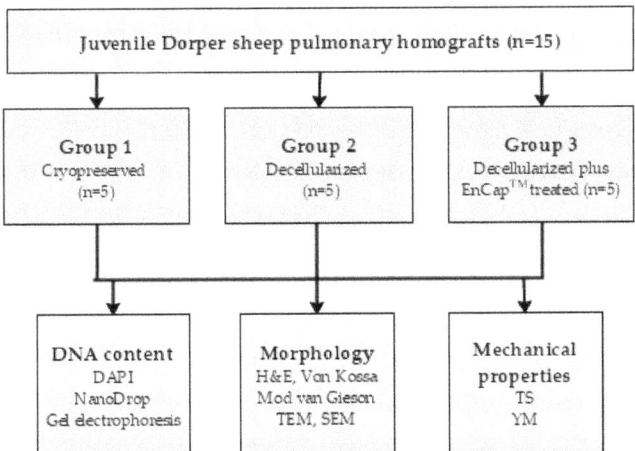

Figure 1. Schematic diagram of the study design.

2.2. Preparation of Homografts

Heart-lung blocks ($n = 15$) from freshly slaughtered juvenile Dorper sheep with a bodyweight of 24–30 kg were collected from a local abattoir and subjected to 48 h ischaemia at 4 °C before dissection and processing, as previously described by Bester et al. [15]. Pulmonary homografts ($n = 15$) were harvested and washed in copious amounts of cold (4 °C) Ringers lactate solution (Fresenius Kabi/Intramed, Midrand, South Africa). All homografts were sterilized overnight at 4 °C in 100 mL Medium199 (Whitehead Scientific, Johannesburg, South Africa) and an antibiotic cocktail consisting of 2.5 mg Amphotericin B (Bristol-Myers Squibb, Bedfordview, South Africa), 50 mg Piperacillin (Brimpharm South Africa (Pty) Ltd., Cape Town, South Africa), 50 mg Vancomycin (Gulf Drug Company, Mount Edgecombe, South Africa), and 25 mg Amikacin sulphate (Bodene (Pty) Ltd., trading as Intramed, Port Elizabeth, South Africa). Valves in Group 1 ($n = 5$) were cryopreserved the next day in 100mL Medium 199 + 11mL dimethyl sulfoxide (DMSO) (Highveld Biological, Johannesburg, South Africa) in a Cryoson controlled rate freezer (Consarctic, Schöllkrippen, Germany) at a rate of -1 °C/min to -140 °C and stored in the vapor phase of liquid nitrogen (LN_2) until evaluation.

Decellularized homografts (Group 2, $n = 5$) were prepared according to our proprietary protocol [24,25] with the addition of Benzonase® (Thermo Fisher Scientific, Johannesburg, South Africa). Briefly, homografts were subjected to osmotic shock in two changes of hypertonic 5% NaCl-solution and distilled water, repeated changes in a multi-detergent solution (0.5% Sodium deoxycholate (SDS) (Sigma-Aldrich, Johannesburg, South Africa)), 1% Sodium deoxycholate (SDC) (Sigma-Aldrich, Johannesburg, South Africa), and 1% Triton-X 100 (Sigma-Aldrich, Johannesburg, South Africa)] in PBS, followed by numerous washings in PBS, and a half-strength antibiotic cocktail, which consisted of the same antibiotic mixture that was used for the sterilization of cryopreserved valves, under constant shaking. These steps were followed by enzymatic treatment with descending concentrations of Benzonase® [26], repeated washing in PBS, delipidation in 70% ethanol, and final storage in PBS with antibiotics.

Additionally, decellularized homografts (Group 3, $n = 5$) were treated with EnCap™ AC technology and stored in propylene oxide (Sigma-Aldrich, Johannesburg, South Africa) [22]. EnCap™ AC technology includes the GA-tanning of the collagen scaffold, and the binding of a polyol, namely, propylene glycol (Sigma-Aldrich, Johannesburg, South Africa) to the free aldehyde groups to lower toxicity, reducing the host inflammatory response and mitigating calcification.

All homografts were confirmed to be culture-negative. To confirm the effective decellularization of the homografts in Groups 2 and 3, DNA quantification was done with 4′,6-diamidino-2-fenielindol (DAPI) staining, gel electrophoresis, and NanoDrop counts by the Cardiovascular Research Unit, UCT Medical School [27].

2.3. Evaluation of Processed Homografts

2.3.1. Histological Analysis

Leaflet and wall-tissue samples were collected in 4% buffered Formalin, embedded in paraffin wax, sectioned and stained according to standard H&E, von Kossa, and modified van Gieson staining protocols for histological evaluation [28].

Pulmonary leaflet and wall-tissue samples for scanning and transmission electron microscopy were collected in 3% GA and processed according to standard protocols for SEM and TEM evaluations [29]. Tissue specimens for SEM were dried using the critical point method (Tousimis critical point dryer, Rockville, MD, USA, ethanol dehydration, carbon dioxide drying gas) and sputter-coated with gold (BIO-RAD, Microscience Division Coating System, London, UK; Au/Ar sputter coating @ 50–60 nm). A Shimadzu SSX 550 scanning electron microscope (Kyoto, Japan, with integral imaging (SDF, TIF and JPG format)) was used to examine and photograph the tissue surface.

Leaflet and wall samples for TEM were fixed in 3% GA overnight, post fixated in Palade's osmium tetroxide, and dehydrated in a graded acetone series. Dehydrated samples were impregnated/embedded in an epoxy [29] to facilitate the creation of ultra-thin sections for the TEM evaluation. Ultra-thin sections were cut from the sample embedded in the epoxy using a Leica ultra-microtome (Leica Ultracut UC7, Vienna, Austria). After sectioning, the samples were stained with uranyl acetate and lead citrate. Sections of the leaflet samples were evaluated by using a Philips (FEI, Eindhoven, The Netherlands) CM100 transmission electron microscope and photographed using an Olympus Soft Imaging System Megaview III digital camera, with Soft Imaging System digital image analysis and documentation software (Olympus, Tokyo, Japan).

2.3.2. Mechanical Properties

The mechanical properties of pulmonary valve leaflet and wall samples were examined using a tensile strength testing apparatus (Lloyds LS100 Plus, IMP, Johannesburg, South Africa), according to the method described by Thubrikar et al. [30]. Tissue samples measured 5 × 10 mm, with leaflets and pulmonary sinus wall samples cut in the circumferential direction. These tissue samples were fixed between clamps at both ends and gradually stretched (0.1 mm/s) by applying constant tension on the two ends, and the data was recorded on a personal computer.

2.3.3. Statistical Analysis

All values are expressed as median values with a corresponding range (first and third quartiles). Statistical analyses were performed using GraphPad Prism version 8.3.1 (GraphPad Software, La Jolla, CA, USA, www.graphpad.com). The overall difference between the groups was assessed using a Kruskal–Wallis (KW) test. The significance was set as $p < 0.05$.

3. Results

The successful decellularization of both the valve leaflets and the pulmonary artery wall tissue in the decellularized (Figure 2C,D) and decellularized plus EnCap^{TM}-treated (Figure 2E,F) groups was shown with DAPI staining.

Gel electrophoresis (fragmented DNA bands < 200 bp) and nanodrop readings of below 50 ng/mg for the same tissue samples were used to confirm and support the successful decellularization. The NanoDrop reading for fresh homograft tissue prior to decellularization was 204.8 ng/mg [200.8–212], and following decellularization no DNA was detectable via NanoDrop. The absence of DNA following the decellularization of homograft tissue was also shown with gel electrophoresis (Figure 3).

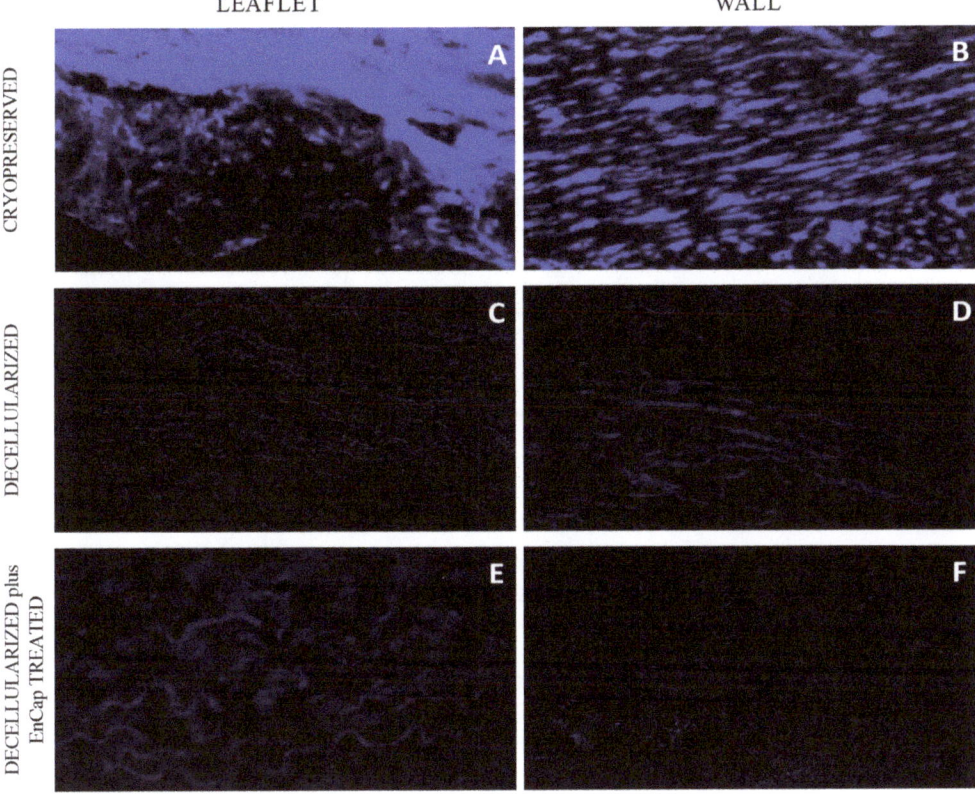

Figure 2. Representative images of DAPI-stained sections of cryopreserved (*n* = 5) (**A**,**B**), decellularized (*n* = 5) (**C**,**D**), and decellularized plus EnCap^{TM}-treated (*n* = 5) (**E**,**F**) leaflet and wall tissue.

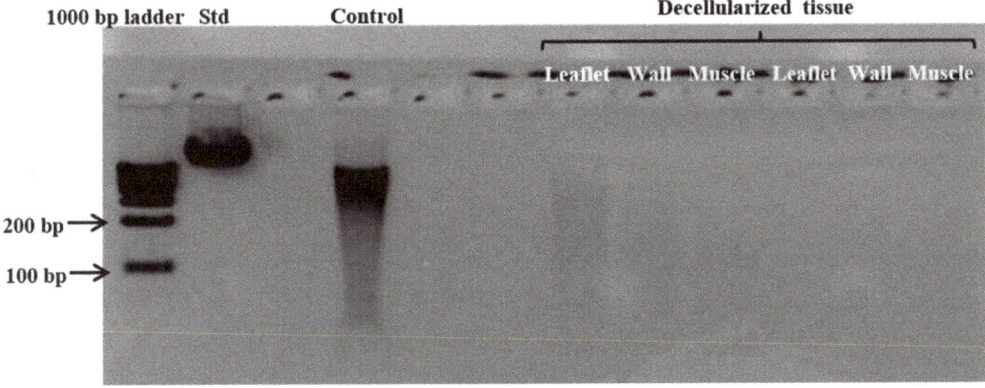

Figure 3. Representative gel electrophoresis image demonstrating the absence of DNA material in leaflet, wall, and muscle tissue of homografts after decellularization. (Standard (Std) = Lamba DNA; control = fresh homograft tissue).

H&E staining of leaflet samples from cryopreserved homografts after 48 h of ischaemia demonstrated the presence of an endothelial layer and the uniform distribution of donor interstitial cells in the ECM (Figure 4A). The leaflets of valves in the decellularized and decellularized plus EnCap^{TM}-treated groups demonstrated well-preserved collagen matrices without any endothelial or interstitial cells (Figure 4D,G). No calcific deposits are visible in any of the samples on von Kossa staining (Figure 4B,E,H), while large numbers of elastin fibers (black arrows) are clearly demonstrated in the ventricularis region of the leaflets in all three groups on modified van Gieson staining (Figure 4C,F,I). The collagen in the cryopreserved group appeared more collapsed compared to fresh untreated tissue [8], and compared to the cryopreserved group, more loosely arranged in the decellularized group and more compact and dense in the decellularized plus EnCap^{TM}-treated group.

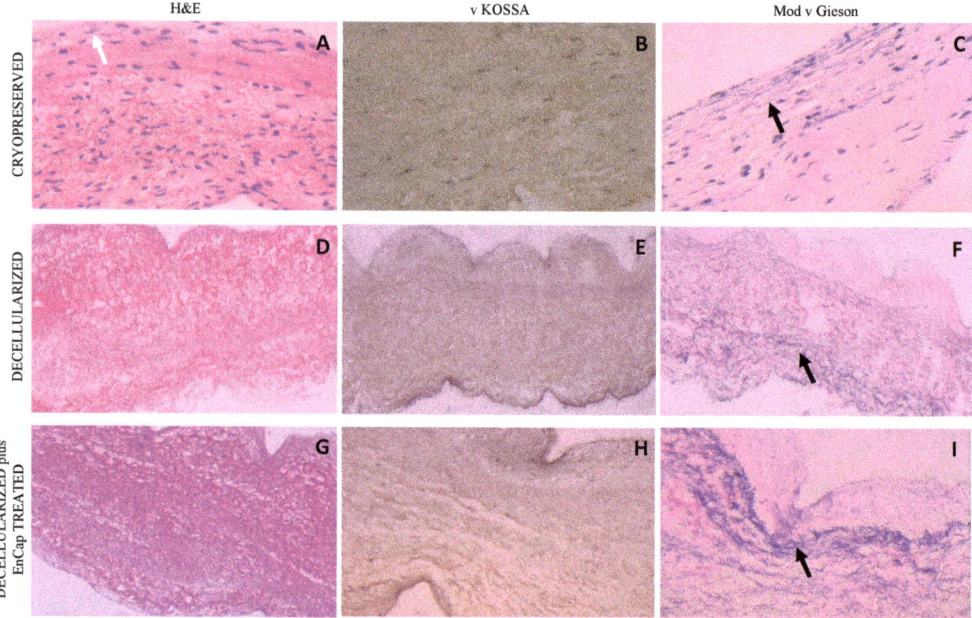

Figure 4. Representative H&E, von Kossa, and modified Verhoeff's van Gieson staining of cryopreserved, decellularized, and decellularized and EnCap^{TM}-treated leaflet samples processed after 48 h cold ischaemia; white arrow points to endothelial layer, and black arrows point to normally distributed elastic fibers ((**A–I**) = 400× magnification).

Similar results were demonstrated for the wall tissue as for the leaflets, with an endothelial cell layer (white arrow) and uniform donor interstitial cell distribution in the cryopreserved group (Figure 5A) and no cells present in the decellularized scaffold or decellularized plus EnCap^{TM}-treated group (Figure 5D,G) on H&E staining. No calcific deposits could be demonstrated with von Kossa staining in any of the three groups (Figure 5B,E,H), and the dense distribution of elastin fibers (black arrows) was demonstrated in all three groups on Modified van Gieson staining (Figure 5C,F,I). Collagen in the walls of the cryopreserved group is again collapsed, loosely arranged in the decellularized group, and dense and compacted in the decellularized plus EnCap^{TM}-treated group.

SEM demonstrated cell dehiscence, although it did demonstrate the uniform coverage of leaflets and wall tissue with endothelial cells in the cryopreserved group (Figure 6A,B). Leaflets and walls in both the decellularized (Figure 6C,D) and decellularized plus EnCap^{TM}-treated groups (Figure 6E,F) demonstrated exposed collagen networks without any endothelial cells and only basal membrane remnants (white arrows).

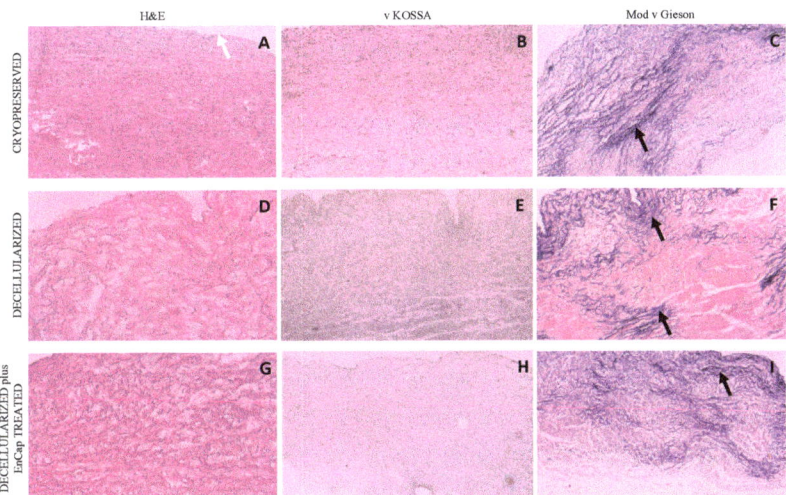

Figure 5. Representative H&E, von Kossa, and modified Verhoeff's van Gieson staining of cryopreserved, decellularized, and decellularized and EnCap™-treated wall tissue processed after 48 h cold ischemia; white arrow points to endothelial layer, and black arrows point to normally distributed elastic fibers ((**A–I**) = 100× magnification).

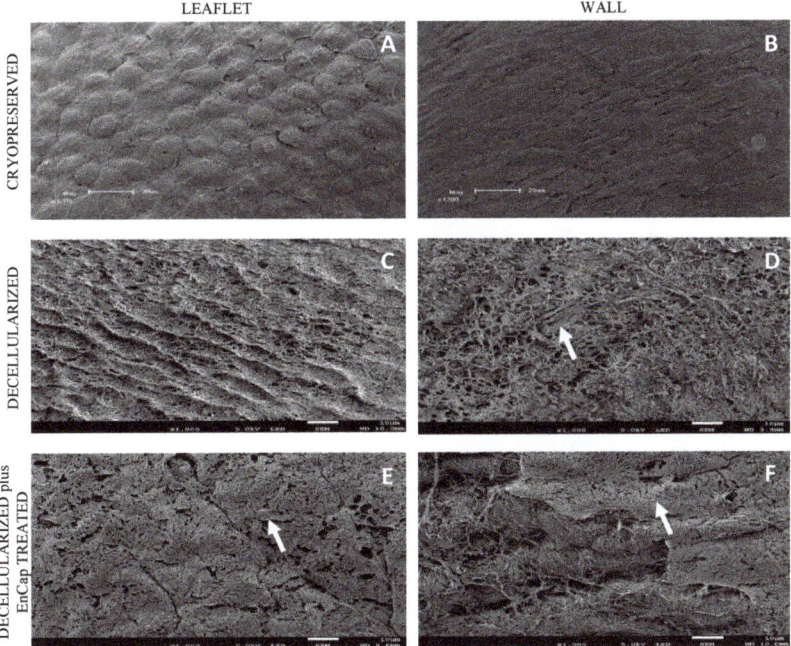

Figure 6. Representative scanning electron microscopy images of cryopreserved, decellularized, and decellularized and EnCap™-treated wall tissue processed after 48 h ischemia; white arrows point to basal membrane remnants ((**A,B**) = 1200× magnification, scale bar 20 µm; and (**C–F**) = 1000× magnification, scale bar 10 µm).

TEM demonstrated the presence of cells and cellular remnants in the cryopreserved group (Figure 7A,B), and no interstitial cells were present in the leaflets and wall tissue in the decellularized group (Figure 7C,D) or decellularized plus EnCap^{TM}-treated group (Figure 7E,F). Collagen bundles appeared more loosely arranged in the decellularized group (Figure 7C,D) and more compressed and dense in the decellularized plus EnCap^{TM}-treated group (Figure 7E,F).

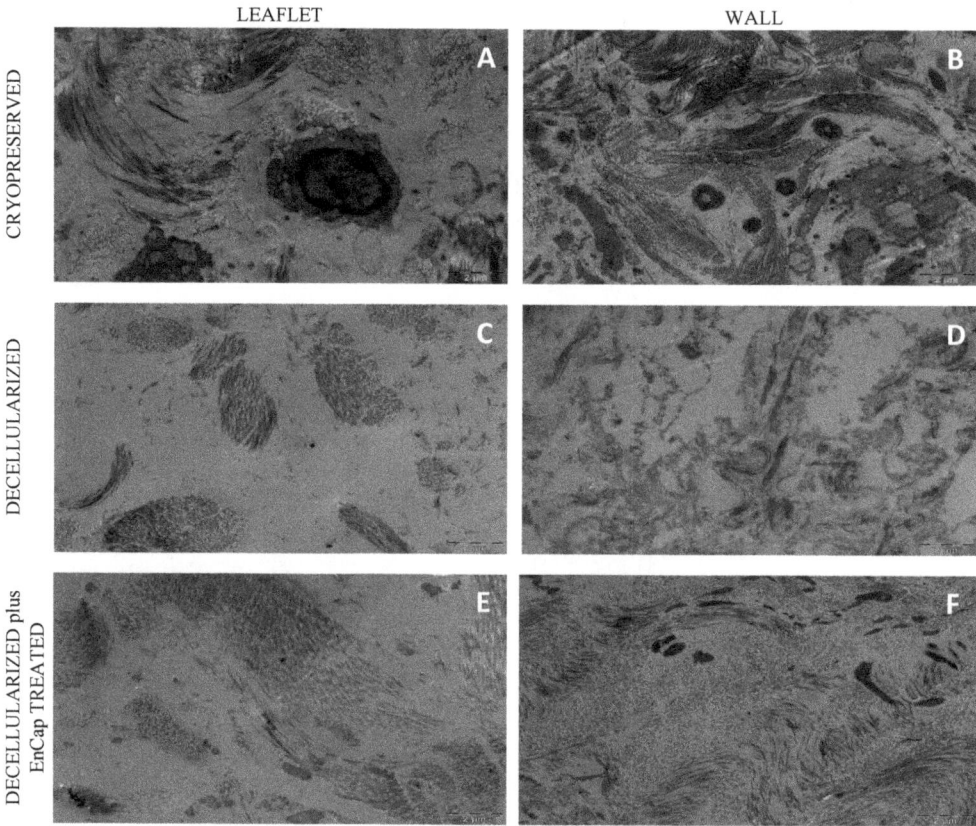

Figure 7. Representative transmission electron microscopy images of cryopreserved, decellularized, and decellularized and EnCap^{TM}-treated wall tissue processed after 48 h ischemia ((**A–F**) = scale bar 2 μm).

The Young's modulus of the decellularized leaflets suggests more extensible leaflet tissue, but no statistically significant differences in the tensile strength or Young's modulus of the leaflet or wall tissue (Table 1) between the three groups could be demonstrated. The stiffness of the leaflets in the decellularized plus EnCap^{TM}-treated group did increase when compared to that in the other two groups; however, the increase was not significant ($p > 0.05$).

Table 1. Baseline TS and YM of cryopreserved, decellularized unfixed, and decellularized plus EnCap^{TM}-treated pulmonary homograft leaflets (n = 5) and wall tissue (n = 5).

	Variable	Cryopreserved	Decellularized	Decellularized Plus EnCapTM-Treated	p-Values
Leaflet	TS (MPa)	3.72 (3.12–5.01)	1.83 (0.01–5.65)	4.96 (4.34–6.76)	0.23
	YM (MPa)	23.34 (14.12–28.87)	22.10 (0.10–55.27)	41.11 (27.76–52.52)	0.40
Wall	TS (MPa)	1.80 (1.35–2.65)	1.89 (1.57–2.38)	1.90 (1.18–2.09)	0.96
	YM (MPa)	1.97 (1.22–3.18)	2.32 (1.82–2.88)	2.21 (1.99–3.58)	0.75

Data presented as median (corresponding range (Q1–Q3)), the difference between groups is determined with the Kruskal–Wallis (KW) test, and no significant differences ($p < 0.05$) are indicated. (TS = tensile strength; YM = Young's modulus).

4. Discussions

Ross (1965) and Barratt-Boyes et al. (1965) were the first to describe the successful use of "fresh" or homovital homografts in the aortic position and pulmonary homografts for RVOT reconstruction, with acceptable homograft performance [31,32]. The superior performance and increased long-term durability of cryopreserved homografts with maintained cellular viability, compared to valves stored at 4 °C [33–35], unfortunately, led to an internationally accepted guideline that homografts from either beating heart or non-beating-heart donors should be harvested and processed within 24 h after death to retain maximum cell viability. These findings not only restricted the available post-mortem donor pool significantly, but the presence of cellular remnants in the ECM evokes immunological reactions from the recipient, causing tissue degeneration and graft failure [36]. Mitchell et al. (1998) reported the early death and loss of endothelial and interstitial cells in cryopreserved homografts following implantation, arguing that valve durability primarily relies on the retention of structural integrity and the preservation of the ECM instead of cell viability [37]. Smit et al. has previously shown that the post-mortem ischaemic time prior to the cryopreservation of homografts can be extended safely to around 48 h [6].

In the current study, the morphology and mechanical properties of standard cryopreserved ovine pulmonary homografts were compared to decellularized pulmonary homografts and decellularized pulmonary homografts treated with EnCap^{TM} AC technology. Homografts were harvested and processed after 48 h cold post-mortem ischaemia.

The histological evaluation of the cryopreserved homografts with H&E demonstrated the presence of an endothelial layer and uniformly distributed interstitial cells, but the collagen scaffold appeared collapsed (Figures 4A and 5A). However, a limitation of H&E staining is the inability to differentiate between morphologically intact cells and non-viable cells [38,39]. SEM demonstrated a confluent endothelial cell layer on the leaflet and wall tissue of cryopreserved homografts, even after 48 h of ischaemia (Figure 6A,B). These endothelial cells presented with prominent nuclei and collapsed extranuclear areas, indicative of non-viable cells [8]. Shenke–Layland and co-workers described the presence of limited collagen-containing structures in the ventricularis of cryopreserved valve leaflets after thawing, with significantly altered and deteriorated collagenous and elastic fiber structures as a result of crystal ice formation in the ECM during cryopreservation [40]. This corresponds with our findings on TEM, where the collagen in the cryopreserved group appeared disrupted and damaged, with interstitial cells and cellular remnants (Figure 7A,B). These cellular remnants that remain in the ECM can activate the immune response from the recipient to the homograft and, together with the damaged and ruptured collagen, will result in early valve degradation [36].

Removing all the cellular content and DNA material from a homograft valve through decellularization results in an implant that significantly reduces the immunologic response from the recipient [41]. Decellularized valve leaflets and wall tissue demonstrated complete acellularity with H&E staining (Figures 4D and 5D); however, the decellularization of leaflets using SDS has been shown to result in a dense ECM network and small pore sizes, which might make the recellularization of the matrix with interstitial cells more difficult [42]. Our devised decellularization process for homografts relies on combining osmotic shock, as an initial step to induce cell lysis, with detergents (SDS, SDC, and TritonX-100) and the addition of Benzonase®. The initial osmotic shock step reduces the concentrations and time of exposure of the tissue to the detergents and the enzyme, thereby reducing the damage to the collagen fibers and the ECM [43–45]. Ethanol was used to remove lipids [46,47]. The decellularization method used in this study resulted in collagen fibers that were more loosely organized with larger spaces between the fibers (Figures 4D and 5D). SEM confirmed the complete removal of the endothelial cells on the leaflet and wall tissue of the decellularized valves, with only exposed collagen layers and remnants of basal membrane remaining (Figure 6), while TEM also demonstrated the loosely arranged collagen network (Figure 7C,D). The decellularization method proved effective in achieving complete acellularity, and the larger interfibrillar spaces could be very beneficial for in vivo recellularization once implanted while maintaining tissue strength when compared to cryopreservation (Table 1).

Converse and co-workers used a multi-detergent and enzymatic washout decellularization protocol, and differential scanning calorimetry (DSC) showed that this method does not reduce the cross-linking of collagen and thereby retains the strength of the tissue. However, the decellularization protocol they used reduced the GAG content, with resultant increased extensibility and changes in the relaxation behavior of the pulmonary valve leaflets [44]. The method currently used led to a reduction of the GAG content when used for the decellularization of bovine pericardium [48]. The GAG content was not evaluated in the current study, and the effect of this decellularization method on the GAG content of the pulmonary homografts should be investigated. However, the decellularization process did not significantly affect the stiffness of the homograft leaflet and the wall tissue compared to the cryopreserved tissue (Table 1). Although the Young's modulus of decellularized leaflets suggests more extensible leaflet tissue after decellularization, the lack of significance may be due to the small number of samples tested. The comparable strength and Young's modulus of the decellularized leaflet and wall tissue to that of the cryopreserved tissue could be attributed to the dehydration effect of ethanol in the delipidation step.

Based on these results, the decellularization of homografts using our proprietary method combined with an enzyme could be used as an alternative processing technique for homografts. Decellularized valves are currently being used in clinical practice. CryoValve® SG (CryoLife, Inc, Kennesaw, GA, USA) pulmonary human heart valves were some of the first decellularized homografts to be used clinically for RVOT reconstruction, and the results compare favourably with cryopreserved homografts [49]; however, improved haemodynamics were observed in the CryoValve® SG group, after four years of implantation, possibly due to decreased antigenicity of these valves following decellularization [50]. A European group led by Haverich has implanted 131 decellularized pulmonary homografts since 2005. A 10-year follow-up study found that decellularized homografts had a 100% freedom from explantation at 10 years follow-up, compared to 84.2% freedom from explantation for cryopreserved homografts and 84.3% freedom from explantation for GA-fixed bovine jugular vein conduits. Additionally, the rate of freedom of infective endocarditis was 100% in patients receiving decellularized homografts [51]. Although the decellularized homografts used by this group provided low gradients in follow-up and exhibited adaptive growth [12], the careful observation of their latest data shows no statistical difference in terms of stenosis and regurgitation when compared to the conventional cryopreserved homografts [52]. An early and midterm study by a group led by da Costa reported that decellularized aortic allografts replacements, the first of their kind, had 100% freedom from reoperation of the graft at 3 years follow-up. The decellularized aortic valves retained structural integrity, low calcification rates, and adequate haemodynamics [53].

Due to the concerns that the decellularization process might negatively affect the strength of the scaffold, the additional fixation and stabilization of the decellularized homografts were investigated. The additional fixation and detoxification of the decellularized valves were done using GA as a fixative and PG for detoxifying the GA-fixed homograft (EnCap^{TM} AC technology) [22]. This process resulted in more compacted and dense collagen layers when compared to the leaflet and wall tissue in the cryopreserved and decellularized groups (Figures 4G, 5G and 7E,F), which might make the recellularization of the matrix with interstitial cells more difficult [42]. Using a juvenile ovine model, Botes et al. (2022) demonstrated that commercial EnCap^{TM}-treated bovine pericardium (not decellularized) had less host cell infiltration than the bovine pericardium processed using our proprietary decellularization process. The EnCap^{TM}-treated tissue showed a loss of collagen fiber integrity, causing the collagen to become densely compacted, contributing to insignificant recipient cell infiltration [54]. The current study showed that even after decellularization, GA fixation led to the collagen layers of the homograft becoming compacted and dense, which might affect cell infiltration. Umashankar et al. (2012) reported the total lack of host tissue incorporation in GA-fixed bovine pericardium, compared to excellent host fibroblast incorporation in decellularized bovine pericardium. The authors conclude that this could be due to GA-treated bovine pericardium being resistant to collagenase, an important enzyme in the body that is responsible for ECM remodeling [55]. The additional fixation and detoxification of decellularized homografts using EnCap^{TM} AC technology added additional cross-links and increased the Young's modulus of leaflets, although not significantly. GA fixation has an effect on the hydration properties of fixated tissue, impairing its ability to rehydrate, which is hypothesized to result from the reaction of GA with hydrophilic amine groups in collagen, reducing the tissue's hydrating capacity and increasing stiffness.

5. Conclusions

In conclusion, the confluent layers of endothelial cells were still present on the leaflet and wall tissue of the cryopreserved group even after 48 h cold ischaemia. Our proprietary decellularization protocol with the additional Benzonase® proved to be effective in removing all endothelial cells, interstitial cells, and DNA from the homograft wall and leaflet tissue, without a significant reduction in strength. The removal of donor cells and nuclear material might reduce recipient immunogenicity of the valve. Cryopreservation disrupted and fractured the collagen fibrils, while decellularization created larger interfibrillar spaces in the collagen network, which might be advantageous for host cell repopulation after transplantation. Elastin appeared to be well preserved in all three groups after processing, with no visible calcific nodules in any of the groups. The additional fixation of the decellularized scaffold created dense collagen networks and increased the leaflet stiffness, which could negatively affect the biomechanical behavior of the homograft. Due to the increased stiffness of the leaflets after fixation, additional fixation with GA following the decellularization of homograft valves is not desirable, but further investigations in this regard might be required. Compared to cryopreservation, the use of our proprietary multi-detergent decellularization method combined with the use of an enzyme is a promising alternative for homograft preservation. The clinical performance of these decellularized homografts should be investigated in pre-clinical (animal-based) studies.

Author Contributions: Conceptualization, J.J.v.d.H., C.J.J., P.M.D. and F.E.S.; data curation, J.J.v.d.H. and F.E.S.; formal analysis, J.J.v.d.H., A.L. and F.E.S.; funding acquisition, F.E.S.; investigation, J.J.v.d.H., C.J.J., J.G. and D.B.; methodology, J.J.v.d.H., P.M.D. and F.E.S.; project administration, J.J.v.d.H., C.J.J., D.B. and L.B.; resources, J.J.v.d.H. and L.B.; supervision, P.M.D. and F.E.S.; validation, J.J.v.d.H., A.L. and F.E.S.; visualization, J.J.v.d.H. and A.L.; writing—original draft, J.J.v.d.H. and A.L.; and writing—review and editing, J.J.v.d.H., A.L. and F.E.S. All authors have read and agreed to the published version of the manuscript.

Funding: This research received no external funding.

Institutional Review Board Statement: The study was approved by the interfaculty Animal Ethics Committee of the University of the Free State (approval number: UFS-AED2016/0101).

Informed Consent Statement: Not applicable.

Data Availability Statement: Data available from corresponding author on reasonable request.

Acknowledgments: We want to thank H Grobler from the Centre for Electron Microscopy at the University of the Free State, Bloemfontein, South Africa for the preparation of the SEM and TEM samples.

Conflicts of Interest: The authors declare no conflict of interest.

References

1. Diaz, R.; Hernandez-Vaquero, D.; Alvarez-Cabo, R.; Avanzas, P.; Silva, J.; Moris, C.; Pascual, I. Long-term outcomes of mechanical versus biological aortic valve prosthesis: Systematic review and meta-analysis. *J. Thorac. Cardiovasc. Surg.* **2019**, *158*, 706–714.e18. [CrossRef]
2. Hechadi, J.; Gerber, B.L.; Coche, E.; Melchior, J.; Jashari, R.; Glineur, D.; Noirhomme, P.; Rubay, J.; El Khoury, G.; de Kerchove, L. Stentless xenografts as an alternative to pulmonary homografts in the Ross operation. *Eur. J. Cardiothorac. Surg.* **2013**, *44*, e32–e39. [CrossRef]
3. Romeo, J.L.R.; Papageorgiou, G.; van de Woestijne, P.C.; Takkenberg, J.J.M.; Westenberg, L.E.H.; van Beynum, I.; Bogers, A.; Mokhles, M.M. Downsized cryopreserved and standard-sized allografts for right ventricular outflow tract reconstruction in children: Long-term single-institutional experience. *Interact. Cardiovasc. Thorac. Surg.* **2018**, *27*, 257–263. [CrossRef]
4. Tierney, E.S.; Gersony, W.M.; Altmann, K.; Solowiejczyk, D.E.; Bevilacqua, L.M.; Khan, C.; Krongrad, E.; Mosca, R.S.; Quaegebeur, J.M.; Apfel, H.D. Pulmonary position cryopreserved homografts: Durability in pediatric Ross and non-Ross patients. *J. Thorac. Cardiovasc. Surg.* **2005**, *130*, 282–286. [CrossRef]
5. Goffin, Y.A.; van Hoeck, B.; Jashari, R.; Soots, G.; Kalmar, P. Banking of cryopreserved heart valves in Europe: Assessment of a 10-year operation in the European Homograft Bank (EHB). *J. Heart Valve Dis.* **2000**, *9*, 207–214.
6. Smit, F.E.; Bester, D.; van den Heever, J.J.; Schlegel, F.; Botes, L.; Dohmen, P.M. Does prolonged post-mortem cold ischaemic harvesting time influence cryopreserved pulmonary homograft tissue integrity? *Cell Tissue Bank.* **2015**, *16*, 531–544. [CrossRef]
7. Kitagawa, T.; Masuda, Y.; Tominaga, T.; Kano, M. Cellular biology of cryopreserved allograft valves. *J. Med. Investig.* **2001**, *48*, 123–132.
8. Bester, D.; Botes, L.; van den Heever, J.J.; Kotze, H.; Dohmen, P.; Pomar, J.L.; Smit, F.E. Cadaver donation: Structural integrity of pulmonary homografts harvested 48 h post mortem in the juvenile ovine model. *Cell Tissue Bank.* **2018**, *19*, 743–754. [CrossRef]
9. Shaddy, R.E.; Hawkins, J.A. Immunology and failure of valved allografts in children. *Ann. Thorac. Surg.* **2002**, *74*, 1271–1275. [CrossRef]
10. Baskett, R.J.; Nanton, M.A.; Warren, A.E.; Ross, D.B. Human leukocyte antigen-DR and ABO mismatch are associated with accelerated homograft valve failure in children: Implications for therapeutic interventions. *J. Thorac. Cardiovasc. Surg.* **2003**, *126*, 232–239. [CrossRef]
11. Ryan, W.H.; Herbert, M.A.; Dewey, T.M.; Agarwal, S.; Ryan, A.L.; Prince, S.L.; Mack, M.J. The occurrence of postoperative pulmonary homograft stenosis in adult patients undergoing the Ross procedure. *J. Heart Valve Dis.* **2006**, *15*, 108–113; discussion 113-4. [PubMed]
12. Cebotari, S.; Tudorache, I.; Ciubotaru, A.; Boethig, D.; Sarikouch, S.; Goerler, A.; Lichtenberg, A.; Cheptanaru, E.; Barnaciuc, S.; Cazacu, A.; et al. Use of fresh decellularized allografts for pulmonary valve replacement may reduce the reoperation rate in children and young adults: Early report. *Circulation* **2011**, *124*, S115–S123. [CrossRef] [PubMed]
13. Erdbrügger, W.; Konertz, W.; Dohmen, P.M.; Posner, S.; Ellerbrok, H.; Brodde, O.E.; Robenek, H.; Modersohn, D.; Pruss, A.; Holinski, S.; et al. Decellularized xenogenic heart valves reveal remodeling and growth potential in vivo. *Tissue Eng.* **2006**, *12*, 2059–2068. [CrossRef] [PubMed]
14. VeDepo, M.C.; Detamore, M.S.; Hopkins, R.A.; Converse, G.L. Recellularization of decellularized heart valves: Progress toward the tissue-engineered heart valve. *J. Tissue Eng.* **2017**, *8*, 2041731417726327. [CrossRef] [PubMed]
15. Bourgine, P.E.; Pippenger, B.E.; Todorov, A.; Tchang, L.; Martin, I. Tissue decellularization by activation of programmed cell death. *Biomaterials* **2013**, *34*, 6099–6108. [CrossRef]
16. Korossis, S.A.; Booth, C.; Wilcox, H.E.; Watterson, K.G.; Kearney, J.N.; Fisher, J.; Ingham, E. Tissue engineering of cardiac valve prostheses II: Biomechanical characterization of decellularized porcine aortic heart valves. *J. Heart Valve Dis.* **2002**, *11*, 463–471.
17. Sheehy, E.J.; Cunniffe, G.M.; O'Brien, F.J. 5 Collagen-based biomaterials for tissue regeneration and repair. In *Peptides and Proteins as Biomaterials for Tissue Regeneration and Repair*; Barbosa, M.A., Martins, M.C.L., Eds.; Woodhead Publishing: Sawston, UK, 2018; pp. 127–150. [CrossRef]
18. Oryan, A.; Kamali, A.; Moshiri, A.; Baharvand, H.; Daemi, H. Chemical crosslinking of biopolymeric scaffolds: Current knowledge and future directions of crosslinked engineered bone scaffolds. *Int. J. Biol. Macromol.* **2018**, *107*, 678–688. [CrossRef] [PubMed]
19. Lee, C.; Lim, H.G.; Lee, C.H.; Kim, Y.J. Effects of glutaraldehyde concentration and fixation time on material characteristics and calcification of bovine pericardium: Implications for the optimal method of fixation of autologous pericardium used for cardiovascular surgery. *Interact. Cardiovasc. Thorac. Surg.* **2017**, *24*, 402–406. [CrossRef] [PubMed]
20. Padala, M. A heart valve is no stronger than its weakest link: The need to improve durability of pericardial leaflets. *J. Thorac. Cardiovasc. Surg.* **2018**, *156*, 207–208. [CrossRef] [PubMed]

21. Strange, G.; Brizard, C.; Karl, T.R.; Neethling, L. An evaluation of Admedus' tissue engineering process-treated (ADAPT) bovine pericardium patch (CardioCel) for the repair of cardiac and vascular defects. *Expert Rev. Med. Devices* **2015**, *12*, 135–141. [CrossRef] [PubMed]
22. Seifter, E.; Frater, R.W.M. *Anticalcification Treatment for Aldehyde-Tanned Biological Tissue*; Albert Einstein College of Medicine of Yeshiva University: Bronx, NY, USA, 1995. Available online: http://www.freepatentsonline.com/5476516.html (accessed on 2 October 2019).
23. Succi, J.E.; Buffolo, E.; Salles, C.A.; Casagrande, I.S.J.; Neto, J.V.; de Mendonça, J.T.; Filho, R.V.; Jaramillo, I.A. Valve replacement with aortic glutaraldehyde preserved homografts: A multicenter study. *Braz. J. Cardiovasc. Surg.* **1986**, *1*, 20–23. [CrossRef]
24. Laker, L.; Dohmen, P.M.; Smit, F.E. Synergy in a detergent combination results in superior decellularized bovine pericardial extracellular matrix scaffolds. *J. Biomed. Mater. Res. B Appl. Biomater.* **2020**, *108*, 2571–2578. [CrossRef] [PubMed]
25. Bester, D.; Smit, F.E.; van den Heever, J.J.; Botes, L.; Dohmen, P.M.C.E. Detoxification and Stabilization of Implantable or Transplantable Biological Material. EU Patent 16792990.0-1455, 28 July 2016.
26. Dijkman, P.E.; Driessen-Mol, A.; Frese, L.; Hoerstrup, S.P.; Baaijens, F.P. Decellularized homologous tissue-engineered heart valves as off-the-shelf alternatives to xeno- and homografts. *Biomaterials* **2012**, *33*, 4545–4554. [CrossRef]
27. Crapo, P.M.; Gilbert, T.W.; Badylak, S.F. An overview of tissue and whole organ decellularization processes. *Biomaterials* **2011**, *32*, 3233–3243. [CrossRef]
28. Bancroft, J.D.; Stevens, A. *Theory and Practice of Histological Techniques*, 3rd ed.; Churchill Livingstone: Edinburgh, UK, 1991.
29. Spurr, A.R. A low-viscosity epoxy resin embedding medium for electron microscopy. *J. Ultrastruct. Res.* **1969**, *26*, 31–43. [CrossRef]
30. Thubrikar, M.J.; Deck, J.D.; Aouad, J.; Nolan, S.P. Role of mechanical stress in calcification of aortic bioprosthetic valves. *J. Thorac. Cardiovasc. Surg.* **1983**, *86*, 115–125. [CrossRef]
31. Ross, D.N. Homograft replacement of the aortic valve. *Lancet* **1962**, *280*, 487. [CrossRef]
32. Barratt-Boyes, B.G. Homograft aortic valve replacement. *N. Z. Med. J.* **1965**, *64*, 41–43. [CrossRef]
33. Angell, W.W.; Oury, J.H.; Lamberti, J.J.; Koziol, J. Durability of the viable aortic allograft. *J. Thorac. Cardiovasc. Surg.* **1989**, *98*, 48–55; discussion 55–56. [CrossRef]
34. O'Brien, M.F.; McGiffin, D.C.; Stafford, E.G.; Gardner, M.A.; Pohlner, P.F.; Mclachlan, G.J.; Gall, K.; Smith, S.; Murphy, E. Allograft aortic valve replacement: Long-term comparative clinical analysis of the viable cryopreserved and antibiotic 4 degrees C stored valves. *J. Card. Surg.* **1991**, *6*, 534–543. [CrossRef]
35. O'Brien, M.F.; Stafford, E.G.; Gardner, M.A.; Pohlner, P.G.; Tesar, P.J.; Cochrane, A.D.; Mau, T.K.; Gall, K.L.; Smith, S.E. Allograft aortic valve replacement: Long-term follow-up. *Ann. Thorac. Surg.* **1995**, *60*, S65–S70. [CrossRef]
36. Lopes, S.A.; da Costa, F.D.; de Paula, J.B.; Dohmen, P.; Phol, F.; Vilani, R.; Roderjan, J.G.; Vieira, E.D. Decellularized heterografts versus cryopreserved homografts: Experimental study in sheep model. *Braz. J. Cardiovasc. Surg.* **2009**, *24*, 15–22. [CrossRef] [PubMed]
37. Mitchell, R.N.; Jonas, R.A.; Schoen, F.J. Pathology of explanted cryopreserved allograft heart valves: Comparison with aortic valves from orthotopic heart transplants. *J. Thorac. Cardiovasc. Surg.* **1998**, *115*, 118–127. [CrossRef]
38. Radosevich, J.A.; Ma, Y.X.; Salwen, H.R.; Rosen, S.T.; Lee, I.; Gould, V.E. Monoclonal antibody 44-3A6 as a probe for a novel antigen found on human lung carcinomas with glandular differentiation. *Cancer Res.* **1985**, *45*, 5808–5812. [CrossRef] [PubMed]
39. Radosevich, J.A.; Haines, G.K.; Elseth, K.M.; Shambaugh, G.E.; Maker, V.K. A new method for the detection of viable cells in tissue sections using 3-[4,5-dimethylthiazol-2-yl]-2,5-diphenyltetrazolium bromide (MTT): An application in the assessment of tissue damage by surgical instruments. *Virchows Archiv. B Cell Pathol. Incl. Mol. Pathol.* **1993**, *63*, 345–350. [CrossRef] [PubMed]
40. Schenke-Layland, K.; Madershahian, N.; Riemann, I.; Starcher, B.; Halbhuber, K.J.; Konig, K.; Stock, U.A. Impact of cryopreservation on extracellular matrix structures of heart valve leaflets. *Ann. Thorac. Surg.* **2006**, *81*, 918–926. [CrossRef]
41. Bibevski, S.; Ruzmetov, M.; Fortuna, R.S.; Turrentine, M.W.; Brown, J.W.; Ohye, R.G. Performance of SynerGraft Decellularized Pulmonary Allografts Compared With Standard Cryopreserved Allografts: Results From Multiinstitutional Data. *Ann. Thorac. Surg.* **2017**, *103*, 869–874. [CrossRef]
42. Liao, J.; Joyce, E.M.; Sacks, M.S. Effects of decellularization on the mechanical and structural properties of the porcine aortic valve leaflet. *Biomaterials* **2008**, *29*, 1065–1074. [CrossRef]
43. Zhou, J.; Fritze, O.; Schleicher, M.; Wendel, H.P.; Schenke-Layland, K.; Harasztosi, C.; Hu, S.; Stock, U.A. Impact of heart valve decellularization on 3-D ultrastructure, immunogenicity and thrombogenicity. *Biomaterials* **2010**, *31*, 2549–2554. [CrossRef]
44. Converse, G.L.; Armstrong, M.; Quinn, R.W.; Buse, E.E.; Cromwell, M.L.; Moriarty, S.J.; Lofland, G.K.; Hilbert, S.L.; Hopkins, R.A. Effects of cryopreservation, decellularization and novel extracellular matrix conditioning on the quasi-static and time-dependent properties of the pulmonary valve leaflet. *Acta Biomater.* **2012**, *8*, 2722–2729. [CrossRef]
45. Somers, P.; de Somer, F.; Cornelissen, M.; Thierens, H.; van Nooten, G. Decellularization of heart valve matrices: Search for the ideal balance. *Artif. Cells Blood Substit. Immobil. Biotechnol.* **2012**, *40*, 151–162. [CrossRef] [PubMed]
46. Flynn, L.E. The use of decellularized adipose tissue to provide an inductive microenvironment for the adipogenic differentiation of human adipose-derived stem cells. *Biomaterials* **2010**, *31*, 4715–4724. [CrossRef] [PubMed]
47. Brown, B.N.; Freund, J.M.; Han, L.; Rubin, J.P.; Reing, J.E.; Jeffries, E.M.; Wolf, M.T.; Tottey, S.; Barnes, C.A.; Ratner, B.D.; et al. Comparison of three methods for the derivation of a biologic scaffold composed of adipose tissue extracellular matrix. *Tissue Eng. Part C Methods* **2011**, *17*, 411–421. [CrossRef] [PubMed]

48. Laker, L.; Dohmen, P.M.; Smit, F.E. The sequential effects of a multifactorial detergent based decellularization process on bovine pericardium. *Biomed. Phys. Eng. Express* **2020**, *6*, 065011. [CrossRef]
49. Burch, P.T.; Kaza, A.K.; Lambert, L.M.; Holubkov, R.; Shaddy, R.E.; Hawkins, J.A. Clinical performance of decellularized cryopreserved valved allografts compared with standard allografts in the right ventricular outflow tract. *Ann. Thorac. Surg.* **2010**, *90*, 1301–1305; discussion 1306. [CrossRef]
50. Brown, J.W.; Elkins, R.C.; Clarke, D.R.; Tweddell, J.S.; Huddleston, C.B.; Doty, J.R.; Fehrenbacher, J.W.; Takkenberg, J.J.M. Performance of the CryoValve SG human decellularized pulmonary valve in 342 patients relative to the conventional CryoValve at a mean follow-up of four years. *J. Thorac. Cardiovasc. Surg.* **2010**, *139*, 339–348. [CrossRef]
51. Sarikouch, S.; Horke, A.; Tudorache, I.; Beerbaum, P.; Westhoff-Bleck, M.; Boethig, D.; Repin, O.; Maniuc, L.; Ciubotaru, A.; Haverich, A.; et al. Decellularized fresh homografts for pulmonary valve replacement: A decade of clinical experience. *Eur. J. Cardiothorac. Surg.* **2016**, *50*, 281–290. [CrossRef]
52. Boethig, D.; Horke, A.; Hazekamp, M.; Meyns, B.; Rega, F.; van Puyvelde, J.; Hübler, M.; Schmiady, M.; Ciubotaru, A.; Stellin, G.; et al. A European study on decellularized homografts for pulmonary valve replacement: Initial results from the prospective ESPOIR Trial and ESPOIR Registry data. *Eur. J. Cardiothorac. Surg.* **2019**, *56*, 503–509. [CrossRef]
53. Da Costa, F.D.A.; Costa, A.C.; Prestes, R.; Domanski, A.C.; Balbi, E.M.; Ferreira, A.D.; Lopes, S.V. The early and midterm function of decellularized aortic valve allografts. *Ann. Thorac. Surg.* **2010**, *90*, 1854–1860. [CrossRef]
54. Botes, L.; Laker, L.; Dohmen, P.M.; Van den Heever, J.J.; Jordaan, C.J.; Lewies, A.; Smit, F.E. Advantages of decellularized bovine pericardial scaffolds compared to glutaraldehyde fixed bovine pericardial patches demonstrated in a 180-day implant ovine study. *Cell Tissue Bank.* **2022**. [CrossRef]
55. Umashankar, P.R.; Mohanan, P.V.; Kumari, T.V. Glutaraldehyde treatment elicits toxic response compared to decellularization in bovine pericardium. *Toxicol. Int.* **2012**, *19*, 51–58. [CrossRef] [PubMed]

Review

Current Advances in the Development of Hydrogel-Based Wound Dressings for Diabetic Foot Ulcer Treatment

Viviana R. Güiza-Argüello [1], Víctor A. Solarte-David [2,*], Angie V. Pinzón-Mora [3], Jhair E. Ávila-Quiroga [3] and Silvia M. Becerra-Bayona [3,*]

[1] Metallurgical Engineering and Materials Science Department, Faculty of Physicochemical Engineering, Universidad Industrial de Santander, Bucaramanga 680002, Colombia; vivragui@uis.edu.co
[2] Program of Biomedical Engineering and Program of Medicine, Faculty of Engineering and Faculty of Health Sciences, Universidad Autónoma de Bucaramanga, Bucaramanga 680003, Colombia
[3] Program of Medicine, Faculty of Health Sciences, Universidad Autónoma de Bucaramanga, Bucaramanga 680003, Colombia; apinzon57@unab.edu.co (A.V.P.-M.); javila28@unab.edu.co (J.E.Á.-Q.)
* Correspondence: sbecerra521@unab.edu.co (V.A.S.-D.); vsolarte@unab.edu.co (S.M.B.-B.)

Abstract: Diabetic foot ulcers (DFUs) are one of the most prevalent complications associated with diabetes mellitus. DFUs are chronic injuries that often lead to non-traumatic lower extremity amputations, due to persistent infection and other ulcer-related side effects. Moreover, these complications represent a significant economic burden for the healthcare system, as expensive medical interventions are required. In addition to this, the clinical treatments that are currently available have only proven moderately effective, evidencing a great need to develop novel strategies for the improved treatment of DFUs. Hydrogels are three-dimensional systems that can be fabricated from natural and/or synthetic polymers. Due to their unique versatility, tunability, and hydrophilic properties, these materials have been extensively studied for different types of biomedical applications, including drug delivery and tissue engineering applications. Therefore, this review paper addresses the most recent advances in hydrogel wound dressings for effective DFU treatment, providing an overview of current perspectives and challenges in this research field.

Keywords: hydrogel; diabetic foot; wound dressing; tissue engineering

1. Introduction

Diabetes mellitus (DM) is one of the most common chronic diseases, which is characterized by high blood glucose levels [1]. This occurs when either the body is unable to make insulin, causing hyperglycemia, or when the body cannot use insulin efficiently. In 2019, it was estimated that approximately 463 million people worldwide were affected by this pathology, suggesting that by 2045 this number will increase to 700 million [2]. The multiple complications associated with DM include cardiovascular diseases, blindness, renal failure, and foot ulcers, which lead to morbidity, amputation, and high mortality rates [3]. Additionally, these complications represent a great economic burden for the healthcare system due to the elevated costs of treatment, which in 2017 rounded at about 327 billion dollars. Overall, medical expenses related to the treatment of a diabetic patient have been estimated to average at around 9601 USD per year [4].

Diabetic foot ulcers (DFUs) are a severe condition derived from DM that frequently carries an increased risk of morbidity and early death. It affects around 40 to 60 million diabetic patients worldwide [1], and it is characterized by the formation of chronic wounds, which involve a combination of metabolic disorders, nerve damage, deficient blood flow and biomechanical changes in the lower limbs. DFUs can eventually lead to chronic trauma, infection, and, in the worst-case scenario, lower limb amputation, which is estimated to occur every 30 s globally [2]. Currently, clinical treatments available for DFU management include debridement, pressure relief ("off-loading"), antibiotics, and revascularization [5].

However, in many cases, these therapies fail to ensure full wound repair [6,7], which in turn reinforces the need for novel alternatives that promote tissue regeneration, while minimizing complications in the wound area.

In this context, wound dressing design has progressively gained attention as a potential avenue for the achievement of clinically effective DFU therapies. The ideal dressing for treatment of chronic wounds, such as DFUs, should fulfill several essential requirements [8–10]: (1) offer a moist environment that stimulates tissue regeneration; (2) prevent bacterial invasion into the wound; (3) display adequate porosity for gas exchange; (4) promote cell migration, proliferation, and neovascularization; (5) be sterile, biocompatible, and easy to change (or biodegrade). In this regard, hydrogels are polymeric three-dimensional (3D) materials that have been shown promising for successful DFU dressing fabrication [11], since they can provide a moist environment that supports cell proliferation and tissue restoration, promoting scar formation [12,13]. These 3D structures can also exhibit soft tissue-like mechanical properties, as well as be chemically or physically modified to serve as carriers of bioactive agents that can aid in effective wound healing. Therefore, the main purpose of this work was to review current advances in the development of hydrogel-based wound dressing systems, specifically focusing on their fabrication technologies and efficacy in the context of DFU treatment.

2. Normal Wound Healing

During normal wound healing, a complex and orchestrated series of events takes place, which results in re-epithelialization and restoration of the injured site. These events must proceed in a regulated fashion through four physiological phases: (i) homeostasis, (ii) inflammation, (iii) proliferation, and (iv) remodeling, as described below:

Homeostasis: upon injury, the immune system swiftly responds via activation of different humoral and cellular cues where antibodies, neutrophils, macrophages, and lymphocytes are recruited to control the infiltration of pathogens that could exacerbate the regular healing process [14]. Additionally, hemorrhaging is mitigated by the formation of a platelet-fibrin clot plug that stops the bleeding and acts as a barrier to avoid interactions between the body system and the outer hostile environment (hemostasis) [15].

Inflammation: the inflammation stage is characterized by swelling and redness around the affected area, due to the removal of damaged cells, growth factors and foreign bodies (e.g., bacteria) from the wound region. Moreover, different inflammatory cytokines, such as interleukin 1 (IL-1), interleukin 6 (IL-6), tumor necrosis factor-alpha (TNF-α), and interferon gamma (IFN-γ) are secreted by neutrophils and macrophages, which further contribute to this inflammatory state as wound repair progresses [16].

Proliferation: when inflammation decreases and wound contraction occurs, revascularization must take place to restore oxygen supply to the wound. At the same time, fibroblasts and epithelial cells migrate to the wound area to induce cell proliferation and differentiation, while simultaneously releasing growth factors, such as epidermal growth factor (EGF), hepatocyte growth factor (HGF), fibroblast growth factor (FGF), and keratinocyte growth factor (KGF), which replace the damaged tissue and the fibrin clot with new extracellular matrix (ECM), promoting re-epithelialization [17,18].

Remodeling: after the proliferation phase, collagen fibers reorganize and epithelial cells proliferate and differentiate, restoring the epithelium; at the same time, granulation tissue becomes mature scar tissue, increasing wound resistance, which eventually leads to scar formation and wound closure. Finally, fibroblasts continue to remodel the underlying dermis over a period of several months [19].

3. Diabetic Foot Ulcers (DFUs)

In DFUs, the normal healing process is disrupted (Figure 1), producing disorders that demand constant treatment [20]. Thus, the broad range of metabolic dysfunctions in the diabetic patient results in deficient repair of the lesion, which primarily stems from reduced production of pro-angiogenic growth factors and chemokines. Moreover, chronic wounds are commonly characterized by an abnormal and prolonged inflammatory phase, with larger production of metalloproteinases (MMPs) that erode the extracellular matrix at a rate that is too high to allow adequate re-epithelialization and remodeling to occur, even inducing senescence in some cases. Furthermore, MMPs inhibit the action of growth factors, which impairs angiogenesis and limits tissue oxygenation, rendering the wound hypoxic and permanently open. Therefore, open wounds are more prone to pathogenic infections, and infections worsen because of the decreased phagocytosis action induced by hyperglycemia. The combination of all these factors results in even more severe health complications for the patient [18,19,21]. In the face of this challenge, tissue engineering approaches have enabled the design of hydrogel-based systems that function as scaffolds that support proper cell interactions and fill the wound with bioactive agents that promote re-epithelialization, decrease inflammation, accelerate healing, and improve scar formation [22,23].

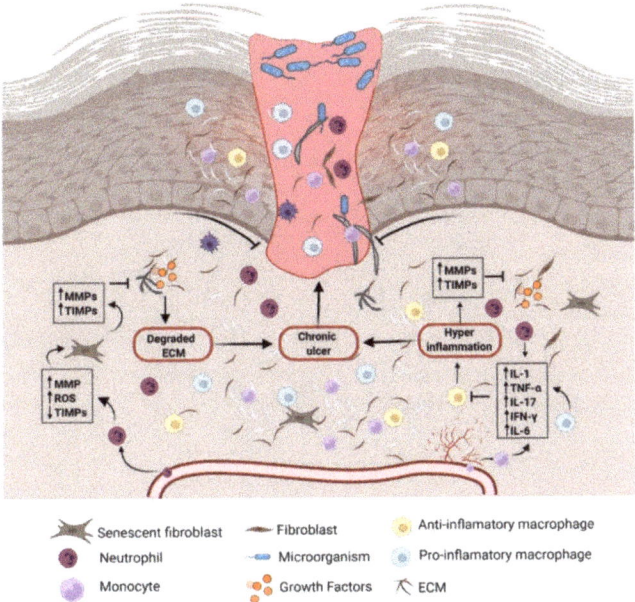

Figure 1. Chronic wound healing process. Diabetic foot ulcers are characterized by continuous inflammation, persistent infections, and necrosis. In these wounds, the balance between Matrix Metalloproteinases (MMPs) and Tissue Inhibitors of Metalloproteinases (TIMPs) is altered, preventing proper remodeling of the extracellular matrix (ECM). Moreover, inflammation is persistent, with high infiltration of immune cells, a response triggered by an increase in interleukins (IL) and pro-inflammatory cytokines associated with cells, such as neutrophils and type I monocytes (pro-inflammatory), which respond to infectious agents. Similarly, there is an increase in reactive oxygen species (ROS), exacerbating the degradation of tissue components. Synergically, these events limit cell migration, angiogenesis, and ECM remodeling, leading to wound chronicity. Created with BioRender.com on 1 June 2022.

4. Hydrogel Wound Dressings for DFU Treatment

The unique potential of hydrogels for effective DFU dressing development lies in their remarkable ability to retain oxygen, absorb wound exudate, and maintain the moist environment that supports normal physiological processes in the wound bed [24–26]. Moreover, hydrogels have shown promising results as antimicrobial substrates [27] that help avoid tissue death and accelerate regular wound repair, and their porous structure promotes adequate exchange of gases, allowing the wound to breathe throughout wound healing and closure (Figure 2). In terms of their fabrication approaches, hydrogels can be made by the assembly of natural polymers such as collagen [28], gelatin [29], cellulose [30], alginate [31], and chitosan [32] (Figure 3); or they can be produced synthetically by using polymers such as polyethylene glycol (PEG) [33], polyvinyl alcohol (PVA) [34], or polyglycolic acid (PGA). Herein, we present an overview of natural and synthetic polymers that have been recently studied for the development of functional hydrogel systems for DFU treatment and repair.

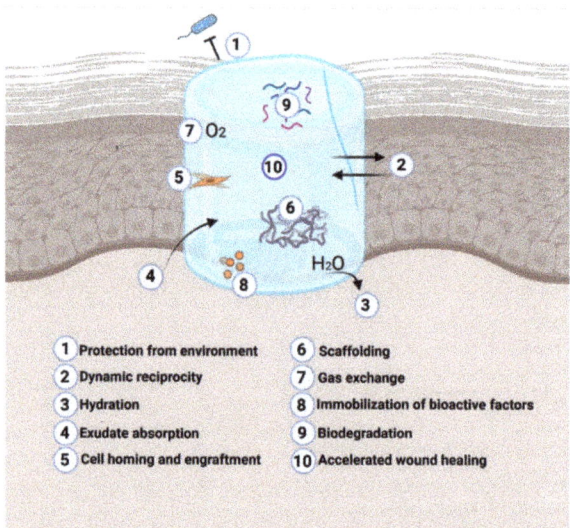

Figure 2. Therapeutic effects of hydrogel dressings during wound healing. Created with BioRender.com on 1 June 2022.

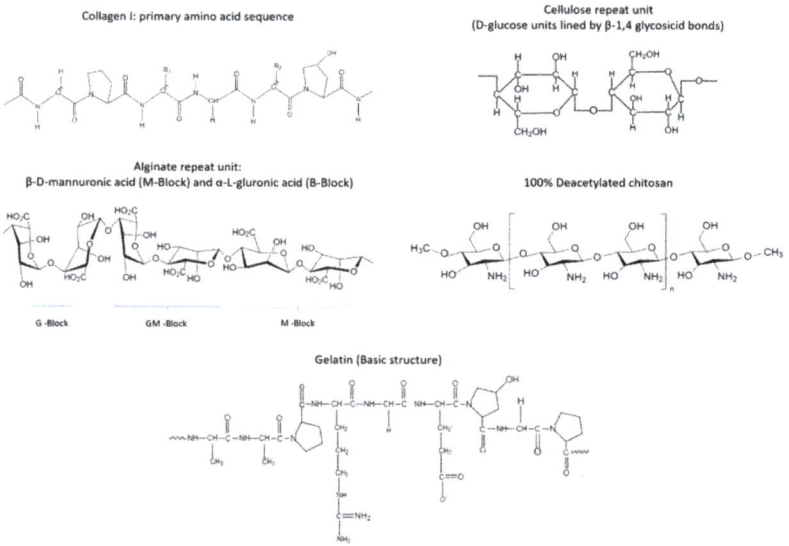

Figure 3. Natural polymers used for the design of hydrogel dressings for DFU treatment.

4.1. Natural DFU Hydrogel Dressings

4.1.1. Collagen and Gelatin Hydrogels

In the context of diabetic wound healing, the therapeutic effect of hydrogels produced from natural polymers such as collagen, alginate or chitosan has been broadly examined in the last few years (Figure 3). Collagen is made of polypeptide chains usually produced by fibroblasts, and it is the most abundant protein found in connective tissue, blood vessels and skin, providing mechanical strength and elasticity [35]. Although there are several types of collagens, collagens type I, II and III are the most predominantly found [36]. Because of its versatile biological activity as well as its biodegradability (thermal or enzymatic) and mechanical properties, collagen has been extensively explored for biomaterial development applications. Specifically, in wound care applications, collagen has been demonstrated to positively impact wound healing by providing structural support and stimulating growth factor release that promotes tissue regeneration. Moreover, its role in MMP deactivation helps maintain favorable biochemical balance and moisture levels in the wound [35].

For instance, Lei et al., evaluated the ability of collagen hydrogels to promote angiogenesis and wound healing in diabetic rat models. Towards this, full-thickness wounds were induced and externally treated with a collagen hydrogel loaded with recombinant human epidermal growth factors. After 14 days of treatment, the rats treated with the fabricated hydrogel showed significantly smaller wound areas, indicating accelerated injury regeneration, relative to the group that did not receive a hydrogel treatment. Regarding angiogenesis, the proposed hydrogel dressing supported endogenous collagen synthesis, as well as the formation of vascularized scar tissue [36]. Moreover, alternative approaches for the development of collagen-based wound dressings have incorporated additional natural or synthetic polymers as well as plant extracts with medicinal properties to enhance the mechanical properties and healing effects of collagen alone in vivo [37,38]. Specifically, the studies performed by Karri et al. reported the fabrication of a hybrid collagen-alginate scaffold that was impregnated with curcumin-loaded chitosan nanoparticles (NPs). Curcumin is a diferuloylmethane found in the Curcuma longa plant, with known antimicrobial, anti-inflammatory, and antioxidant properties [39]. Using a diabetic rat model, the authors evaluated the wound healing effect of the collagen-alginate scaffolds, which were produced by a freeze-drying method. It was observed that the wounds treated with the proposed scaffold displayed improved wound contraction and reduced inflammation, as well as the formation of thick granulation tissue. Positive wound remodeling was further confirmed by proper collagen deposition and epithelial tissue formation [39]. In addition, collagen/chitosan hydrogels have been used as treatment for patients with neuropathic DFU, leading to higher healing rates compared to standard treatment (saline gauze) [38]. Specifically, a Kaplan–Meier survival analysis showed that ulcers treated with the collagen/chitosan hydrogels healed completely after around 12 weeks, whereas ulcers treated with the control took about 21 weeks to heal. In a similar manner, Nilforoushzadeh et al. found that the use of pre-vascularized collagen/fibrin hydrogels on DFU patients resulted in increased hypodermis thickness and an accelerated wound healing process [37].

On the other hand, gelatin is a natural polymer obtained from the thermal hydrolysis of collagen, and it has also been studied for the design of wound dressings (Figure 3). Relative to collagen, gelatin exhibits lower antigenicity [40] and higher degradability, as it can be cleaved by most proteases. Moreover, gelatin is a less expensive alternative to collagen, which preserves the bioactivity required to elicit proper tissue regeneration responses [41,42]. For example, Hsu et al. developed a biodegradable composite hydrogel by combining gelatin and hyaluronic acid (HA) through chemical crosslinking with 1-(3-dimethylaminopropyl)-3-ethylcarbodiimide hydrochloride (EDC chemistry) followed by lyophilization and impregnation with recombinant thrombomodulin (rhTM) rich in epidermal growth factor-like domains. The therapeutic effect of the synthesized hydrogel was examined in full-thickness diabetic wounds on a murine model. After 10 days of treatment, it was observed that a hydrogel formulation containing 0.1% HA and 9 µg rhTM, which also displayed 70% rhTM loading capacity, significantly stimulated collagen production, re-epithelialization, and the formation of granulation tissue, as well as angiogenic response [43].

Further research on gelatin-based dressings, such as the study reported by Zheng et al., explored the use of this biopolymer for the design of injectable scaffolds that could be fabricated in situ to produce dressings that could successfully fill the irregular volume of a lesion through minimally invasive surgical treatments. Specifically, the authors prepared an injectable hydrogel precursor from a gelatin solution loaded with tannic acid, a bactericidal agent, which could then be gelled at body temperature upon blending with gellan through electrostatic complexation [44]. The self-recovery property of this gelatin hydrogel relied on the reversibility of its crosslinks, which rendered a versatile material with in situ moldability to wound beds and self-healing traits. Additionally, the microporous structure of the resulting gel scaffold facilitated cell adhesion and migration, as well as tannic acid delivery into the wound, favoring wound healing and providing a rapid wound closure effect [44].

4.1.2. Chitosan Hydrogels

Chitosan is a natural amino polysaccharide obtained from the alkaline N-deacetylation of chitin [45], which has been used in a wide variety of applications as a scaffold in tissue engineering, especially for the fabrication of hydrogel systems due to its low cost and toxicity, ease of production, antimicrobial activity, biocompatibility, and hemostatic potential [46,47] (Figure 3). In addition to this, chitosan is a highly versatile material that can be chemically modified and processed in different forms, such as nanofibers, nanoparticles, nanocomposites, hydrogels, and films, among others. This flexibility favors the development of scaffolds, hydrogels, and bioactive films with tunable properties [48,49]. Overall, chitosan has yielded promising results when implemented for pharmaceutical applications, particularly as drug delivery vehicles with unique potential for wound care applications, as is the case of DFUs [50]. In this sense, chitosan has been recently shown to be an attractive material for the development of hydrogels through additive manufacturing technologies to produce porous 3D scaffolds with specific shapes and bioactivity towards the successful reconstruction of complex tissues.

For instance, Intini et al., employed a 3D printing approach for the fabrication of chitosan scaffolds with controlled and reproducible structures, as well as mechanical properties that resembled those of native skin [46]. The hydrogel precursor solution consisted of 6% w/v chitosan in acetic acid, containing 290 mM sucrose to increase hydrophilicity and elasticity of the final scaffold [51]. Upon successful printing, the hydrogel was gelled with 8% w/v KOH solution and further tested on a rat diabetic wound model. Although no significant differences in wound closure rate were observed relative to a commercial dressing, the proposed hydrogel appeared to provide an enhanced antibacterial effect, a result that was ascribed to chitosan's intrinsic antimicrobial properties [46]. Moreover, Thangavel and coworkers [52] investigated the influence of natural dressings based on pure chitosan and L-glutamic acid on the wound healing process in diabetic rats. The authors hypothesized that since L-glutamic acid is a known precursor of proline synthesis, its delivery in the wound could stimulate collagen synthesis, and thus, skin regeneration. The proposed hydrogel was prepared through physical crosslinking in 1 M NaOH, in the presence of glycerol as a plasticizer. Comparison of rats treated with gauze dressing, pure chitosan hydrogel, or chitosan hydrogel + 1% L-glutamic acid, indicated the significant therapeutic effect of the latter, as evidenced by complete re-epithelialization after 16 days of treatment (2 cm × 2 cm wounds), in addition to the enhanced levels of collagen deposition and crosslinking that were observed for this group. Furthermore, the positive results from CD31 staining revealed that, indeed, L-glutamic acid promoted new blood vessel formation, whereas a reduction in CD68 levels after 12 days of treatment indicated that the chitosan-L-glutamic acid hydrogel helped regulate the inflammatory response, and therefore, contributed to proper wound healing. Table 1 presents a compilation of recent in vivo studies that have addressed DFU dressing development using natural polymeric sources.

4.1.3. Alginate Hydrogels

Alginate is a natural block copolymer containing blocks of (1-4)-linked β-D-mannuronic acid (M) and α-L-guluronic acid (G) monomers, typically arranged in three different forms: sequential M residues (MMMMMM), sequential G residues (GGGGGG) and regions of combined M-G units (GMGMGM) (Figure 3). The physical and mechanical properties of alginate can vary depending on how these block sequences are distributed [53].

Alginate has been widely explored for tissue repair applications involving cell, enzyme, and peptide immobilization, as well as the fabrication of supporting matrices for drug delivery systems, because of its intrinsic ability to allow platelets and erythrocytes to adhere and trigger wound healing responses [54]. Moreover, its hydrophilic nature provides a moist environment that helps reduce scar formation and improves wound re-epithelialization while minimizing bacterial infection. In this regard, Tellechea et al., examined the design of alginate hydrogels (Figure 4) for the encapsulation of human umbilical cord-derived outgrowth endothelial cells (OECs) and the neuropeptides Substance P and Neurotensin, two biological agents with anti-inflammatory attributes. The results from testing in a diabetes-induced mouse model revealed that the alginate hydrogels enabled sustained neuropeptide release during the 10-day period of the study, promoting a remarkable wound size reduction of around 80%. Moreover, the incorporation of vascular endothelial growth factor (VEGF) into the proposed hydrogel further enhanced and accelerated wound healing, as evidenced by a 40% wound size reduction after 4 days of treatment, which suggested a synergistic effect supporting of neovascularization [55].

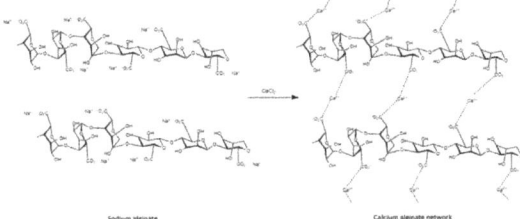

Figure 4. Ionic crosslinking of sodium alginate to produce calcium alginate hydrogels.

More recently, the use of polydeoxyribonucleotide (PDRN)-based therapies for wound healing has increasingly gained attention, since PDRN has been shown to effectively induce angiogenesis, collagen synthesis, as well as anti-inflammatory responses in the wound bed [56]. In this sense, the studies conducted by Shin et al. demonstrated that PDRN-loaded alginate hydrogels prepared by ionic crosslinking yielded full re-epithelialization of a murine diabetic wound after 14 days of treatment, a process characterized by reduced inflammation and controlled PDRN release [57]. In a similar fashion, therapeutic treatments with edaravone have been elucidated to accelerate DFU wound healing. Edaravone is an antioxidant agent with free-radical scavenging properties that reduces the amount of reactive oxygen species (ROS) present in the wound bed, while preventing vascular endothelial cell injury and inducing angiogenesis in diabetic patients [58].

Table 1. Recent in vivo studies on natural hydrogels for diabetic wound healing.

Ref.	Year	Polymer Source and Material	Additional Functional Component (s)	Synthesis Method	Diabetic Model	Therapeutic Effect
[59]	2022	Sodium alginate (2% w/v) hydrogel	Deferoxamine (560 µg/mL) and copper nanoparticles (200 µg/mL)	Ionic crosslinking with 0.1 M $CaCl_2$	STZ-induced male C57BL/6 mice	Enhanced antimicrobial effect as well as angiogenesis by upregulation of HIF-1α and VEGF. Reduced inflammatory response.

Table 1. Cont.

Ref.	Year	Polymer Source and Material	Additional Functional Component (s)	Synthesis Method	Diabetic Model	Therapeutic Effect
[60]	2021	Sodium alginate/pectin (5% w/w) composite hydrogel	Simvastatin (20 mg/mL)	Combined solvent-casting and ionic crosslinking with 0.5% w/v $CaCl_2$	STZ-induced male Wistar rats	Accelerated wound closure due to the presence of SIM, which promoted re-epithelialization, fibroblast proliferation and collagen production.
[61]	2021	Silk nanofiber (1 wt%) hydrogel	Deferoxamine (60 µM and 120 µM)	Concentration-dilution-thermal incubation method	STZ-induced male Sprague–Dawley rats	Enhanced collagen deposition and wound healing rates: 80% on day 14, and 100% on day 21. Improved angiogenic and inflammatory responses.
[57]	2020	Sodium alginate (2–5% w/v) hydrogel	Polydeoxyribonucleotide (100 µg/mL)	Ionic crosslinking with $CaCO_3$	Male C57BLKS/J-db/db mice	Improved re-epithelialization and granulation tissue formation. Increased collagen production and angiogenesis.
[62]	2019	Sodium alginate (1.5% w/w) hydrogel	Edaravone-loaded Eudragit nanoparticles	Ionic crosslinking with 0.5% w/w $CaCl_2$	STZ-induced male C57BL/6 mice	Downregulation of reactive oxygen species favored accelerated wound healing.
[43]	2019	Gelatin (4% w/v)/hyaluronic acid (0.1% w/v) composite hydrogel	Thrombomodulin (9 and 15 µg)	Chemical crosslinking (0.05% EDC)	STZ-induced male C57BL/6JNarl mice	Enhanced granulation tissue formation, re-epithelialization, collagen deposition, and angiogenesis.
[46]	2018	Chitosan (6% w/v) hydrogel	D-(+) raffinose pentahydrate (290 mM)	Physical crosslinking in alkaline solution (8% w/v KOH)	STZ-induced female Wistar rats	Increased bactericidal effect and accelerated wound healing.
[52]	2017	Chitosan (2 wt. %) hydrogel	L-glutamic acid (0.25–1.0%)	Physical crosslinking in alkaline solution (1 M NaOH)	STZ-induced male Wistar rats	Enhanced re-epithelialization, collagen deposition, and neovascularization.
[39]	2016	Chitosan/starch hydrogel	Chitosan silver nanoparticles (5 ppm Ag in 6.9 mg/mL chitosan)	Reductive alkylation crosslinking	Alloxan-induced male albino rats	Significantly improved wound healing rate. Increased bactericidal response.
[39]	2016	Collagen/alginate (50/50 w/w) hydrogel	Curcumin (1 wt.%)-loaded chitosan nanoparticles	Chemical crosslinking (EDC)	STZ-induced male Wistar rats	Reduced inflammation. Enhanced cell adhesion and proliferation. Accelerated wound closure.
[63]	2016	Gelatin/hydroxyphenyl propionic acid hydrogel (5 wt%)	Interleukin-8 (IL-8, 0.5 µg/mL) or macrophage inflammatory protein-3α (MIP-3α, 1 µg/mL)	Horseradish peroxidase (HRP)-catalyzed cross-linking	STZ-induced male ICR mice	Increased cell infiltration, re-epithelialization, neovascularization, and collagen deposition.

Fan et al., presented a pioneer work in which alginate hydrogels were used to entrap Eudragit nanoparticles loaded with edaravone. This composite dressing was used on mouse diabetic wounds, and the results after two weeks of treatment indicated ROS downregulation and wound closure, relative to the controls, i.e., free edaravone application (without dressing) or untreated diabetic wounds [62]. Nonetheless, the authors found that significantly positive outcomes could only be attained at low edaravone concentrations in the hydrogel, because high doses of this drug removed too much ROS, which appeared detrimental for wound healing.

4.2. Synthetic and Semi-Synthetic DFU Hydrogel Dressings

The design of a successful DFU dressing inherently demands the fulfillment of biofunctional as well as structural requirements. Despite the outstanding bioactivity exhibited by natural hydrogels, their mechanical performance and reproducibility still need to be considerably improved. In this context, synthetic polymers are more versatile materials, which display unique properties that can be tailored through controlled physical or chemical processes. As opposed to their natural counterparts, synthetic polymers can be more easily produced at an industrial scale, and their tunability allows for their use in different configurations that favor desired tissue growth [64]. Moreover, because their hydrophilic and hydrophobic domains can be tightly controlled, synthetic polymers can also exhibit more homogeneous structures and higher water absorption capacities. Furthermore, semi-synthetic hydrogels are prepared from natural and synthetic polymers (Figure 5), which combined, result in composite matrices with enhanced biological and material properties. While the natural element provides the desired bioactivity and biocompatibility, the synthetic component facilitates control of hydrogel properties, such as biodegradability as well as mechanical and swelling behavior. This synergistic effect has strengthened the therapeutic potential of hydrogel dressings, and for this reason, semi-synthetic hydrogel systems have been at the core of DFU dressing research for the past few years (Table 2).

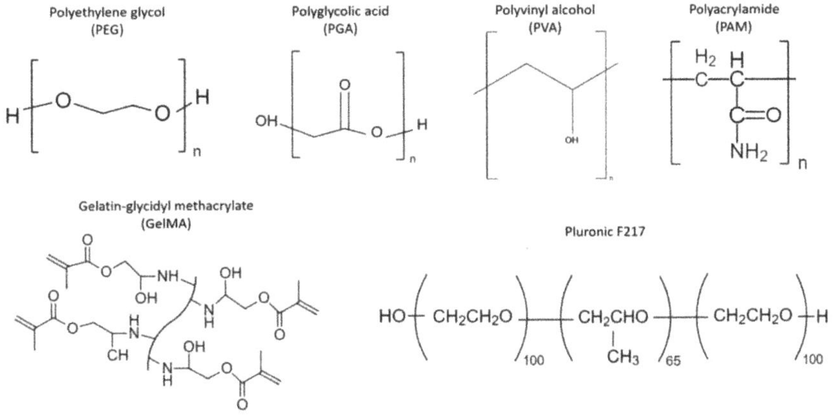

Figure 5. Synthetic polymers used for the design of hydrogel dressings for DFU treatment.

4.2.1. Polyethylene Glycol (PEG)-Based Systems

PEG-based hydrogels have been employed for the fabrication of biological systems due to their exceptional biocompatibility and resistance to protein adhesion [65] (Figure 5). The introduction of functional groups can yield PEG derivatives such as polyethylene glycol diacrylate (PEGDA) (Figure 6) and polyethylene glycol dimethacrylate (PEGDM), which can be chemically crosslinked to produce durable matrices [66,67] that allow tethering or embedding of biomolecules that promote proper tissue regeneration [68] (Figure 6).

Table 2. Recent in vivo studies on synthetic/semi-synthetic hydrogels for diabetic wound healing.

Ref.	Year	Polymer Source and Material	Additional Functional Component(s)	Synthesis Method	Diabetic Model	Therapeutic Effect
[69]	2022	Methacrylate gelatin (GelMA)/PEGDA microneedle patch	Tazarotene (1 mg/10 mL) and exosomes (100 µg/mL) from human umbilical vein endothelial cells (HUVECs)	Photopolymerization with lithium acylphosphinate salt (LAP 0.05%, g/mL)	STZ-induced male C57BL mice	Accelerated collagen deposition, epithelial regeneration, and angiogenesis.
[70]	2022	PLGA-PEG-PLGA thermosensitive hydrogel	Copper-based MOFs containing curcumin and metformin hydrochloride	Thermal gelation	STZ-induced male BALB/c mice	Significant reduction of oxidative stress; enhanced cell migration, neovascularization, and collagen formation.
[71]	2022	Injectable hydrogel prepared from 4,5-imidazoledicarboxylic acid, zinc nitrate hexahydrate, deferoxamine mesylate and glucose oxidase (GOX)	Deferoxamine mesylate (DFO, 8.3 µg/mL)	Phase-transfer-mediated programmed GOX loading	STZ-induced female BALB/c mice	Release of zinc ions and DFO resulted in enhanced antibacterial and angiogenic effect. Significant induction of re-epithelialization and collagen deposition.
[72]	2022	PDLLA-PEG-PDLLA (25% w/v) thermosensitive hydrogel	Prussian blue nanoparticles (PBNPs, 333.3 µg/mL and 666.6 µg/mL)	Thermal gelation	STZ-induced C57BL/6J mice	Decreased reactive oxygen species (ROS) production as well as IL-6 and TNF-α levels. PBNPs dose-dependent accelerated wound closure.
[73]	2022	pH/glucose dual responsive hydrogel prepared from dihydrocaffeic acid and L-Arginine co-grafting chitosan, phenylboronic acid and benzaldehyde difunctional polyethylene glycol-co-poly(glycerol sebacic acid) and polydopamine-coated graphene oxide (GO)	Metformin (2 mg/mL)	Double dynamic bond of a Schiff-base and phenylboronate ester	STZ-induced Sprague–Dawley rats	Antibacterial properties, tissue adhesion, hemostasis. Decreased inflammatory response. Increased wound closure ratio, re-epithelialization, and regeneration of blood vessels.
[74]	2022	Supramolecular guanosine-quadruplex hydrogel	Hemin (0.36–0.54 mg) and GOX (0.125–0.5 mg)	Self-assembled gelation	STZ-induced male BABL/c mice	Significantly faster antibacterial effect, relative to commercial antibiotic. Decreased glucose concentration in the wound.
[75]	2022	Chitosan/polyvinyl acetate heterogeneous hydrogel	Human epidermal growth factor (EGF)-loaded nanoparticles, polyhexamethylene biguanide, and perfluorocarbon emulsions	Freeze-thaw cycling	STZ-induced Sprague-Dawley rats	High antibacterial and anti-inflammatory effect. Enhanced collagen production and wound closure efficiency, relative to commercial dressings.
[76]	2022	Double-layered GelMA-PLL hydrogel	Vascular endothelial growth factor (VEGF)-mimetic peptide	Photopolymerization with lithium acylphosphinate salt (LAP)	STZ-induced Sprague-Dawley rats	Enhanced antibacterial and wound-healing effect. Improved collagen deposition, angiogenesis, and re-vascularization.
[77]	2022	Oxidized alginate/platelet-rich plasma (PRP) fibrin hydrogel		Ionic crosslinking with 1.22 M $CaSO_4 \cdot 2H_2O$	Male db/db (BKS.Cg-m+/+Leprdb/J) mice	Accelerated wound maturation and closure.

Table 2. Cont.

Ref.	Year	Polymer Source and Material	Additional Functional Component(s)	Synthesis Method	Diabetic Model	Therapeutic Effect
[78]	2022	PTFE/PU patch	Calcium-alginate hydrogel microparticles (MPs) containing *Chlorella vulgaris* and *Bacillus licheniformis*	MP encapsulation in porous PTFE membrane (inner lining) and a transparent PU film (back lining)	STZ-induced mice	Enhanced wound healing effect: 50% wound closure by day 3, and full wound closure on day 12.
[79]	2021	GelMA (10% w/v) hydrogel	Bioactive glass particles loaded with cerium (1% w/v)	Photopolymerization with LAP (0.1% w/v)	STZ-induced Sprague-Dawley rats	Wound closure of almost 95% on day 21.
[80]	2021	Cecropin-modified hyaluronic acid/oxidized dextran/PRP composite hydrogel		Schiff base reaction	Male db/db mice	Accelerated healing of infected wounds. Shortened inflammatory stage. Increased mature collagen content.
[81]	2021	Pluronic F-127 (20%) hydrogel	Ag nanocubes with virus-like mesoporous silica containing gentamicin	Thermal gelation	STZ-induced Kunming mice	Bacterial infected wounds were fully healed by day 20, with enhanced collagen production.
[82]	2021	Carboxymethyl chitosan/poly(dextran-g-4-formylbenzoic acid) hydrogel	Peptide-modified PAN nanofibers	Schiff base reaction	Diabetic ICR mice	Enhanced antibacterial and angiogenic effect. Reduced inflammatory response. Wound closure > 96% at day 14.
[83]	2021	Hydroxyl propyl methyl cellulose (2% w/w) hydrogel	Lipid nanoparticles loaded with Valsartan (1% w/w)	Thermal gelation	STZ-induced male Sprague-Dawley rats	Enhanced healing response mediated through COX-2, NF-κB, NO, TGF-β, MMPs and VEGF pathways.
[84]	2021	Polyacrylamide/gelatin/ε-polylysine composite hydrogel		Free-radical polymerization	STZ-induced male Sprague-Dawley rats	Increased granulation tissue formation, collagen deposition, and angiogenesis. Enhanced antibacterial effect.
[85]	2021	Conductive hydrogel made from acrylamide-co-polymerized ionic liquid (VAPimBF4) and konjac glucomannan		Chemical crosslinking (EDC/NHS chemistry)	STZ-induced male Kunming mice	Highest wound healing rate when coupled with electrical stimulation. Increased antibacterial effect, Col-1 production, and new vessel growth.
[86]	2021	N-carboxyethyl chitosan/adipic acid dihydrazide pH responsive hydrogel	Insulin (0.67 U/mL)	Crosslinking by hyaluronic acid-aldehyde (imine and acylhydrazone bonds)	STZ-induced male Sprague-Dawley rats	Significant reduction of glucose levels in the wound. Decreased inflammation phase. Increased granulation tissue formation, collagen deposition, and re-epithelialization.
[87]	2021	Quaternized chitosan/oxidized hyaluronic acid self-healing hydrogel	α-lipoic acid-loaded MOFs	Schiff base reaction	STZ-induced male Sprague-Dawley rats	Increased collagen deposition, cell proliferation and neovascularization. Accelerated wound healing.
[88]	2021	Chitosan/polyvinyl acetate hydrogel	Chitosan nanoparticles loaded with human epidermal growth factor (EGF, 60 μg/mL) and Ag^+ ions	Freeze-thaw cycling	STZ-induced Sprague-Dawley rats	Remarkable antibacterial effect. Enhanced tissue maturation and wound closure: 40% on day 3, and 97% on day 14.

Table 2. Cont.

Ref.	Year	Polymer Source and Material	Additional Functional Component(s)	Synthesis Method	Diabetic Model	Therapeutic Effect
[89]	2021	Pluronic F-127 (20% w/v) hydrogel	Sodium ascorbyl phosphate (400 µM) and Wharton's jelly mesenchymal stem cells (WJMSC)	Thermal gelation	STZ-induced male Sprague-Dawley rats	Shortened inflammatory response. Improved dermis regeneration, neovascularization, and collagen deposition.
[90]	2020	Supramolecular hydrogel based on ferrocene, hyaluronic acid, β-cyclodextrin, and rhein		Intermolecular $\pi-\pi$ interactions and hydrogen bonds	STZ-induced C57 mice	Anti-inflammatory properties of rhein facilitated transition from the inflammatory phase into the proliferation phase, thus, favoring normal wound healing.
[91]	2020	Pluronic F-127 hydrogel	Exosomes derived from human umbilical cord MSCs (300 µg/mL)	Thermal gelation	STZ-induced male Sprague-Dawley rats	Increased vascularization of wound granulation tissue, shortening wound healing time. Improved epithelial regeneration.
[92]	2020	4-carboxybenzaldehyde-PEG/glycol chitosan/silk fibroin/PRP self-healing hydrogel		Schiff base reaction + crosslinking with 10% calcium gluconate	STZ-induced Sprague-Dawley rats	Enhanced angiogenesis, re-epithelialization, nerve repair, and wound healing rate.
[93]	2020	Chitosan/polyurethane hydrogel membrane	Bone marrow mononuclear cells (1×10^6) injected into the edge of the wound prior to hydrogel application	Chemical crosslinking (urea/urethane bonds)	STZ-induced female Wistar rats	Hemostatic and anti-inflammatory effect. Wound closure > 90% after 14 days.
[94]	2020	Stimuli-responsive supramolecular hydrogel made from polyvinyl alcohol/N-carboxyethyl chitosan/agarose/Ag nanowires		Hydrogen bonding	STZ-induced male Sprague-Dawley rats	Bactericidal effect. Promoted angiogenesis and collagen deposition. Accelerated wound healing rate.
[95]	2020	Poly(N-isopropylacrylamide)/poly(γ-glutamic acid) hydrogel (20 mg/mL total concentration)	Superoxide dismutase (2 mg/mL)	Thermal gelation	STZ-induced male Sprague-Dawley rats	Reduced inflammation. Enhanced collagen production and epidermal formation.
[96]	2020	N-carboxyethyl chitosan (7.5% w/v)/adipic acid dihydrazide (7.5% w/v)/hyaluronic acid-aldehyde (5% w/v) composite hydrogel	Encapsulated bone marrow mesenchymal stem cells (2×10^5)	Crosslinking by hyaluronic acid-aldehyde (imine and acylhydrazone bonds)	STZ-induced male Sprague-Dawley rats	Inhibited chronic inflammation. Enhanced formation of granulation tissue, cell proliferation and neovascularization.
[97]	2020	γ-polyglutamic acid (0.5 g/mL) hydrogel	Human cell-free fat extract (5 mg/mL)	Chemical crosslinking (EDC/NHS chemistry)	Male BKS-Leprem2Cd479/Nju mice	Improved cell proliferation, collagen deposition and continuous epidermal formation. Significant angiogenesis.
[98]	2020	Silk fibroin-polyvinyl pyrrolidone hydrogel	L-carnosine and curcumin	Mixing/vortex shearing (physical crosslinking)	STZ-induced BALB/c mice	Significant antibacterial and anti-inflammatory effect. Enhanced wound healing.

Table 2. Cont.

Ref.	Year	Polymer Source and Material	Additional Functional Component(s)	Synthesis Method	Diabetic Model	Therapeutic Effect
[99]	2020	[2-(methacryloloxy)ethyl] dimethyl-(3-sulfopropyl) ammonium hydroxide (SBMA)/2-Hydroxyethyl methacrylate (HEMA) and 3-[[2-(Methacryloyloxy)ethyl] dimethylammonio] propionate (CBMA)/HEMA zwitterionic cryogels	miRNA146a-conjugated cerium oxide nanoparticles	Free-radical polymerization with 13.6 mg/mL ammonium persulfate	Db/Db female mice	Full wound healing on day 14. Downregulation of inflammatory markers. Increased Col1a2 expression.
[100]	2020	Polyvinyl alcohol (8% w/v)/sodium alginate (1% w/v) hydrogel	Green tea polyphenol nanoparticles	Ionic crosslinking ($CaCl_2$, 100 µg/mL) and hydrogen bonding	STZ-induced female Sprague-Dawley rats	Increased granulation tissue formation and collage deposition. Accelerated wound healing.
[101]	2019	Chitosan/PEG hydrogel	Ag nanoparticles	Chemical crosslinking with glutaraldehyde	Alloxan-induced rabbits	Increased bactericidal effect. Accelerated re-epithelialization and collagen deposition. Full wound closure on day 14.
[102]	2018	A5G81-modified poly(polyethylene glycol cocitric acid-co-N-isopropylacrylamide) hydrogel		Thermal gelation	B6.BKS(D)-Lepr[db]/J mice	Enhanced re-epithelialization and granulation tissue formation. Faster wound closure than that achieved with commercial dressings.
[103]	2018	Hyperbranched PEG/thiolated hyaluronic acid injectable hydrogel	Encapsulated adipose-derived stem cells (2.5 × 10^6 cell/mL)	thiol-ene click reaction	STZ-induced male Sprague-Dawley rats	Reduced inflammatory response. Increased angiogenesis and re-epithelialization.
[104]	2017	Polymethyl methacrylate/Polyvinyl alcohol hydrogel particles	Collagen, Ag nanowires, and chitosan	UV photocrosslinking (Irgacure 184)	STZ-induced male Wistar rats STZ-induced Landrace pigs	Enhanced collagen production and epidermal cell migration. Reduced inflammatory response.
[105]	2017	Phenylboronic-modified chitosan (1.2 wt%)/poly(vinyl alcohol) (0.6 wt%)/benzaldehyde-capped PEG (0.6 wt%) hydrogel	Insulin (0.3 wt%) and L929 fibroblasts (1.2 × 10^6 cells/mL)	Schiff base reaction	STZ-induced Sprague-Dawley rats	Improved control of glucose levels in wound. Increased neovascularization and collagen deposition. Enhanced wound closure rate.
[106]	2016	Sodium carboxymethyl-cellulose/propylene glycol hydrogel	*Blechnum orientale* extract (2–4% wt)	Hydrogen bonding	STZ-induced male Sprague-Dawley rats	Significant bactericidal and antioxidative effect. Enhanced re-epithelialization, fibroblast proliferation, collagen synthesis, and angiogenesis.
[107]	2016	Gelatin methacrylate (15% w/v) hydrogel	Desferrioxamine (1% w/v)	UV photocrosslinking with Irgacure 2959 (0.5% w/v)	STZ-induced male Sprague-Dawley rats	Accelerated neovascularization, granulation tissue remodeling, and wound closure.

In the context of wound healing, Xu et al., fabricated an in situ polymerizable hydrogel encapsulating adipose-derived stem cells to promote skin regeneration in murine diabetic wounds. By means of a reversible addition−fragmentation chain-transfer (RAFT) polymerization mechanism, hyperbranched multi-acrylated PEG macromers were synthesized and further combined with thiolated hyaluronic acid to produce a hydrogel via thiol-ene click reaction [103]. The results from in vivo testing revealed the therapeutic effect of the proposed system, as evidenced by accelerated healing rates and granular tissue formation, in addition to reduced inflammation.

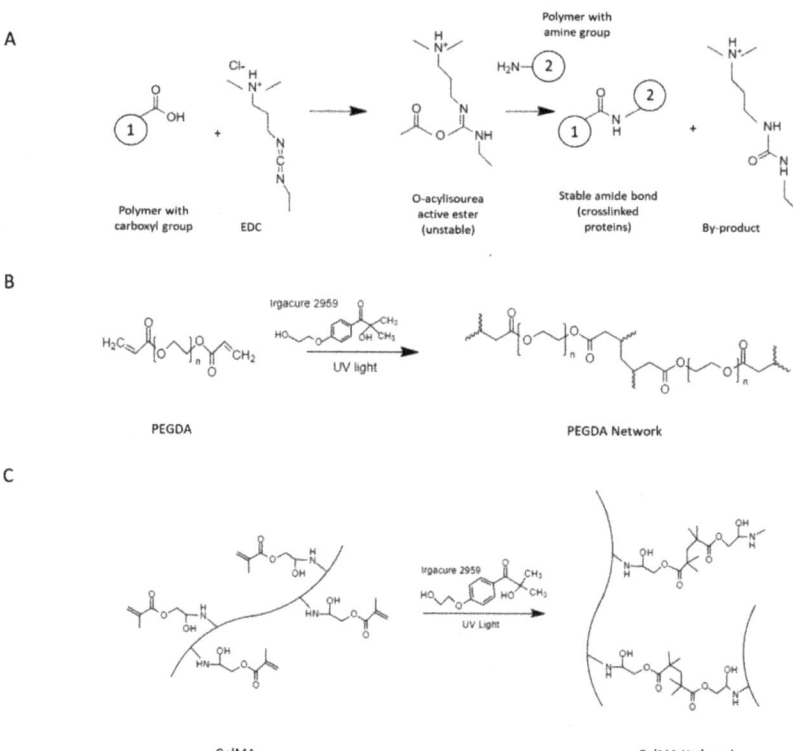

Figure 6. Hydrogel fabrication through chemical crosslinking via: (**A**) 1-(3-dimethylaminopropyl)-3-ethylcarbodiimide hydrochloride (EDC chemistry), and (**B**) and (**C**) UV photopolymerization using Irgacure 2959.

Several factors contribute to the chronicity associated with DFUs, being oxidative stress one of the most detrimental elements for diabetic wound healing. For this reason, a lot of research efforts are currently focusing on the development of materials with scavenging properties that can protect the wound bed from the mitochondrial damage inflicted by reactive oxygen species (ROS). In this sense, the recent work conducted by Xu et al. involved the fabrication of a poly(d,L-lactide)-PEG-poly(d,L-lactide) (PLGA-PEG-PLGA) hydrogel for the encapsulation and controlled release of Prusian blue nanoparticles (PBNPs), a synthetic material that can mimic the ROS scavenging activity of naturally occurring antioxidant enzymes, such as peroxidase, catalase, and superoxide dismutase [72]. The thermosensitive properties of the prepared hydrogel allowed in situ gelation at physiological temperature. Furthermore, in vivo experiments on murine diabetic wound models demonstrated the ROS scavenging features of PBNPs, as confirmed by dihydroethidium staining. This in turn favored an anti-inflammatory response that mitigated chronic inflammation and thus, enhanced proper wound healing and neovascular remodeling, relative to the control treatment without PBNPs.

Moreover, the studies conducted by Yang and coworkers aimed at the design of a multifunctional hydrogel system that simultaneously addressed the regulation of blood glucose and oxidative stress levels in diabetic ulcers [70]. To this end, the authors developed copper-based metal-organic frameworks (MOFs) by a solvothermal method, which could be loaded with metformin hydrochloride (MH), a hypoglycemic and hydrophilic drug, as well as curcumin, a natural lipophilic agent that has been previously shown to have ROS scavenging properties [108]. The resulting MOFs exhibited ~65% drug entrapment efficiency and were further embedded into PLGA-PEG-PLGA thermosensitive hydrogels for the controlled and sustained release of the therapeutic agents, using a type 1 diabetes wound model. It was observed that blood glucose levels showed a continuously declining trend from day 5 until the last day of treatment (day 20), whereas the antioxidative capacity of the hydrogel was validated through a sensitive chemiluminescent of L-012. Wound analysis revealed that the proposed composite dressing promoted cell proliferation, neovascularization, and collagen formation, reaching >90% wound closure after 20 days.

Finally, Pluronics or Poloxamers are a family of polyethylene oxide/polypropylene oxide/polyethylene oxide (PEO-PPO-PEO) triblock copolymers which are non-toxic and FDA-approved for biomedical use (Figure 5). The PPO portion acts as the hydrophobic component, whereas the PEO segments provide hydrophilicity to the material. Within this family of polymers, Pluronic F-127 hydrogels have increasingly drawn attention, as they can be safely used as synthetic platforms for the delivery of drugs and cells through in situ gelation at body temperature [109–111]. For instance, Jiao et al. recently evaluated the effect of Wharton's jelly mesenchymal stem cells (WJMSCs) encapsulated in a Pluronic F-127 hydrogel (20% w/v) on the healing process of a type 2 diabetes wound model [89]. Prior to encapsulation, the hydrogel precursor was supplemented with sodium ascorbyl phosphate to increase WJMSC viability. The results showed that the proposed hydrogel dressing significantly stimulated collagen deposition, re-epithelialization, and anti-inflammatory response 14 days post-transplantation. Moreover, WJMSC engraftment in the wound was also evidenced, which confirmed the immunomodulatory and regenerative potential of this hydrogel system for DFU treatment.

4.2.2. Polyvinyl Alcohol (PVA)-Based Systems

PVA is a hydrophilic polymer with biocompatible, biodegradable, and semi-crystalline features that have been of great interest in the biomedical field [112] (Figure 5). PVA can be physically crosslinked by repeated freeze-thaw cycles (cryogelation), or chemically crosslinked using glutaraldehyde or epichlorohydrin. Both synthesis routes render PVA hydrogels with remarkable hydrophilicity and chemical stability. In cryogelation, as the temperature of the PVA solution decreases, more ice crystals are formed, which in turn causes the PVA polymer chains to be "pushed out" as impurities, increasing their concentration in the unfrozen liquid phase. This phenomenon enables physical interactions among PVA molecules and the eventual creation of PVA crystallites. Such crystallites are not soluble in water and, therefore, remain after thawing, so that at the end of repeated freeze-thaw cycles, this yields a porous hydrogel [113,114]. In this regard, Takei et al. used a cryogelation technique to incorporate PVA into a chitosan-gluconic acid conjugate (GC) hydrogel, towards improving the mechanical properties, water retention capacity, and resistance to enzyme degradation of pure GC hydrogels. The produced dressings were also loaded with basic fibroblast growth factor (bFGF) and then evaluated on partial-thickness diabetic wounds. The authors reported that hydrogel implantation contributed to a high expression of inflammatory cells and the secretion of chemical mediators that accelerated wound healing, avoiding the formation of fibrotic tissue [115].

Furthermore, PVA-based hydrogels have been recently tested for the stabilization and sustained release of natural biomolecules with antioxidant potential, such as green tea-derived polyphenols (TPs). In their studies, Chen et al., explored the fabrication of a PVA-alginate composite system, whose synthesis involved a combination of ionic crosslinking and hydrogen bonding to encapsulate TPs nanoparticles [100]. The therapeutic effect of the proposed dressing was evaluated on excisional linear wounds in diabetic rats for seven days. As reported by the authors, the designed hydrogel system promoted higher collagen deposition and maturation as well as granulation tissue formation, relative to the controls (hydrogel without TPs and hydrogel with powdered TPs). Finally, an analysis of the PI3K/AKT protein pathway, which has been associated with cell proliferation, angiogenesis and glucose metabolism [116,117] allowed to confirm that TPs helped regulate such pathway, which translated into a significant improvement of the wound healing process [118].

Another remarkable example of the development of multifunctional semi-synthetic hydrogels for DFU applications is the study conducted by Zhao et al. The authors set out to create a dressing that enabled tissue restoration, while allowing controlled in situ degradation, avoiding the neo tissue damage and patient discomfort that is commonly observed upon dressing replacement [94]. In brief, stimuli-responsive 3D supramolecular structures were fabricated through hydrogen bonding using PVA, agarose, N-carboxyethyl chitosan and silver nanowires. The incorporation of $Na_2B_4O_7$ into the hydrogel precursor solution granted the final material the ability to degrade upon exposure to heat, pH reduction, ultrasound, or diol-containing molecules, such as glucose. This biodegradability was attributed to the presence of reversible boronate-based reactions between PVA and $Na_2B_4O_7$, which can be affected by the factors mentioned above. Furthermore, in vivo testing of this scaffold revealed strong bactericidal traits, and after 20 days of treatment, foot wounds in diabetic rats were almost fully healed, versus approximately 35% of wound closure observed in the control group (untreated).

4.2.3. Gelatin-Based Systems

Gelatin has been studied as a delivery vehicle for the transport of bioactive agents, such as cytokines and growth factors secreted by mesenchymal stem cells, which contribute to tissue repair and play an important role in wound healing by regulating the wound environment and stimulating cell migration, adhesion, and proliferation. This results in enhanced chemotactic responses, which are very much needed in a diseased cell environment such as DFUs. For instance, Yoon et al. developed an enzyme-catalyzed gelatin hydrogel as a sprayable dressing material capable of delivering cell-attracting chemokines for diabetic wound healing [63]. The gelatin-hydroxyphenyl propionic acid hydrogels were prepared in situ by crosslinking the phenol moieties with horseradish peroxidase and H_2O_2, while simultaneously loading the material with chemotactic cytokines (IL-8, 5 µg/mL or MIP-3α, 10 µg/mL). Diabetic mice were treated with the fabricated dressing, and the observed results showed accelerated cell infiltration into the wound area, which in turn promoted wound healing, neovascularization, and increased collagen deposition.

Additionally, the studies reported by Chen et al. described the development of a gelatin methacrylate (GelMA) system (Figures 5 and 6) using in situ photo-crosslinking to obtain hydrogels coupled with desferrioxamine (DFO), an angiogenic drug that supports wound healing. As evidenced by the results, the proposed hydrogel dressings enhanced blood vessel network formation, which in turn ensured sufficient oxygen and nutrient delivery into the wound area. Ultimately, these GelMA scaffolds promoted granulation and epithelial tissue generation, as well as a favorable cell microenvironment that led to effective diabetic wound treatment [107]. More recently, Yuan and coworkers designed a novel microneedle patch system based on a GelMA/PEGDA composite matrix [69]. The rationale behind their studies was that the incorporation of gel microneedles on the inner side of the wound dressing could significantly improve transdermal delivery of target biological agents, without the pain associated with local injection approaches. Towards this end, tazarotene, a medication known for its ability to stimulate collagen production and angiogenesis, was mixed with the GelMA/PEGDA precursor solution by means of grafted isocyanatoethyl acrylate-modified β-cyclodextrin (β-CD-AOI$_2$) to improve the water solubility of the drug. After this, exosomes (100 µg/mL) derived from human umbilical vein endothelial cells (HUVECs) were added to the mixture and further poured into a vacuum mold for microneedle formation. Finally, the hydrogel dressing was obtained through photopolymerization with lithium acylphosphinate salt (0.05%, g/mL). Following in vitro assays that confirmed the sustained and controlled delivery of exosomes and tazarotene, in vivo testing was performed on a murine diabetic wound model. The results showed that the hydrogel loaded with both, tazarotene and HUVEC-exos, elicited a significantly enhanced wound healing response in terms of cell migration, angiogenesis, and wound closure rate, relative to the blank hydrogel (pure GelMA/PEGDA), or the hydrogel loaded with tazarotene only. The presence of PEGDA in the matrix ensured increased crosslink density, and thus, desired mechanical stability for the developed system.

4.2.4. Chitosan-Based Systems

Because of its remarkable antibacterial, anti-inflammatory, and hemostatic features [119–121], chitosan has been continuously employed as a base material for the synthesis of semi-synthetic hydrogel dressings with advanced biological and physical properties. For instance, the experiments conducted by Mirhamed et al., demonstrated the potential of chitosan-derived hydrogels for preserving the bioactivity of key growth factors, ensuring their sustained release into a wound. Through carbodiimide conjugation reaction, recombinant human epidermal growth factor (rhEGF) was conjugated to sodium carboxymethyl chitosan (NaCMCh), with the aim of protecting rhEGF from the proteolytic degradation that occurs in chronic wound environments. These conjugates were further embedded into a NaCMCh/polyvinylpyrrolidone (PVP) hydrogel. The authors were able to show that the developed dressing maintained rhEGF functionality over time, and its controlled release and degradation significantly increased rhEGF therapeutic effect. Moreover, in vivo evaluation of this hydrogel in diabetic rats proved that it indeed stimulated fast wound closure, re-epithelialization, as well as granulation tissue formation [122].

Additional studies on chitosan-derived 3D wound dressings have sought to improve the mechanical behavior of the synthesized material as well. In this sense, Kamel et al. prepared hydrogel-based scaffolds from the combination of hydroxypropyl methylcellulose (HPMC) with high-viscosity chitosan. This mixture was then enriched with pioglitazone HCl (PG), a known anti-diabetic drug. The hydrogel was freeze-dried and further used to treat diabetic wounds in rats. The authors highlighted the improved flexibility and adhesiveness of the scaffold, which could be attributed to the presence of HPMC. Moreover, relative to the control (untreated) group, the positive effect of the HPMC-chitosan dressing was evidenced by a faster wound healing rate, as well as a reduction in matrix metalloproteinase-9 (MMP9) and tumor necrosis factor (TNF-α) levels in the wound [123].

More recently, Lee and coworkers proposed the fabrication of a novel multifunctional DFU hydrogel system to simultaneously address several issues of the diabetic wound environment, and thus, provide a more effective treatment alternative [75]. By means of a freeze-thaw cycling method (-20 to 25 °C), the authors prepared heterogeneous chitosan/polyvinyl acetate dressings containing the following: (1) a perfluorocarbon emulsion to favor oxygen transport into the wound; (2) EGF-loaded nanoparticles to stimulate cell adhesion and proliferation; and (3) polyhexamethylene biguanide (PHMB), an antimicrobial and antiviral agent to prevent infection. Material characterization of the synthesized hydrogel revealed that the nanoparticles and the emulsion also served as mechanical reinforcement elements for the scaffold, whereas in vitro evaluation demonstrated a remarkable bactericidal effect against *S. aureus* and *S. epidermidis*. Upon performing in vivo experiments in diabetic rats, the developed composite dressing offered an improved wound closure rate and reduced inflammatory response, relative to treatment with gauze or a commercially available dressing (HeraDerm).

4.2.5. Plasma-Based Systems

Blood plasma has been reported to be a promising material source for the design of bioactive matrices, due primarily to its biochemical composition and its autologous nature [124]. Blood plasma contains platelets, glycoproteins, immunoglobulins, and growth factors that swiftly adhere and aggregate at the injury site, preventing excessive bleeding and subsequent vascular complications [125]. Specifically, platelets play a key role in the hemostatic system since they are degranulated of their secretory granules, which grants them the capacity to stimulate tissue repair, blood vessel regeneration, and cell differentiation [126].

Platelet-rich plasma (PRP) is a type of autologous plasma with high platelet concentration [127], which is considered a natural gelling sealant that works as a drug vehicle for the release of essential platelet-derived growth factors, such as: platelet-derived growth factor (PDGF) (AA, BB, and AB isomers), transforming growth factor-β (TGF-β), platelet factor 4 (PF4), interleukin-1 (IL-1), platelet-derived angiogenesis factor (PDAF), vascular endothelial growth factor (VEGF), epidermal growth factor (EGF), epithelial cell growth factor (ECGF), and insulin-like growth factor (IGF), among others [128,129]. These factors support the wound healing process by (i) stimulating cells around the wound area to participate in the restoration of the affected tissue; and (ii) promoting cell differentiation through the control of cytokine release and macrophage activity, reducing inflammation and accelerating tissue regeneration in chronic environments such as DFUs.

Current studies have reported the successful incorporation of PRP into the development of hydrogel dressings for DFU treatment. For instance, Qian et al., designed a composite system from silk fibroin, glycol chitosan (GCTS), and PRP, which was eventually tested in a type 2 diabetes rat model. In short, PEG was modified with 4-carboxybenzaldehyde (CB) through the esterification reaction of carboxyl with hydroxyl to produce CBPEG, which could crosslink GCTS through a reversible Schiff base reaction [92]. Moreover, silk fibroin was added to provide mechanical reinforcement to the scaffold. The authors evaluated the in vitro and in vivo performance of this composite hydrogel, finding that it allowed for a more controlled degradation of the PRP present in the matrix, as well as a more sustained delivery of the bioactive factors associated with PRP when compared to a pure PRP hydrogel. This was attributed to the resistance to enzymatic degradation of chitosan and silk fibroin. Altogether, the biological properties of this novel hydrogel resulted in accelerated wound healing, characterized by increased collagen deposition, granulation tissue formation, re-epithelialization, and nerve regeneration.

More recently, the experiments conducted by Wei et al., focused on the creation of a hydrogel with enhanced bactericidal features towards regulating chronic inflammation in diabetic wound environments [80]. In this sense, dextran and hyaluronic acid (HA) were chosen as the base materials, which underwent chemical modification prior to hydrogel synthesis. Essentially, cecropin, an antimicrobial peptide (AMP), was tethered to HA through EDC/NHS chemistry (amide bond formation) to yield an HA-AMP polymer, while dextran was oxidized (ODEX) with sodium periodate ($NaIO_4$) and diethylene glycol to add aldehyde groups. Final hydrogel fabrication was achieved through a Schiff's base reaction between the aldehyde groups in ODEX and the amino groups in HA-AMP and PRP. A control hydrogel was also prepared, which contained everything but PRP. The authors observed that both types of hydrogels exhibited significant antibacterial properties against *P. aeruginosa* and *S. aureus* in infected diabetic wounds (mouse) and also promoted downregulation of inflammatory factors, such as TGF-β1, TNF-α, IL-6 and IL-1β. Nonetheless, the hydrogel containing PRP exhibited greater VEGF expression, confirming the angiogenic potential of PRP. Furthermore, following a similar oxidation procedure to modify alginate, Garcia-Orue and coworkers set out to develop a biodegradable matrix that could deliver the relevant growth factors present in PRP to promote wound healing without the need to change the dressing after use [77]. Oxidation degree, and thus, the biodegradability of the final hydrogel could be tightly controlled by adjusting the amount of $NaIO_4$ utilized for alginate oxidation. Oxidized alginate (2.5%)-PRP hydrogels were produced by ionic crosslinking using CaSO$_4$, along with a control hydrogel lacking PRP. They were then applied to wounds in diabetic mice for 15 days. Interestingly, even though wound healing appeared enhanced in both sample groups (hydrogels with or without PRP), no significant differences were found between them. This result disagreed with what had been previously observed during in vitro testing with human fibroblasts and keratinocytes, in which the PRP-containing hydrogels displayed increased cell adhesion and proliferation, relative to the control hydrogel. The authors attributed this inconsistency between in vitro and in vivo tests to the fact that the PRP used was of human origin, and therefore, had a positive effect when tested with human cells, but such effect could have been hindered by its xenogeneic nature when applied to mice.

5. Conclusions

Diabetic foot ulcers are a serious health condition that poses a significant risk to the patient and a great financial burden to the healthcare system. Impaired healing in DFUs is a complex and multifactorial problem since diabetic wounds are characterized by high levels of blood glucose and reactive oxygen species, infection, abnormal inflammatory response, as well as hindered angiogenesis. Currently, the development of a successful clinical treatment for DFUs remains a challenge for the scientific community. Nonetheless, extensive research efforts are continuously made towards elucidating the key combination of elements that could allow for the fabrication of a truly effective DFU treatment.

Due to their inherent properties, hydrogels are attractive materials for the design of optimal DFU wound dressing systems. Hydrogels are 3D platforms that not only can provide skin-like mechanical properties and a moist environment for the wound bed, but also can serve as vehicles for the delivery of biological cues that are essential for adequate wound repair. In this context, natural and synthetic polymers have been explored for the preparation of pure or composite hydrogel dressings that can additionally transport a myriad of relevant bioactive agents, such as antibiotics, antioxidants, stem cells, PRP, growth factors, and insulin, among others. Even though natural polymers such as chitosan, alginate, and gelatin can intrinsically help ameliorate the defective conditions of the diabetic wound environment, there currently appears to be an agreement on the need to design semi-synthetic hydrogels that can strategically take advantage of the traits offered by natural polymers (biodegradability, biocompatibility, bioactivity, and antibacterial properties), while incorporating the unique versatility and reproducibility provided by synthetic polymers. For instance, nano-hydrogels embedded with quercetin and oleic acid have recently shown promising results in a clinical trial, by reducing wound healing time without adverse effects in 28 DFU patients [130].

Furthermore, because DFUs are a multifactorial problem, the growing body of scientific data in the field seems to indicate that the answer to this problem could be the development of multifunctional hydrogel dressings; in other words, 3D polymeric systems that can simultaneously address several issues of the diabetic wound, ideally in the absence of additional chemicals (crosslinkers) that could compromise the cytocompatibility of the scaffold. Moreover, based on the cumulative results from multiple in vivo studies recently reported, the delivery of antioxidants and antibacterial agents, as well as bioactivity to stimulate desired cell infiltration and proliferation, appear to be determinant factors in the success of clinically viable DFU hydrogel treatments.

Author Contributions: All authors listed have made a substantial, direct, and intellectual contribution to the work. Conceptualization: V.A.S.-D. and S.M.B.-B. Writing original draft: V.R.G.-A., A.V.P.-M. and J.E.Á.-Q. Editing: V.R.G.-A. and S.M.B.-B. Critical feedback to the final version: V.R.G.-A., S.M.B.-B. and V.A.S.-D. All authors have read and agreed to the published version of the manuscript.

Funding: This research was funded by Minciencias, contract number 880 of 2019 and grant number 124777757672".

Conflicts of Interest: The authors declare no conflict of interest.

References

1. American Diabetes Association. Diagnosis and Classification of Diabetes Mellitus. *Diabetes Care* **2013**, *37* (Suppl. S1), S81–S90.
2. International Diabetes Federation. *IDF Diabetes Atlas*; International Diabetes Federation: Brussels, Belgium, 2019.
3. Schmidt, A.M. Highlighting Diabetes Mellitus: The Epidemic Continues. *Arterioscler. Thromb. Vasc. Biol.* **2018**, *38*, e1–e8. [CrossRef] [PubMed]
4. American Diabetes Association. Economic Costs of Diabetes in the U.S. in 2017. *Diabetes Care* **2018**, *41*, 917–928. [CrossRef] [PubMed]
5. Fard, A.S.; Esmaelzadeh, M.; Larijani, B. Assessment and treatment of diabetic foot ulcer. *Int. J. Clin. Pract.* **2007**, *61*, 1931–1938. [CrossRef]
6. Hinchliffe, R.J.; Andros, G.; Apelqvist, J.; Bakker, K.; Friederichs, S.; Lammer, J.; Lepantalo, M.; Mills, J.; Reekers, J.; Shearman, C.P.; et al. A systematic review of the effectiveness of revascularization of the ulcerated foot in patients with diabetes and peripheral arterial disease. *Diabetes/Metab. Res. Rev.* **2012**, *28*, 179–217. [CrossRef]
7. Lebrun, E.; Tomic-Canic, M.; Kirsner, R.S. The role of surgical debridement in healing of diabetic foot ulcers. *Wound Repair Regen.* **2010**, *18*, 433–438. [CrossRef]
8. Dhivya, S.; Padma, V.V.; Santhini, E. Wound dressings—A review. *Biomedicine* **2015**, *5*, 22. [CrossRef]
9. Han, G.; Ceilley, R. Chronic Wound Healing: A Review of Current Management and Treatments. *Adv. Ther.* **2017**, *34*, 599–610. [CrossRef]
10. Oliveira, A.; Simões, S.; Ascenso, A.; Reis, C.P. Therapeutic advances in wound healing. *J. Dermatol. Treat.* **2020**, *33*, 2–22. [CrossRef]
11. Dumville, J.C.; O'Meara, S.; Deshpande, S.; Speak, K. Hydrogel dressings for healing diabetic foot ulcers. *Cochrane Database Syst. Rev.* **2013**, *7*, CD009101. [CrossRef]
12. Morton, L.M.; Phillips, T.J. Wound healing update. *Semin. Cutan. Med. Surg.* **2012**, *31*, 33–37. [CrossRef] [PubMed]
13. David, V.A.S.; Güiza-Argüello, V.R.; Arango-Rodríguez, M.L.; Sossa, C.L.; Becerra-Bayona, S.M. Decellularized Tissues for Wound Healing: Towards Closing the Gap Between Scaffold Design and Effective Extracellular Matrix Remodeling. *Front. Bioeng. Biotechnol.* **2022**, *10*, 821852. [CrossRef] [PubMed]
14. Park, J.E.; Barbul, A. Understanding the role of immune regulation in wound healing. *Am. J. Surg.* **2004**, *187*, 11S–16S. [CrossRef]
15. Goldberg, S.R.; Diegelmann, R.F. Wound healing primer. *Surg. Clin. N. Am.* **2010**, *90*, 1133–1146. [CrossRef] [PubMed]
16. Landén, N.X.; Li, D.; Ståhle, M. Transition from inflammation to proliferation: A critical step during wound healing. *Cell. Mol. Life Sci.* **2016**, *73*, 3861–3885. [CrossRef] [PubMed]
17. Stevens, L.J.; Page-McCaw, A. A secreted MMP is required for reepithelialization during wound healing. *Mol. Biol. Cell* **2012**, *23*, 1068–1079. [CrossRef] [PubMed]
18. Caley, M.P.; Martins, V.L.; O'Toole, E.A. Metalloproteinases and Wound Healing. *Adv. Wound Care* **2015**, *4*, 225–234. [CrossRef] [PubMed]
19. Perez-Favila, A.; Martinez-Fierro, M.L.; Rodriguez-Lazalde, J.G.; Cid-Baez, M.A.; Zamudio-Osuna, M.D.J.; Martinez-Blanco, M.d.R.; Mollinedo-Montaño, F.E.; Rodriguez-Sanchez, I.P.; Castañeda-Miranda, R.; Garza-Veloz, I. Current Therapeutic Strategies in Diabetic Foot Ulcers. *Medicina* **2019**, *55*, 714. [CrossRef]
20. Guo, S.; DiPietro, L.A. Factors Affecting Wound Healing. *J. Dent. Res.* **2010**, *89*, 219–229. [CrossRef]
21. Martin, P.; Nunan, R. Cellular and molecular mechanisms of repair in acute and chronic wound healing. *Br. J. Dermatol.* **2015**, *173*, 370–378. [CrossRef]
22. Straccia, M.C.; D'Ayala, G.G.; Romano, I.; Oliva, A.; Laurienzo, P. Alginate Hydrogels Coated with Chitosan for Wound Dressing. *Mar. Drugs* **2015**, *13*, 2890–2908. [CrossRef]
23. Koehler, J.; Verheyen, L.; Hedtrich, S.; Brandl, F.P.; Goepferich, A.M. Alkaline poly(ethylene glycol)-based hydrogels for a potential use as bioactive wound dressings. *J. Biomed. Mater. Res. Part A* **2017**, *105*, 3360–3368. [CrossRef]
24. Kumar, A.; Wang, X.; Nune, K.C.; Misra, R. Biodegradable hydrogel-based biomaterials with high absorbent properties for non-adherent wound dressing. *Int. Wound J.* **2017**, *14*, 1076–1087. [CrossRef]
25. Ribeiro, M.; Espiga, A.; Silva, D.; Henriques, J.; Ferreira, C.; Silva, J.; Pires, E.; Chaves, P. Development of a new chitosan hydrogel for wound dressing. *Wound Repair Regen.* **2009**, *17*, 817–824. [CrossRef]
26. Paladini, F.; Meikle, S.T.; Cooper, I.R.; Lacey, J.; Perugini, V.; Santin, M. Silver-doped self-assembling di-phenylalanine hydrogels as wound dressing biomaterials. *J. Mater. Sci. Mater. Med.* **2013**, *24*, 2461–2472. [CrossRef]
27. Qu, J.; Zhao, X.; Liang, Y.; Zhang, T.; Ma, P.X.; Guo, B. Antibacterial adhesive injectable hydrogels with rapid self-healing, extensibility and compressibility as wound dressing for joints skin wound healing. *Biomaterials* **2018**, *183*, 185–199. [CrossRef]
28. Helary, C.; Zarka, M.; Giraud-Guille, M.M. Fibroblasts within concentrated collagen hydrogels favour chronic skin wound healing. *J. Tissue Eng. Regen. Med.* **2011**, *6*, 225–237. [CrossRef]
29. Thi, P.L.; Lee, Y.; Tran, D.L.; Thi, T.T.H.; Kang, J.I.; Park, K.M.; Park, K.D. In situ forming and reactive oxygen species-scavenging gelatin hydrogels for enhancing wound healing efficacy. *Acta Biomater.* **2020**, *103*, 142–152. [CrossRef]
30. Basu, A.; Celma, G.; Strømme, M.; Ferraz, N. In Vitro and in Vivo Evaluation of the Wound Healing Properties of Nanofibrillated Cellulose Hydrogels. *ACS Appl. Bio Mater.* **2018**, *1*, 1853–1863. [CrossRef]

31. Kang, J.I.; Park, K.M.; Park, K.D. Oxygen-generating alginate hydrogels as a bioactive acellular matrix for facilitating wound healing. *J. Ind. Eng. Chem.* **2019**, *69*, 397–404. [CrossRef]
32. Xie, Y.; Liao, X.; Zhang, J.; Yang, F.; Fan, Z. Novel chitosan hydrogels reinforced by silver nanoparticles with ultrahigh mechanical and high antibacterial properties for accelerating wound healing. *Int. J. Biol. Macromol.* **2018**, *119*, 402–412. [CrossRef]
33. Jafari, A.; Hassanajili, S.; Azarpira, N.; Karimi, M.B.; Geramizadeh, B. Development of thermal-crosslinkable chitosan/maleic terminated polyethylene glycol hydrogels for full thickness wound healing: In vitro and in vivo evaluation. *Eur. Polym. J.* **2019**, *118*, 113–127. [CrossRef]
34. Sung, J.H.; Hwang, M.R.; Kim, J.O.; Lee, J.H.; Kim, Y.I.; Kim, J.H.; Chang, S.W.; Jin, S.G.; Kim, J.A.; Lyoo, W.S.; et al. Gel characterisation and in vivo evaluation of minocycline-loaded wound dressing with enhanced wound healing using polyvinyl alcohol and chitosan. *Int. J. Pharm.* **2010**, *392*, 232–240. [CrossRef] [PubMed]
35. Haycocks, S.; Chadwick, P.; Cutting, K. Collagen matrix wound dressings and the treatment of DFUs. *J. Wound Care* **2013**, *22*, 369–375. [CrossRef] [PubMed]
36. Lei, J.; Chen, P.; Li, Y.; Wang, X.; Tang, S. Collagen hydrogel dressing for wound healing and angiogenesis in diabetic rat models. *Int. J. Clin. Exp. Med.* **2017**, *10*, 16319–16327.
37. Nilforoushzadeh, M.A.; Sisakht, M.M.; Amirkhani, M.A.; Seifalian, A.M.; Banafshe, H.R.; Verdi, J.; Nouradini, M. Engineered skin graft with stromal vascular fraction cells encapsulated in fibrin–collagen hydrogel: A clinical study for diabetic wound healing. *J. Tissue Eng. Regen. Med.* **2019**, *14*, 424–440. [CrossRef] [PubMed]
38. Djavid, G.E.; Tabaie, S.M.; Tajali, S.B.; Totounchi, M.; Farhoud, A.; Fateh, M.; Ghafghazi, M.; Koosha, M.; Taghizadeh, S. Application of a collagen matrix dressing on a neuropathic diabetic foot ulcer: A randomised control trial. *J. Wound Care* **2020**, *29*, S13–S18. [CrossRef] [PubMed]
39. Karri, V.V.S.R.; Kuppusamy, G.; Talluri, S.V.; Mannemala, S.S.; Kollipara, R.; Wadhwani, A.D.; Mulukutla, S.; Raju, K.R.S.; Malayandi, R. Curcumin loaded chitosan nanoparticles impregnated into collagen-alginate scaffolds for diabetic wound healing. *Int. J. Biol. Macromol.* **2016**, *93*, 1519–1529. [CrossRef] [PubMed]
40. Zhao, X.; Lang, Q.; Yildirimer, L.; Lin, Z.Y.; Cui, W.; Annabi, N.; Ng, K.W.; Dokmeci, M.R.; Ghaemmaghami, A.M.; Khademhosseini, A. Photocrosslinkable Gelatin Hydrogel for Epidermal Tissue Engineering. *Adv. Health Mater.* **2016**, *5*, 108–118. [CrossRef]
41. Jahan, I.; George, E.; Saxena, N.; Sen, S. Silver-Nanoparticle-Entrapped Soft GelMA Gels as Prospective Scaffolds for Wound Healing. *ACS Appl. Bio Mater.* **2019**, *2*, 1802–1814. [CrossRef]
42. Ikada, Y.; Tabata, Y. Protein release from gelatin matrices. *Adv. Drug Deliv. Rev.* **1998**, *31*, 287–301.
43. Hsu, Y.-Y.; Liu, K.-L.; Yeh, H.-H.; Lin, H.-R.; Wu, H.-L.; Tsai, J.-C. Sustained release of recombinant thrombomodulin from cross-linked gelatin/hyaluronic acid hydrogels potentiate wound healing in diabetic mice. *Eur. J. Pharm. Biopharm.* **2018**, *135*, 61–71. [CrossRef]
44. Zheng, Y.; Liang, Y.; Zhang, D.; Sun, X.; Liang, L.; Li, J.; Liu, Y.-N. Gelatin-Based Hydrogels Blended with Gellan as an Injectable Wound Dressing. *ACS Omega* **2018**, *3*, 4766–4775. [CrossRef]
45. Kumar, M.N.R. A review of chitin and chitosan applications. *React. Funct. Polym.* **2000**, *46*, 1–27. [CrossRef]
46. Intini, C.; Elviri, L.; Cabral, J.; Mros, S.; Bergonzi, C.; Bianchera, A.; Flammini, L.; Govoni, P.; Barocelli, E.; Bettini, R.; et al. 3D-printed chitosan-based scaffolds: An in vitro study of human skin cell growth and an in-vivo wound healing evaluation in experimental diabetes in rats. *Carbohydr. Polym.* **2018**, *199*, 593–602. [CrossRef]
47. Jiang, T.; James, R.; Kumbar, S.G.; Laurencin, C.T. Chitosan as a Biomaterial: Structure, Properties, and Applications in Tissue Engineering and Drug Delivery. In *Natural and Synthetic Biomedical Polymers*; Kumbar, S.G., Laurencin, C.T., Deng, M., Eds.; Elsevier: Oxford, UK, 2014; pp. 91–113.
48. Thapa, B.; Narain, R. Mechanism, current challenges and new approaches for non viral gene delivery. In *Polymers and Nanomaterials for Gene Therapy*; Narain, R., Ed.; Woodhead Publishing: Sawston, UK, 2016; pp. 1–27.
49. Ahmad, M.; Zhang, B.; Manzoor, K.; Ahmad, S.; Ikram, S. Chitin and chitosan-based bionanocomposites. In *Bionanocomposites*; Zia, K.M., Jabeen, F.; Anjum, M.N., Ikram, S., Eds.; Elsevier: Amsterdam, The Netherlands, 2020; pp. 145–156.
50. El-Naggar, M.Y.; Gohar, Y.M.; Sorour, M.A.; Waheeb, M.G. Hydrogel Dressing with a Nano-Formula against Methicillin-Resistant Staphylococcus aureus and Pseudomonas aeruginosa Diabetic Foot Bacteria. *J. Microbiol. Biotechnol.* **2016**, *26*, 408–420. [CrossRef]
51. Bettini, R.; Romani, A.A.; Morganti, M.M.; Borghetti, A.F. Physicochemical and cell adhesion properties of chitosan films prepared from sugar and phosphate-containing solutions. *Eur. J. Pharm. Biopharm.* **2008**, *68*, 74–81. [CrossRef]
52. Thangavel, P.; Ramachandran, B.; Chakraborty, S.; Kannan, R.; Lonchin, S.; Muthuvijayan, V. Accelerated Healing of Diabetic Wounds Treated with L-Glutamic acid Loaded Hydrogels Through Enhanced Collagen Deposition and Angiogenesis: An In Vivo Study. *Sci. Rep.* **2017**, *7*, 1–15. [CrossRef]
53. Sun, J.; Tan, H. Alginate-Based Biomaterials for Regenerative Medicine Applications. *Materials* **2013**, *6*, 1285–1309. [CrossRef]
54. Sharmeen, S.; Rahman, M.S.; Islam, M.M.; Islam, M.S.; Shahruzzaman, M.; Mallik, A.K.; Haque, P.; Rahman, M.M. Application of polysaccharides in enzyme immobilization. In *Functional Polysaccharides for Biomedical Applications*; Maiti, S., Jana, S., Eds.; Woodhead Publishing: Sawston, UK, 2019; pp. 357–395.
55. Tellechea, A.; Silva, E.A.; Min, J.; Leal, E.; Auster, M.E.; Pradhan-Nabzdyk, L.; Shih, W.; Mooney, D.; Veves, A. Alginate and DNA Gels Are Suitable Delivery Systems for Diabetic Wound Healing. *Int. J. Low. Extrem. Wounds* **2015**, *14*, 146–153. [CrossRef]

56. Kwon, T.-R.; Han, S.W.; Kim, J.H.; Lee, B.C.; Kim, J.M.; Hong, J.Y.; Kim, B.J. Polydeoxyribonucleotides Improve Diabetic Wound Healing in Mouse Animal Model for Experimental Validation. *Ann. Dermatol.* **2019**, *31*, 403–413. [CrossRef] [PubMed]
57. Shin, D.Y.; Park, J.-U.; Choi, M.-H.; Kim, S.; Kim, H.-E.; Jeong, S.-H. Polydeoxyribonucleotide-delivering therapeutic hydrogel for diabetic wound healing. *Sci. Rep.* **2020**, *10*, 1–14. [CrossRef] [PubMed]
58. Naito, R.; Nishinakamura, H.; Watanabe, T.; Nakayama, J.; Kodama, S. Edaravone, a free radical scavenger, accelerates wound healing in diabetic mice. *Wounds* **2014**, *26*, 163–171. [PubMed]
59. Li, S.; Wang, X.; Chen, J.; Guo, J.; Yuan, M.; Wan, G.; Yan, C.; Li, W.; Machens, H.-G.; Rinkevich, Y.; et al. Calcium ion cross-linked sodium alginate hydrogels containing deferoxamine and copper nanoparticles for diabetic wound healing. *Int. J. Biol. Macromol.* **2022**, *202*, 657–670. [CrossRef]
60. Rezvanian, M.; Ng, S.-F.; Alavi, T.; Ahmad, W. In-vivo evaluation of Alginate-Pectin hydrogel film loaded with Simvastatin for diabetic wound healing in Streptozotocin-induced diabetic rats. *Int. J. Biol. Macromol.* **2021**, *171*, 308–319. [CrossRef]
61. Ding, Z.; Zhang, Y.; Guo, P.; Duan, T.; Cheng, W.; Guo, Y.; Zheng, X.; Lu, G.; Lu, Q.; Kaplan, D.L. Injectable Desferrioxamine-Laden Silk Nanofiber Hydrogels for Accelerating Diabetic Wound Healing. *ACS Biomater. Sci. Eng.* **2021**, *7*, 1147–1158. [CrossRef]
62. Fan, Y.; Wu, W.; Lei, Y.; Gaucher, C.; Pei, S.; Zhang, J.; Xia, X. Edaravone-Loaded Alginate-Based Nanocomposite Hydrogel Accelerated Chronic Wound Healing in Diabetic Mice. *Mar. Drugs* **2019**, *17*, 285. [CrossRef]
63. Yoon, D.S.; Lee, Y.; Ryu, H.A.; Jang, Y.; Lee, K.-M.; Choi, Y.; Choi, W.J.; Lee, M.; Park, K.M.; Park, K.D.; et al. Cell recruiting chemokine-loaded sprayable gelatin hydrogel dressings for diabetic wound healing. *Acta Biomater.* **2016**, *38*, 59–68. [CrossRef]
64. Gunatillake, P.A.; Adhikari, R.; Gadegaard, N. Biodegradable synthetic polymers for tissue engineering. *Eur. Cell Mater.* **2003**, *5*, 1–16. [CrossRef]
65. Alcantar, N.A.; Aydil, E.S.; Israelachvili, J.N. Polyethylene glycol–coated biocompatible surfaces. *J. Biomed. Mater. Res.* **2000**, *51*, 343–351. [CrossRef]
66. Beamish, J.A.; Zhu, J.; Kottke-Marchant, K.; Marchant, R.E. The effects of monoacrylated poly(ethylene glycol) on the properties of poly(ethylene glycol) diacrylate hydrogels used for tissue engineering. *J. Biomed. Mater. Res. Part A* **2009**, *92*, 441–450. [CrossRef]
67. Zhu, J. Bioactive modification of poly(ethylene glycol) hydrogels for tissue engineering. *Biomaterials* **2010**, *31*, 4639–4656. [CrossRef]
68. Sokic, S.; Christenson, M.; Larson, J.; Papavasiliou, G. In Situ Generation of Cell-Laden Porous MMP-Sensitive PEGDA Hydrogels by Gelatin Leaching. *Macromol. Biosci.* **2014**, *14*, 731–739. [CrossRef]
69. Yuan, M.; Liu, K.; Jiang, T.; Li, S.; Chen, J.; Wu, Z.; Li, W.; Tan, R.; Wei, W.; Yang, X.; et al. GelMA/PEGDA microneedles patch loaded with HUVECs-derived exosomes and Tazarotene promote diabetic wound healing. *J. Nanobiotechnol.* **2022**, *20*, 1–18. [CrossRef]
70. Yang, L.; Liang, F.; Zhang, X.; Jiang, Y.; Duan, F.; Li, L.; Ren, F. Remodeling microenvironment based on MOFs-Hydrogel hybrid system for improving diabetic wound healing. *Chem. Eng. J.* **2021**, *427*, 131506. [CrossRef]
71. Yang, J.; Zeng, W.; Xu, P.; Fu, X.; Yu, X.; Chen, L.; Leng, F.; Yu, C.; Yang, Z. Glucose-responsive multifunctional metal–organic drug-loaded hydrogel for diabetic wound healing. *Acta Biomater.* **2021**, *140*, 206–218. [CrossRef]
72. Xu, Z.; Liu, Y.; Ma, R.; Chen, J.; Qiu, J.; Du, S.; Li, C.; Wu, Z.; Yang, X.; Chen, Z.; et al. Thermosensitive Hydrogel Incorporating Prussian Blue Nanoparticles Promotes Diabetic Wound Healing via ROS Scavenging and Mitochondrial Function Restoration. *ACS Appl. Mater. Interfaces* **2022**, *14*, 14059–14071. [CrossRef]
73. Liang, Y.; Li, M.; Yang, Y.; Qiao, L.; Xu, H.; Guo, B. pH/Glucose Dual Responsive Metformin Release Hydrogel Dressings with Adhesion and Self-Healing via Dual-Dynamic Bonding for Athletic Diabetic Foot Wound Healing. *ACS Nano* **2022**, *16*, 3194–3207. [CrossRef]
74. Li, Y.; Su, L.; Zhang, Y.; Liu, Y.; Huang, F.; Ren, Y.; An, Y.; Shi, L.; van der Mei, H.C.; Busscher, H.J. A Guanosine-Quadruplex Hydrogel as Cascade Reaction Container Consuming Endogenous Glucose for Infected Wound Treatment—A Study in Diabetic Mice. *Adv. Sci.* **2022**, *9*, 2103485. [CrossRef]
75. Lee, Y.-H.; Lin, S.-J. Chitosan/PVA Hetero-Composite Hydrogel Containing Antimicrobials, Perfluorocarbon Nanoemulsions, and Growth Factor-Loaded Nanoparticles as a Multifunctional Dressing for Diabetic Wound Healing: Synthesis, Characterization, and In Vitro/In Vivo Evaluation. *Pharmaceutics* **2022**, *14*, 537. [CrossRef]
76. Hu, Y.; Liu, M.; Zhou, D.; Chen, F.; Cai, Q.; Yan, X.; Li, J. Gelatine methacrylamide-based multifunctional bilayer hydrogels for accelerating diabetic wound repair. *Mater. Des.* **2022**, *218*, 110687. [CrossRef]
77. Garcia-Orue, I.; Santos-Vizcaino, E.; Sanchez, P.; Gutierrez, F.B.; Aguirre, J.J.; Hernandez, R.M.; Igartua, M. Bioactive and degradable hydrogel based on human platelet-rich plasma fibrin matrix combined with oxidized alginate in a diabetic mice wound healing model. *Biomater. Adv.* **2022**, *135*, 112695. [CrossRef]
78. Chen, H.; Guo, Y.; Zhang, Z.; Mao, W.; Shen, C.; Xiong, W.; Yao, Y.; Zhao, X.; Hu, Y.; Zou, Z.; et al. Symbiotic Algae–Bacteria Dressing for Producing Hydrogen to Accelerate Diabetic Wound Healing. *Nano Lett.* **2021**, *22*, 229–237. [CrossRef]
79. Chen, Y.-H.; Rao, Z.-F.; Liu, Y.-J.; Liu, X.-S.; Liu, Y.-F.; Xu, L.-J.; Wang, Z.-Q.; Guo, J.-Y.; Zhang, L.; Dong, Y.-S.; et al. Multifunctional Injectable Hydrogel Loaded with Cerium-Containing Bioactive Glass Nanoparticles for Diabetic Wound Healing. *Biomolecules* **2021**, *11*, 702. [CrossRef]
80. Wei, S.; Xu, P.; Yao, Z.; Cui, X.; Lei, X.; Li, L.; Dong, Y.; Zhu, W.; Guo, R.; Cheng, B. A composite hydrogel with co-delivery of antimicrobial peptides and platelet-rich plasma to enhance healing of infected wounds in diabetes. *Acta Biomater.* **2021**, *124*, 205–218. [CrossRef]

81. Wang, P.; Jiang, S.; Li, Y.; Luo, Q.; Lin, J.; Hu, L.; Liu, X.; Xue, F. Virus-like mesoporous silica-coated plasmonic Ag nanocube with strong bacteria adhesion for diabetic wound ulcer healing. *Nanomed. Nanotechnol. Biol. Med.* **2021**, *34*, 102381. [CrossRef]
82. Qiu, W.; Han, H.; Li, M.; Li, N.; Wang, Q.; Qin, X.; Wang, X.; Yu, J.; Zhou, Y.; Li, Y.; et al. Nanofibers reinforced injectable hydrogel with self-healing, antibacterial, and hemostatic properties for chronic wound healing. *J. Colloid Interface Sci.* **2021**, *596*, 312–323. [CrossRef]
83. El-Salamouni, N.S.; Gowayed, M.A.; Seiffein, N.L.; Moneim, R.A.A.; Kamel, M.A.; Labib, G.S. Valsartan solid lipid nanoparticles integrated hydrogel: A challenging repurposed use in the treatment of diabetic foot ulcer, in-vitro/in-vivo experimental study. *Int. J. Pharm.* **2020**, *592*, 120091. [CrossRef]
84. Liu, H.; Li, Z.; Zhao, Y.; Feng, Y.; Zvyagin, A.V.; Wang, J.; Yang, X.; Yang, B.; Lin, Q. Novel Diabetic Foot Wound Dressing Based on Multifunctional Hydrogels with Extensive Temperature-Tolerant, Durable, Adhesive, and Intrinsic Antibacterial Properties. *ACS Appl. Mater. Interfaces* **2021**, *13*, 26770–26781. [CrossRef]
85. Liu, P.; Jin, K.; Wong, W.; Wang, Y.; Liang, T.; He, M.; Li, H.; Lu, C.; Tang, X.; Zong, Y.; et al. Ionic liquid functionalized non-releasing antibacterial hydrogel dressing coupled with electrical stimulation for the promotion of diabetic wound healing. *Chem. Eng. J.* **2021**, *415*, 129025. [CrossRef]
86. Li, Z.; Zhao, Y.; Liu, H.; Ren, M.; Wang, Z.; Wang, X.; Liu, H.; Feng, Y.; Lin, Q.; Wang, C.; et al. pH-responsive hydrogel loaded with insulin as a bioactive dressing for enhancing diabetic wound healing. *Mater. Des.* **2021**, *210*, 110104. [CrossRef]
87. Li, Q.; Liu, K.; Jiang, T.; Ren, S.; Kang, Y.; Li, W.; Yao, H.; Yang, X.; Dai, H.; Chen, Z. Injectable and self-healing chitosan-based hydrogel with MOF-loaded α-lipoic acid promotes diabetic wound healing. *Mater. Sci. Eng. C* **2021**, *131*, 112519. [CrossRef] [PubMed]
88. Lee, Y.-H.; Hong, Y.-L.; Wu, T.-L. Novel silver and nanoparticle-encapsulated growth factor co-loaded chitosan composite hydrogel with sustained antimicrobility and promoted biological properties for diabetic wound healing. *Mater. Sci. Eng. C Mater. Biol. Appl.* **2021**, *118*, 111385. [CrossRef] [PubMed]
89. Jiao, Y.; Chen, X.; Niu, Y.; Huang, S.; Wang, J.; Luo, M.; Shi, G.; Huang, J. Wharton's jelly mesenchymal stem cells embedded in PF-127 hydrogel plus sodium ascorbyl phosphate combination promote diabetic wound healing in type 2 diabetic rat. *Stem Cell Res. Ther.* **2021**, *12*, 559. [CrossRef]
90. Zhao, W.; Zhang, X.; Zhang, R.; Zhang, K.; Li, Y.; Xu, F.-J. Self-Assembled Herbal Medicine Encapsulated by an Oxidation-Sensitive Supramolecular Hydrogel for Chronic Wound Treatment. *ACS Appl. Mater. Interfaces* **2020**, *12*, 56898–56907. [CrossRef]
91. Yang, J.; Chen, Z.; Pan, D.; Li, H.; Shen, J. Umbilical Cord-Derived Mesenchymal Stem Cell-Derived Exosomes Combined Pluronic F127 Hydrogel Promote Chronic Diabetic Wound Healing and Complete Skin Regeneration. *Int. J. Nanomed.* **2020**, *15*, 5911–5926. [CrossRef]
92. Qian, Z.; Wang, H.; Bai, Y.; Wang, Y.; Tao, L.; Wei, Y.; Fan, Y.; Guo, X.; Liu, H. Improving Chronic Diabetic Wound Healing through an Injectable and Self-Healing Hydrogel with Platelet-Rich Plasma Release. *ACS Appl. Mater. Interfaces* **2020**, *12*, 55659–55674. [CrossRef]
93. Viezzer, C.; Mazzuca, R.; Machado, D.C.; Forte, M.M.D.C.; Ribelles, J.L.G. A new waterborne chitosan-based polyurethane hydrogel as a vehicle to transplant bone marrow mesenchymal cells improved wound healing of ulcers in a diabetic rat model. *Carbohydr. Polym.* **2020**, *231*, 115734. [CrossRef]
94. Zhao, Y.; Li, Z.; Li, Q.; Yang, L.; Liu, H.; Yan, R.; Xiao, L.; Liu, H.; Wang, J.; Yang, B.; et al. Transparent Conductive Supramolecular Hydrogels with Stimuli-Responsive Properties for On-Demand Dissolvable Diabetic Foot Wound Dressings. *Macromol. Rapid Commun.* **2020**, *41*, e2000441. [CrossRef]
95. Dong, Y.; Zhuang, H.; Hao, Y.; Zhang, L.; Yang, Q.; Liu, Y.; Qi, C.; Wang, S. Poly(N-Isopropyl-Acrylamide)/Poly(gamma-Glutamic Acid) Thermo-Sensitive Hydrogels Loaded with Superoxide Dismutase for Wound Dressing Application. *Int. J. Nanomed.* **2020**, *15*, 1939–1950. [CrossRef]
96. Bai, H.; Kyu-Cheol, N.; Wang, Z.; Cui, Y.; Liu, H.; Liu, H.; Feng, Y.; Zhao, Y.; Lin, Q.; Li, Z. Regulation of inflammatory microenvironment using a self-healing hydrogel loaded with BM-MSCs for advanced wound healing in rat diabetic foot ulcers. *J. Tissue Eng.* **2020**, *11*, 204173142094724. [CrossRef]
97. Yin, M.; Wang, X.; Yu, Z.; Wang, Y.; Wang, X.; Deng, M.; Zhao, D.; Ji, S.; Jia, N.; Zhang, W. γ-PGA hydrogel loaded with cell-free fat extract promotes the healing of diabetic wounds. *J. Mater. Chem. B* **2020**, *8*, 8395–8404. [CrossRef]
98. Sonamuthu, J.; Cai, Y.; Liu, H.; Kasim, M.S.M.; Vasanthakumar, V.R.; Pandi, B.; Wang, H.; Yao, J. MMP-9 responsive dipeptide-tempted natural protein hydrogel-based wound dressings for accelerated healing action of infected diabetic wound. *Int. J. Biol. Macromol.* **2020**, *153*, 1058–1069. [CrossRef]
99. Sener, G.; Hilton, S.A.; Osmond, M.J.; Zgheib, C.; Newsom, J.P.; Dewberry, L.; Singh, S.; Sakthivel, T.S.; Seal, S.; Liechty, K.W.; et al. Injectable, self-healable zwitterionic cryogels with sustained microRNA—Cerium oxide nanoparticle release promote accelerated wound healing. *Acta Biomater.* **2019**, *101*, 262–272. [CrossRef]
100. Chen, G.; He, L.; Zhang, P.; Zhang, J.; Mei, X.; Wang, D.; Zhang, Y.; Ren, X.; Chen, Z. Encapsulation of green tea polyphenol nanospheres in PVA/alginate hydrogel for promoting wound healing of diabetic rats by regulating PI3K/AKT pathway. *Mater. Sci. Eng. C* **2020**, *110*, 110686. [CrossRef]

101. Masood, N.; Ahmed, R.; Tariq, M.; Ahmed, Z.; Masoud, M.; Ali, I.; Asghar, R.; Andleeb, A.; Hasan, A. Silver nanoparticle impregnated chitosan-PEG hydrogel enhances wound healing in diabetes induced rabbits. *Int. J. Pharm.* **2019**, *559*, 23–36. [CrossRef]
102. Zhu, Y.; Cankova, Z.; Iwanaszko, M.; Lichtor, S.; Mrksich, M.; Ameer, G.A. Potent laminin-inspired antioxidant regenerative dressing accelerates wound healing in diabetes. *Proc. Natl. Acad. Sci. USA* **2018**, *115*, 6816–6821. [CrossRef]
103. Xu, Q.; Sigen, A.; Gao, Y.; Guo, L.; Creagh-Flynn, J.; Zhou, D.; Greiser, U.; Dong, Y.; Wang, F.; Tai, H.; et al. A hybrid injectable hydrogel from hyperbranched PEG macromer as a stem cell delivery and retention platform for diabetic wound healing. *Acta Biomater.* **2018**, *75*, 63–74. [CrossRef]
104. Hsieh, H.-T.; Chang, H.-M.; Lin, W.-J.; Hsu, Y.-T.; Mai, F.-D. Poly-Methyl Methacrylate/Polyvinyl Alcohol Copolymer Agents Applied on Diabetic Wound Dressing. *Sci. Rep.* **2017**, *7*, 9531. [CrossRef]
105. Zhao, L.; Niu, L.; Liang, H.; Tan, H.; Liu, C.; Zhu, F. pH and Glucose Dual-Responsive Injectable Hydrogels with Insulin and Fibroblasts as Bioactive Dressings for Diabetic Wound Healing. *ACS Appl. Mater. Interfaces* **2017**, *9*, 37563–37574. [CrossRef]
106. Lai, J.C.-Y.; Lai, H.-Y.; Rao, N.K.; Ng, S.-F. Treatment for diabetic ulcer wounds using a fern tannin optimized hydrogel formulation with antibacterial and antioxidative properties. *J. Ethnopharmacol.* **2016**, *189*, 277–289. [CrossRef]
107. Chen, H.; Guo, L.; Wicks, J.; Ling, C.; Zhao, X.; Yan, Y.; Qi, J.; Cui, W.; Deng, L. Quickly promoting angiogenesis by using a DFO-loaded photo-crosslinked gelatin hydrogel for diabetic skin regeneration. *J. Mater. Chem. B* **2016**, *4*, 3770–3781. [CrossRef]
108. Kamar, S.S.; Abdel-Kader, D.H.; Rashed, L.A. Beneficial effect of Curcumin Nanoparticles-Hydrogel on excisional skin wound healing in type-I diabetic rat: Histological and immunohistochemical studies. *Ann. Anat.-Anat. Anz.* **2018**, *222*, 94–102. [CrossRef]
109. Yap, L.-S.; Yang, M.-C. Thermo-reversible injectable hydrogel composing of pluronic F127 and carboxymethyl hexanoyl chitosan for cell-encapsulation. *Colloids Surf. B Biointerfaces* **2019**, *185*, 110606. [CrossRef]
110. Li, H.; Jia, Y.; Liu, C. Pluronic(R) F127 stabilized reduced graphene oxide hydrogel for transdermal delivery of ondansetron: Ex vivo and animal studies. *Colloids Surf B Biointerfaces* **2020**, *195*, 111259. [CrossRef]
111. Cao, J.; Su, M.; Hasan, N.; Lee, J.; Kwak, D.; Kim, D.Y.; Kim, K.; Lee, E.H.; Jung, J.H.; Yoo, J.W. Nitric Oxide-Releasing Thermoresponsive Pluronic F127/Alginate Hydrogel for Enhanced Antibacterial Activity and Accelerated Healing of Infected Wounds. *Pharmaceutics* **2020**, *12*, 926. [CrossRef]
112. Zahid, A.A.; Ahmed, R.; Rehman, S.R.U.; Augustine, R.; Tariq, M.; Hasan, A. Nitric oxide releasing chitosan-poly (vinyl alcohol) hydrogel promotes angiogenesis in chick embryo model. *Int. J. Biol. Macromol.* **2019**, *136*, 901–910. [CrossRef]
113. Yokoyama, F.; Masada, I.; Shimamura, K.; Ikawa, T.; Monobe, K. Morphology and structure of highly elastic poly(vinyl alcohol) hydrogel prepared by repeated freezing-and-melting. *Colloid Polym. Sci.* **1986**, *264*, 595–601. [CrossRef]
114. Figueroa-Pizano, M.D.; Vélaz, I.; Martínez-Barbosa, M.E. A Freeze-Thawing Method to Prepare Chitosan-Poly(vinyl alcohol) Hydrogels Without Crosslinking Agents and Diflunisal Release Studies. *J. Vis. Exp.* **2020**, *155*, e59636. [CrossRef] [PubMed]
115. Takei, T.; Nakahara, H.; Tanaka, S.; Nishimata, H.; Yoshida, M.; Kawakami, K. Effect of chitosan-gluconic acid conjugate/poly(vinyl alcohol) cryogels as wound dressing on partial-thickness wounds in diabetic rats. *J. Mater. Sci. Mater. Med.* **2013**, *24*, 2479–2487. [CrossRef] [PubMed]
116. Xie, Y.; Shi, X.; Sheng, K.; Han, G.; Li, W.; Zhao, Q.; Jiang, B.; Feng, J.; Li, J.; Gu, Y. PI3K/Akt signaling transduction pathway, erythropoiesis and glycolysis in hypoxia (Review). *Mol. Med. Rep.* **2018**, *19*, 783–791. [CrossRef]
117. Keppler-Noreuil, K.M.; Parker, V.E.; Darling, T.; Martinez-Agosto, J.A. Somatic overgrowth disorders of the PI3K/AKT/mTOR pathway & therapeutic strategies. *Am. J. Med. Genet. Part C Semin. Med. Genet.* **2016**, *172*, 402–421. [CrossRef]
118. Lao, G.; Yan, L.; Yang, C.; Zhang, L.; Zhang, S.; Zhou, Y. Controlled Release of Epidermal Growth Factor from Hydrogels Accelerates Wound Healing in Diabetic Rats. *J. Am. Podiatr. Med. Assoc.* **2012**, *102*, 89–98. [CrossRef]
119. Mohan, K.; Ganesan, A.R.; Muralisankar, T.; Jayakumar, R.; Sathishkumar, P.; Uthayakumar, V.; Chandirasekar, R.; Revathi, N. Recent insights into the extraction, characterization, and bioactivities of chitin and chitosan from insects. *Trends Food Sci. Technol.* **2020**, *105*, 17–42. [CrossRef]
120. Li, X.; Bai, H.; Yang, Y.; Yoon, J.; Wang, S.; Zhang, X. Supramolecular Antibacterial Materials for Combatting Antibiotic Resistance. *Adv. Mater.* **2018**, *31*, e1805092. [CrossRef]
121. Li, Z.; Yang, F.; Yang, R. Synthesis and characterization of chitosan derivatives with dual-antibacterial functional groups. *Int. J. Biol. Macromol.* **2015**, *75*, 378–387. [CrossRef]
122. Hajimiri, M.; Shahverdi, S.; Esfandiari, M.A.; Larijani, B.; Atyabi, F.; Rajabiani, A.; Dehpour, A.R.; Amini, M.; Dinarvand, R. Preparation of hydrogel embedded polymer-growth factor conjugated nanoparticles as a diabetic wound dressing. *Drug Dev. Ind. Pharm.* **2015**, *42*, 707–719. [CrossRef]
123. Kamel, R.; El-Batanony, R.; Salama, A. Pioglitazone-loaded three-dimensional composite polymeric scaffolds: A proof of concept study in wounded diabetic rats. *Int. J. Pharm.* **2019**, *570*, 118667. [CrossRef]
124. Gil Park, Y.; Lee, I.H.; Park, E.S.; Kim, J.Y. Hydrogel and Platelet-Rich Plasma Combined Treatment to Accelerate Wound Healing in a Nude Mouse Model. *Arch. Plast. Surg.* **2017**, *44*, 194–201. [CrossRef]
125. Ekblad, T.; Faxälv, L.; Andersson, O.; Wallmark, N.; Larsson, A.; Lindahl, T.L.; Liedberg, B. Patterned Hydrogels for Controlled Platelet Adhesion from Whole Blood and Plasma. *Adv. Funct. Mater.* **2010**, *20*, 2396–2403. [CrossRef]

126. Samberg, M.; Stone, R.; Natesan, S.; Kowalczewski, A.; Becerra, S.; Wrice, N.; Cap, A.; Christy, R. Platelet rich plasma hydrogels promote in vitro and in vivo angiogenic potential of adipose-derived stem cells. *Acta Biomater.* **2019**, *87*, 76–87. [CrossRef]
127. Eppley, B.L.; Woodell, J.E.; Higgins, J. Platelet Quantification and Growth Factor Analysis from Platelet-Rich Plasma: Implications for Wound Healing. *Plast. Reconstr. Surg.* **2004**, *114*, 1502–1508. [CrossRef]
128. El-Sharkawy, H.; Kantarci, A.; Deady, J.; Hasturk, H.; Liu, H.; Alshahat, M.; Van Dyke, T.E. Platelet-Rich Plasma: Growth Factors and Pro- and Anti-Inflammatory Properties. *J. Periodontol.* **2007**, *78*, 661–669. [CrossRef]
129. Chicharro-Alcántara, D.; Rubio-Zaragoza, M.; Damiá-Giménez, E.; Carrillo-Poveda, J.M.; Cuervo-Serrato, B.; Peláez-Gorrea, P.; Sopena-Juncosa, J.J. Platelet Rich Plasma: New Insights for Cutaneous Wound Healing Management. *J. Funct. Biomater.* **2018**, *9*, 10. [CrossRef]
130. Gallelli, G.; Cione, E.; Serra, R.; Leo, A.; Citraro, R.; Matricardi, P.; Di Meo, C.; Bisceglia, F.; Caroleo, M.C.; Basile, S.; et al. Nano-hydrogel embedded with quercetin and oleic acid as a new formulation in the treatment of diabetic foot ulcer: A pilot study. *Int. Wound J.* **2019**, *17*, 485–490. [CrossRef]

Article

In Vitro Tissue Reconstruction Using Decellularized Pericardium Cultured with Cells for Ligament Regeneration

Mika Suzuki [1], Tsuyoshi Kimura [1,*], Yukina Yoshida [1], Mako Kobayashi [1], Yoshihide Hashimoto [1], Hironobu Takahashi [2], Tatsuya Shimizu [2], Shota Anzai [3], Naoko Nakamura [1,3] and Akio Kishida [1]

[1] Department of Material-Based Medical Engineering, Institute of Biomaterials and Bioengineering, Tokyo Medical and Dental University, Tokyo 101-0062, Japan; mika.mbme@tmd.ac.jp (M.S.); ma200041@tmd.ac.jp (Y.Y.); mako.mbme@tmd.ac.jp (M.K.); hashimoto.atrm@tmd.ac.jp (Y.H.); naoko@shibaura-it.ac.jp (N.N.); kishida.mbme@tmd.ac.jp (A.K.)
[2] Institute of Advanced Biomedical Engineering and Science, Tokyo Women's Medical University, Tokyo 162-8666, Japan; takahashi.hironobu@twmu.ac.jp (H.T.); shimizu.tatsuya@twmu.ac.jp (T.S.)
[3] Department of Bioscience and Engineering, Shibaura Institute of Technology, Saitama 337-8570, Japan; bn16228@shibaura-it.ac.jp
* Correspondence: kimurat.mbme@tmd.ac.jp; Tel.: +81-3-5280-8110

Abstract: Recent applications of decellularized tissues have included the ectopic use of their sheets and powders for three-dimensional (3D) tissue reconstruction. Decellularized tissues are fabricated with the desired functions to employ them to a target tissue. The aim of this study was to develop a 3D reconstruction method using a recellularized pericardium to overcome the difficulties in cell infiltration into tight and dense tissues, such as ligament and tendon tissues. Decellularized pericardial tissues were prepared using the high hydrostatic pressurization (HHP) and surfactant methods. The pericardium consisted of bundles of aligned fibers. The bundles were slightly disordered in the surfactant decellularization method compared to the HHP decellularization method. The mechanical properties of the pericardium were maintained after the HHP and surfactant decellularizations. The HHP-decellularized pericardium was rolled up into a cylindrical formation. Its mechanical behavior was similar to that of a porcine anterior cruciate ligament in tensile testing. NIH3T3, C2C12, and mesenchymal stem cells were adhered with elongation and alignment on the HHP- and surfactant-decellularized pericardia, with dependences on the cell type and decellularization method. When the recellularized pericardium was rolled up into a cylinder formation and cultured by hanging circulation for 2 days, the cylinder formation and cellular elongation and alignment were maintained on the decellularized pericardium, resulting in a layer structure of cells in a cross-section. According to these results, the 3D-reconstructed decellularized pericardium with cells has the potential to be an attractive alternative to living tissues, such as ligament and tendon tissues.

Keywords: decellularization; extracellular matrix; porcine pericardium; high hydrostatic pressurization method; surfactant method; 3D fabrication; ligament

1. Introduction

Tissue engineering and regenerative medicine are emerging fields focused on curing complex and chronic diseases. Decellularized tissue prepared from a living tissue is an extracellular matrix, which can be used as a scaffold in tissue engineering, owing to its low immunogenicity and high compatibility. Numerous decellularized tissues, such as the dermis [1], urinary bladder matrix [2], small intestinal submucosa [3], aorta [4,5], ligament, and tendon [6–8], have been developed using various physical and chemical decellularization methods. Mainly, decellularized tissues have been employed as alternative tissues. Particularly, for tissues where mechanical compliance is important, such as vascular, ligament, and tendon tissues, decellularized tissues are implanted orthotopically [9] by retaining the histological structure to express typical biomechanical properties. Numerous decellularization methods have been developed to remove cells while maintaining the tissue structure [10,11]. Although decellularization has been achieved effectively for tight, rigid, and dense tissues such as ligament, tendon, and vascular tissues, recipient cells cannot easily infiltrate these tissues [8,12]. Moreover, complications such as a repair-site rupture and adhesion formation slow the healing because of the hypocellular composition of tissues [13]. Thus, decellularized tissues are fabricated by creating holes and slits to promote cell infiltration while maintaining their shape [12]. Decellularized tissues have been formed into powders [14], sheets [15], and gels [16] for ectopic use and three-dimensional (3D) printing [17]. Several decellularized tissues are commercially available and used ectopically, e.g., for wound healing and adhesion prevention [18,19]. We reported that a sheet of decellularized aortic intermedia was formed into a tube and applied as an alternative small-diameter vascular graft [20,21].

In this study, we propose a 3D tissue fabrication method using a decellularized membranous tissue recellularized to apply it as a ligament and tendon alternative. Decellularized pericardia were prepared by physical and chemical decellularization methods as decellularized membranous tissues, and then were recellularized by several types of cells: fibroblast, myoblast, and mesenchymal stem cells. The recellularized pericardium was rolled up, formed cylindrically, and cultured in vitro as a reconstructed ligament-like tissue (Figure 1).

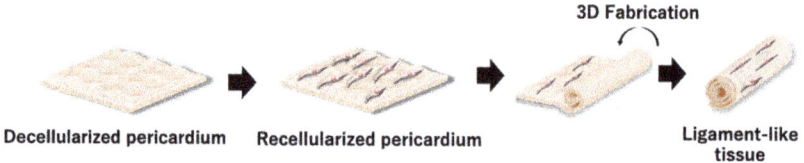

Figure 1. Scheme of a 3D tissue reconstructed using the recellularized pericardium.

2. Materials and Methods

2.1. Materials

Fresh porcine pericardium and anterior cruciate ligament (ACL) were obtained from a local slaughterhouse (Tokyo Shibaura Zouki, Tokyo, Japan) and stored at 4 °C until use. DNase I was purchased from Roche Diagnostics (Tokyo, Japan). Dulbecco's modified Eagle medium (DMEM), magnesium chloride hexahydrate ($MgCl_2$), neutral-buffered (pH = 7.4) solution of 10% formalin, protease-K, phosphate-buffered saline (PBS), NaCl, sodium deoxycholate (SDC), sodium dodecyl sulfate (SDS), and tert-butyl alcohol were purchased from FUJIFILM Wako Pure Chemical Corp. (Osaka, Japan). Bovine serum albumin and tris(hydroxymethyl)aminomethane (Tris) were obtained from Sigma-Aldrich Inc., (St. Louis, MO, USA). An isodine solution was purchased from Shionogi Healthcare (Osaka, Japan). Gentamicin sulfate was procured from Tokyo Chemical Industry Co. Ltd. (Tokyo, Japan). Glutaraldehyde (25%) was obtained from TAAB Laboratories Equipment, Ltd. (Aldermaston, UK). Phenol/chloroform was purchased from Nippon Gene (Tokyo, Japan). Ethylenediaminetetraacetic acid (EDTA) and Calcein-AM solution (0.5 mg/mL) were purchased from Dojindo Laboratories (Kumamoto, Japan). NIH3T3 and C2C12 cells were purchased from Riken Bioresource Research Center (Tsukuba, Japan). Mesenchymal stem cells were purchased from JCRB Cell Bank, National Institute of Biomedical Innovation, Health and Nutrition (Osaka, Japan). A mesenchymal stem cell medium (Powered by 10) was purchased from GlycoTechnica Ltd. (Yokohama, Japan).

2.2. Preparation of a Decellularized Porcine Pericardium

The porcine pericardium was separated into a fibrous pericardium and serous pericardium. The serous pericardium was trimmed by removing fat off the underside of the parietal pericardium. The pericardium was packed into a plastic bag containing saline and sealed. The pericardium was subjected to a high hydrostatic pressure of 1000 MPa at 30 °C for 10 min using a high-hydrostatic-pressure machine (Dr. Chef; Kobelco, Kobe, Japan), and then washed with DNase (0.1 mg/mL) and MgCl$_2$ (50 mM) in saline at 4 °C for 4 days, 80% ethanol in saline at 4 °C for 3 days, and 0.1 M citric acid in a saline solution at 4 °C for 3 days. Surfactant decellularization was carried out according to reported procedures [22]. The pericardium was treated with a solution of 0.5% SDS and 0.5% SDC for 24 h, and then washed with PBS at room temperature for 12 h six times.

2.3. Deoxyribonucleic Acid (DNA) Quantification

Residual DNA quantification was performed to evaluate the quality of the decellularization. The decellularized pericardium was freeze-dried and incubated in a lysis buffer containing 50-mgmL^{-1} protease K, 50 mM Tris–HCl, 1% SDS, 10 mM NaCl, and 20 mM EDTA at 55 °C for 12 h. The DNA was purified with phenol/chloroform extraction and ethanol precipitation. The residual DNA content was quantified at 260 nm using a microvolume spectrophotometer (Nanodrop; Thermo Fisher Scientific K.K., Tokyo, Japan) and normalized to the tissue dry weight of 20 mg.

2.4. Hematoxylin–Eosin (H&E) Staining

To evaluate the efficiency of cell removal, decellularized pericardia were fixed with a neutral-buffered (pH = 7.4) solution of 10% formalin in PBS at 25 °C for 24 h and dehydrated with 70%, 80%, 90%, and 100% ethanol. The samples were then replaced with xylene and embedded in paraffin. Paraffin sections were cut into 4 μm thick sections and stained with hematoxylin and eosin.

2.5. Scanning Electron Microscopy (SEM) Observation of the Pericardium

The decellularized pericardia were fixed for 24 h at room temperature in 2.5% glutaraldehyde, dehydrated through a graded ethanol series, soaked for 72 h in t-butyl alcohol, sputter-coated with Au, and imaged with SEM (S-3400N, Hitachi Ltd., Tokyo, Japan).

2.6. Mechanical Properties

The mechanical strengths of the untreated pericardium and high-hydrostatic-pressurization (HHP)- and surfactant-decellularized pericardia were evaluated. All samples were cut into dumbbell-shaped pieces. The test pieces were 35 mm long and 2 mm wide. Wall thicknesses were measured using a creep meter (RE2-33005 B, Yamaden Co., Ltd., Tokyo, Japan) for each sample. The HHP-decellularized pericardium (5 cm × 13 cm) was tightly rolled up into a cylinder form. Stress–strain curves of the pieces and cylinder of the pericardium were obtained using a universal testing machine (AGS-X, Shimadzu Co., Ltd., Kyoto, Japan). Each sample was strained at a rate of 5 mm/min.

2.7. Cell Culture

DMEM containing 10% FBS and 1% penicillin/streptomycin was used for cell culture of C2C12 and NIH3T3. Poweredby 10 was used for cell culture of human mesenchymal stem cells (hMSCs). C2C12, NIH3T3, and hMSCs were seeded on the HHP- and surfactant-decellularized pericardia (2 × 10^4 cells/cm^2) and incubated at 37 °C under 5% CO$_2$ for 4 days. The culture medium was changed every 2 days. After 2 and 4 days, the cells were incubated with Calcein-AM (1 μg/mL) at 37 °C for 30 min and observed using a fluorescent microscope (BZ-X710, Keyence Corp., Osaka, Japan) after washing with PBS twice. The cell growth was assessed by a cell counting kit-8 (Dojindo Labolratories, Kumamoto, Japan). After culturing for 1, 2, 3, and 4 days, the absorbance of WST-8 at 450 nm was measured with a multispectrometer (Cytation, BioTek Japan, Tokyo, Japan). The cell density was calculated using the standard curve of absorbance.

2.8. Cell Culture in the Cylindrically Formed Pericardium

A silicone mold (5 cm × 6.5 cm) with an empty space (3 cm × 4 cm) was placed on the HHP-decellularized pericardium (5 cm × 7 cm). The C2C12 cells were seeded on the HHP-decellularized pericardium at a cell density of 2 × 10^4 cells/cm^2 and cultured for

1 day. A medical-grade suture was placed on one end of the recellularized pericardium. The pericardium was rolled up into a cylinder form so that its suture was in the center. Both ends of the suture were fixed at the edge of the plastic cup and the rolled pericardium was hanged. After filling it with the culture medium, the rolled pericardium was cultured for 2 days with a stirring culture medium. After cutting it in half, one half was subjected to H&E staining. The other half was partially peeled, incubated with Calcein-AM (1 μg/mL) at 37 °C for 30 min, and observed using a fluorescent microscope after washing with PBS twice to evaluate the cell adhesion and survival.

2.9. Statistical Analysis

Each experiment was performed six times. The results are expressed as mean ± standard deviation. One-way analysis of variance (ANOVA) and Tukey's post hoc multiple comparison tests were carried out to evaluate statistical significance. A p-value < 0.05 was considered statistically significant. For the cell growth analysis, the samples were compared each day.

3. Results

3.1. Preparation of the Decellularized Pericardium

The porcine pericardia were decellularized with HHP and surfactant methods. Figure 2A–F show the H&E staining of the untreated pericardium, HHP-decellularized pericardium, and surfactant-decellularized pericardium. There were no large changes in the shape and thickness of the pericardium upon the decellularization. For the HHP- and surfactant-decellularized pericardia, no blue spots of cells were observed. Gaps between the bundles of fibers occurred in the HHP-decellularized pericardium, which were expanded in the surfactant-decellularized pericardium. The residual DNA was 1052.9 ± 61.6 ng/mg of tissue weight and significantly decreased to 33.1 ± 0.8 and 46.6 ± 0.9 ng/mg of tissue weight for the HHP and surfactant pericardia, respectively.

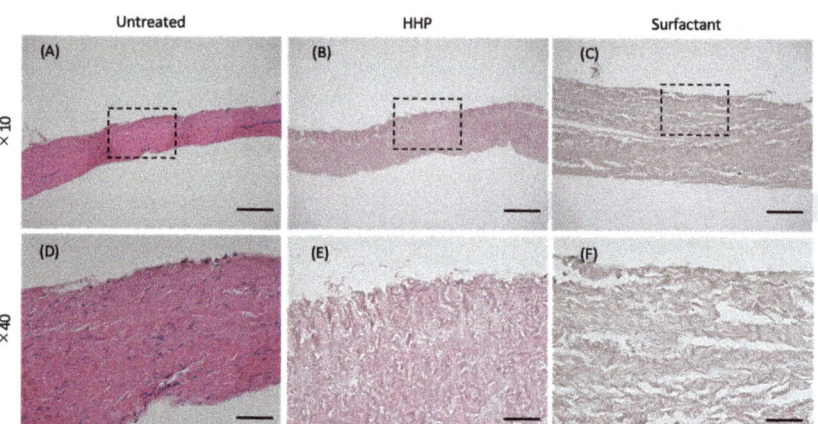

Figure 2. H&E staining of the (**A,D**) untreated, (**B,E**) HHP-decellularized, and (**C,F**) surfactant-decellularized pericardia. Scale bars: (**A–C**) 200 μm, (**D–F**) 50 μm. Nuclei can be observed in black in images (**A,D**), but are absent after the decellularization (**B,C,E,F**).

Figure 3A–F show SEM images of the surfaces of the untreated pericardium, HHP-decellularized pericardium, and surfactant-decellularized pericardium. Adhered and spread cells were observed on the untreated pericardium. In gaps between the cells, aligned fibers were also observed. On the other hand, for the HHP-decellularized and surfactant-decellularized pericardia, no cells were observed, while wavy fibers were observed along the longitudinal direction of the fibers. In addition, melted and disordered fibers were partially observed for the surfactant-decellularized pericardium.

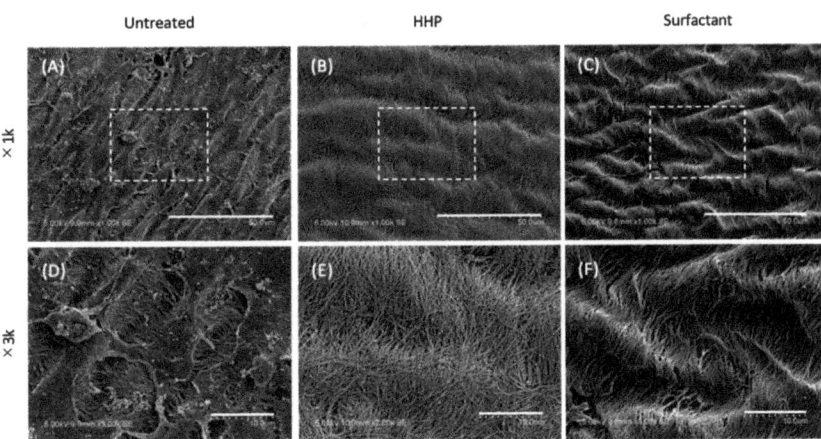

Figure 3. SEM observation of the (**A,D**) untreated, (**B,E**) HHP-decellularized, and (**C,F**) surfactant-decellularized pericardia. Scale bars: (**A–C**) 50 µm, (**D–F**) 10 µm.

3.2. Mechanical Properties of the Decellularized Pericardium and 3D Reconstruction

Figure 4 shows the mechanical properties of the untreated pericardium and pericardia decellularized by the HHP and surfactant methods. Their stress–strain curves were J-curves, typical for biological tissues. There were no large differences in the mechanical parameters, such as the ultimate tensile strength, failure strain, and elastic modulus.

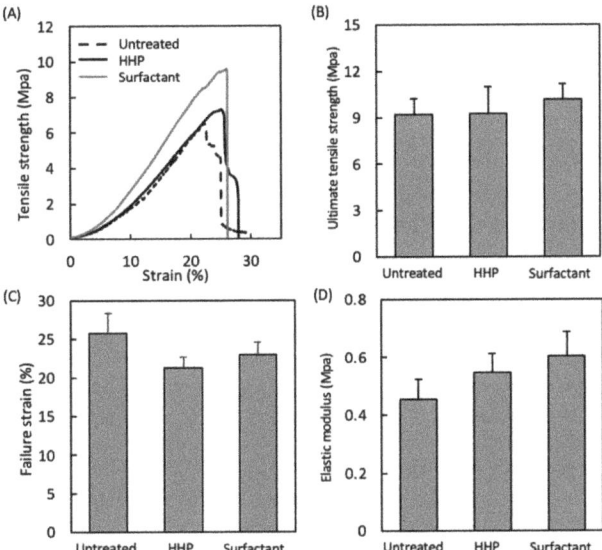

Figure 4. (**A**) Typical stress–strain curves of the untreated pericardium, HHP-decellularized pericardium, and surfactant-decellularized pericardium. Mechanical properties of the untreated pericardium, HHP-decellularized pericardium, and surfactant-decellularized pericardium: (**B**) ultimate tensile strength, (**C**) failure strain, and (**D**) elastic modulus. Mean ± standard deviation, $n = 6$; no significant differences were identified between groups (one-way ANOVA, $p < 0.05$; Tukey's post hoc comparison).

The cylindrically formed pericardium decellularized by the HHP method was subjected to a tensile test. Porcine ACL was used as a control. The porcine ACL and rolled pericardium were stretched, partially ruptured, and broken during the tensile test (Figure 5A,B). The typical stress–strain curves are shown in Figure 5C. The J-curve was observed in the early phase of the porcine ACL, and then the tensile strength was decreased slowly after the maximum tensile strength was reached, where the porcine ACL was partially ruptured. On the other hand, for the rolled pericardium, the shape of the S–S curve was similar to that of the porcine ACL, although the tensile strength increased without the J-shape in the early phase. There was no difference between their ultimate tensile strengths.

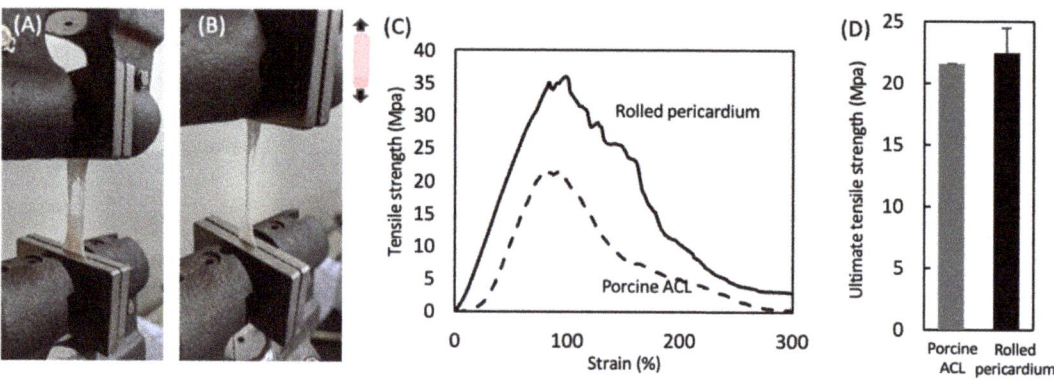

Figure 5. Photographs of the (**A**) porcine ACL and (**B**) rolled pericardium decellularized with the HHP method during a tensile test. Mechanical properties of the porcine ACL and rolled pericardium decellularized with the HHP method: (**C**) typical S–S curves and (**D**) ultimate tensile strength. Mean ± standard deviation, n = 6; no significant differences were identified between groups (one-way ANOVA, $p < 0.05$; Tukey's post hoc comparison).

3.3. Recellularization of the Decellularized Pericardium and Hanging Circulation Culture of the 3D-Reconstructed Pericardium

Figure 6 shows fluorescent microscopy images of various types of cells seeded on the HHP- and surfactant-decellularized pericardia. TCPS was used as a control. All types of cells were adhered in random directions on the TCPS at day 2 and grew. On the other hand, on the HHP- and surfactant-decellularized pericardia, the cells were aligned along the fiber direction at day 2 and grew while maintaining the cellular direction for 4 days of cultivation. Elongated cells, whose longitudinal axis was parallel with the fiber direction, were observed on the HHP- and surfactant-decellularized pericardia at day 2 and kept for 4 days of culture. There were no large differences in the alignment and morphology of cells between the HHP- and surfactant-decellularized pericardia, although the cellular densities differed between them for NIH3T3 cells and MSCs. The cellular growth is shown in Figure 7. For the NIH3T3 cells, the growth was inhibited on the surfactant-decellularized pericardium, while an effective growth was observed on the HHP-decellularized pericardium. For the C2C12 cells, a slow growth was observed, irrespective of the substrate. For the MSCs, compared to TCPS, the cellular growth was suppressed on the HHP- and surfactant-decellularized pericardia.

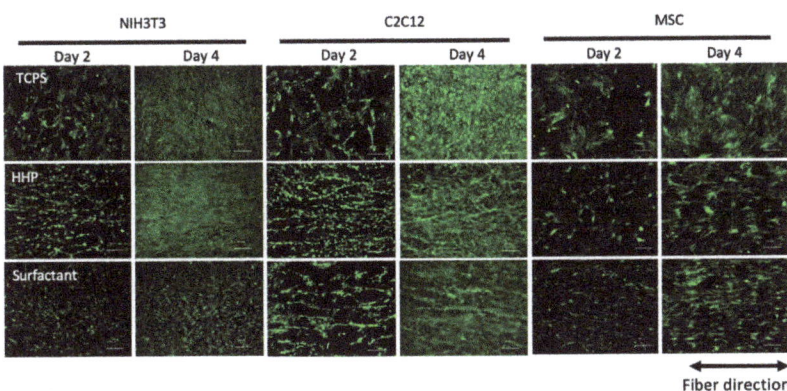

Figure 6. Fluorescent microscopy observation of various types of cells seeded on TCPS, HHP-decellularized pericardium, and surfactant-decellularized pericardium. The cells were stained with Calcein-AM, which shows living cells. Scale bar: 200 μm.

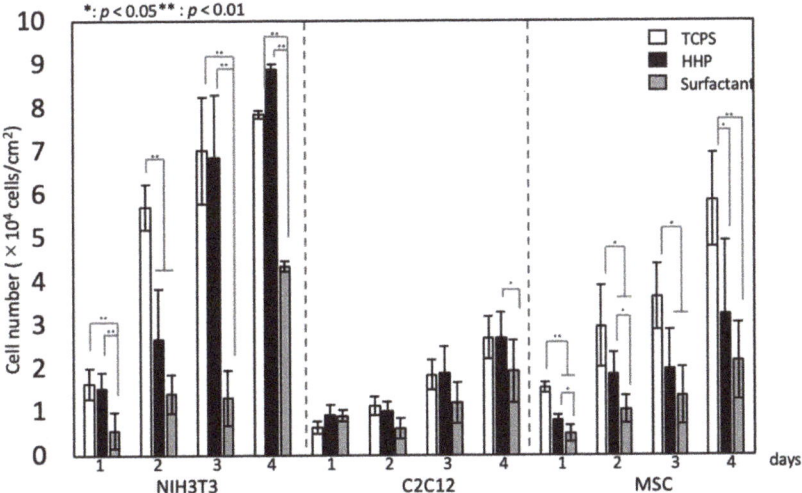

Figure 7. Cell growth on TCPS, HHP-decellularized pericardium, and surfactant-decellularized pericardium. Mean ± standard deviation, $n = 6$. One-way ANOVA, *: $p < 0.05$, **: $p < 0.01$; Tukey's post hoc comparison.

The HHP-decellularized pericardium was recellularized with C2C12 cells, formed cylindrically, and then hanging-cultured (Figure 8A). After 2 days of culture, the cylinder formation was kept. The cross-section of the cylindrically formed pericardium stained with H&E is shown in Figure 8B,C. The layer of the pericardia and numerous cells between layers were observed. Fluorescent microscopy images of cells stained with Calcein-AM, which can stain living cells, are shown in Figure 8D–F. The living cells were observed on the entire pericardium, although the cellular density varied with the location (Figure 8D). The living cells were oriented along the fiber direction (Figure 8E,F).

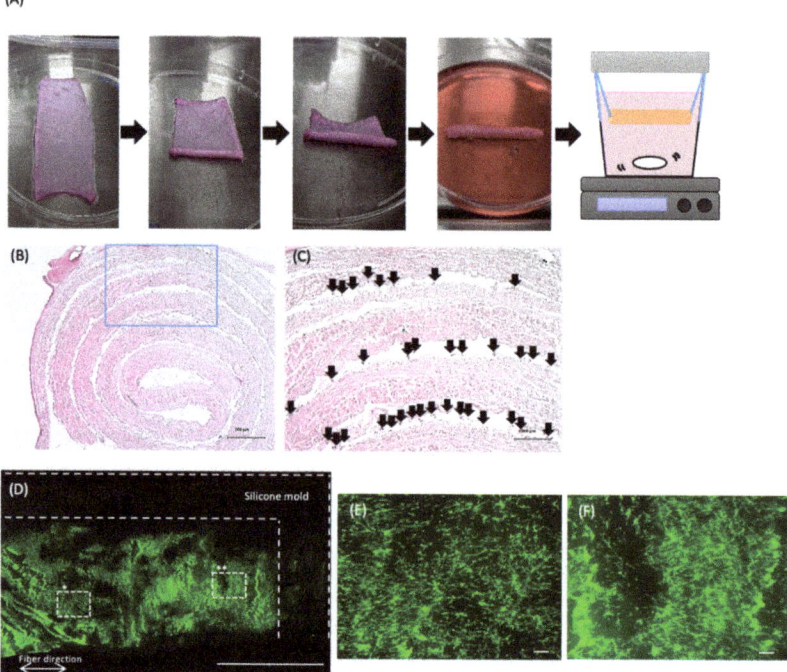

Figure 8. (**A**) Cylinder formation of the recellularized pericardium with C2C12 cells at a cell density of 2×10^3 cells/cm^2 and hanging culture. (**B,C**) H&E staining of the cross-section of the hanging-cultured pericardium for 2 days. The arrows show cells. (**D–F**) Fluorescent images of the Calcein-AM-stained cells on the partially peeled pericardium from the cylindrically formed pericardium after hanging culturing. (**E,F**): * and ** of hash square regions. Scale bars: (**D**) 1 cm, (**E,F**) 200 μm.

4. Discussion

Decellularized tissues are applied as alternative tissues. Numerous decellularization methods have been developed and an effective decellularization method with retained mechanical properties has been achieved. However, for tight and dense tissues, such as ligament, tendon, and vascular tissues, cells cannot easily infiltrate them. Recently, decellularized tissues have been fabricated into sheet and powder forms. They are used ectopically and induce recipient cells early when a mechanical contribution is not required [23]. We previously reported that the aortic intima-medium was formed into a tube and applied as a small-diameter vascular graft, which indicated that the mechanical compliance was adjusted by the appropriate fabrication of the decellularized tissue [20,21]. In this study, we developed a 3D tissue reconstruction of a ligament-like tissue with cells inside and similar mechanical properties to those of native tissues through the recellularization and fabrication of a decellularized pericardium.

Decellularized pericardia were prepared by the HHP and surfactant methods. The decellularization was achieved effectively while maintaining the mechanical properties, irrespective of the decellularization method. We reported that HHP decellularization could decellularize various tissues while retaining the tissue structure and mechanical properties [9]. On the other hand, the surfactant decellularization was an effective method, while the strong detergent destroyed the tissue structure, which contributed to the typical biological mechanical properties. We also reported that the SDS decellularization of the intima-media of aorta induced histological disordering and remarkable decreases in ultimate tensile strength and elastic modulus from 4.0 ± 1.9 to 1.5 ± 0.4 MPa and from

6.0 ± 2.9 to 2.4 ± 0.4 MPa, respectively [24]. However, the surfactant-decellularized pericardium was obtained with retained mechanical properties (Figure 4). It is considered that the combination of SDS and SDC could decellularize the pericardium [25]. A cylindrically formed pericardium was obtained, which exhibited a behavior similar to that of the porcine ACL in tensile testing (Figure 5D). This suggests that the appropriate fabrication could fit a target tissue even for the ectopic use of the cardiovascular tissue to a musculoskeletal tissue from a mechanics perspective. On the other hand, the ultimate tensile strength of a natural human ACL is approximately 36 MPa [26]. Thus, it is required to improve the rolling-up of the decellularized pericardium (e.g., number and strength) to apply it clinically.

The recellularization of various types of cells, NIH3T3, C2C12, and MSCs, was investigated in HHP- and surfactant-decellularized pericardia. Cellular adhesion and growth were shown on both decellularized pericardia. All types of cells were elongated and aligned on both decellularized pericardia (Figure 6). The percentages of NIH3T3, C2C12, and hMSCs with angles below 30° against the fiber direction were approximately 70%, 74%, and 49%, respectively, higher than those on TCPS (approximately 33%, 35%, and 33%, respectively) (data not shown). The cellular shape and alignment depend on the physical, chemical, and morphological properties of the substrate [27]. Particularly, the large elongation and alignment of cells were observed on the aligned fibers with nanoscale diameters compared to those with microscale diameters [28]. The SEM observation revealed that the surfaces of the HHP- and surfactant-decellularized pericardia consisted of a bundle of nanoscale fibers (71 ± 18 nm). The fiber bundles were aligned (Figure 3B,C,E,F). Thus, it is considered that the aligned fibrillar structures of the HHP- and surfactant-decellularized pericardia affected the cellular elongation and alignment along the fiber direction [29]. For the recellularization of NIH3T3 cells, the cells were effectively aligned with the fiber direction of the HHP-decellularized pericardium compared to the surfactant-decellularized pericardium. This may have been caused by the partially melted and disordered fibers for the surfactant-decellularized pericardium due to the sensitive nature of the fibroblast [30,31]. The cell growth was inhibited on the surfactant-decellularized pericardium compared to the HHP-decellularized pericardium (Figure 7). The residual surfactant in the surfactant-decellularized pericardium probably inhibited the cellular growth because of the high cytotoxicity of SDS [32]. The growth suppression of MSCs on the HHP-decellularized pericardium was also observed, compared to that on TCPS, which suggests that the cellular elongation on the fibrillar structure of the HHP-decellularized pericardium was related to the proliferation [33].

Finally, the recellularized pericardium was rolled up into a cylinder formation and cultivated with the hanging circulation culture method for 2 days (Figure 8). The cylinder formation was maintained without disordering. The cells were still alive in the cylindrically formed pericardium. The cellular features, elongation and alignment, were still maintained on the decellularized pericardium in a cylinder formation, which indicates that the fibrous membranous tissue could contribute to the cellular shape in the tissue, irrespective of the rolling process. These results suggest that the 3D tissue with cells reconstructed by this method could be adapted as a ligament-like tissue, although further studies are needed.

We expect that the proposed tissue-engineered product can be clinically applied. Artificial synthetic grafts and autologous tissue grafts are used in replacement surgery. The deterioration, due to the long-term implantation and loss of function at the site of tissue harvesting, needs to be overcome. In this regard, we proposed a method using tissue engineering techniques. The proposed tissue-engineered product is one of the candidates to overcome the above problems because of its potential for long-term compatibility, owing to the presence of cells inside the tissue. In addition, it may be adapted not only as a graft, but also to the site of harvesting in autologous transplantation. Furthermore, the decellularized pericardium used in this study was flexible, simple to fabricate, and could be processed into various shapes, which could provide various clinical applications, including hand surgery and orthopedic surgery. To widely apply the proposed method, the selection of cells to be recellularized and in vivo studies are necessary.

5. Conclusions

We successfully developed a 3D tissue reconstruction method, through which the decellularized pericardium was recellularized and fabricated into a cylinder formation. The cells were elongated and aligned on the decellularized pericardium. The reconstructed decellularized pericardium expressed a mechanical behavior similar to that of porcine ACL. Further in vivo experiments, such as compatibility and mechanical compliance experiments, are needed to apply the 3D-reconstructed decellularized pericardium as an alternative to ligaments and tendons.

Author Contributions: Conceptualization, T.K., N.N. and A.K.; methodology, M.S., Y.Y. and S.A.; validation, M.S., Y.Y., M.K., Y.H. and H.T.; writing—original draft preparation, M.S. and M.K.; writing—review and editing, T.S. and T.K.; project administration, T.K.; funding acquisition, A.K. All authors have read and agreed to the published version of the manuscript.

Funding: This study was partly supported by the Japan Society for the Promotion of Science KAKENHI (grant numbers 16H03180, 19H04465, and 21H04954), Creative Scientific Research of the Viable Material via Integration of Biology and Engineering from Ministry of Education, Culture, Sports, Science and Technology (MEXT), and Cooperative Research Project of the Research Center for Biomedical Engineering from MEXT.

Institutional Review Board Statement: Not applicable.

Informed Consent Statement: Not applicable.

Data Availability Statement: Not applicable.

Conflicts of Interest: The authors declare no conflict of interest.

Abbreviations

HHP	High hydrostatic pressurization
H&E	Hematoxylin–eosin
SDS	Sodium dodecyl sulfate
SDC	Sodium deoxycholate
FBS	Fatal bovine serum
ACL	Anterior cruciate ligament
TCPS	Tissue culture polystylene

References

1. Reing, J.E.; Brown, B.N.; Daly, K.A.; Freund, J.M.; Gilbert, T.W.; Hsiong, S.X.; Huber, A.; Kullas, K.E.; Tottey, S.; Wolf, M.T.; et al. The effects of processing methods upon mechanical and biologic properties of porcine dermal extracellular matrix scaffolds. *Biomaterials* **2010**, *31*, 8626–8633. [CrossRef] [PubMed]
2. Zhu, Y.; Hideyoshi, S.; Jiang, H.B.; Matsumura, Y.; Dziki, J.L.; LoPresti, S.T.; Huleihel, L.; Faria, G.N.F.; Fuhrman, L.C.; Lodono, R.; et al. Injectable, Porous, Biohybrid Hydrogels Incorporating Decellularized Tissue Components for Soft Tissue Applications. *Acta Biomater.* **2018**, *73*, 112–126. [CrossRef] [PubMed]
3. Andrée, B.; Bär, A.; Haverich, A.; Hilfiker, A. Small Intestinal Submucosa Segments as Matrix for Tissue Engineering: [Review]. *Tissue Eng. Part B Rev.* **2013**, *19*, 279–291. [CrossRef] [PubMed]
4. Quinta, C.; Kondob, Y.; Mansonc, R.L.; Lawsonc, J.H.; Dardikb, A.; Niklason, L.E. Decellularized tissue-engineered blood vessel as an arterial conduit. *Proc. Natl. Acad. Sci. USA* **2011**, *108*, 9214–9219. [CrossRef]
5. Lin, C.H.; Hsia, K.; Ma, H.; Lee, H.; Lu, J.H. In Vivo Performance of Decellularized Vascular Grafts: A Review Article. *Int. J. Mol. Sci.* **2018**, *19*, 2101. [CrossRef]
6. Woods, T.; Gratzer, P.F. Effectiveness of Three Extraction Techniques in the Development of a Decellularized Bone-Anterior Cruciate Ligament-Bone Graft. *Biomaterials* **2005**, *26*, 7339–7349. [CrossRef]
7. Jones, G.; Herbert, A.; Berry, H.; Edwards, J.H.; Fisher, J.; Ingham, E. Decellularization and Characterization of Porcine Superflexor Tendon: A Potential Anterior Cruciate Ligament Replacement. *Tissue Eng. Part A* **2017**, *23*, 124–134. [CrossRef]
8. Schulze-Tanzil, G.; Al-Sadi, O.; Ertel, W.; Lohan, A. Decellularized Tendon Extracellular Matrix-A Valuable Approach for Tendon Reconstruction? *Cells.* **2012**, *1*, 1010–1028. [CrossRef]
9. Nakamura, N.; Kimura, T.; Kishida, A. Overview of the Development, Applications, and Future Perspectives of Decellularized Tissues and Organs. *ACS Biomater. Sci. Eng.* **2017**, *3*, 1236–1244. [CrossRef]

10. Hashimoto, Y.; Funamoto, S.; Sasaki, S.; Honda, T.; Hattori, S.; Nam, K.; Kimura, T.; Mochizuki, M.; Fujisato, T.; Kobayashi, H.; et al. Preparation and Characterization of Decellularized Cornea Using High-Hydrostatic Pressurization for Corneal Tissue Engineering. *Biomaterials* **2010**, *31*, 3941–3948. [CrossRef]
11. Iablonskii, P.; Cebotari, S.; Tudorache, I.; Granados, M.; Morticelli, L.; Goecke, T.; Klein, N.; Korossis, S.; Hilfiker, A.; Haverich, A. Tissue-Engineered Mitral Valve: Morphology and Biomechanics. *Interact. Cardiovasc. Thorac. Surg.* **2015**, *20*, 712–719; discussion 719. [CrossRef] [PubMed]
12. Lu, C.C.; Zhang, T.; Amadio, P.C.; An, K.N.; Moran, S.L.; Gingery, A.; Zhao, C. Lateral Slit Delivery of Bone Marrow Stromal Cells Enhances Regeneration in the Decellularized Allograft Flexor Tendon. *J. Orthop. Transl.* **2019**, *19*, 58–67. [CrossRef] [PubMed]
13. Tozer, S.; Duprez, D. Tendon and Ligament: Development, Repair and Disease. *Birth Defects Res. C Embryo Today* **2005**, *75*, 226–236. [CrossRef] [PubMed]
14. Edgar, L.; Altamimi, A.; García Sánchez, M.G.; Tamburrinia, R.; Asthana, A.; Gazia, C.; Orlando, G. Utility of Extracellular Matrix Powders in Tissue Engineering. *Organogenesis* **2018**, *14*, 172–186. [CrossRef] [PubMed]
15. Ning, L.J.; Jiang, Y.L.; Zhang, C.H.; Zhang, Y.; Yang, J.L.; Cui, J.; Zhang, Y.J.; Yao, X.; Luo, J.C.; Qin, T.W. Fabrication and Characterization of a Decellularized Bovine Tendon Sheet for Tendon Reconstruction. *J. Biomed. Mater. Res. A* **2017**, *105*, 2299–2311. [CrossRef]
16. Spang, M.T.; Christman, K.L. Extracellular Matrix Hydrogel Therapies: In Vivo Applications and Development. *Acta Biomater.* **2018**, *68*, 1–14. [CrossRef]
17. Pati, F.; Jang, J.; Ha, D.H.; Won Kim, S.; Rhie, J.W.; Shim, J.H.; Kim, D.H.; Cho, D.W. Printing Three-Dimensional Tissue Analogues with Decellularized Extracellular Matrix Bioink. *Nat. Commun.* **2014**, *5*, 3935. [CrossRef]
18. Badylak, S.F.; Freytes, D.O.; Gilbert, T.W. Extracellular Matrix as a Biological Scaffold Material: Structure and Function. *Acta Biomater.* **2009**, *5*, 1–13. [CrossRef]
19. Heath, D.E. A Review of Decellularized Extracellular Matrix Biomaterials for Regenerative Engineering Applications. *Regen. Eng. Transl. Med.* **2019**, *5*, 155–166. [CrossRef]
20. Negishi, J.; Hashimoto, Y.; Yamashita, A.; Zhang, Y.; Kimura, T.; Kishida, A.; Funamoto, S. Evaluation of Small-Diameter Vascular Grafts Reconstructed from Decellularized Aorta Sheets. *J. Biomed. Mater. Res. A* **2017**, *105*, 1293–1298. [CrossRef]
21. Wu, P.; Nakamura, N.; Morita, H.; Nam, K.; Fujisato, T.; Kimura, T.; Kishida, A. A Hybrid Small-Diameter Tube Fabricated from Decellularized Aortic Intima-Media and Electrospun Fiber for Artificial Small-Diameter Blood Vessel. *J. Biomed. Mater. Res. A* **2019**, *107*, 1064–1070. [CrossRef] [PubMed]
22. Lichtenberg, A.; Tudorache, I.; Cebotari, S.; Ringes-Lichtenberg, S.; Sturz, G.; Hoeffler, K.; Hurscheler, C.; Brandes, G.; Hilfiker, A.; Haverich, A. In Vitro Re-Endothelialization of Detergent Decellularized Heart Valves under Simulated Physiological Dynamic Conditions. *Biomaterials* **2006**, *27*, 4221–4229. [CrossRef] [PubMed]
23. Taylor, D.A.; Sampaio, L.C.; Ferdous, Z.; Gobin, A.S.; Taite, L.J. Decellularized Matrices in Regenerative Medicine. *Acta Biomater.* **2018**, *74*, 74–89. [CrossRef] [PubMed]
24. Wu, P.; Nakamura, N.; Kimura, T.; Nam, K.; Fujisato, T.; Funamoto, S.; Higami, T.; Kishida, A. Decellularized Porcine Aortic Intima-Media as a Potential Cardiovascular Biomaterial. *Interact. Cardiovasc. Thorac. Surg.* **2015**, *21*, 189–194. [CrossRef]
25. Laker, L.; Dohmen, P.M.; Smit, F.E. Synergy in a Detergent Combination Results in Superior Decellularized Bovine Pericardial Extracellular Matrix Scaffolds. *J. Biomed. Mater. Res. A* **2020**, *108*, 2571–2578. [CrossRef]
26. Butler, D.L.; Guan, Y.; Kay, M.D.; Cummings, J.F.; Feder, S.M.; Levy, M.S. Location-Dependent Variations in the Material Properties of the Anterior Cruciate Ligament. *J. Biomech.* **1992**, *25*, 511–518. [CrossRef]
27. Nikkhah, M.; Eshak, N.; Zorlutuna, P.; Annabi, N.; Castello, M.; Kim, K.; Dolatshahi-Pirouz, A.; Edalat, F.; Bae, H.; Yang, Y.; et al. Directed Endothelial Cell Morphogenesis in Micropatterned Gelatin Methacrylate Hydrogels. *Biomaterials* **2012**, *33*, 9009–9018. [CrossRef]
28. Li, X.; Wang, X.; Yao, D.; Jiang, J.; Guo, X.; Gao, Y.; Li, Q.; Shen, C. Effects of Aligned and Random Fibers with Different Diameter on Cell Behaviors. *Colloids Surf. B Biointerfaces* **2018**, *171*, 461–467. [CrossRef]
29. Humphrey, J.D.; Dufresne, E.R.; Schwartz, M.A. Mechanotransduction and Extracellular Matrix Homeostasis. *Nat. Rev. Mol. Cell Biol.* **2014**, *15*, 802–812. [CrossRef]
30. Wang, B.; Shi, J.; Wei, J.; Wang, L.; Tu, X.; Tang, Y.; Chen, Y. Fabrication of elastomer pillar arrays with height gradient for cell culture studies. *Microelectron. Eng.* **2017**, *175*, 50–55. [CrossRef]
31. Kimura, T.; Kondo, M.; Hashimoto, Y.; Fujisato, T.; Nakamura, N.; Kishida, A. Surface Topography of PDMS Replica Transferred from Various Decellularized Aortic Lumens Affects Cellular Orientation. *ACS Biomater. Sci. Eng.* **2019**, *5*, 5721–5726. [CrossRef] [PubMed]
32. Gilbert, T.W.; Sellaro, T.L.; Badylak, S.F. Decellularization of Tissues and Organs. *Biomaterials* **2006**, *27*, 3675–3683. [CrossRef] [PubMed]
33. Xie, J.; Bao, M.; Bruekers, S.M.C.; Huck, W.T.S. Collagen Gels with Different Fibrillar Microarchitectures Elicit Different Cellular Responses. *ACS Appl. Mater. Interfaces* **2017**, *9*, 19630–19637. [CrossRef] [PubMed]

Article

In-Vitro Endothelialization Assessment of Heparinized Bovine Pericardial Scaffold for Cardiovascular Application

My Thi Ngoc Nguyen [1,2,3] and Ha Le Bao Tran [1,2,3,*]

1. Laboratory of Tissue Engineering and Biomedical Materials, University of Science, Ho Chi Minh City 700000, Vietnam; ntnmy@hcmus.edu.vn
2. Department of Physiology and Animal Biotechnology, Faculty of Biology—Biotechnology, University of Science, Ho Chi Minh City 700000, Vietnam
3. Vietnam National University, Ho Chi Minh City 700000, Vietnam
* Correspondence: tlbha@hcmus.edu.vn

Abstract: (1) Background: Hemocompatibility is a critical challenge for tissue-derived biomaterial when directly contacting the bloodstream. In addition to surface modification with heparin, endothelialization of the grafted material is suggested to improve long-term clinical efficacy. This study aimed to evaluate the ability to endothelialize in vitro of heparinized bovine pericardial scaffolds. (2) Methods: bovine pericardial scaffolds were fabricated and heparinized using a layer-by-layer assembly technique. The heparinized scaffolds were characterized for heparin content, surface morphology, and blood compatibility. Liquid extraction of the samples was prepared for cytotoxicity testing on human endothelial cells. The in-vitro endothelialization was determined via human endothelial cell attachment and proliferation on the scaffold. (3) Results: The heparinized bovine pericardial scaffold exhibited a heparin coating within its microfiber network. The scaffold surface immobilized with heparin performed good anti-thrombosis and prevented platelet adherence. The proper cytotoxicity impact was observed for a freshly used heparinized sample. After 24 h washing in PBS 1X, the cell compatibility of the heparinized scaffolds was improved. In-vitro examination results exhibited human endothelial cell attachment and proliferation for 7 days of culture. (4) Conclusions: Our in-vitro analysis provided evidence for the scaffold's ability to support endothelialization, which benefits long-term thromboresistance.

Keywords: heparinize; bovine pericardium; decellularization; scaffold; hemocompatibility; endothelialization

1. Introduction

Bovine pericardium, first utilized for bioprosthetic heart valve fabrication, has been widely used in clinical practice [1,2]. Decellularization is considered the critical factor that significantly promotes bovine pericardium and other tissue-derived extracellular matrix (ECM) for biomedical research and application. This technique focuses on eliminating xenogenic antigenicity while preserving the native ECM structural, mechanical, and biologic properties. For long-term stability in vivo, further fixation or crosslinking, i.e., glutaraldehyde treatment, was adopted to prepare tissue ECM-derived biomaterials. Decellularized and glutaraldehyde treated bovine pericardium are now commonly used as commercially available products proven for their durability, biocompatibility, and an appropriate effect on host tissue remodeling. In the field of cardiovascular practice, it has been found that bovine pericardial patches revealed significantly less suture line bleeding (at 14%) than a prosthetic patch material (at 55% in the Dacron patches) [3,4]. Somehow, there is still a demand for a non-thrombogenic strategy due to the particular risk of coagulation on this biomaterial when directly contacting the bloodstream [4–7].

To prevent early thrombosis after implantation, patients are treated with anticoagulation drugs via injection or oral use. Another potential solution is to improve the hemocompatibility of blood-contacting biomaterials. Material-surface modification has been taken into account, including immobilization of bioactive anticoagulants. Heparin is the most commonly used in systemic anticoagulant therapy. Heparin belongs to the family of glycosaminoglycans and is characterized as a highly sulfated linear polysaccharide, with large quantities of hydroxyl, amino, carboxyl, and sulfonyl. Typically, heparin is linked to the extracellular matrix (ECM) using an active crosslinker such as glutaraldehyde and (1-(3-dimethylaminopropyl)-3-ethylcarbodiimide hydrochloride) (so-called EDC) [5–7]. These chemical immobilization methods have shown effectiveness in grafting heparin onto ECM materials with considerable stability, whereas the residual amount of crosslink reagent presents potential cytotoxicity and calcification. In addition, a heparin incorporation method by layer-by-layer (LBL) assembly was reported, in which complex multilayers of iron-heparin were deposited onto pericardial ECM fibers [8,9]. This procedure effectively reduced blood-clot formation on the material surface. It also revealed heparin release and maintained an antithrombotic effect during a 30-day washing in phosphate buffer saline [9]. However, when reaching the end-point of heparin release, thrombus formation would likely occur, which could diminish the clinical efficacy of this material. Thus, there is a demand for a long-term strategy to improve the hemocompatibility of this material.

Indeed, the principal player in thrombosis prevention is the endothelial layer on blood-contacting surfaces. The endothelial layer was shown to not occur on any cardiovascular device immediately after implantation, causing thrombosis [10]. For early endothelialization, biochemical modification of the surfaces and immobilization of biomolecules to the surfaces have been useful. The present publications have provided evidence for the attachment of endothelial cells on heparin-modified material surfaces [11–13]. At the same time, pericardial scaffolds were also proven to support endothelial cell adherence and proliferation [14]. Therefore, it could be predicted that a heparinized bovine pericardial scaffold might exhibit a positive endothelialization effect. Our previous study successfully incorporated heparin onto a bovine pericardial scaffold, which indicated good thrombosis resistance. In the current research, we aimed to investigate further the ability of the scaffold to support endothelialization in vitro. The heparinized bovine pericardial scaffolds were examined for their cytotoxicity and effect on human endothelial cell attachment and proliferation.

2. Materials and Methods

2.1. Preparation of Bovine Pericardial Scaffold with Heparin Modification

Bovine pericardial scaffold (BPS) was prepared and heparinized according to our previous publications [9,15]. Briefly, the bovine pericardium was decellularized with 10 mM Tris-HCl (Sigma Aldrich, St. Louis, MI, USA) and 0.15% SDS (Sigma Aldrich, USA), followed by further stabilizing in 0.1% glutaraldehyde (Sigma Aldrich, USA). The BPS was obtained after lyophilization and sterilization by gamma irradiation. The BPS underwent a heparin modification, in which the BPS was incubated in DHI ($[Fe(OH)_2]^+$) (Sigma Aldrich, USA) and heparin solution (Sigma Aldrich, USA). The procedure includes seven cycles of layer-by-layer (LbL) assembly for heparin deposition on the scaffold. In each cycle, BPS was immersed in DHI solution for 5 min, then washed in 0.9% NaCl (Sigma Aldrich, USA) solution for 5 min, followed by heparin condition for 5 min at room temperature. The heparinized BPS (HepBPS) were examined freshly after fabrication.

2.2. Toluidine Blue O (TBO) Assay

Heparin content in HepBPS was measured by TBO assay, which colorimetrically measures glucosamine concentration. HcpBPS samples were lyophilized and cut into 1×1 cm and incubated in 0.04% TBO solution (Sigma-Aldrich, USA) for 4 h. The samples were washed in deionized water to eliminate unreacted TBO solution and immersed in 80% Ethanol/0.1 M NaOH mixture to extract the heparin/TBO complex. The finalized solution was measured for absorbance at 530 nm using a microplate reader (EZ Read 400, Biochrom, Cambridge, UK). Heparin content on samples was evaluated by comparing with a calibration curve.

2.3. Structural Characteristics

2.3.1. Hematoxyline and Eosin Staining

The BPS and HepBPS were fixed with 10% formalin, embedded in paraffin, and processed for staining with hematoxylin and eosin (H&E). Images were taken using Olympus CKX-RCD microscope (Tokyo, Japan) equipped with a DP2-BSW microscope digital camera.

2.3.2. Scanning Electron Microscopy

The BPS and HepBPS were washed once by PBS 1X and fixed in a homemade fixation solution (1.35% Paraformaldehyde, 3% Glutaraldehyde, 4.45% Succrose in PBS 1X, pH 7.4) for SEM evaluation. All samples were fixed under 4 °C for 24 h and dehydrated through a graded series of ethanol. Samples were sputter coated with a layer of gold for 60 s to observe the high-resolution images. The ultrastructure of all samples was investigated using scanning electron microscopy (SEM, JSM-6510, JEOL, Tokyo, Japan).

2.4. Thrombosis and Platelet Attachment

To evaluate the potential antithrombogenicity, the scaffolds were immersed in citrated whole blood. After incubation at 37 °C for 2 h, blood coagulation was activated by adding with 0.1 M $CaCl_2$ solution (ration 1:10, v/v) [9]. The scaffolds were briefly flushed with distilled water and examined for the presence of blood clots on the surface.

In platelet attachment test, platelet rich plasma (PRP) was collected from whole blood. Each scaffold was incubated with PRP at 37 °C for 2 h, followed by washing with PBS 1X. The attachment of platelets on the scaffolds was examined by SEM.

2.5. In-Vitro Cytotoxicity Tests

In-Vitro Cell Culture

HUVEC were purchased from the American Type Culture Collection (ATCC, Manassas, VA, USA). HUVEC were cultured in vascular cell basal medium (ATCC, USA) supplemented with Endothelial Cell Growth Kit-VEGF and Penicillin-Streptomycin-Amphotericin B Solution (ATCC, USA) (so-called complete culture medium). The cells were cultured at 37 °C and 5% CO_2, with a constant humidity of 95% (Panasonic, Osaka, Japan). Culture medium was freshly changed every 2 days. HUVEC at 4th passage were used for the assessment. HUVEC suspension was prepared by trypsinization and seeded into each well of a 96-well plate at 1×10^4 cells/mL. The culture plate was incubated in 5% CO_2 humid atmosphere at 37 °C overnight for cell attachment and spreading.

2.6. Sample Extraction and Testing

The HepBPS samples underwent either non-wash or 24 h wash in PBS 1X solution before the extraction process. The extract of the samples was prepared following the ISO 10993-12 standard. Complete culture medium was used as an extraction medium. Each sample was extracted in the extraction medium (1 cm^2 per 1 mL) at 37 °C for 24 h to obtain a test liquid extract. The culture medium was removed completely on the test day and the test extract (100 μL) was added. Culture medium and culture medium containing 20% DMSO were used as the blank and positive control, respectively. The plate was incubated in 5% CO_2 humid atmosphere at 37 °C for further 24 h. Cell viability was determined using MTT assay.

2.7. Cytotoxicity Determination

After 24 h incubation, tested liquid extract and controls were removed, followed by adding 100 μL of salt (MTT) solution (Sigma-Aldrich, St. Louis, MI, USA) (0.5 mg/mL) into each well. The resultant mixture was incubated in 5% CO_2 humid atmosphere at 37 °C for 4 h for the formation of formazan crystals. The MTT solution was removed and stopping solution of DMSO and Ethanol (Sigma-Aldrich, USA) (ratio 1:1, v/v) was added to the wells. Stopping solution was mixed in wells to completely dissolve the formazan crystal. The optical density (OD) was measured at 570 nm with the ELISA Microplate Reader (Biochrom EZ Read 400, Cambridge, UK).

Sample cytotoxicity was determined according to the relative growth rate (RGR, %). Formular for RGR of test sample (%) was as $(OD_{test}/OD_{blank}) \times 100\%$. The liquid extract is confirmed as no-toxic to the cells if the RGR value is higher than 70 %, according to ISO 10993-5 guidance [2].

2.8. Cell Attachment on the Scaffolds

To examine cell attachment, the HepBPS was cut into round specimens and placed into the wells of 96-well plate. HUVEC were seeded into each well at a density of 3×10^4 cells per well and cultured for 24 h. The specimen seeded with HUVEC were collected, rinsed once in PBS 1X, and incubated in Calcein A solution (Thermo Scientific, Waltham, MA, USA) (ration 1:1000 in PBS, v/v) for 35 min at 37 °C. Samples were briefly rinsed in PBS and imaged using a fluorescence microscope (Olympus IX81, Olympus, Tokyo, Japan). HUVEC adherence and morphology were also observed by scanning electron microscope as per earlier described procedure.

2.9. Cell Proliferation Assay

For the proliferation evaluation, HUVEC were seeded into the well containing HepBPS specimen with a density of 3×10^4 cells. HUVEC proliferation was examined by a Cell Counting Kit-8 (CCK-8) (Sigma-Aldrich, USA) on day 1, 4, and 7. At each time point, medium containing 10% CCK-8 was added to each well and incubated for 4 h at 37 °C. The medium was transferred into a new 96-well plate for measuring the optical absorbance at 450 nm with ELISA Microplate Reader.

2.10. Statistical Analysis

All data sets were analyzed by Student's *t*-test for comparison between two groups using the GraphPad 8.0 software. Data were expressed as mean ± standard division of the mean (SD), and statistical significance was set at $p < 0.05$. All experiments were conducted in triplicate.

3. Results

3.1. Characterization of the Heparinized Bovine Pericardial Scaffold (HepBPS)

Macroscopical observation presented a significant difference between the BPS (Figure 1A) and HepBPS (Figure 1C), in which HepBPS adopted a diffuse brown appearance. After incubation with the TBO solution, the BPS sample obtained a light blue color (Figure 1B), whereas a homogeneous presence of purple crystals was clearly detected on the HepBPS sample (Figure 1D). This result indicated a successful deposition of heparin on the scaffold using the LbL assemble technique. Quantitatively, heparin content in the HepPBS was determined as 169.5 ± 17.31 mg/cm^2 (Figure 1E).

The pericardium derived scaffold was composed of extracellular matrix fibers, which are basic and was stained with eosin as an acid dye (Figure 2A). After heparin modification, the outer layer of sections was stained dark blue (Figure 2B). Heparin is highly acidic because of sulfate and carboxylic acid groups [16,17], which could generate selective reactions with hematoxylin as a basic dye. SEM images confirmed this variation in surface morphology between these surfaces before and after heparin modification (Figure 2C,D). After modification with heparin by the LbL technique, SEM illustration showed the deposition of the DHI/heparin complex as coatings around the fibrils of the HepBPS structure (Figure 2D).

There was the apparent formation of a blood clot on the surface of the BPS sample compared with the thrombus-free surface of HepBPS sample. On the BPS membrane, some adhered platelets were found (Figure 2E), which corresponded to its thrombus formation. HepBPS performed better hemocompatibility, with nearly no platelets attached (Figure 2F).

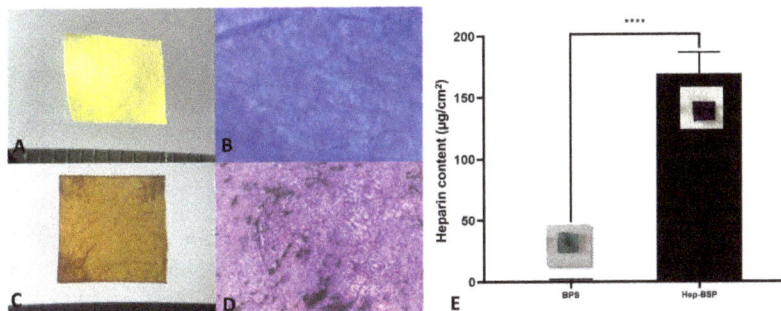

Figure 1. Preparation of heparinized bovine pericardial scaffold. (**A**)—Macroscopic observation of Bovine pericardial scaffold (BPS). (**B**)—Stereo microscope image of BPS after toluidine blue incubation. (**C**)—Heparinized bovine pericardial scaffold (HepBPS). (**D**)—Stereo microscope image of HepBPS after toluidine blue incubation. (**E**)—Heparin quantification of the scaffolds. ****: p value < 0.0001.

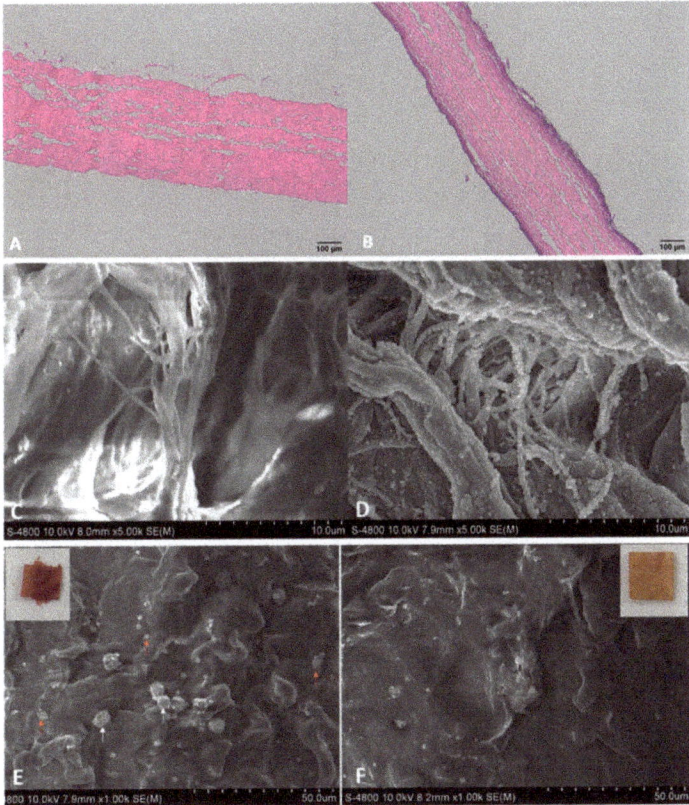

Figure 2. Characterization of heparinized bovine pericardial scaffold. (**A**)—Hematoxylin and Eosin (H&E) staining of Bovine pericardial scaffold (BPS). (**B**)—H&E staining of Heparinized bovine pericardial scaffold (HepPBS). (**C**)—Scanning electron microscope (SEM) of BPS. (**D**)—SEM of HepPBS. (**E**)—Platelet attachment and SEM examination of BPS. (**F**)—Platelet attachment and SEM examination of HepBPS. H&E staining: All scale bars are 10 μm. SEM: All scale bars are 50 μm. Red arrows indicate platelets. White arrows indicate white blood cells.

3.2. In-Vitro Cytotoxicity Tests

In-vitro cytotoxicity tests aimed to provide predictive evidence of biocompatibility. HUVEC were cultured in the condition of the HepBPS liquid extract. After 24 h incubation, no cell death and changes in cell pattern were observed in the complete medium as a negative control (Figure 3A). Meanwhile, 20% DMSO solution as a positive control was highly toxic to the cells, resulting in cell death and detachment (Figure 3B). For the non-washed sample, which imitated the immediate effect of HepBPS when implanted, there was a proper negative effect on cell viability indicated by the decrease in cell adherence density (Figure 3C) and relatively low RGR percentage (75.83%). After 24 h washing in PBS 1X, the cytotoxicity of the HepBPS liquid extract was pretty low, showing as high a level of RGR as 96.56% (Figure 3E). The liquid extract of DMSO caused harsh affects on cell viability, with RGR values at 3%.

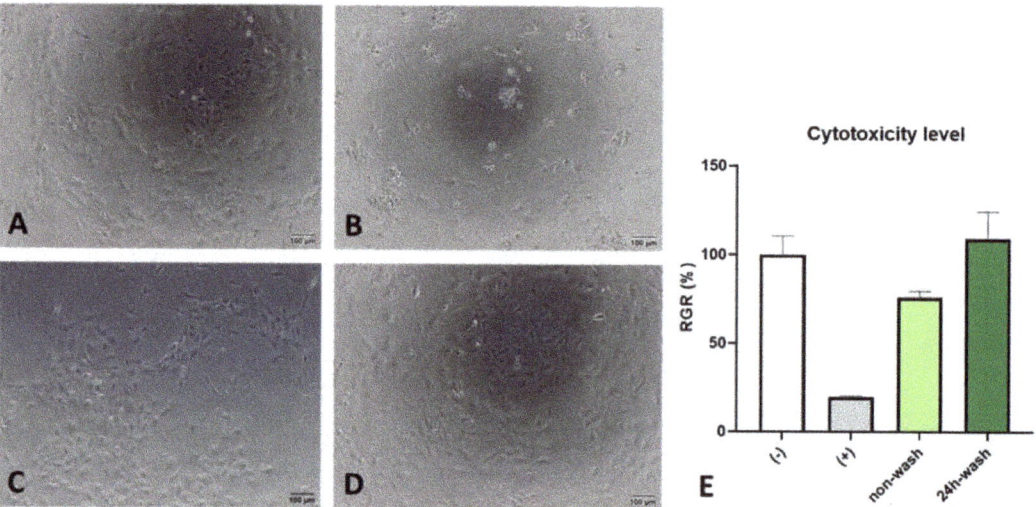

Figure 3. Observation of HUVEC cultured in different solutions. (**A**)—Culture medium as a negative control. (**B**)—Liquid extract from latex as a positive control. (**C**)—Liquid extract from a non-washed HepBPS. (**D**)—Liquid extract from HepBPS after 24 h wash in PBS 1X. (**E**) Cytotoxicity determination (Magnification 10×). All scale bars are 100 µm.

3.3. Cell Attachment on the Scaffolds

HUVEC were used to examine the cell attachment support of the scaffolds. HUVEC were seeded onto either BPS or HepBPS and visualized by calcein staining (Figure 4) and SEM (Figure 5). HUVEC attachment on both scaffolds was detected after culturing for 24 h. Somehow, the density of endothelial cells on the HepBPS membrane was similar to that on the surface of the BPS membrane (Figure 4B,D). SEM images also exhibited HUVEC spreading morphology with cellular projections interacting with scaffold components.

3.4. Cell Proliferation on the Scaffolds

Cell proliferation was determined by CCK8 assay at different time points, on day 1, 4, and 7, as shown in the chart (Figure 5E). In both types of scaffolds, the proliferation rate of HUVEC within the first four days was recorded. During the next 4 and 7 days, the optical absorbance of the incubated solution did not significantly increase, indicating a limited growth rate after 7 days. However, data indicated a slower cell growth rate in the scaffolds after heparinization, which was shown as a significantly higher in OD value of the BPS group compared to the HepBPS group.

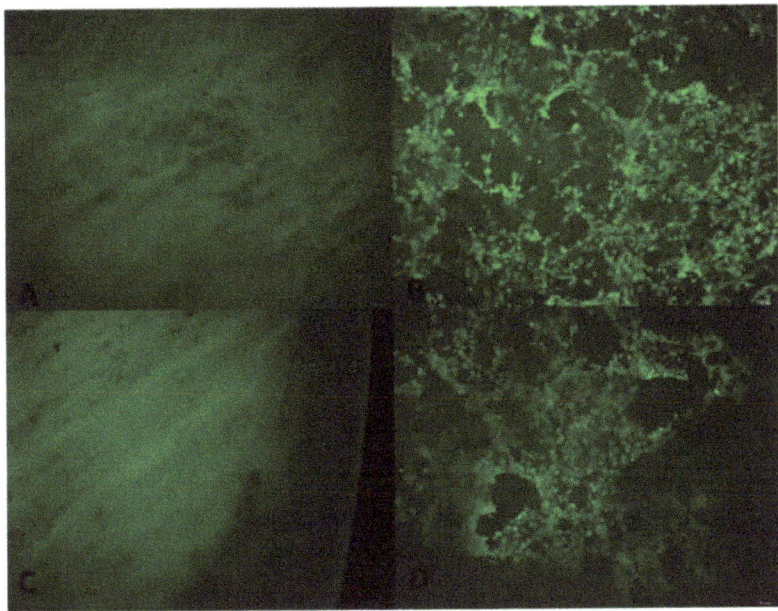

Figure 4. Calcein staining of BPS and HepBPS after HUVEC seeding. (**A**)—BPS without cell seeding. (**B**)—BPS seeded with HUVEC. (**C**)—HepBPS without cell seeding. (**D**)—HepBPS seeded with HUVEC. All scale bars represent 100 μm.

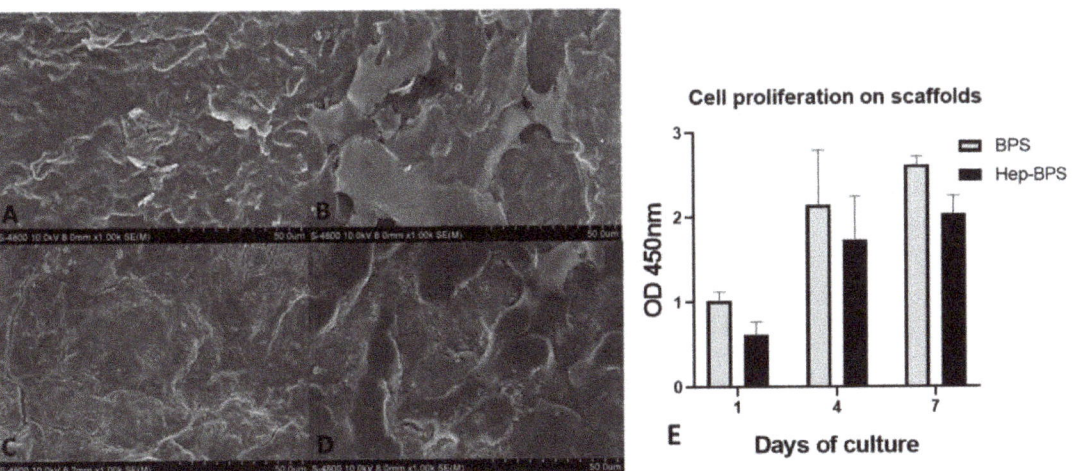

Figure 5. Assessment of cell attachment by SEM and cell proliferation on scaffolds. (**A**)—BPS without cell seeding. (**B**)—BPS seeded with HUVEC. (**C**)—HepBPS without cell seeding. (**D**)—HepBPS seeded with HUVEC. (**E**)—CCK8 assay for cell proliferation. All scale bars represent 50 μm.

4. Discussion

Acellular-tissue-matrix-derived biomaterials provide extensive applications in clinical surgery due to their ready availability and potential tissue regeneration. In the case of cardiovascular patch fabrication, bovine pericardium is the most well-known material for its suitable thickness, low rate of suture bleeding, good biocompatibility, and mechanical properties. As the cardiovascular patch is always in direct contact with the bloodstream, besides the mentioned advantages, this type of material still has a drawback regarding its hemocompatibility. A hemocompatible material should not cause any adverse interactions with blood components, especially not activating blood coagulation or blood clots. However, the pericardium is a biologic tissue; it is mainly composed of collagen fibers, which create an attractive surface for the absorption of plasma proteins and platelet attachment. Therefore, blood clots or thrombosis on the material surface is an obvious consequence. Therefore, there is a demand to improve the hemocompatibility of this material.

To prevent early thrombosis after implantation, surface modification strategies have been developed to improve blood compatibility, including the immobilization of bioactive anticoagulants. In this case, heparin is the most commonly used in systemic anticoagulant therapy. Our previous study successfully established a heparinization of the bovine pericardial scaffold using the layer-by-layer assembly technique [9]. The results showed that heparin could be incorporated into the bovine pericardial scaffold, as proven via SEM images, histological analysis, and heparin amount assay. The seven cycles selected for the LbL technique was determined due to the high level of heparin accumulation, which indicated seven assembly cycles as the heparin immobilization threshold.

In this current study, we also used this technique to immobilize heparin to the scaffolds and further tested for their anti-thrombotic activity and endothelialization support. The results confirmed that the preparation technique effectively deposited heparin in the bovine pericardial scaffold. The macroscopic analysis demonstrated adequate uniform coverage over the surface of the scaffold. Compared with the untreated scaffolds, heparinization resulted in a visually clot-free surface and empty platelet attachment. This observation was similar to the present studies on adopting heparin for surface modification, including decellularized vascular graft and live matrix [8,9,18,19]. Accordingly, surface modification with heparin could provide anti-thrombosis with two effects. In a direct impact, a surface with heparin would prevent the adsorption of plasma proteins and platelet adherence, therefore successfully creating an anti-thrombosis surface. In the other way, after modification, the material can release heparin which, in turn, suppresses the activity of thrombin, thus creating an anti-thrombosis microenviroment within the material and keeping the material surface in a free-clotting condition [9,18].

Endothelial cells form a consensus layer in the blood vessel and maintain the normal physiology condition of blood vessel. One of the most essential functions of the endothelial cell is to prevent thrombosis. Therefore, from a material perspective, endothelialization is a process in which the endothelial cells can form a layer on the material surface. If the cardiovascular patch could achieve this ideal stage, long-term anti-thrombosis could be guaranteed. Our published data showed that the heparinized bovine pericardial scaffolds could release heparin and maintain the anti-thrombus stage for 30 days [9]. This result indicated that the blood clotting on the scaffold surface possibly happens as a consequence of complete heparin release. Therefore, the ability to support endothelialization could, in turn, ensure the blood compatibility of the scaffold. Overall, our study demonstrated that HepBPS could provide an appropriate attachment and proliferation of human endothelial cells. However, before this performance, the HepBPS should undergo 24 h washing in PBS 1X solution; otherwise, potential cytotoxicity could be a certain (as shown in cytotoxicity assay, Figure 3). Although heparin is frequently used as an anticoagulant, there were findings of the toxic effects of heparin in cell cultures [20]. Additionally, there was the presence of DHI ions in the HepBPS, which was also unloaded during incubation and possibly caused a decrease in cell viability. Meanwhile, the in-vitro cytotoxicity of the releasing heparin and DHI ions in our specific case and present publications [9,10,20] remains unsolved, thus demanding a detailed investigation in further study. Endothelialization was detected on HepBPS via endothelial cell attachment and proliferation on the scaffold (as shown in Figures 4 and 5). This implied the two-phase anti-thrombotic activity of the HepBPS when implanted. In the early time of implantation, heparin release and heparin on the graft surface would take the main role in anti-thrombus. Then, the scaffold itself could support endothelialization, which benefits the long-term thromboresistance by preventing the sub-endothelial matrix from blood contact.

As mentioned earlier, endothelial cells were confirmed to attach to a heparin-modified material surface [11–13,17], which was also found in our current result. The results also showed that the endothelialization effect on the heparinized scaffolds was not comparable to the un-modified one, which might raise a contradiction on the initial purpose of employing heparin on the bovine pericardial scaffolds. In fact, SEM images revealed completely coated microfibers with the heparin/DHI complex after the LbL assembly procedure. Consequently, cell attachment would be dismissed due to reduced or a lack of cell–matrix interaction. Endothelial cells were shown not to express direct receptors for heparin. The positive effects of the heparin immobilized scaffold on endothelial cells were described via "bridging" molecules, e.g., chitosan, gelatin [11,17], or binding and stabilizing cell growth factors, e.g., VEGF [12,13]. Therefore, the decrease in cell adherence or proliferation on HepBPS could be explainable. Another mechanism for endothelialization relates to the protein adsorption level on the scaffold. In-vivo transplantation clearly demonstrates that protein adsorption is the first event that happens at the interface, leading to a later material associated with cell attachment and growth [11,13]. The heparinized scaffold is negatively charged, possibly minimizing negatively charged protein adsorption [18,21] and endothelialization. Taken together, our in-vitro results on HepBPS remained the predictive evidence for the scaffold's ability to support endothelial cell attachment and proliferation in both cases, including the later period of heparin release in vitro/in vivo, long-term interaction, and uptaking of plasma proteins and growth factors in vivo. Therefore, the mechanisms for interactions between the bovine pericardium, heparin and cells need to be performed. Further investigations on these undefined behaviors of the heparinized bovine pericardial scaffold in vitro and in vivo are also recommended to provide a better understanding of the clinical application.

5. Conclusions

In this study, heparinized bovine pericardial scaffolds were constructed via LbL assembly successfully. The results demonstrated that the use of the heparinization technique provided anti-thrombosis and prevented platelet adhesion. After an adequate heparin release by immersing in PBS 1X, the scaffolds could support the attachment and proliferation of endothelial cells, which exhibited a promising effect on endothelialization and long-term hemocompatibility. However, in order to achieve complete understanding of its behaviors, additional investigation should be performed.

Author Contributions: M.T.N.N. prepared the material, conducted experimental work, wrote the draft paper, and performed the theatrical analysis; H.L.B.T. mentored, reviewed, and revised the papers. All authors have read and agreed to the published version of the manuscript.

Funding: This research is funded by Vietnam National University Ho Chi Minh City (VNU-HCM) under grant number B2021-18-03.

Institutional Review Board Statement: Not applicable.

Data Availability Statement: Data presented in this study are available in the article.

Acknowledgments: All authors acknowledge funding received for this project from Vietnam National University Ho Chi Minh City (VNU-HCM) under grant number B2021-18-03. We acknowledge the contributing advice on the blood-contacting test and cardiovascular applications from Thang Quoc BUI, Pediatric Cardiovascular Surgery, Doctor of Medicine: Cho Ray Hospital, Ho Chi Minh City, Vietnam.

Conflicts of Interest: The authors declare no conflict of interest.

References

1. Dimitrievska, S.; Cai, C.; Weyers, A.; Balestrini, J.L.; Lin, T.; Sundaram, S.; Hatachi, G.; Spiegel, D.A.; Kyriakides, T.R.; Miao, J.; et al. Click-coated, heparinized, decellularized vascular grafts. *Acta Biomater.* **2015**, *13*, 177–187. [CrossRef] [PubMed]
2. Shklover, J.; McMasters, J.; Alfonso-Garcia, A.; Higuita, M.L.; Panitch, A.; Marcu, L.; Griffiths, L. Bovine pericardial extracellular matrix niche modulates human aortic endothelial cell phenotype and function. *Sci. Rep.* **2019**, *9*, 16688. [CrossRef] [PubMed]
3. Li, X.; Guo, Y.; Ziegler, K.R.; Model, L.S.; Eghbalieh, S.D.; Brenes, R.A.; Kim, S.T.; Shu, C.; Dardik, A. Current usage and future directions for the bovine pericardial patch. *Ann. Vasc. Surg.* **2011**, *25*, 561–568. [CrossRef] [PubMed]
4. Ren, S.; Li, X.; Wen, J.; Zhang, W.; Liu, P. Systematic Review of Randomized Controlled Trials of Different Types of Patch Materials during Carotid Endarterectomy. *PLoS ONE* **2013**, *8*, e55050. [CrossRef] [PubMed]
5. Cai, Z.; Gu, Y.; Cheng, J.; Li, J.; Xu, Z.; Xing, Y.; Wang, C.; Wang, Z. Decellularization, cross-linking and heparin immobilization of porcine carotid arteries for tissue engineering vascular grafts. *Cell Tissue Bank.* **2019**, *20*, 569–578. [CrossRef]
6. Zhou, M.; Liu, Z.; Liu, C.; Jiang, X.; Wei, Z.; Qiao, W.; Ran, F.; Wang, W.H.; Qiao, T.; Liu, C. Tissue engineering of small-diameter vascular grafts by endothelial progenitor cells seeding heparin-coated decellularized scaffolds. *J. Biomed. Mater. Research. Part B Appl. Biomater.* **2012**, *100*, 111–120. [CrossRef] [PubMed]
7. Lee, W.K.; Park, K.D.; Keun Han, D.; Suh, H.; Park, J.C.; Kim, Y.H. Heparinized bovine pericardium as a novel cardiovascular bioprosthesis. *Biomaterials* **2000**, *21*, 2323–2330. [CrossRef]
8. Tao, Y.; Hu, T.; Wu, Z.; Tang, H.; Hu, Y.; Tan, Q.; Wu, C.H. Heparin nanomodification improves biocompatibility and biomechanical stability of decellularized vascular scaffolds. *Int. J. Nanomed.* **2012**, *7*, 5847–5858. [CrossRef]
9. Nguyen, M.T.N.; Tran, H.L.B. Heparinization of the bovine pericardial scaffold by layer-by-layer (LBL) assembly technique. *J. Sci. Adv. Mater. Devices* **2022**, *7*, 100405. [CrossRef]
10. Jana, S. Endothelialization of cardiovascular devices. *Acta Biomater.* **2019**, *99*, 53–71. [CrossRef]
11. Leijon, J.; Carlsson, F.; Brännström, J.; Sanchez, J.; Larsson, R.; Nilsson, B.; Magnusson, P.U.; Rosenquist, M. Attachment of Flexible Heparin Chains to Gelatin Scaffolds Improves Endothelial Cell Infiltration. *Tissue Eng. Part A* **2013**, *19*, 1336–1348. [CrossRef] [PubMed]
12. Xie, J.; Wan, J.; Tang, X.; Li, W.; Peng, B. Heparin modification improves the re-endothelialization and angiogenesis of decellularized kidney scaffolds through antithrombosis and anti-inflammation in vivo. *Transl. Androl. Urol.* **2021**, *10*, 3656–3668. [CrossRef] [PubMed]
13. Nie, C.; Ma, L.; Cheng, C.; Deng, J.; Zhao, C. Nanofibrous heparin and heparin-mimicking multilayers as highly effective endothelialization and antithrombogenic coatings. *Biomacromolecules* **2015**, *16*, 992–1001. [CrossRef] [PubMed]
14. Liu, Z.Z.; Wong, M.L.; Griffiths, L.G. Effect of bovine pericardial extracellular matrix scaffold niche on seeded human mesenchymal stem cell function. *Sci. Rep.* **2016**, *6*, 37089. [CrossRef]
15. Nguyen, M.T.N.; Tran, H.L.B. Effect of Modified Bovine Pericardium on Human Gingival Fibroblasts in vitro. *Cells Tissues Organs* **2018**, *206*, 296–307. [CrossRef]
16. Liu, M.; Yue, X.; Dai, Z.; Ma, Y.; Xing, L.; Zha, Z.; Liu, S.; Li, Y. Novel Thrombo-Resistant Coating Based on Iron—Polysaccharide Complex Multilayers. *ACS Appl. Mater. Interfaces* **2009**, *1*, 113–123. [CrossRef]
17. Wacker, M.; Riedel, J.; Walles, H.; Scherner, M.; Awad, G.; Varghese, S.; Schürlein, S.; Garke, B.; Veluswamy, P.; Wippermann, J.; et al. Comparative Evaluation on Impacts of Fibronectin, Heparin-Chitosan, and Albumin Coating of Bacterial Nanocellulose Small-Diameter Vascular Grafts on Endothelialization In Vitro. *Nanomaterials* **2021**, *11*, 1952. [CrossRef]
18. Patel, H. Blood biocompatibility enhancement of biomaterials by heparin immobilization: A review. *Blood Coagul. Fibrinolysis* **2021**, *32*, 237–247. [CrossRef]
19. Bruinsma, B.G.; Kim, Y.; Berendsen, T.A.; Ozer, S.; Yarmush, M.L.; Uygun, B.E. Layer-by-layer heparinization of decellularized liver matrices to reduce thrombogenicity of tissue engineered grafts. *J. Clin. Transl. Sci.* **2015**, *1*, 48–56. [CrossRef]
20. Gurbuz, H.A.; Durukan, A.B.; Sevim, H.; Ergin, E.; Gurpinar, A.; Yorgancioglu, C. Heparin toxicity in cell culture: A critical link in translation of basic science to clinical practice. *Blood Coagul. Fibrinolysis* **2013**, *24*, 742–745. [CrossRef]
21. Biran, R.; Pond, D. Heparin coatings for improving blood compatibility of medical devices. *Adv. Drug Deliv. Rev.* **2017**, *112*, 12–23. [CrossRef] [PubMed]

Article

In Vitro Biocompatibility and Degradation Analysis of Mass-Produced Collagen Fibers

Kiran M. Ali, Yihan Huang, Alaowei Y. Amanah, Nasif Mahmood, Taylor C. Suh and Jessica M. Gluck *

Department of Textile Engineering, Chemistry and Science, Wilson College of Textiles, NC State University, Raleigh, NC 27695, USA; kmmumtaz@ncsu.edu (K.M.A.); yhuang38@ncsu.edu (Y.H.); aya8@duke.edu (A.Y.A.); nmahmoo3@ncsu.edu (N.M.); tacook3@ncsu.edu (T.C.S.)
* Correspondence: jmgluck@ncsu.edu; Tel.: +1-919-515-6637

Abstract: Automation and mass-production are two of the many limitations in the tissue engineering industry. Textile fabrication methods such as electrospinning are used extensively in this field because of the resemblance of the extracellular matrix to the fiber structure. However, electrospinning has many limitations, including the ability to mass-produce, automate, and reproduce products. For this reason, this study evaluates the potential use of a traditional textile method such as spinning. Apart from mass production, these methods are also easy, efficient, and cost-effective. This study uses bovine-derived collagen fibers to create yarns using the traditional ring spinning method. The collagen yarns are proven to be biocompatible. Enzymatic biodegradability was also confirmed for its potential use in vivo. The results of this study prove the safety and efficacy of the material and the fabrication method. The material encourages higher cell proliferation and migration compared to tissue culture-treated plastic plates. The process is not only simple but is also streamlined and replicable, resulting in standardized products that can be reproduced.

Keywords: collagen; tissue engineering; biomaterials; biocompatibility; scaffolds

1. Introduction

Disease, injury, and trauma can damage and degenerate tissues in the human body that can be treated via the repair, replacement, or regeneration of the tissue [1]. Often, the tissues are either taken from the patient's own body (autograft) or from a donor (allograft). An example of an autograft is a bone graft, which is considered the "golden standard" treatment for spinal surgery [2]. Bone, tendons, cartilage, skin, heart valves, and veins are some of the tissues that can be treated using these methods [3]. This type of treatment has been revolutionary and has saved numerous lives. However, the treatment has many drawbacks, including cost and a sometimes difficult recovery, whereas a transplant may result in an immune response or even organ rejection [4].

For this purpose, tissue engineering is beneficial, as tissues are regenerated instead of being replaced. It involves the expansion of cells from a patient's biopsy with the help of ex vivo cell culture that function together to create tissues and organs. Tissue regeneration is carried out with the help of biological support in the form of engineered scaffolds, which restore or improve cell function [1]. Scaffolds play an important role in the manipulation of cell function and guidance toward new tissue development [5].

Various biomaterials have been used for the fabrication of scaffolds for regenerative medicine. Among them, collagen is one of the most prominent proteins due to its excellent biocompatibility [6]. Its high flexibility, optimal mechanical strength, and ability to absorb fluids, makes it an ideal biopolymer. Collagen is also one of the most prominent proteins found in the extracellular matrix (ECM) of humans and is expressed in tissues throughout the human body such as tendons, ligaments, bones, the dermis, dentins, and blood vessels [7–9]. Additionally, bovine and porcine collagens have shown degradable properties and have been used for the fabrication of biodegradable medical devices/scaffolds [10,11], such as drug delivery systems [12], and in bone tissue repair [13]. In a clinical study by Charriere et al. [14], the immunological response to bovine collagen implants was evaluated. Out of 705 participants, only 2.3% of the patients showed an adverse reaction to the collagen implant. Another study by Keefe et al. [15] also showed that the adverse effects were only observed in 1–2% of the treated patients, which were absolved when the material was resorbed by the hosts. For this reason, collagen polymers, especially porcine and bovine collagen polymers, are widely used in regenerative medicine.

There are multiple fabrication methods for the creation of scaffolds, such as solvent casting/particle leaching, thermally induced phase separation (TIPS), and three-dimensional (3D) printing [16]. Although these techniques have been very advantageous for the creation of scaffolds, they all have disadvantages such as a lack of reproducibility, low production efficiency, and reduced mechanical strength [16,17]. The most common problems with these methods are reproducibility, repeatability, and standardization.

Textile fabrication methods, such as ring spinning, knitting, weaving, and braiding, have been used for centuries. Most of these fabrication techniques, including for yarn and fabric production, have now been automated and standardized, resulting in efficiently reproducible products. Using these methods to produce scaffolds will provide us with the ability to mass-produce products that are identical and that can be created in a standardized manner [18].

Traditional textile manufacturing methods, including knitting, weaving, and electrospinning, have been used for scaffold fabrication. There is a resemblance between the structure of native ECM and textile fibers [18]. The yarn-like structure of ECM resembles the fibrous morphology of textile materials, which has been proven to be beneficial for cellular migration and proliferation [19,20]. The fibrous architecture of ECM helps to provide support to the cell by creating a mesh of collagen, elastin, and other proteins. It is because of the fibrous structure of ECM that the cells are able to form a communication/signaling network [20]. Additionally, textile methods also create 3D structures, which are desirable for cellular scaffolds in terms of cell proliferation and migration [21].

Different methods are available for spinning yarns, including the ring spinning, rotor spinning, wrap-spinning, and core-spinning methods [22] (Table 1). Among these, ring spinning is the most popular method for creating yarns using staple fibers (short fibers ranging from 10–500 mm in length). The spun yarns can be used to create knit, woven, or braided structures for tissue engineering applications.

Table 1. The advantages and disadvantages of different spinning methods.

Method	Advantages	Disadvantages	Ref
Ring Spinning	The yarns that are produced have high strength. This method is applicable to a wide variety of fibers.	The method consumes high amounts of energy, and therefore, production costs are also high.	[22,23]
Rotor Spinning	The production costs of using this method are relatively low.	The resultant strength of the yarns is low.	[22,23]
Wrap-Spinning	The yarns that are produced through this method are highly absorbent. The method has high production efficiency.	The yarns have low strength.	[22–24]
Core-Spinning	This method uses two or more fibers, providing excellent properties to the resultant composite yarn.	There are limited applications for this method.	[22,23,25]

Weaving provides strength to the yarn. A study by Gilmore et al. [26] used round and grooved cross-sectional fibers to create plain weave and satin weave scaffolds. The results suggested that the fabric structure has a significant influence on the resultant properties, especially on the permeability of the scaffold. Similarly, knitting provides elasticity and strength to fibrous scaffolds. A study by Lieshout et al. [27] shows a comparison of knitted and electrospun scaffolds for aortic valves. Human myofibroblasts were cultured on both scaffolds over the course of 23 days. The results compared tissue formation, which was evaluated via confocal laser scanning microscopy. The study showed that the electrospun scaffold tore within 6 h, while the knit scaffold remained intact. Another study by Zhang et al. [28] explored the idea of using a circular knit machine to produce a small-caliber vascular graft. The scaffold had excellent mechanical properties (bursting strength, suture retention strength, and compliance) that were comparable to the coronary artery under normotensive pressure.

This study evaluates an effort to mass-produce collagen fiber scaffolds that can be fabricated via traditional textile processes (Figure 1). In this study, collagen yarns are produced using the ring-spinning method, which is a process of creating yarns by twisting fibers together and winding them on a bobbin. The method comprises four processes: carding, drawing, roving, and ring spinning (Figure 2). The purpose of carding and drawing is to clean and align the fibers. After drawing, the fibers are passed to a roving machine, which fuses the fibers together through twisting, creating a roving. A roving is a sliver of fibers that is larger in diameter than yarn. The roving is then passed to a ring spinning machine, which provides the fibers with more twist, reducing the diameter to convert them into a yarn [22]. As a result, yarns are stronger when they are fabricated via ring spinning than other methods, such as rotor spinning. Ring spinning also provides a high production rate and can be used for any type of fiber [23].

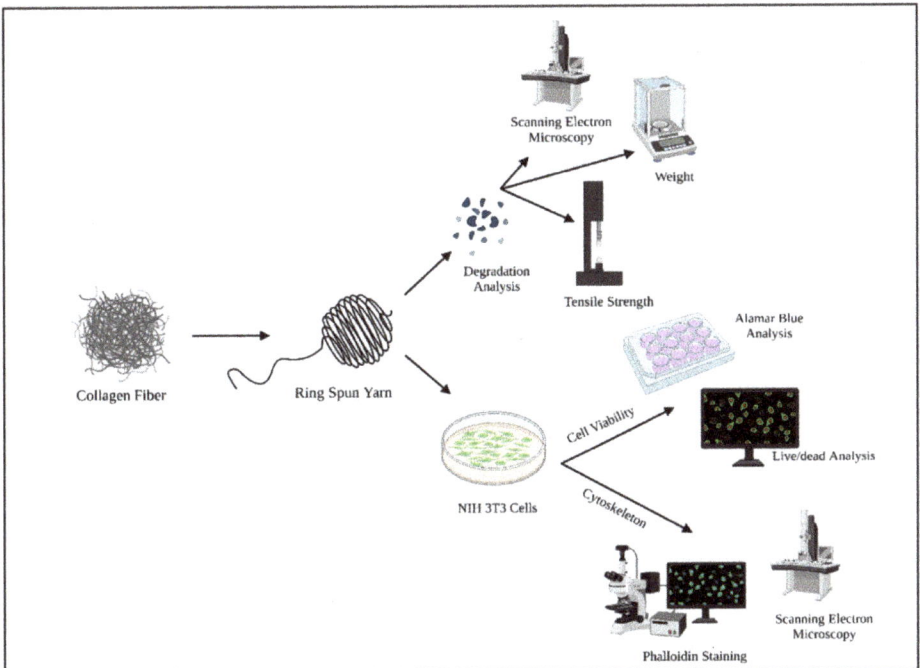

Figure 1. Experimental Design Schematic. Collagen fibers were provided by the Kaneka Corporation and spun to yarns using the traditional ring spinning method. The resulting yarns were analyzed for biodegradability and biocompatibility. The biodegradability of the material was analyzed via weight changes, tensile strength changes, and morphological analysis. The biocompatibility was checked with Alamar blue and live/dead analysis for cell viability, and SEM and phalloidin staining were used for cytoskeleton analysis.

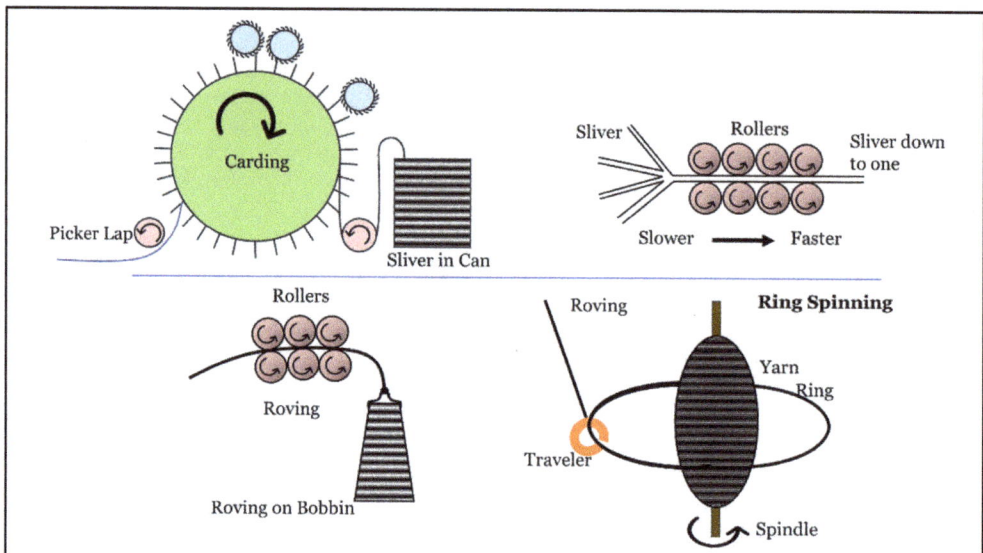

Figure 2. Ring spinning process for collagen yarns. The collagen fibers were provided by the Kaneka Corporation. The fibers were first run through the card and drawer with multiple sets of rollers. The rollers combed the fibers to clean and aligned them before they were gathered to form a sliver in a can. The sliver then passed through a roving frame, where the sliver reduced in diameter to form a roving, which was wound on a bobbin. Finally, the roving bobbins were passed through a ring spinning machine, where their diameters were further reduced via twisting with the help of a traveler and a ring.

After the yarns were fabricated, they were subjected to biocompatibility and biodegradation analysis to evaluate the perspective use of this technique in the future to massproduce scaffolds. The biocompatibility analysis provides knowledge about the safety and efficacy of the fabricated when encountering cells. This is crucial for the application of this process in tissue engineering products, especially for dermal [29], cardiovascular [28], tendon [30], and ligament [18] applications.

2. Materials and Methods

Collagen fibers were provided by the Kaneka Corporation. The fibers were processed through a card chute feed system (Rieter Card C4) followed by two rounds of drawing (Rieter RSB851). The aligned fibers were then converted to a roving with a roving machine (Rieter Fly F4/1) and finally spun into yarn on a ring spinning machine (Rieter G5/2) (Figure 2). The yarns had a fineness of 20 Ne, which is the thickness or coarseness of a yarn, and were measured according to the English Cotton Count (Ne) method, which is defined as "the number of 840-yard length per pound" [23]. Tensile strength and ultrastructure characterization were carried out via scanning electron microscopy (SEM), and an in vitro biocompatibility analysis was performed on the resulting yarns (Figure 1).

2.1. Degradation Study

A degradation study was conducted to check the ability of the material to degrade when implanted in the body or when used as an in vivo application. Collagenase is normally found in almost all mammalian tissues, including pig pancreas, beef pancreas, and human tissues. It helps in wound healing by moving the keratinocytes over the collagen-rich dermis during re-epithelialization.

A collagenase solution was prepared according to the protocol by Alberti et al. [31]. Collagenase isolated from *Clostridium histolyticum* cleaves the bonds between neutral amino acids and glycine, which is found with a high frequency in collagen. This type of degradation is a simplistic method to predict the in vivo degradation profile of our ring-spun collagen yarns. First, a collagen yarn sample was dried, weighed, and placed in 12-well plates. In a separate glassware, 0.5 mL of 0.1 M TRIS-HCl and 0.005 M of $CaCl_2$ were mixed with 2 mg/mL of collagenase (Type I, powder, Gibco, >125 units/mg). A 2 mL amount of the prepared solution was added to each well such that each sample was immersed in the solution, and the samples were placed in an incubator at 37 °C for a total of 8 weeks. Their weight, tensile strength, and morphology were monitored biweekly. The ultrastructure and morphology of the yarns were assessed using SEM to evaluate the degradation pattern in the collagen yarns.

2.1.1. Weight Change

A total of 6 samples were prepared, all of which were 20 cm each in length. It was determined that the length would be longer to increase the accuracy. The samples were weighed and placed in a 12-well plate and immersed in collagenase solution for the entire length of the study. The samples were then placed in the incubator at 37 °C. The solution was maintained at a pH of 7–7.4 throughout the study. The samples were dried every 14 days by aspirating the solution and drying in the desiccator for 24 h. After drying, their weight was measured, and the samples were placed in the original well plates again and submerged in collagenase solution at 37 °C. Average mass loss was calculated with the following formula:

$$\text{Mass Loss} = \frac{\text{Original Mass} - \text{Final Mass}}{\text{Original Mass}}$$

2.1.2. Tensile Test

A total of 24 samples were prepared, each of which were 20 cm in length, and the samples were placed in 12-well plates. Collagenase solution was added to each well, and the samples were stored in the incubator at 37 °C. The pH of the samples was maintained at 7–7.4. Every 14 days, 6 samples were dried and were placed in the desiccator for 24 h. When they had dried completely, the samples were mounted on cardboard (as shown in Figure 3a). An MTS Criterion 43 Tensile tester (MTS Systems, Eden Prairie, MN USA) was used for the tensile test, and a 50 N load cell, a gauge length of 1 cm, and a crosshead speed of 10 mm/min were maintained for the tensile test.

2.1.3. Scanning Electron Microscopy

The samples were dried and mounted on stubs using double-sided carbon tape. They were sputter-coated with gold/palladium using a SC7620 Mini Sputter Coater (Quorum Technologies, East Sussex, UK) for 45 s, resulting in a 10 nm coating. Samples were imaged with a Phenom G1 desktop SEM (Phenom, ThermoFisher, Eindhoven, The Netherlands). A total of 9 samples were analyzed per time point, and 4 images were taken from each sample. The morphology of the yarns was observed at biweekly intervals. The images were analyzed for surface and bulk degradation.

2.2. Biocompatibility

2.2.1. Sample Preparation

The collagen yarns were sterilized by immersing in 70% ethanol for 20 min. The ethanol was then aspirated followed by washing with phosphate-buffered solution (PBS, Hyclone, Cytiva, Long, UT, USA).

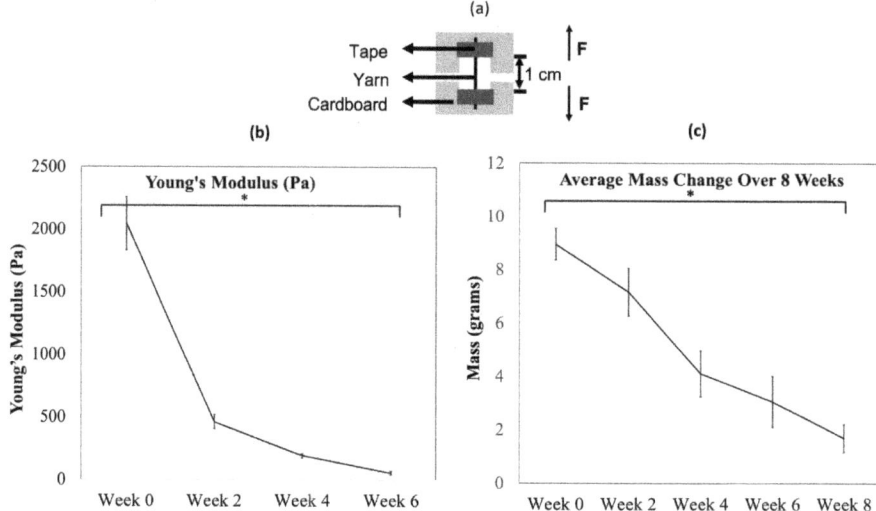

Figure 3. Mechanical analysis over time: Collagen yarns were soaked in a collagenase solution for 8 weeks (n = 3 samples per period). Samples were prepared and mounted using a cardboard holder (**a**) for tensile testing. The samples showed a significant loss (97%, * $p < 0.0001$) in strength over 6 weeks. After 6 weeks, the samples lost mechanical integrity and therefore could not be tested further. There was a significant mass reduction * ($p < 0.0001$) from week 0 to week 8. (**b**) Mass change was calculated and recorded over time for the material degradation analysis. In the study, 20/s yarns were soaked in collagenase solution for 8 weeks, (n = 6 per period) and an average mass loss of 80 ± 6% was recorded over 8 weeks. Standard deviation is shown by error bars (**c**).

2.2.2. Cell Culture

Mouse fibroblast cells (NIH-3T3) were used for all of the biocompatibility studies. Cells were maintained in Dulbecco's Modified Eagle Medium (DMEM, Invitrogen, Waltham, MA, USA) supplemented with 10% fetal bovine serum (FBS, Atlas Biologicals, Fort Collins, CO, USA) and 1% penicillin-streptomycin (10,000 U/mL, Gibco, Invitrogen, Waltham, MA, USA). Cells were seeded onto the collagen yarns at a passage of 9–11 and at a density of 2×10^6 cells/sample. Cell-only controls were seeded at a density of 0.5×10^6 cells/well. The cell-seeded scaffolds were then maintained in culture for 7 days. The medium was changed every 72 h. Instead of aspirating the medium, the samples were transferred into another well, and more medium was slowly added over the seeded samples. This was to avoid cells detaching from the surface of the collagen yarn surface. Cell proliferation, functioning, and morphology were checked.

The biocompatibility was then evaluated via a Live/Dead™ assay (Invitrogen, Waltham, MA, USA), alamarBlue proliferation assay (Invitrogen, Waltham, MA, USA), immunocytochemistry, and SEM analysis of the seeded scaffolds. The live/dead assay and alamarBlue were evaluated on days 1, 3, 5, and 7, whereas phalloidin (F-actin) staining and SEM were carried out on days 1 and 7.

2.2.3. Cell Proliferation

The live/dead assay was carried out using an Invitrogen Live/Dead™ Cell Imaging kit following the manufacturer's instructions. The solution from the kit was diluted to a ratio of 1:1 with PBS. After mixing the solution, 250 µL of the prepared working solution was added to each scaffold sample and placed in the incubator at 37 °C and 10% CO_2 for 20 min. The samples were then observed under a fluorescent microscope (EVOS™FL Auto 2 Imaging System, ThermoFisher, Waltham, MA, USA) to check for green (live) and red (dead) fluorescence. The kit contains calcein; AM, a permanent dye that is an alive cell indicator (ex/em 488 nm/515 nm); and BOBO-3 Iodide, which is a dead cell indicator (ex/em 570 nm/602 nm). This test was performed on cell-seeding days 1, 3, 5, and 7.

The alamarBlue assay was also performed to quantify the proliferation of the cells that had been seeded on collagen fiber scaffolds. This assay provides quantitative data on cell metabolism (proliferation) by measuring the changes in fluorescence generated by resazurin, which reduces to resorufin in response to chemical reductions due to cell growth. The dye changes its color and fluorescence from blue/non-fluorescent (oxidized) to red/fluorescent (unoxidized).

The cells were seeded on the yarn for 7 days. The alamarBlue assay was carried out on days 1, 3, 5, and 7 with the positive control cells only and with the negative controls of the yarn in media only (no cells). The samples were incubated with alamarBlue Reagent (Fisher Scientific, Waltham, MA, USA) for 1 h. Later, the samples were read with a plate reader (Synergy HT, BioTek, Santa Clara, CA, USA) set to 540/25 λ excitation, 590/35 λ emission, and maintained at 37 °C.

2.2.4. Phalloidin Staining

Phalloidin staining was carried out to label and identify F-actin (cytoskeleton) expression in the cells to observe the cellular morphology. On day 7, the samples were fixed with 4% paraformaldehyde for 20 min and washed with PBS. The samples were then permeabilized with 0.1% TritonX-100 for 30 min and 0.1% Tween-20 for 15 min. Blocking buffer was prepared with 0.1%Tween-20, 2% bovine serum albumin, and 2% goat serum. Phalloidin (Invitrogen™ ActinGreen™ 488 ReadyProbes™ Reagent) was used in the concentration of 2 drops/mL of blocking buffer and incubated for 1 h. After incubation, the samples were washed with PBS and stained with Hoechst (Invitrogen™ Hoechst 33342, Eugene, OR, USA) (1:1000 concentration Hoechst: PBS) for 5 min. The samples were then imaged using fluorescence microscopy (EVOS™ FL Auto 2 imaging system, ThermoFisher, Waltham, MA, USA).

2.2.5. Ultrastructure Analysis (Scanning Electron Microscopy)

Collagen fiber yarns seeded with NIH 3T3 fibroblasts were fixed in buffered formalin. The samples were washed 3 × 10 min with 0.15 M sodium phosphate buffer and at a pH of 7.4 followed by post-fixation in 1% osmium tetroxide/0.15 M sodium phosphate buffer at a pH of 7.4 for 1 h. After washing with deionized water (3 × 10 min), the samples were treated with 1% tannic acid in water for 30 min. The scaffolds were washed in deionized water and dehydrated through an increasing ethanol series (30%, 50%, 75%, 90%, 100%, 100%, and 100%—15 min each). The samples were transferred in 100% ethanol to a Samdri-795 critical point dryer (Tousimis Research Corporation, Rockville, MD, USA) and dried using liquid carbon dioxide as the transitional solvent. The scaffolds were mounted onto 13 mm diameter aluminum stubs with carbon adhesive tabs and were sputter-coated with 8 nm of gold/palladium alloy (60Au: 40Pd) using a Cressington 208HR Sputter Coater (Ted Pella, Inc., Redding, CA, USA). Images were taken using a Zeiss Supra 25 FESEM (Carl Zeiss Microscopy, Jena, Germany) operating at 5 kV or 10 kV using the SE2 detector, 30 μm aperture, and approximate working distances from 15 to 25 mm. The samples were imaged for any cell structures that were growing horizontally or vertically over the fiber and yarn surfaces.

2.3. Statistical Analysis

Data were collected from 3 to 9 samples and expressed as means ± standard deviation. Statistical analysis was performed using the Student's t test, and significance was determined at $p < 0.0001$.

3. Results

This study evaluated the biodegradability and biocompatibility of mass-produced ring spun collagen fiber yarns (Figure 1). The collagenase (Type I) enzyme was used to assess yarn degradation. Mass changes, tensile strength, and the morphology of the yarns were checked at biweekly intervals. The biocompatibility of the yarns was also checked using mouse fibroblast cells. The cell viability, cell metabolic activity, and the morphology of the cytoskeleton was observed for biocompatibility analysis.

3.1. Collagen Yarns Degrade in the Presence of Enzymes in an 8-Week Study

During the enzymatic degradation with collagenase Type I, the collagen yarns lost most of their mass and mechanical integrity over 8 weeks. On average, the samples lost 80 ± 6% of their mass over 8 weeks (Figure 3c), which indicates that the material is degradable, and the remnants were dispersed in the solution. The pH of the solution remained the same throughout the study. The degradation of collagen fibers was confirmed via the assessment of the tensile strength and morphology of the samples, as indicated by the losses observed in the measured Young's modulus (Figure 3b). The modulus loss indicates the degradation of the collagen fibers in the presence of collagenase solution. The samples lost 97% of their tensile strength in 6 weeks. Fiber samples were mechanically unstable by week 6 and therefore could not be tested for tensile strength in week 8.

Gradual degradation was evident in the yarns for eight weeks (Figure 4) when week 0 and week 8 are compared. A polymer undergoes bulk degradation if the diffusion of water/fluids into the polymer is faster than the bonds break [32]. The SEM image analysis of the collagen yarns indicated that the length of the fibers broken down over 8 weeks (Figure 4 marked with arrows), suggesting bulk degradation. On the other hand, the surface of the fiber also changed from smooth to rough and beady (marked with stars), which is a sign of surface degradation in the fibers.

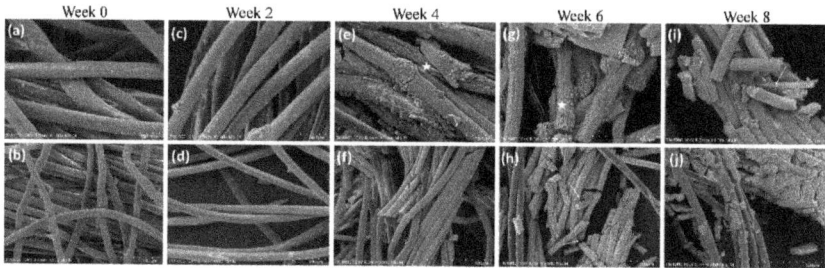

Figure 4. Bulk and surface degradation is observed over 8 weeks. The samples were subjected to collagenase for 8 weeks. The morphology of the fibers was observed via SEM. The fibers had broken down and degraded significantly at week 8 (images **i,j**) when compared to week 0 (images **a,b**). The surface of the fibers also changed, which shows that there is surface degradation in the fibers ($n = 6$). The top images (**a,c,e,g,i**) show a higher magnification (50 um), whereas the bottom images (**b,d,f,h,j**) show a lower magnification (100 um). Arrows indicate broken fiber lengths; stars indicate a surface transition from smooth to rough and beady.

3.2. The Collagen Material Is Biocompatible with Mouse Fibroblast (NIH 3T3) Cell Line

High cell viability on the collagen yarns was observed over 7 days (Figure 5). After only 24 h of cell culture, the cells were observed to form colonies, which was a strong indication that the material provides a heterogeneous environment where the cells are more attracted to some areas, forming clusters and migrating to a suitable area. The cells adhered to the yarn/fiber surface and proliferated along the length (Figure 5c,d).

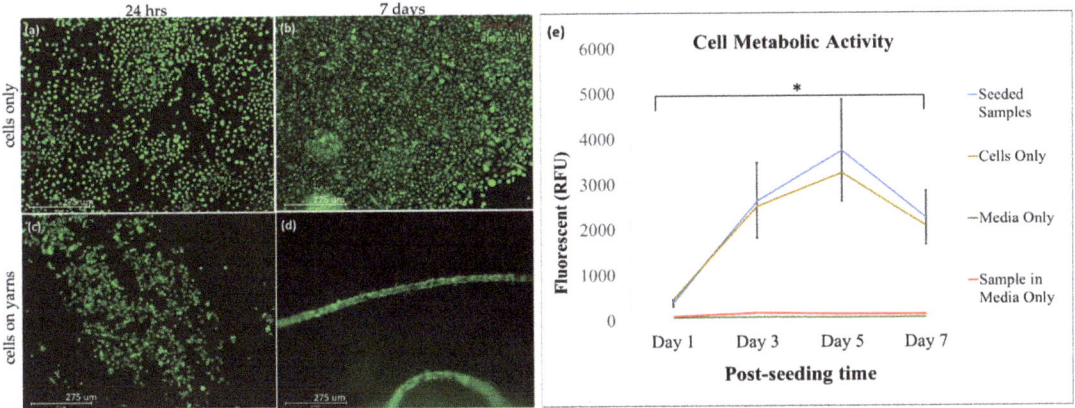

Figure 5. High biocompatibility of collagen yarns is observed. Viability of NIH 3T3 cells (**a**) seeded on collagen yarns. The green signal indicates live cells, and the red signal indicates dead cells. The live/dead assay of the seeded samples at 24 h (**c**) and after 7 days (**d**). The cell-only controls at 24 h and 7 days are also shown (**a**,**b**). We see evidence of the cells adhering and proliferating along the length of the fiber for at least 7 days, indicating the biocompatibility of the collagen yarns ($n = 10$). The results from the live/dead assay were confirmed and quantified with the alamarBlue assay (**e**). We observed an increased in metabolic activity that was correlated to increased proliferation in the cells that were seeded on our collagen yarns (**e**, blue positive control (cells only in yellow) and in the negative controls (medium-only in green and sample in medium in red)). A significant increase in metabolic activity was observed in the cells seeded on collagen yarns from day 1 to day 5 (* $p < 0.0001$). The error bars show the standard deviation ($n = 9$).

The collagen yarns support high cellular viability (Figure 5), and the corresponding metabolic activity confirmed the viability (Figure 5e). The metabolic activity results show an 8x increase in metabolic activity from day 1 to day 5, only to reduce after that. The reduction in proliferation may be due to limited space for the cells to proliferate any further. A study by Streitchan et al. [33] has shown that cells slow down or pause their proliferation cycle due to spatial constraints in response to contact inhibition. This was highlighted on day 5, where we observe a reduction in metabolic activity. Cells proliferated more on the collagen yarns compared to standard tissue culture plastic dishes, indicating their superior biocompatibility.

The morphology of the cells seeded on the collagen yarns is observed to be a bright ellipsoid shape that wraps around the strands of the collagen fibers, as evidenced by f-actin detection (Figure 6a,b). The ultrastructure analysis from the SEM images (Figure 6c–h) highlights that cell growth can be observed along the direction of the collagen fiber.

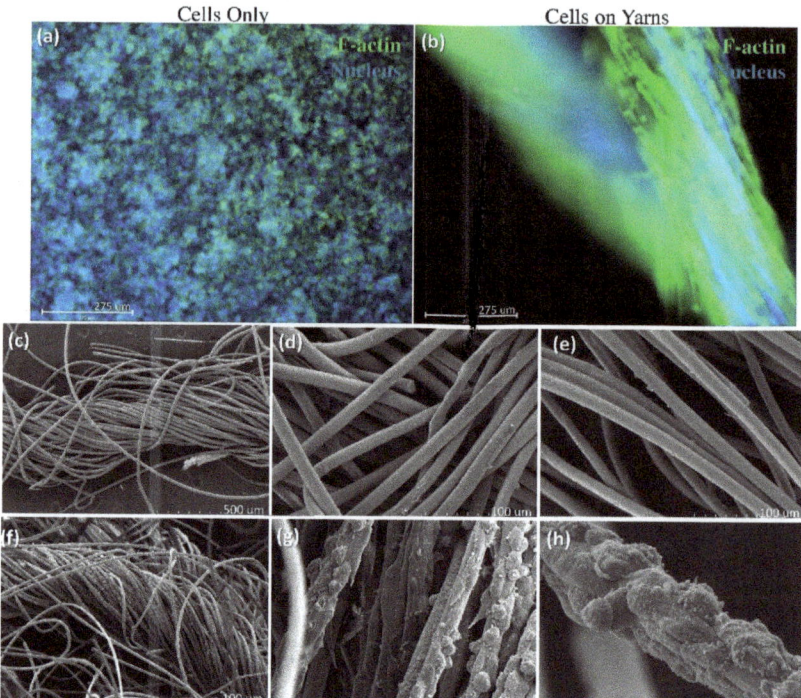

Figure 6. Cells proliferate along the length of the collagen yarns. The samples were stained with phalloidin to identify the cytoskeleton of the cells on the collagen yarns at day 7 (**a**,**b**). F-actin, a cytoskeleton protein, highlights the typical morphology of the cells (**a**) and highlights the adhesion of the cells along the length of the collagen yarns (**b**), scale bar = 275 μm (n = 6). Representative SEM images of collagen fibers alone (**c**–**e**) at varying magnifications. NIH 3T3 cells were seeded and maintained on collagen fiber yarns for 7 days (**f**–**h**). Scale bar (**a**,**b**) = 275 μm, (**c**) = 500 μm, (**d**–**f**) = 100 μm, (**g**) = 10 μm, (**h**) = 2 μm (n = 3).

4. Discussion

This study demonstrates that textile-based mass-production techniques may be used for the fabrication of biocompatible tissue-engineered scaffolds. The yarns in this study were produced using a traditional ring spinning technique facility where general textile yarns are produced. These yarns can be used to produce woven, knit, or braided structures that can be used for tissue engineering applications. For example, Learn et al. [30] used collagen yarns to fabricate a woven scaffold for rotator cuff tendon regeneration, whereas Ruan et al. [34] used a collagen-silk blend to create knit scaffolds for anterior cruciate ligament reconstruction. The data demonstrate that collagen fiber yarns are biocompatible. The biocompatibility of a material is defined as the properties of a material that are safe and effective for cells to proliferate, migrate, and function [35]. The results from our viability and proliferation assays combined with our fluorescent and electron microscopy image analysis show that both the material and the fabrication method support cellular proliferation and maintain their expected cellular morphology (Figures 5 and 6). It is interesting to observe that the cells migrated from the tissue culture-treated plate to the fiber surface, providing compelling evidence of the biocompatibility of the material.

Similarly, the fabrication method demonstrates the ability of the collagen fibers to degrade over time in the presence of the collagenase enzyme. Collagenase is normally present in most mammalian tissues, including those in humans [36]. Collagenase provides a good imitation of the in vivo environment, where enzymatic activity is observed to cleave native fibrillar collagen. In this study, degradation is seen by a reduction in mass and mechanical integrity over time, which signifies that the fabrication methods provide additional mechanical strength without changing the degradable nature of the polymer. Degradation is an important property for implantable biopolymers in situation where the scaffold is only required for mechanical support until the resulting tissue develops in vivo. This prevents the need for surgical procedures to remove the scaffold when it is no longer required [36].

Previously, conventional fabrication techniques such as fiber melts, fiber bonding, melt molding, and fiber composite foam have been used, which can provide the necessary morphology. However, these methods have multiple disadvantages, such as a lack of structural stability or the solvent residue being toxic or harmful [17]. On the other hand, traditional textile fabrication methods have many advantages, such as good mechanical stability, a 3D structure, a fibrous nature, and most importantly, the ability to fabricate products that are identical and that can be produced in bulk [18].

Collagen is one of the most extensively used materials in tissue engineering due to its abundant presence in the native ECM, where it provides biological and structural integrity to cells [7]. However, when used as a scaffold, the material does not provide mechanical strength, which limits its applications [37]. Collagen is normally used as a blended material or is chemically cross-linked to provide the required strength to the material [37]. However, the use of ring spinning technology solves this problem due to the twisting of fibers by providing excellent mechanical support.

Additionally, the fibrous structure of textile yarns replicates the thread-like morphology of ECM [18]. Therefore, using collagen in a textile form provides the required support to the cells whiles also providing a medium for networking and signaling [20]. Moreover, the 3D structure of textiles is also beneficial for cell proliferation and migration. Biomimetic fibrous scaffolds are crucial in tissue engineering when the goal is to best recapitulate the native ECM. Creating fibrous scaffolds from collagen has been repeatedly demonstrated to be beneficial for a variety of tissue engineering applications [38], including scaffolds for heart valve repair [39] and cardiac tissue engineering [40], with bone [34,41] and skin tissue engineering applications being the most common examples.

A study by Xie et al. [42] demonstrated the potential use of wet spun yarn in tissue engineering applications. However, collagen is sensitive to temperature, and therefore, wet spinning requires the yarns to be cross-linked to acquire the required thermal and chemical stability. On the other hand, the ring spinning method is dry and does not require elevated temperatures. Therefore, it is likely to be a more stable form of spinning for collagen fibers. Another common method using for processing collagen for tissue engineering applications includes chemical cross-linking, which is destructive to the natural helical structure of collagen. Thus, we showed that ring-spun collagen yarns represent an exciting new opportunity for the scalable production of biomimetic biomaterials that can be used for a variety of tissue engineering applications.

Regardless of the excellent results of the in vitro analysis of the fabricated textile yarns, further studies are required to create compact structures that are either woven, knit, or braided. We are currently working on fabricating knit structures for in vitro studies.

5. Conclusions

Collagen fiber yarns are a highly biocompatible material/technique, as they promote cell adhesion and proliferation. The structure of the material encourages cell adhesion and proliferation without requiring any intricate and complex fabrication technique, which not only saves time but also simplifies and streamlines the fabrication process. The process also provides a standardized fabrication technique for creating identical scaffolds. Furthermore, this proof-of-concept study sets the foundation for further analyses of the underlying textile and biological properties of mass-produced collagen fiber scaffolds that support cell behavior and that can be used for further tissue engineering and regenerative medicine applications.

Author Contributions: Conceptualization, K.M.A. and J.M.G.; methodology, K.M.A., Y.H. and A.Y.A.; formal analysis, K.M.A., A.Y.A. and J.M.G.; resources, J.M.G.; data curation, K.M.A., N.M. and T.C.S.; writing—original draft preparation, K.M.A.; writing—review and editing, K.M.A., N.M., T.C.S. and J.M.G.; visualization, K.M.A., Y.H., N.M. and T.C.S.; supervision, J.M.G.; project administration, K.M.A. and J.M.G.; funding acquisition, J.M.G. All authors have read and agreed to the published version of the manuscript.

Funding: The research was supported by the Kaneka Corporation (K.M.A., J.M.G.), the NCSU Laboratory Research Equipment Program (LREP) (J.M.G.), the Wilson College of Textiles (J.M.G.), the Department of Textile Engineering, Chemistry and Science (J.M.G., K.M.A., T.C.S.), and The Provost's Fellowship at NC State University (T.C.S.).

Institutional Review Board Statement: Not Applicable.

Informed Consent Statement: Not Applicable.

Data Availability Statement: The data presented in this study are available within this publication.

Acknowledgments: The authors wish to thank Jeff Vercellone and Dennis Mater from the Kaneka Corporation; Wilson College of Textiles, North Carolina State University; Judy Elson from the Chemistry and Microscopy Laboratory, Textile Engineering, Chemistry, and Science, Wilson College of Textiles, North Carolina State University, Physical Testing Laboratory; Timothy Pleasants from Zeiss Textiles Extension, North Carolina State University; and the Microscopy Services Laboratory, University of North Carolina at Chapel Hill.

Conflicts of Interest: The authors declare no conflict of interest.

References

1. O'Brien, F.J. Biomaterials & scaffolds for tissue engineering. *Mater. Today* **2011**, *14*, 88–95. [CrossRef]
2. Zileli, M.; Benzel, E.C.; Bell, G.R. Chapter 93—Bone graft harvesting. In *Spine Surgery*, 3rd ed.; Benzel, E.C., Ed.; Churchill Livingstone: Philadelphia, PA, USA, 2005; pp. 1253–1261. [CrossRef]
3. *Transplanting Human Tissue: Ethics, Policy, and Practice*; Youngner, S.J.; Youngner, S.J., Anderson, M.W.; Schapiro, R., Eds.; Oxford University Press: Oxford, UK, 2004.
4. Atala, A. Tissue engineering and regenerative medicine: Concepts for clinical application. *Rejuvenation Res.* **2004**, *7*, 15–31. [CrossRef] [PubMed]
5. Chen, G.; Ushida, T.; Tateishi, T. Scaffold design for tissue engineering. *Macromol. Biosci.* **2002**, *2*, 67–77. [CrossRef]
6. Rahmanian-Schwarz, A.; Held, M.; Knoeller, T.; Stachon, S.; Schmidt, T.; Schaller, H.E.; Just, L. In vivo biocompatibility and biodegradation of a novel thin and mechanically stable collagen scaffold. *J. Biomed. Mater. Res.* **2014**, *102*, 1173–1179. [CrossRef]
7. Gelse, K.; Pöschl, E.; Aigner, T. Collagens—Structure, function, and biosynthesis. *Adv. Drug Deliv. Rev.* **2003**, *55*, 1531–1546. [CrossRef]
8. Glowacki, J.; Mizuno, S. Collagen scaffolds for tissue engineering. *Biopolym. Orig. Res. Biomol.* **2008**, *89*, 338–344. [CrossRef]
9. Cen, L.; Liu, W.; Cui, L.; Zhang, W.; Cao, Y. Collagen tissue engineering: Development of novel biomaterials and applications. *Pediatr. Res.* **2008**, *63*, 492–496. [CrossRef]
10. George, J.; Onodera, J.; Miyata, T. Biodegradable honeycomb collagen scaffold for dermal tissue engineering. *J. Biomed. Mater. Res.* **2008**, *87*, 1103–1111. [CrossRef]
11. Roßbach, B.P.; Gülecyüz, M.F.; Kempfert, L.; Pietschmann, M.F.; Ullamann, T.; Ficklscherer, A.; Niethammer, T.R.; Zhang, A.; Klar, R.M.; Müller, P.E. Rotator cuff repair with autologous tenocytes and biodegradable collagen scaffold: A histological and biomechanical study in sheep. *Am. J. Sports Med.* **2020**, *48*, 450–459. [CrossRef]
12. Khan, R.; Khan, M.H. Use of collagen as a biomaterial: An update. *J. Indian Soc. Periodontol.* **2013**, *17*, 539–542. [CrossRef]
13. Wei, S.; Ma, J.; Xu, L.; Gu, X.; Ma, X. Biodegradable materials for bone defect repair. *Mil. Med. Res.* **2020**, *7*, 54. [CrossRef] [PubMed]
14. Charriere, G.; Bejot, M.; Schnitzler, L.; Ville, G.; Hartmann, D.J. Reactions to a bovine collagen implant: Clinical and immunologic study in 705 patients. *J. Am. Acad. Dermatol.* **1989**, *21*, 1203–1208. [CrossRef]
15. Keefe, J.; Wauk, L.; Chu, S.; DeLustro, F. Clinical use of injectable bovine collagen: A decade of experience. *Clin. Mater.* **1992**, *9*, 155–162. [CrossRef]
16. Dutta, R.C.; Dey, M.; Dutta, A.K.; Basu, B. Competent processing techniques for scaffolds in tissue engineering. *Biotechnol. Adv.* **2017**, *35*, 240–250. [CrossRef] [PubMed]
17. Yang, S.; Leong, K.; Du, Z.; Chua, C. The design of scaffolds for use in tissue engineering. Part I. Traditional factors. *Tissue Eng.* **2001**, *7*, 679–689. [CrossRef] [PubMed]
18. Jiao, Y.; Li, C.; Liu, L.; Wang, F.; Liu, X.; Mao, J.; Wang, L. Construction and application of textile-based tissue engineering scaffolds: A review. *Biomater. Sci.* **2020**, *8*, 3574–3600. [CrossRef]

19. Nivedhitha Sundaram, M.; Deepthi, S.; Mony, U.; Shalumon, K.T.; Chen, J.P.; Jayakumar, R. Chitosan hydrogel scaffold reinforced with twisted poly(l lactic acid) aligned microfibrous bundle to mimic tendon extracellular matrix. *Int. J. Biol. Macromol.* **2019**, *122*, 37–44. [CrossRef]
20. Benjamin, M.; Kaiser, E.; Milz, S. Structure-function relationships in tendons: A review. *J. Anat.* **2008**, *212*, 211–228. [CrossRef]
21. Badekila, A.K.; Kini, S.; Jaiswal, A.K. Fabrication techniques of biomimetic scaffolds in three-dimensional cell culture: A review. *J. Cell. Physiol.* **2021**, *236*, 741–762. [CrossRef]
22. Alagirusamy, R.; Das, A. Chapter 8—Conversion of fibre to yarn: An overview. In *Textiles and Fashion*; Sinclair, R., Ed.; Woodhead Publishing: Sawston, UK, 2015; pp. 159–189. [CrossRef]
23. Lawrence, C.A. *Advances in Yarn Spinning Technology*; Elsevier Science: Amsterdam, The Netherlands, 2010.
24. Subramaniam, V.; Mohammad, P. Wrap spinning technology—A critical review of yarn properties. *Indian J. Fibre Text. Res.* **1992**, *17*, 252–254.
25. Das, A.; Alagirusamy, R. 3—Fundamental principles of open end yarn spinning. In *Advances in Yarn Spinning Technology*; Lawrence, C.A., Ed.; Woodhead Publishing: Sawston, UK, 2010; pp. 79–101. [CrossRef]
26. Gilmore, J.; Yin, F.; Burg, K.J.L. Evaluation of permeability and fluid wicking in woven fiber bone scaffolds. *J. Biomed. Mater. Res. Part B Appl. Biomater.* **2019**, *107*, 306–313. [CrossRef] [PubMed]
27. Van Lieshout, M.I.; Vaz, C.M.; Rutten, M.C.M.; Peters, G.W.M.; Baaijens, F.P.T. Electrospinning versus knitting: Two scaffolds for tissue engineering of the aortic valve. *J. Biomater. Sci. Polym. Ed.* **2006**, *17*, 77–89. [CrossRef] [PubMed]
28. Zhang, F.; Bambharoliya, T.; Xie, Y.; Liu, L.; Celik, H.; Wang, L.; Akkus, O.; King, M.W. A hybrid vascular graft harnessing the superior mechanical properties of synthetic fibers and the biological performance of collagen filaments. *Mater. Sci. Eng. C Mater. Biol. Appl.* **2021**, *118*, 111418. [CrossRef] [PubMed]
29. Wang, X.; Li, Q.; Hu, X.; Ma, L.; You, C.; Zheng, Y.; Sun, H.; Han, C.; Gao, C. Fabrication and characterization of poly(l-lactide-co-glycolide) knitted mesh-reinforced collagen–chitosan hybrid scaffolds for dermal tissue engineering. *J. Mech. Behav. Biomed. Mater.* **2012**, *8*, 204–215. [CrossRef] [PubMed]
30. Learn, G.D.; McClellan, P.E.; Knapik, D.M.; Cumsky, J.L.; Webster-Wood, V.; Anderson, J.M.; Gillespie, R.J.; Akkus, O. Woven collagen biotextiles enable mechanically functional rotator cuff tendon regeneration during repair of segmental tendon defects in vivo. *J. Biomed. Mater. Res. Part B Appl. Biomater.* **2019**, *107*, 1864–1876. [CrossRef]
31. Alberti, K.A.; Xu, Q. Biocompatibility and degradation of tendon-derived scaffolds. *Regen. Biomater.* **2016**, *3*, 1–11. [CrossRef]
32. Von Burkersroda, F.; Schedl, L.; Göpferich, A. Why degradable polymers undergo surface erosion or bulk erosion. *Biomaterials* **2002**, *23*, 4221–4231. [CrossRef]
33. Streichan, S.J.; Hoerner, C.R.; Schneidt, T.; Holzer, D.; Hufnagel, L. Spatial constraints control cell proliferation in tissues. *Proc. Natl. Acad. Sci. USA* **2014**, *111*, 5586–5591. [CrossRef]
34. Rico-Llanos, G.; Borrego-González, S.; Moncayo-Donoso, M.; Becerra, J.; Visser, R. Collagen type I biomaterials as scaffolds for bone tissue engineering. *Polymers* **2021**, *13*, 599. [CrossRef]
35. Cvrček, L.; Horáková, M. Chapter 14—Plasma modified polymeric materials for implant applications. In *Non-Thermal Plasma Technology for Polymeric Materials*; Thomas, S., Mozetič, M., Cvelbar, U., Špatenka, P., Praveen, K.M., Eds.; Elsevier: Amsterdam, The Netherlands, 2019; pp. 367–407. [CrossRef]
36. Ge, Z.; Jin, Z.; Cao, T. Manufacture of degradable polymeric scaffolds for bone regeneration. *Biomed. Mater.* **2008**, *3*, 022001. [CrossRef]
37. Dong, C.; Lv, Y. Application of collagen scaffold in tissue engineering: Recent advances and new perspectives. *Polymers* **2016**, *8*, 42. [CrossRef] [PubMed]
38. Tonndorf, R.; Aibibu, D.; Cherif, C. Isotropic and anisotropic scaffolds for tissue engineering: Collagen, conventional, and textile fabrication technologies and properties. *Int. J. Mol. Sci.* **2021**, *22*, 9561. [CrossRef] [PubMed]
39. Saidy, N.T.; Wolf, F.; Bas, O.; Keijdener, H.; Hutmacher, D.W.; Mela, P.; De-Juan-Pardo, E.M. Biologically inspired scaffolds for heart valve tissue engineering via melt electrowriting. *Small* **2019**, *15*, 1900873. [CrossRef] [PubMed]
40. Roshanbinfar, K.; Vogt, L.; Ruther, F.; Roether, J.A.; Boccaccini, A.R.; Engel, F.B. Nanofibrous composite with tailorable electrical and mechanical properties for cardiac tissue engineering. *Adv. Funct. Mater.* **2020**, *30*, 1908612. [CrossRef]
41. Ma, C.; Wang, H.; Chi, Y.; Wang, Y.; Jiang, L.; Xu, N.; Wu, Q.; Feng, Q.; Sun, X. Preparation of oriented collagen fiber scaffolds and its application in bone tissue engineering. *Appl. Mater. Today* **2021**, *22*, 100902. [CrossRef]
42. Xie, Y.; Chen, J.; Celik, H.; Akkus, O.; King, M.W. Evaluation of an electrochemically aligned collagen yarn for textile scaffold fabrication. *Biomed. Mater.* **2021**, *16*, 025001. [CrossRef]

Article

Synthesis and In Vitro Characterization of Ascorbyl Palmitate-Loaded Solid Lipid Nanoparticles

Maja Ledinski [1], Ivan Marić [2], Petra Peharec Štefanić [1], Iva Ladan [3], Katarina Caput Mihalić [1], Tanja Jurkin [2], Marijan Gotić [4] and Inga Urlić [1,*]

1. Faculty of Science, Division of Molecular Biology, Department of Biology, University of Zagreb, 10 000 Zagreb, Croatia; maja.ledinski@biol.pmf.hr (M.L.); petra.peharec.stefanic@biol.pmf.hr (P.P.Š.); katarina.caput.mihalic@biol.pmf.hr (K.C.M.)
2. Division of Materials Chemistry, Radiation Chemistry and Dosimetry Laboratory, Ruđer Bošković Institute, 10 000 Zagreb, Croatia; imaric@irb.hr (I.M.); tanja.jurkin@irb.hr (T.J.)
3. Croatian Institute of Public Health, Department of Health Ecology, 10 000 Zagreb, Croatia; iva.ladan@hzjz.hr
4. Division of Materials Physics, Laboratory for Molecular Physics and Synthesis of New Materials, Ruđer Bošković Institute, 10 000 Zagreb, Croatia; gotic@irb.hr
* Correspondence: ingam@biol.pmf.hr

Abstract: Antitumor applications of ascorbic acid (AA) and its oxidized form dehydroascorbic acid (DHA) can be quite challenging due to their instability and sensitivity to degradation in aqueous media. To overcome this obstacle, we have synthesized solid lipid nanoparticles loaded with ascorbyl palmitate (SLN-AP) with variations in proportions of the polymer Pluronic F-68. SLNs were synthesized using the hot homogenization method, characterized by measuring the particle size, polydispersity, zeta potential and visualized by TEM. To investigate the cellular uptake of the SLN, we have incorporated coumarin-6 into the same SLN formulation and followed their successful uptake for 48 h. We have tested the cytotoxicity of the SLN formulations and free ascorbate forms, AA and DHA, on HEK 293 and U2OS cell lines by MTT assay. The SLN-AP in both formulations have a cytotoxic effect at lower concentrations when compared to ascorbate applied the form of AA or DHA. Better selectivity for targeting tumor cell line was observed with 3% Pluronic F-68. The antioxidative effect of the SLN-AP was observed as early as 1 h after the treatment with a small dose of ascorbate applied (5 μM). SLN-AP formulation with 3% Pluronic F-68 needs to be further optimized as an ascorbate carrier due to its intrinsic cytotoxicity.

Keywords: ascorbate; ascorbyl palmitate; drug delivery; cellular uptake; nanoparticles; antitumor effect

1. Introduction

Ascorbate has been extensively studied as an antitumor agent, but its instability remains the major problem for its use. Ascorbic acid (AA) and its oxidized form, dehydroascorbic acid (DHA), are readily degraded in aqueous media in the presence of oxygen and metal ions and are also sensitive to light. To improve the stability of ascorbate, salts and lipophilic derivatives of AA are often used [1]. The alkyl esters of AA are of particular interest because they can penetrate cell membranes, however, they are poorly soluble in aqueous environments [2]. Ascorbyl palmitate (Figure 1) is an alkyl ester already used in the cosmetics, dermatology, and food industries as a more stable lipophilic derivative of AA because it retains the antioxidant properties of ascorbate [3]. Therefore, it is important to develop formulations that could allow its administration under hydrophilic conditions.

In order to apply AP in hydrophilic conditions, researchers have already tried various formulations. Kristl et al. [4] tested the incorporation of AP into microemulsions, liposomes, and solid lipid nanoparticles (SLNs), which were evaluated as the second most stable out of five formulations. On the other hand, Gopinath et al. [5] developed a method for the formation of ascorbyl palmitate bilayer vesicles (aspasomes) using the film hydration

method and sonication. They had found that AP could assemble into bilayer vesicles by adding cholesterol and diacetyl phosphate. Teeranachaideekul et al. [6] also investigated the incorporation of AP into lipid nanoparticles by synthesizing both SLN and nanostructured lipid carriers by high-pressure homogenization. They found that AP in nanoparticles with glycerol monostearate produced stable formulations with high (100%) encapsulation efficiency.

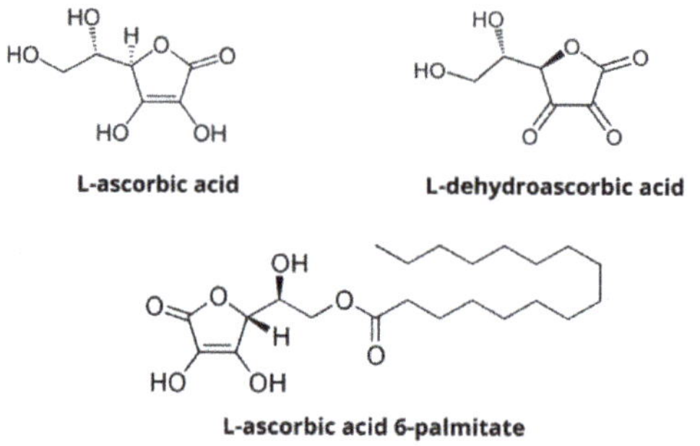

Figure 1. Chemical structures of hydrophilic compounds ascorbic acid and dehydroascorbic acid and lipophilic derivative, ascorbyl palmitate.

Ascorbate is a known antioxidant, however, depending on the concentration it can have a prooxidant effect that selectively affects tumor cells [7]. Although this antitumor effect is mostly explained by prooxidant activity of ascorbate, it can also affect tumor cells by inducing epigenetic modifications, affecting glycolysis metabolism and surviving in hypoxia [7,8]. Therefore, AP, as a carrier of ascorbate, was recognized as an antitumor agent and researchers have incorporated it into nanoparticles with various chemotherapeutic agents. Zhou et al. [9] incorporated AP and paclitaxel into the SLN, Li et al. [10] synthesized liposomes for co-administration of AP and docetaxel, and Jukanti et al. [11] PEGylated liposomes for co-administration of AP and doxorubicin. All of these formulations were designed to combine the antitumor activity reported for ascorbate with the cytotoxic effect of chemotherapeutic agents, which is referred to as combination chemotherapy.

Several sources indicate that the polymers that go into these formulations must be carefully selected because of their effect on the stability of the nanoparticle formulations [5,11]. In this work, we synthesized and characterized ascorbyl palmitate-loaded SLNs (Figure 2) with different proportions of the polymer Pluronic F-68, which is commonly used as a surfactant in the synthesis of nanoparticles due to its amphiphilic properties and good biocompatibility [12].

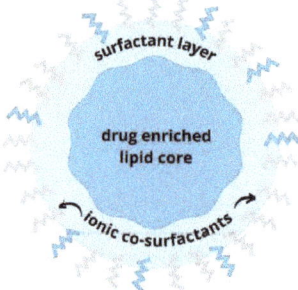

Figure 2. The structure of solid lipid nanoparticle. SLNs have lipid core with an incorporated lipid compound of interest and a surfactant layer on the surface which enables solubility in hydrophilic solutions. Ionic co-surfactants provide electrical stability and prevent the possible aggregation of nanoparticles.

In the study of drug delivery systems, confirmation of cellular uptake is crucial before further experiments are performed. We followed nanoparticle uptake by visualizing the uptake of SLN loaded with fluorescent coumarin-6. We also analyzed the cytotoxicity of SLN on HEK 293 and U2OS cell lines to test whether these formulations could have an antitumor potential.

2. Materials and Methods

2.1. Materials

Glyceryl monostearate (GMS), dimethyldioctadecylammonium bromide (DDAB), ascorbyl palmitate (AP), ascorbic acid (AA), dehydroascorbic acid (DHA), coumarin-6, MTT (3-(4,5-Dimethylthiazol-2-yl)-2,5-Diphenyltetrazolium Bromide), 0.25% Trypsin/EDTA, 2′,7′-Dichlorofluorescin diacetate and Pluronic F-68 were purchased from Sigma Aldrich, Germany. Dulbecco's Modified Eagle Medium High glucose (4.5 g/L), fetal bovine serum (FBS) and penicillin/streptomycin were purchased from Capricorn, Germany.

2.2. SLN Synthesis and Characterization

Ascorbyl palmitate-loaded SLNs were prepared by ultrasonic method [9]. GMS (50 mg), DDAB (5 mg) and AP (10 mg) were heated to 85 °C, just above the melting point of GMS. At the same time, 5 mL of Pluronic-F68 (either 3% or 10%) was heated to the same temperature, in a water bath. When both phases reached 85 °C, an aqueous phase was added to the lipid phase and pre-emulsion was sonicated for 25 min in a water bath at 85 °C (Figure 3). SLNs were formed after cooling in an ice bath and were stored at 4 °C. Size, polydispersity and zeta potential of SLNs were measured by multiple angle dynamic light scattering (MADLS) using Zetasizer Ultra (Malvern Panalytical). The size distributions given by MADLS were reported as distributions by intensity, and results are presented as the mean value of at least 3 measurements each, performed at three angles. Zeta potential was taken as the mean value of three measurements. For visualization using TEM, SLNs were firstly incubated on a Formvar®/carbon copper grids and washed in ultra-pure water (Milli-Q, 18.2 MΩ·cm, Merck Millipore, Billerica, MA, USA). Sample was then incubated with uranyl acetate for 5 min and after staining washed in ultra-pure water 3 times. After the sample had dried, SLNs were visualized using a transmission electron microscope Morgagni 268D (Philips/FEI) and operating at 70 kV.

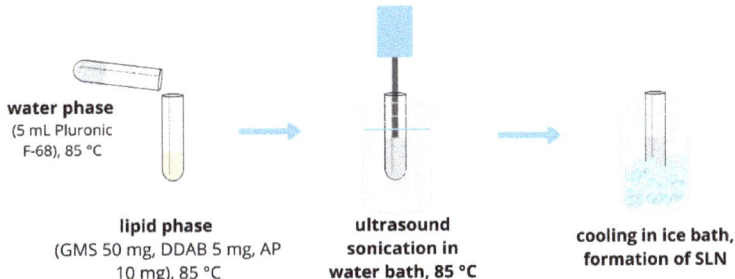

Figure 3. Synthesis of solid lipid nanoparticles.

2.3. Cell Culture

Experiments were performed using human embryonic kidney HEK 293 and human osteosarcoma U2OS cell lines, which were kindly provided by Vjekoslav Tomaić (Ruđer Bošković Institute, Zagreb, Croatia). HEK 293 and U2OS cell lines were cultured in DMEM High glucose (4.5 g/L) supplemented with 10% FBS and 1% penicillin/streptomycin. Cells were cultured at 37 °C in the presence of 5% CO_2.

2.4. Cellular Uptake of Solid Lipid Nanoparticles

SLNs containing coumarin-6 were synthesized using the same method as mentioned above. Synthesis was carried out in the dark. To remove free coumarin-6 from the SLN suspension, SLN-coumarin-6 were centrifuged at $20,000 \times g$, 30 min (Hettich universal 320R) and SLN-coumarin-6 pellet was resuspended in ultra-pure water. Cells were seeded at a density of 30,000 cells/mL in a Petri dish and allowed to adhere overnight at 37 °C, 5% CO_2. They were treated with SLN-coumarin-6 and visualized by using fluorescent microscopy (Olympus BX51; software DPController, Olympus Optical) at time points 1 h, 3 h, 6 h, 24 h and 48 h.

2.5. Cytotoxicity Analysis Using MTT Test

HEK 293 and U2OS cells were seeded in 96-well plate (1×10^4 cells/well) and allowed to adhere overnight at 37 °C, 5% CO_2. Treatments (SLN-AP, blank SLN, AA and DHA) were prepared in cell culture medium. Cells were treated with blank SLNs in the same amount as SLN-AP for each concentration to serve as a control for the effect of the empty nanoparticles. In addition, all treatment concentrations were normalized to ascorbate composition to account for differences in molecular masses between AA, DHA and AP. Cells were treated in triplicates and after 24 h, treatment medium was aspirated, and cells were washed three times with 200 µL PBS. Cells were incubated 4 h with MTT solution in cell culture medium at concentration 0.5 mg/mL, and 170 µL DMSO was added to dissolve formazan crystals. Absorbance was measured at 570 nm using GloMax microplate reader (Promega), and cell viability was calculated as a percentage of untreated control after reducing blank absorbance to all the samples.

2.6. Analysis of Reactive Oxygen Species

U2OS cells were seeded in black 96-well plates (2×10^4 cells/well) and incubated for 24 h at 37 °C, 5% CO_2. Treatments (AA in free form, SLN-AP, blank SLN) were prepared in cell culture medium and cells were treated in triplicates for 1 h and 6 h. After the incubation, cells were washed 3 times with PBS to remove any treatment residue. Then, 20 µM 2′,7′-dichlorofluorescin diacetate (DCHF-DA) was prepared in PBS and cells were incubated for 30 min at 37 °C, 5% CO_2. Fluorescence was measured at excitation 490 nm, emission 510–570 nm using GloMax microplate reader (Promega).

3. Results

3.1. SLN Synthesis and Characterization

The SLN were characterized by measuring their size, polydispersity and zeta potential. All nanoparticles have nanometer size as shown in Table 1 and Figure 4. SLN-APs prepared in 3% Pluronic are larger in size and have a smaller polydispersity index when compared to blank SLNs of the same formulation, which are smaller in size but have a higher polydispersity index. The SLN-APs prepared in 10% Pluronic are smaller than the blank SLNs of the same formulation, but the polydispersity indices do not differ greatly. Both formulations of the empty SLNs have a positive zeta potential, while SLN-APs have a very low positive for 3% Pluronic F-68 formulation and slightly negative zeta potential for 10% Pluronic F-68 formulation.

Table 1. Characterization of blank solid lipid nanoparticles (blank SLN) and ascorbyl palmitate-loaded solid lipid nanoparticles (SLN-AP). Before measurement, samples were diluted 10× in ultra-pure water. Average size, polydispersity and zeta potential were measured on Zetasizer, Malvern Panalytical. Data are expressed as mean ± ($n \geq 3$).

Measurement	3% Pluronic F-68		10% Pluronic F-68	
	Blank SLN	SLN-AP	Blank SLN	SLN-AP
Diameter/nm	18.2 ± 3.21 125 ± 4.52	337 ± 8.52	445 ± 14.3	414 ± 32.1
PDI	0.291 ± 0.0224	0.137 ± 0.0336	0.182 ± 0.0219	0.198 ± 0.0451
zeta potential/mV	+38.1 ± 3.36	+1.45 ± 0.337	+21.5 ± 0.308	−4.18 ± 0.251

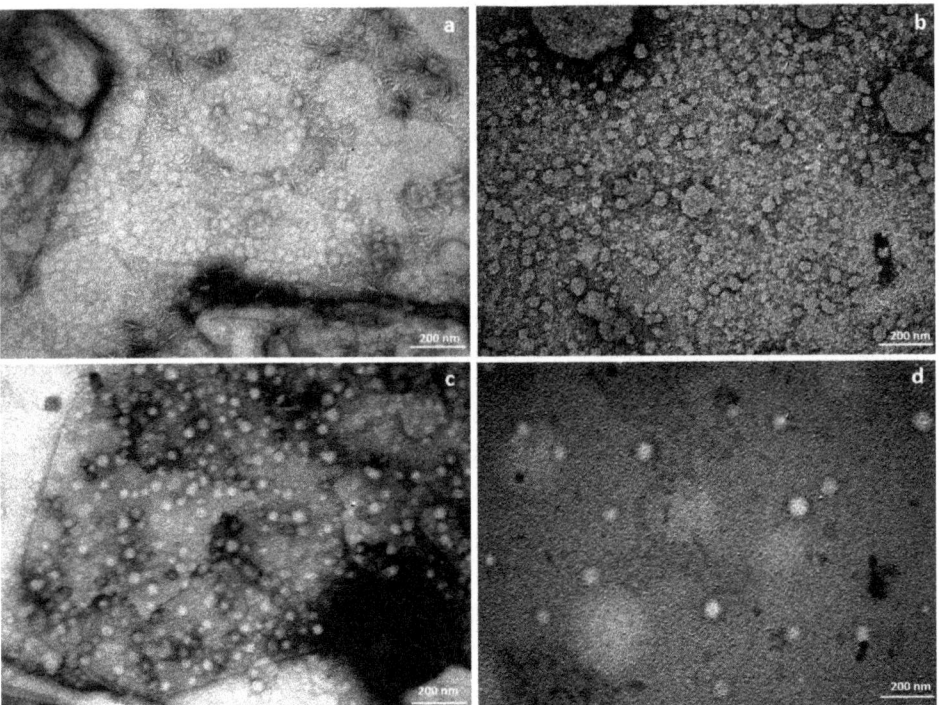

Figure 4. TEM images of SLN. SLN suspensions were contrasted with uranyl acetate and washed in ultra-pure water before visualization. Scale bar = 200 nm (**a**) SLN-AP 3% Pluronic; (**b**) blank SLN 3% Pluronic; (**c**) SLN-AP 10% Pluronic; and (**d**) blank SLN 10% Pluronic.

Morphological evaluation of all the SLN suspensions using TEM revealed that both suspensions contain round SLNs with a size in the nanometer range. Rod-like shapes are seen in the formulation SLN-AP (Figure 4a) prepared with 3% Pluronic. The image of empty SLNs prepared in 3% Pluronic (Figure 4b) confirms the polydispersity data with different sizes of SLNs. Likewise, the images of SLNs prepared in 10% Pluronic (Figure 4c,d) are consistent with the DLS data.

3.2. Cellular Uptake of SLN

HEK 293 and U2OS cell lines were incubated with SLN-coumarin-6 and visualized under a fluorescence microscope. Fluorescence is detected in both cell types and increases with incubation time, as shown in Figure 5.

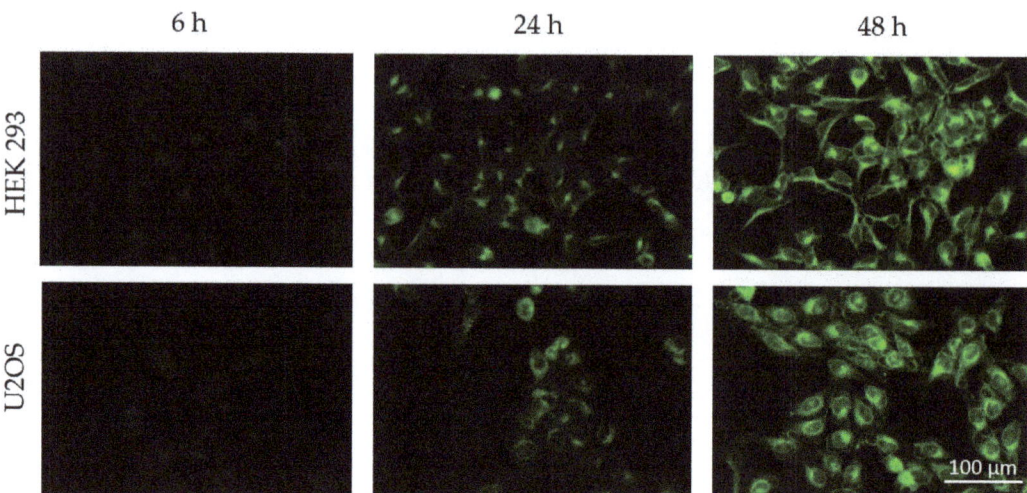

Figure 5. Cellular uptake of SLN-coumarin-6 during 48 h in cell lines HEK 293 and U2OS. Cells were treated with SLN-coumarin-6 and photographed at time points 6 h, 24 h and 48 h. Green fluorescence represents coumarin-6 that entered the cells. Scale bar = 100 µm.

3.3. Cytotoxicity Analysis

We used the MTT assay to evaluate the cytotoxicity of the two SLN formulations, prepared in 3% and 10% Pluronic F-68.

Regarding the SLN prepared in 3% Pluronic F-68, blank SLN presented mild cytotoxicity in HEK 293 cells, and stronger cytotoxicity in tumor U2OS cells. Likewise, SLN-AP also presented mild cytotoxicity on HEK 293 cell and stronger in U2OS cells. For concentrations higher than 5 µM of ascorbate, SLN-AP show cytotoxic effect on both cell lines, HEK 293 and U2OS, but the data suggest that tumor cells are slightly more sensitive to the treatment (Figure 6a).

When analyzing data regarding the SLN prepared in 10% Pluronic F-68, blank SLN presented selective cytotoxicity in HEK 293 and U2OS cell, significantly reducing viability in tumor U2OS cell. After 50 µM ascorbate, SLN-AP of this formulation begin to exert a cytotoxic effect, as shown in Figure 6b. U2OS cell are more sensitive to the treatment with SLN-AP then HEK293 cells (Figure 6b).

Altogether, cytotoxicity is higher for the SLN prepared with 3% Pluronic F-68, and tumor cells are slightly more sensitive to the SLN and SLN-AP treatments than HEK 293.

To evaluate the difference in cytotoxicity between the SLN formulations and ascorbate in AA and DHA, we performed a cytotoxicity assay and treated the cells in the same way with AA and DHA. While DHA shows no cytotoxic effect on HEK 293 and U2OS cell lines, AA has an IC50 of 1.94 mM for HEK 293 and 4.17 mM for U2OS, as shown in Figure 6c. Therefore, SLN-APs, prepared in 3% and 10% Pluronic, are effective in much lower concentrations than AA and DHA in the free form.

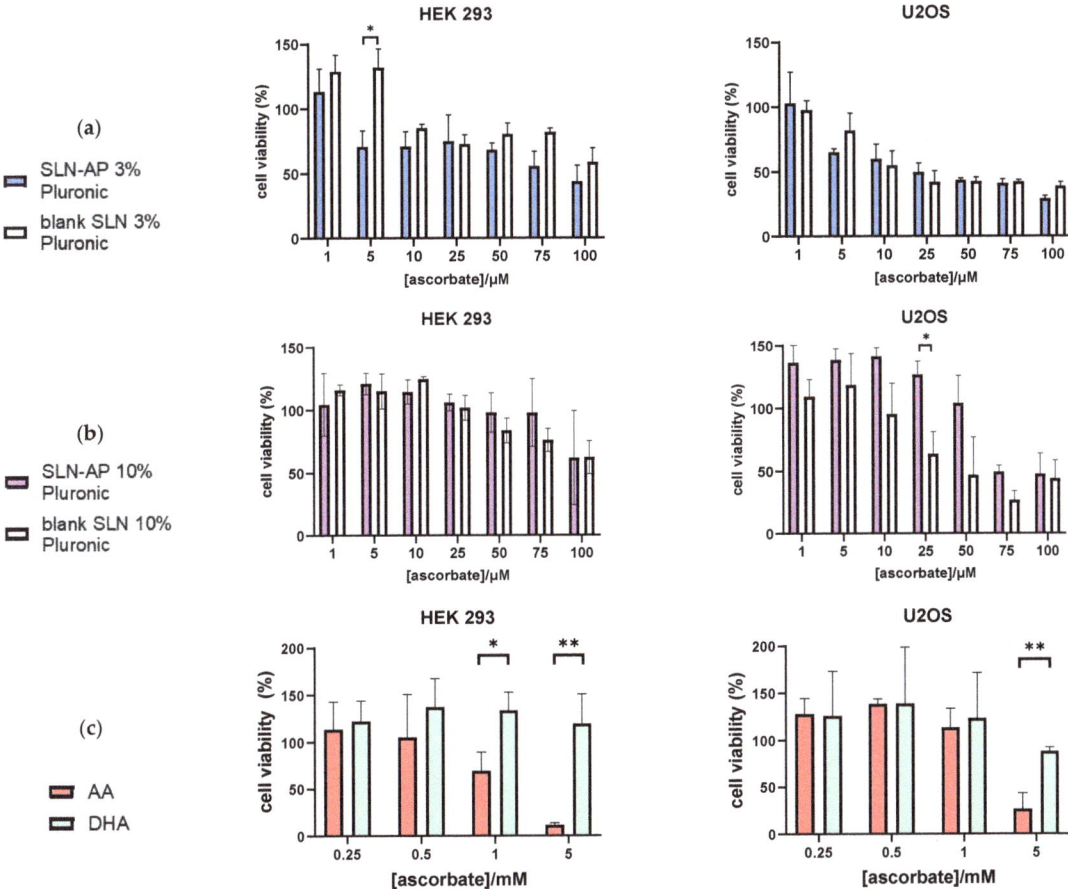

Figure 6. Cytotoxicity of SLN AP and blank SLN synthesized with either 3% (**a**) or 10% (**b**) Pluronic F-68, as well as AA and DHA in free form (**c**), on HEK 293 and U2OS cell lines was tested by MTT test during 24 h. All treatments were normalized to ascorbate composition. Data are expressed as percentage of negative control, mean ± sd, $n = 3$. Holm–Šídák multiple unpaired t-tests were performed. Statistical significance: * $p \leq 0.05$, ** $p \leq 0.01$.

3.4. Analysis of Reactive Oxygen Species

Analysis of reactive oxygen species has been performed on U2OS cells following 1 h and 6 h of incubation with SLN-AP and blank SLN. Fluorescence is proportional to the amount of ROS detected after the treatment. As visible from Figure 7, AA in free form did not affect ROS both 1 h and 6 h after the treatment. On the other hand, one hour after the treatment, both SLN-AP and blank SLN lower the basal amount of ROS in the U2OS cells, with SLN-AP having significant antioxidative effect when applied in concentrations of 5–100 µM (normalized to ascorbate composition). Six hours after the treatment, SLN-AP still show the antioxidative effect, while blank SLN did not significantly affect the amount of ROS in the cells.

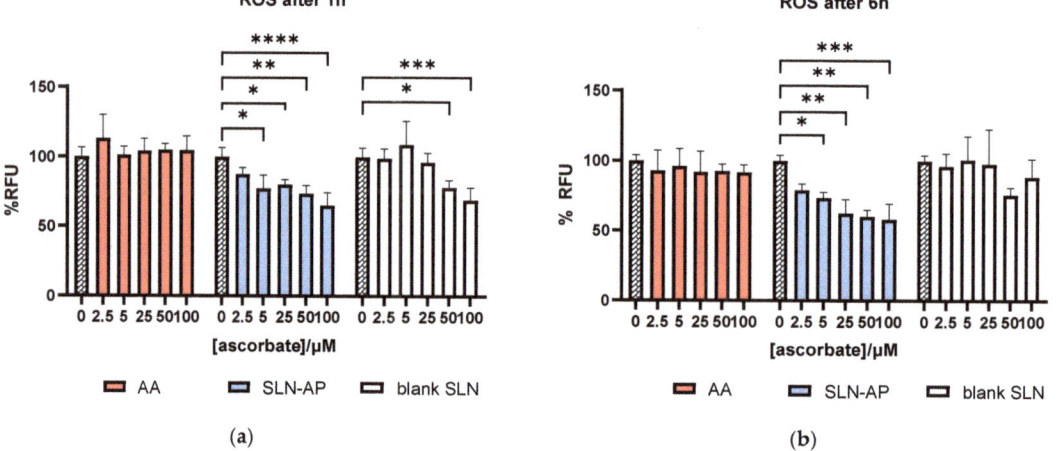

Figure 7. In vitro ROS analysis in cell line U2OS. Cells were treated with AA in free form, SLN-AP and blank SLN for 1 h (**a**) and 6 h (**b**). Fluorescence was measured at 490 nm/570 nm. All treatments were normalized to ascorbate composition. Data are expressed as percent of negative control, mean ± sd, n = 3. Two-way ANOVA was performed with Dunnet's correction. Statistical significance: * $p \leq 0.05$, ** $p \leq 0.01$, *** $p \leq 0.001$, **** $p \leq 0.0001$.

4. Discussion

It is widely known that AA and DHA are readily applicable sources of ascorbate because they are soluble in water as well as in the cell culture medium. However, since they are sensitive to light, oxygen and interaction with metal ions, they are easily degraded [1,9]. To overcome this problem, we have decided to use AP as the ascorbate source. As a lipophilic compound, AP should be able to pass the cell membrane [2]. It can be dissolved in water at a very low concentration (179.48 nM at 25 °C), but is readily soluble in higher concentrations in ethanol and some other organic compounds [13]. However, when dissolving AP in ethanol in cell culture medium, AP interacts with various salts and precipitates. To overcome these obstacles, we incorporated AP into the solid lipid nanoparticles.

We prepared two formulations of SLNs, with the only difference being the wt/vol % of Pluronic F-68. We found that this did not have much effect on the zeta potential, as both formulations of the blank SLN had a positive zeta potential (due to the co-surfactant DDAB) and SLN-AP had almost neutral or negative zeta potential when prepared with 3% and 10% Pluronic F-68, respectively. The measured almost neutral and slightly negative zeta potential is likely due to the orientation of the hydrophilic ascorbate on the surface of SLN-AP, which was previously shown by Jukanti et al. for AP liposomes [11]. Formulations prepared with 10% Pluronic F-68 have larger nanoparticles and lower PDI, suggesting it is beneficial for the synthesis of stable formulations. TEM images are consistent with DLS data, however, rod-like shapes are visible in SLN-AP formulation prepared in 3% Pluronic F-68. This is not surprising, as lipids in lipid-based formulations can aggregate into many structures due to varying temperatures, water concentrations, or the presence of different components [14]. For example, Yakimova et al. [15] showed that solid lipid nanoparticles (SLNs) loaded with luminescent markers such as monosubstituted pillar [5] arenes with terminal OH group produced SLNs with elongated shape, i.e., spindle- and rod-shaped SLNs. In our case, incorporation of AP into the formulation may had lead to slight changes in the morphology of SLNs, as shown in Figure 4a.

To evaluate the cellular uptake of the synthesized nanoparticles, we used coumarin-6, a hydrophobic fluorophore commonly used for visualization of drug delivery systems, and incorporated it into the same SLN formulation [16,17]. Zhou et al. [9] and Shi et al. [18] showed the uptake of nanoparticles with coumarin-6 within 4 h; we followed the gradual uptake of SLN-coumarin-6 during 48 h. The fluorescence observed upon visualization of cells incubated with SLN (Figure 5) is partly due to the SLN-coumarin-6 taken up by the cells and partly due to the coumarin-6 released. The gradual increase in fluorescence confirmed the successful uptake of SLN as well as the probable release of the compound contained in the formulation.

When we examined the cytotoxicity of the SLN-AP, we found that U2OS cells are more sensitive to the treatment then HEK 293 cells. Our results show that SLN-AP prepared in 3% and 10% Pluronic F-68 are cytotoxic in much lower concentrations than AA and DHA in the free form. However, overall cytotoxicity is slightly higher for SLNs prepared using 3% Pluronic F-68 (Figure 6) with U2OS cells being more sensitive to the treatment then HEK 293 cells. It is proven that ascorbate can have a selective cytotoxic effect on tumor cells [19–21]. Few mechanisms are suggested to explain how can ascorbate have such a selective cytotoxic effect on tumor cells when applied in free form. Ascorbate has a prooxidative effect when interacting with metal ions in cell culture media because of the H_2O_2 generation [7,22], it downregulates the transcription factors important for adjusting to hypoxic conditions [23], inhibits GAPDH by generating ROS in *KRAS* and *BRAF* mutant colorectal cancer cells [20] and induces caspase-independent cell death pathway in breast cancer cells [21]. Ascorbate was also investigated when applied in small concentrations together with various chemotherapeutics and was shown to enhance their cytotoxic effect [7]. Several clinical trials are exploring administration of ascorbate in combination chemotherapy [22]. Regarding delivery systems, ascorbate derivative AP is usually explored in combination chemotherapy, where it acts both as a carrier and as a therapeutic—either as an antioxidant [11] or as an enhancer of chemotherapeutic effect [9,10]. Sawant et al. have synthesized AP loaded micelles in order to investigate the effect of AP itself and found that AP loaded micelles have cytotoxic effect due to ROS generation, mostly in the µM range of applied AP [24].

Analysis of reactive oxygen species was performed using the fluorescent probe DCHF-DA, which is de-esterified upon oxidation in the cell [25]. While AA in free form does not affect the level of ROS, SLN-AP significantly decreases them (Figure 7) both 1 h and 6 h after the treatment. Since SLN have antioxidative properties, we conclude that cytotoxicity of SLN is not due to ROS generation. While some studies have reported prooxidant effects of ascorbyl palmitate nanoparticles [18,26], our results show that this formulation can be applied at µM concentrations in which it has no cytotoxic effect and exhibits antioxidant properties both 1 h and 6 h after application. In this formulation, the antioxidant effect of low dose of ascorbate is achieved through a delivery system that should be able to overcome physiological barriers in the organism [27].

5. Conclusions

In conclusion, ascorbyl palmitate is a stable source of ascorbate. To ensure successful delivery of AP to the cells, we loaded AP into solid lipid nanoparticles (SLN-AP) using the hot homogenization method with 3% and 10% polymer Pluronic F-68 and their characterization confirmed similar properties. The cytotoxicity assay performed on normal and tumor cell lines showed the appropriate concentration range for the use of both formulations, which depends on whether we want to study antioxidant properties or selective antitumor properties of the SLN-AP. However, both formulations need to be optimized to reduce their intrinsic cytotoxicity.

Author Contributions: Conceptualization, M.G. and I.U.; methodology, M.G. and I.U.; software, M.L. and I.M.; formal analysis, M.L., I.M., K.C.M., P.P.Š., I.L. and T.J.; investigation, M.G., I.U. and M.L.; data curation, M.L., I.M. and P.P.Š.; writing—original draft preparation, M.L.; writing—review and editing, M.L., I.M., K.C.M., P.P.Š., I.L., T.J., M.G. and I.U.; visualization, M.L. and P.P.Š.; supervision, M.G. and I.U.; project administration, I.U.; funding acquisition, M.G. and I.U. All authors have read and agreed to the published version of the manuscript.

Funding: This research was funded by Croatian Science Foundation, grant number IP-2018-01-7590, project "Ascorbic acid for selective targeting of sarcoma stem cells (ASTar)". Zetasizer Ultra (Malvern Panalytical) was purchased by the project UIP-2017-05-7337 financially supported by the Croatian Science Foundation.

Acknowledgments: This work was financially supported by the Croatian Science Foundation under the project IP-2018-01-7590 Ascorbic acid for selective targeting of sarcoma stem cells (ASTar).

Conflicts of Interest: The authors declare no conflict of interest.

References

1. Caritá, A.C.; Fonseca-Santos, B.; Shultz, J.D.; Michniak-Kohn, B.; Chorilli, M.; Leonardi, G.R. Vitamin C: One compound, several uses. Advances for delivery, efficiency and stability. *Nanomed. Nanotechnol. Biol. Med.* **2020**, *24*, 102117. [CrossRef] [PubMed]
2. Moribe, K.; Limwikrant, W.; Higashi, K.; Yamamoto, K. Drug Nanoparticle Formulation Using Ascorbic Acid Derivatives. *J. Drug Deliv.* **2011**, *2011*, 138929. [CrossRef] [PubMed]
3. Pokorski, M.; Marczak, M.; Dymecka, A.; Suchocki, P. Ascorbyl palmitate as a carrier of ascorbate into neural tissues. *J. Biomed. Sci.* **2003**, *10*, 193–198. [CrossRef]
4. Kristl, J.; Volk, B.; Gasperlin, M.; Sentjurc, M.; Jurkovic, P. Effect of colloidal carriers on ascorbyl palmitate stability. *Eur. J. Pharm. Sci.* **2003**, *19*, 181–189. [CrossRef]
5. Gopinath, D.; Ravi, D.; Rao, B.R.; Apte, S.S.; Renuka, D.; Rambhau, D. Ascorbyl palmitate vesicles (Aspasomes): Formation, characterization and applications. *Int. J. Pharm.* **2004**, *271*, 95–113. [CrossRef] [PubMed]
6. Teeranachaideekul, V.; Souto, E.B.; Müller, R.H.; Junyaprasert, B. Release studies of ascorbyl palmitate-loaded semi-solid nanostructured lipid carriers (NLC gels). *J. Microencapsul.* **2008**, *2048*, 110–120. [CrossRef]
7. Vissers, M.C.M.; Das, A.B. Potential mechanisms of action for vitamin C in cancer: Reviewing the evidence. *Front. Physiol.* **2018**, *9*, 809. [CrossRef]
8. Ngo, B.; Van Riper, J.M.; Cantley, L.C.; Yun, J. Targeting cancer vulnerabilities with high-dose vitamin C. *Nat. Rev. Cancer* **2019**, *19*, 271–282. [CrossRef]
9. Zhou, M.; Li, X.; Li, Y.; Yao, Q.; Ming, Y.; Li, Z.; Lu, L.; Shi, S. Ascorbyl palmitate-incorporated paclitaxel-loaded composite nanoparticles for synergistic anti-tumoral therapy. *Drug Deliv.* **2017**, *24*, 1230–1242. [CrossRef]
10. Li, J.; Guo, C.; Feng, F.; Fan, A.; Dai, Y.; Li, N.; Zhao, D.; Chen, X.; Lu, Y. Co-delivery of docetaxel and palmitoyl ascorbate by liposome for enhanced synergistic antitumor efficacy. *Nat. Publ. Gr.* **2016**, *6*, 38787. [CrossRef]
11. Jukanti, R.; Devraj, G.; Shashank, A.S.; Devraj, R. Biodistribution of ascorbyl palmitate loaded doxorubicin pegylated liposomes in solid tumor bearing mice. *J. Microencapsul.* **2011**, *28*, 142–149. [CrossRef] [PubMed]
12. Yu, J.; Qiu, H.; Yin, S.; Wang, H.; Li, Y. Polymeric drug delivery system based on pluronics for cancer treatment. *Molecules* **2021**, *26*, 3610. [CrossRef]
13. Ascorbyl Palmitate | C22H38O7–PubChem. Available online: https://pubchem.ncbi.nlm.nih.gov/compound/Ascorbyl-palmitate (accessed on 10 August 2021).
14. Hallan, S.S.; Sguizzato, M.; Esposito, E.; Cortesi, R. Challenges in the physical characterization of lipid nanoparticles. *Pharmaceutics* **2021**, *13*, 549. [CrossRef] [PubMed]
15. Yakimova, L.S.; Guralnik, E.G.; Shurpik, D.N.; Evtugyn, V.G.; Osin, Y.N.; Subakaeva, E.V.; Sokolova, E.A.; Zelenikhin, P.V.; Stoikov, I.I. Morphology, structure and cytotoxicity of dye-loaded lipid nanoparticles based on monoamine pillar[5]arenes. *Mater. Chem. Front.* **2020**, *4*, 2962–2970. [CrossRef]
16. Rivolta, I.; Panariti, A.; Lettiero, B.; Sesana, S.; Gasco, P.; Gasco, M.R.; Masserini, M.; Miserocchi, G. Cellular uptake of coumarin-6 as a model drug loaded in solid lipid nanoparticles. *J. Physiol. Pharmacol.* **2011**, *62*, 45–53. [PubMed]
17. Finke, J.H.; Richter, C.; Gothsch, T.; Kwade, A.; Büttgenbach, S.; Müller-Goymann, C.C. Coumarin 6 as a fluorescent model drug: How to identify properties of lipid colloidal drug delivery systems via fluorescence spectroscopy? *Eur. J. Lipid Sci. Technol.* **2014**, *116*, 1234–1246. [CrossRef]
18. Shi, S.; Yang, L.; Yao, Q.; Li, X.; Ming, Y.; Zhao, Y. Ascorbic palmitate as a bifunctional drug and nanocarrier of paclitaxel for synergistic anti-tumor therapy. *J. Biomed. Nanotechnol.* **2018**, *14*, 1601–1612. [CrossRef] [PubMed]
19. Chen, Q.; Espey, M.G.; Krishna, M.C.; Mitchell, J.B.; Corpe, C.P.; Buettner, G.R.; Shacter, E.; Levine, M. Pharmacologic ascorbic acid concentrations selectively kill cancer cells: Action as a pro-drug to deliver hydrogen peroxide to tissues. *Proc Natl Acad Sci USA* **2005**, *102*, 13604–13609. [CrossRef] [PubMed]
20. Yun, J.; Mullarky, E.; Lu, C.; Bosch, K.N.; Kavalier, A.; Rivera, K.; Roper, J.; Chio, I.I.C.; Giannopoulou, E.G.; Rago, C.; et al. Vitamin C selectively kills KRAS and BRAF mutant colorectal cancer cells by targeting GAPDH. *Science* **2015**, *350*, 1391–1396. [CrossRef]
21. Hong, S.W.; Jin, D.H.; Hahm, E.S.; Yim, S.H.; Lim, J.S.; Kim, K.I.; Yang, Y.; Lee, S.S.; Kang, J.S.; Lee, W.J.; et al. Ascorbate (vitamin C) induces cell death through the apoptosis-inducing factor in human breast cancer cells. *Oncol. Rep.* **2007**, *18*, 811–815. [CrossRef]
22. Satheesh, N.J.; Samuel, S.M.; Büsselberg, D. Combination Therapy with Vitamin C Could Eradicate Cancer Stem Cells. *Biomolecules* **2020**, *10*, 79. [CrossRef] [PubMed]

23. Carr, A.C.; Cook, J.; Carr, A.C. Intravenous Vitamin C for Cancer Therapy–Identifying the Current Gaps in Our Knowledge. *Front. Physiol.* **2018**, *9*, 1182. [CrossRef] [PubMed]
24. Sawant, R.R.; Vaze, O.; Souza, G.G.M.D.; Rockwell, K.; Vladimir, P. Palmitoyl Ascorbate-Loaded Polymeric Micelles: Cancer Cell Targeting and Cytotoxicity. *Pharm. Res.* **2011**, *28*, 301–308. [CrossRef]
25. Armstrong, D. Advanced protocols in oxidative stress III, Methods in Molecular Biology. *Adv. Protoc. Oxidative Stress III* **2014**, *594*, 1–477. [CrossRef]
26. Sawant, R.R.; Vaze, O.S.; Rockwell, K.; Torchilin, V.P. Palmitoyl ascorbate-modified liposomes as nanoparticle platform for ascorbate-mediated cytotoxicity and paclitaxel co-delivery. *Eur. J. Pharm. Biopharm.* **2010**, *75*, 321–326. [CrossRef]
27. Vaiserman, A.; Koliada, A.; Zayachkivska, A.; Lushchak, O. Nanodelivery of Natural Antioxidants: An Anti-aging Perspective. *Front. Bioeng. Biotechnol.* **2020**, *7*, 447. [CrossRef]

Article

A Study on the Correlation between the Oxidation Degree of Oxidized Sodium Alginate on Its Degradability and Gelation

Hongcai Wang [1,2,3], Xiuqiong Chen [2,3], Yanshi Wen [1,2,3], Dongze Li [2,3], Xiuying Sun [2,3], Zhaowen Liu [2,3,4], Huiqiong Yan [1,2,3,*] and Qiang Lin [1,2,3,*]

1. Key Laboratory of Tropical Medicinal Resource Chemistry of Ministry of Education, College of Chemistry and Chemical Engineering, Hainan Normal University, Haikou 571158, China; whcz0505@163.com (H.W.); wenyanshi2022@163.com (Y.W.)
2. Key Laboratory of Natural Polymer Functional Material of Haikou City, College of Chemistry and Chemical Engineering, Hainan Normal University, Haikou 571158, China; chenxiuqiongedu@163.com (X.C.); dongzeli2019@163.com (D.L.); sunxiuying2019@163.com (X.S.); liuzhaowenyifan@126.com (Z.L.)
3. Key Laboratory of Water Pollution Treatment & Resource Reuse of Hainan Province, College of Chemistry and Chemical Engineering, Hainan Normal University, Haikou 571158, China
4. College of Pharmacy, Gannan Medical University, Ganzhou 341000, China
* Correspondence: yanhqedu@163.com (H.Y.); linqianggroup@163.com (Q.L.);
  Tel.: +86-0898-65884995 (H.Y.); +86-0898-65889422 (Q.L.)

Abstract: Oxidized sodium alginate (OSA) is selected as an appropriate material to be extensively applied in regenerative medicine, 3D-printed/composite scaffolds, and tissue engineering for its excellent physicochemical properties and biodegradability. However, few literatures have systematically investigated the structure and properties of the resultant OSA and the effect of the oxidation degree (OD) of alginate on its biodegradability and gelation ability. Herein, we used $NaIO_4$ as the oxidant to oxidize adjacent hydroxyl groups at the C-2 and C-3 positions on alginate uronic acid monomer to obtain OSA with various ODs. The structure and physicochemical properties of OSA were evaluated by Fourier transform infrared spectroscopy (FT-IR), ^{1}H nuclear magnetic resonance (^{1}H NMR), X-ray Photoelectron Spectroscopy (XPS), X-ray Diffraction (XRD), and thermogravimetric analysis (TGA). At the same time, gel permeation chromatography (GPC) and a rheometer were used to determine the hydrogel-forming ability and biodegradation performance of OSA. The results showed that the two adjacent hydroxyl groups of alginate uronic acid units were successfully oxidized to form the aldehyde groups; as the amount of $NaIO_4$ increased, the OD of OSA gradually increased, the molecular weight decreased, the gelation ability continued to weaken, and degradation performance obviously rose. It is shown that OSA with various ODs could be prepared by regulating the molar ratio of $NaIO_4$ and sodium alginate (SA), which could greatly broaden the application of OSA-based hydrogel in tissue engineering, controlled drug release, 3D printing, and the biomedical field.

Keywords: oxidized sodium alginate; oxidation degree; biodegradation gelation ability; rheological properties

1. Introduction

Sodium alginate (SA), an irregular linear natural polysaccharide polymer derived from algae and bacteria, comprises 1,4-linked β-D-mannuronic acid (M Block) and 1,4-linked α-L-guluronic acid (G Block) with a homogeneous (poly-G, poly-M) or heterogeneous (MG) block composition [1,2], as illustrated in Figure 1. Owing to the advantages of good biocompatibility, low toxicity, non-immunogenicity, reproducibility, and plasticity, alginate is widely used in wound dressing, tissue engineering, pharmaceutical industries, food, and cosmetics [3–6]. However, as a result of the numerous carboxyl and hydroxyl groups on its molecular backbone, the hydrophilic alginate suffers from uncontrollable degradation and massive swelling properties, leading to its weak stability in biological buffers, which significantly restricts its practical application in the biomedical field [7,8]. In addition, SA with high-molecular-weight alginate hardly degrades in the body because of the lack of alginate-degrading enzymes [9,10]. The low-molecular-weight alginate with molecular weight lower than 50 kDa can be removed from the body through the kidney [11–13]. It is worth noting that the oxidation of alginate can strengthen its biodegradability and significantly reduce its molecular weight [14,15]. Simultaneously, the low molecular weight of oxidized sodium alginate (OSA) hydrogels is conducive to be used as the degradable hydrogel scaffolds for drug delivery system and tissue engineering for its functional groups (such as aldehyde groups) that can be quickly degraded in the body compared with natural alginate [14,16,17].

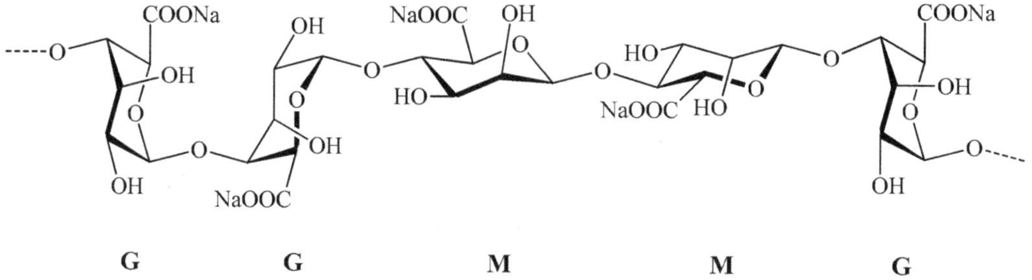

Figure 1. Molecular structure of sodium alginate.

In general, the oxidation of SA can be achieved via oxidizing agents including potassium permanganate, ozone, hydrogen peroxide, and periodate. Lu et al. [18] reported that SA could be oxidized by potassium permanganate under acidic conditions, and the degradation of SA increased with the increase in the amount of potassium permanganate and the decrease in pH value of the solution. Wu et al. [19] illustrated that the free radicals produced by ozone self-decomposition could cause the degradation of SA, thereby obtaining the low molecular weight of OSA. Mao et al. [20] studied the oxidation of SA using hydrogen peroxide, which could achieve the desired molecular weight of alginate by adjusting the reaction temperature, hydrogen peroxide concentration, initial concentration of SA, and pH value in the system. To note, it can be observed that there are two adjacent -OH groups at the C-2 and C-3 positions on the repetitive unit of the alginate chain, which could be specifically oxidized by the periodate to generate the aldehyde groups [21,22]. Among the above oxidation methods, using sodium periodate ($NaIO_4$) to oxidize SA is regarded as the most commonly used method to endow alginate with active functional groups and easy degradation in drug controlled delivery [23]. The oxidation of alginate with $NaIO_4$ to form OSA involved C2–C3 bond breakage, thus transforming the uronic acid into an open chain adduct containing aldehyde groups. Subsequently, the generated aldehyde groups will react spontaneously with hydroxyl groups present on the adjacent uronic acid in the alginate chain to form a cyclic hemiacetal [24], as shown in Figure 2. Moreover, the carboxyl groups of alginate can be completely retained, and the oxidation degree (OD) of alginate can be also controlled by adjusting the concentration of $NaIO_4$ during the oxidation process [25].

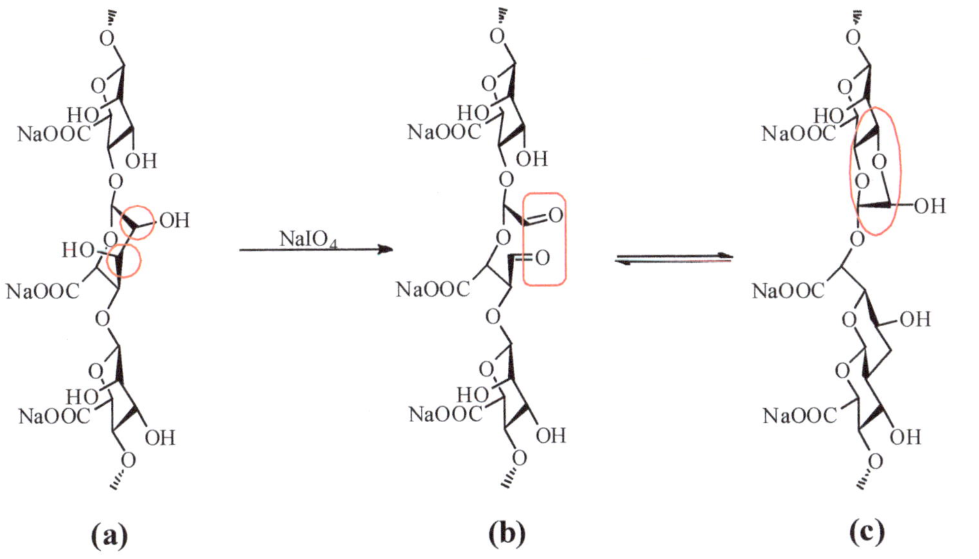

Figure 2. Schematic presentation of (**a**) the oxidation reaction of sodium alginate by NaIO$_4$, (**b**) the formation of OSA, and (**c**) the formation of hemiacetal.

Oxidation degree (OD) is the basic parameter of OSA, which significantly affects the physical and chemical properties of OSA-based materials such as hydrogels [26], microspheres [27], and electrospinning materials [28]. The increase in OD may make OSA-based materials possess a low crosslinking degree and network density. Additionally, OD is also related to the gelling properties of alginate. Generally, alginate and its derivatives exhibit gelling ability with divalent cations [29–32] (such as Ca^{2+}, Ba^{2+}, Mg^{2+}, Sr^{2+}, etc.). However, the strength of the ionic crosslink between the OSA polymer chain and the divalent ion are weakened with the destruction of the cooperative interaction during oxidation process [23]. Fortunately, alginate with a lower OD can still form the hydrogel [25]. Therefore, it is generally necessary to select an OSA with the suitable OD before the actual application. Although some works about oxidation reactions with NaIO$_4$ on the hydroxyl groups of alginate uronic units have been reported [23–25], the systematic characterization of the resultant OSA by multiple testing methods and the correlation between the OD on the degradability and gelation of OSA have rarely been studied at present. A large number of studies have shown a great reduction in the molecular weight of SA after being oxidized by NaIO$_4$, resulting in the formation of aldehyde groups with higher reactivity [24,27,28]. The resultant OSA not only retained the water solubility, low, and biocompatibility of alginate, but also exhibited good biodegradability and better molecular flexibility. Therefore, research on the correlation between the OD of OSA on its biodegradability and gelation is beneficial to exploring the application potential of OSA-based hydrogels in the biomedical field.

Herein, to broaden the applicability of alginate, we prepared OSA with various theoretical ODs using NaIO$_4$ as the oxidant. The structure and physicochemical properties of OSA was evaluated by Fourier transform infrared spectroscopy (FT-IR), ^{1}H nuclear magnetic resonance (^{1}H NMR), X-ray Photoelectron Spectroscopy (XPS), X-ray Diffraction (XRD), and thermogravimetric analysis (TGA). Meanwhile, gel permeation chromatography (GPC) and rheometer (DHR) were used to determine the correlation between the ODs of oxidized sodium alginate on its biodegradability and gelation. Additionally, the cytotoxicity of the formed OSA hydrogel against the MC3T3-E1 cells was also evaluated.

2. Materials and Methods

2.1. Materials

Sodium alginate (SA, M_W = 1,106,340, G/M = 2:1), sodium periodate ($NaIO_4$), D-Glucono-δ-lactone (GDL), nanosized hydroxyapatite (HAP), and sodium dihydrogen phosphate (NaH_2PO_4) were bought from Aladdin Chemical Reagent Co., Ltd., Shanghai, China. Anhydrous ethanol, hydrochloric acid, potassium chloride, sodium chloride (NaCl), sodium hydroxide (NaOH), and pH 7.4 phosphate buffer solution (PBS) were obtained from Zhongyao Chemical Reagent Co., Ltd., Beijing, China. Starch and ethylene glycol were bought from Macklin Co. Ltd., Shanghai, China. All chemicals were analytical grade and were used without further purification.

2.2. Methods

2.2.1. Synthesis of Oxidized Sodium Alginate

The oxidation of SA was carried out in aqueous solution using $NaIO_4$ as the oxidant to prepare oxidized sodium alginate (OSA) with various theoretical oxidation degrees based on the reported methods [23,24]. Briefly, 5 g of SA was fully dissolved in 200 mL of distilled water in a dark bottle; then, 50 mL of anhydrous ethanol was added under vigorous stirring to obtain 2.0% (w/v) SA solution. Afterwards, a certain amount of $NaIO_4$ was added to initiate the oxidation reaction at 25 °C in N_2 atmosphere to obtain OSA. Various ratios of $NaIO_4$ to the number of repetitive uronic acid of alginate (5, 10, 15, or 20 mol %) were used, as shown in Table 1. Each reaction was performed for a period of 24 h until about 10 mL of ethylene glycol was incorporated to reduce the unreacted $NaIO_4$. Subsequently, 4 times the volume of anhydrous ethanol and 4 g of NaCl were added to precipitate OSA; then, they were placed in the dialysis bag with a molecular weight cutoff of 3500 to dialyze against deionized water for 5 days. Finally, the dried OSA was obtained by freeze-drying the dialyzed OSA solution.

Table 1. Oxidation reaction parameters with $NaIO_4$ for OSA with various OD.

Theoretical OD	Absorbance	Mole of Alginate Uronic Acid (mmol)	$NaIO_4$ (g)	Actual OD
5%	0.002	1.2625	0.27	4.99%
10%	0.020	2.5250	0.54	9.90%
15%	0.016	3.7875	0.81	14.91%
20%	0.018	5.0500	1.08	19.88%

2.2.2. Determination of Oxidation Degree of OSA by UV–Vis Absorption Spectroscopy

The oxidation degree (OD) of OSA could be determined by UV–Vis absorption spectroscopy based on the difference between the initial and final amount of $NaIO_4$ during the oxidation reaction. Before adding the ethylene glycol to quench the oxidation reaction, 1 mL of reaction solution was transferred and diluted to 250 mL with distilled water. Then, 3.5 mL of this diluted solution was mixed with 1.50 mL of indicator solution that was prepared by mixing equal volumes of 20% (w/v) KI and 1% (w/v) soluble starch solutions, using pH 7.0 PBS as the solvent. The absorbance of diluted mixed solution was rapidly measured with a Shimadzu UV-1800 (Shimadzu, Kyoto, Japan) UV–visible spectrophotometer at 290 nm. The concentration of $NaIO_4$ in the solution was determined with a calibration curve whose linear regression equation was fitted as $y = 0.2188 \times x(10^{-5} \text{ mol/L}) + 0.0146$ ($R^2 = 0.9990$), which was prepared by standard concentration of $NaIO_4$ in PBS in the range of $0.5 \sim 2.5 \times 10^{-5}$ mol/L, as presented in Figure 3. Therefore, the actual OD of OSA was calculated by the amount of $NaIO_4$ according to the following equation:

$$OD = \frac{N_i - N_r}{N_0} \times 100\% \tag{1}$$

where N_i, N_r, and N_0, respectively, represent the initial moles of $NaIO_4$, the residual moles of $NaIO_4$, and initial moles of uronic acid of alginate.

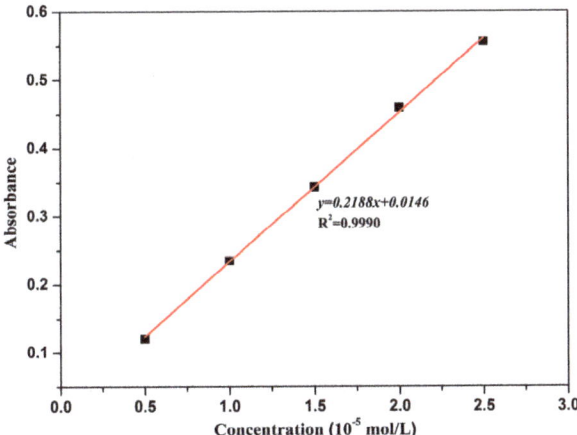

Figure 3. Calibration curve of NaIO$_4$ as a function of its concentration.

2.2.3. Characterization of Oxidized Sodium Alginate

The successful synthesis of OSA was confirmed by Fourier transform infrared spectrophotometer (FT-IR), ^{1}H nuclear magnetic resonance (^{1}H NMR), and X-ray Photoelectron Spectroscopy (XPS). The FT-IR measurement was performed on disks that were prepared by compressing the mixture of KBr and a slight amount of sample using a Nicolet-6700 FT-IR spectrophotometer (Thermo Scientific, Waltham, MA, USA). The FT-IR spectra of the sample were recorded in the range of wavenumbers between 4000 and 400 cm^{-1} for 64 scans with a spectral resolution of 4.0 cm^{-1}. The ^{1}H NMR spectroscopy of the sample that was dissolved in D$_2$O (99%) with a concentration of 10 mg/mL was recorded on an AV 400 NMR nuclear magnetic resonance spectrometer (Bruker, Ettlingen, Germany) at 25 °C. The surface elemental composition of the sample was examined by XPS using an axis ultra DLD apparatus (Kratos, Manchester, UK), which was equipped with a monochromatic Al KR X-ray source operating at 15 kV, 10 mA (150 W). The spectra were collected in fixed analyzer transmission mode (FAT): survey scans at 1200~0 eV with 1.0 eV steps at an analyzer pass energy of 160 eV; narrow scans at 0.1 eV steps at an analyzer pass energy of 20 eV. The crystalline structure and thermal stability of OSA were examined by X-ray Diffraction (XRD) and thermogravimetric analysis (TGA). The XRD pattern was performed on an AXS/D8 X-ray diffractometer (Bruker, Cambridge, UK) with Cu-Kα radiation (λ = 0.154 nm). The measurement was conducted in a step scan mode at a scanning speed of 0.025°/s over a 2θ range of 5°~60°. The thermal property of OSA was determined by a Q600 thermogravimetric analyzer (TA Instrument, New Castle, DE, USA). The TGA test was carried out under a N$_2$ atmosphere at a heating rate of 10 K/min with the range of 30~800 °C.

2.2.4. In Vitro Biodegradation of Oxidized Sodium Alginate

The in vitro biodegradation of OSA was performed at 37 °C using PBS buffer containing 10,000 U/mL lysozyme to simulate the physiological conditions. A total 0.2 g of OSA with various ODs was immersed in the 100-mL PBS buffer for 60 days. At different time intervals (6 days, 12 days, 20 days, 28 days, 38 days, 48 days, and 60 days), 2 mL aliquot of samples were withdrawn and their molecular weights were determined by an e2695 Gel Permeation Chromatography equipped with an Ultrahydrogel^{TM}120 (7.8 × 300 mm^2) column (Waters, Milford, MA, USA). A total 0.05% sodium azide was used as the mobile phase with a flow rate of 0.6 mL/min at 40 °C. Similarly, each time interval for the sample was tested in parallel for 3 times to take the average value. Within the time interval, the biodegradation of OSA could be evaluated by its weight-average molecular weight (M_W).

2.2.5. Gelation Ability of Oxidized Sodium Alginate

To examine the gelation ability of OSA, the internal gelation of OSA was conducted based on our previous work [33]. As shown in Scheme 1, this method allowed homogeneous alginate hydrogel to be obtained by the control of the release of Ca^{2+} from the insoluble hydroxyapatite (HAP) in the presence of D-glucono-δ-lactone (GDL), which avoided the unbalanced crosslinking density, thus enhancing the mechanical property and homogeneity of the hydrogel [34,35]. In detail, a certain amount of HAP was ultrasonically dispersed in 2% (w/v) OSA solution and stirred vigorously until a homogenous solution was formed. Then, the ion cross-linking of OSA was initiated by the addition of a certain amount of GDL under the magnetic stirring. In this study, HAP–GDL complex with the molar ratio of 1:10 was used as the cross-linking system, and the molar ratio of Ca^{2+} from HAP and carboxyl from OSA was fixed at 0.18. After stirring at high speed for 3 min, the mixture was quickly transferred into a 12-well tissue culture plate and physically cross-linked at 4 °C for 24 h to obtain homogeneous OSA hydrogel. The resultant OSA hydrogel was left for 30 min to eliminate any air bubbles for further rheological measurement. The rheological properties of the OSA hydrogel with various ODs were analyzed by steady shear test and dynamic sweep measurements using a rotational rheometer with parallel-plate geometry (DHR TA Instruments, New Castle, DE, USA) at 25 °C. Steady shear measurements were carried out to record the apparent viscosity (η) with the shear rate ranging from 0.1 to 1000 s^{-1}, while the oscillation frequency sweep measurements were conducted to record storage modulus G' and loss modulus G'' with the angular frequencies (ω) ranging from 0.1 to 100 rad/s and the strain amplitude was fixed at 1%.

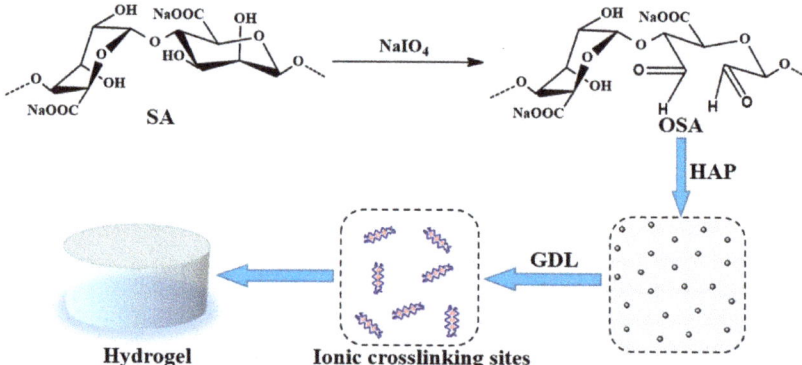

Scheme 1. Schematic representation of the preparation of OSA and its ionic crosslinking by HAP–GDL.

2.2.6. In Vitro Cytotoxicity of Oxidized Sodium Alginate Hydrogel

To verify the application potential of OSA hydrogels in the biomedical field, the osteoblastic MC3T3-E1 cells cultured with the culture medium containing 90% MEM-α, 10% fetal bovine serum, 100 U/mL penicillin, and 100 μg/mL streptomycin were applied to examine their cytocompatibility by using the Cell Counting Kit-8 (CCK-8) assay. The OSA—10% hydrogels were cut into circular disks with a diameter of 20 mm and a height of 10 mm; then, they were sterilized by cobalt 60 radiation with the irradiation intensity of 8 kGy. The MC3T3-E1 cells were seeded on the OSA—10% hydrogels in the 24-well tissue culture plates at a density of 5×10^4 per well, while the same cells were seeded on the tissue culture plates as a blank control. Afterwards, the culture medium was replenished to make the total amount of medium per well reach up to 500 μL. Subsequently, they were transferred to an incubator containing 5% CO_2, 95% air, and 100% relative humidity at 37 °C and their culture media were replaced every 2 days. After 2 days and 5 days incubation, 50 μL of CCK-8 reagent was added to 500 μL of medium in each well, and they were placed in the incubator at 37 °C for 4 h. Finally, 100 μL of solution from each well was transferred to a 96-well plate, whose absorbance value (OD) was determined by an X-mark microplate reader (Bio-rad, Hercules, CA, USA) at a wavelength of 450 nm. Since the OD of the cell medium on the OSA—10% hydrogel is proportional to the number of living cells, the cell viability of the MC3T3-E1 cells on the OSA—10% hydrogel could be judged by comparing the OD values.

3. Results and Discussion

3.1. Oxidation of Alginate via $NaIO_4$

Since there are a large number of hydrophilic groups such as hydroxyl and carboxyl groups on the backbone of alginate molecules, they can easily form intramolecular hydrogen bonds, resulting in the strong and rigid molecular structure of alginate, which may restrict its scope of application in drug delivery [36,37]. However, the inert dihydroxy groups on the alginate backbone can be oxidized to generate the reactive dialdehyde groups by exploiting the oxidative properties of $NaIO_4$, which significantly improved its biodegradability [23,38]. The cleavage of the C2–C3 bond of alginate uronic acid monomer could occur in the oxidation reaction of $NaIO_4$, which strictly destroyed the rigid structure of the alginate molecular backbone, thus enhancing its molecular flexibility [24]. In this work, the SA was partially oxidized to a theoretical extent (5%, 10%, 15%, and 20%) and the oxidation degree (OD)—defined as the percentage of oxidized uronic acid groups in the alginate—was determined by the consumption of $NaIO_4$ with the results shown in Table 1. It was obvious that the OD of OSA increased with the increase in the amount of $NaIO_4$, and the actual OD was close to the theoretical OD. This result directly indicated that most of the added $NaIO_4$ participated in the oxidation reaction, and the oxidation reaction of alginate with $NaIO_4$ was feasible and active.

3.2. Molecular Structure and Thermal Stability of OSA

The molecular structure and specific functional groups of the resultant OSA could be confirmed by FT-IR and ^1H NMR spectroscopy. As shown in Figure 4a, both SA and OSA revealed the broad hydroxyl stretching vibration absorption peaks in the range of 4000~3000 cm^{-1} [39]. In detail, SA exhibited main characteristic peaks at 2926, 1616, and 1417 cm^{-1} owing to the C-H stretching vibration of the polysaccharide structure and the asymmetric and symmetric stretching vibration of $-COO^-$ [40]. Additionally, the absorption peaks at 1030 cm^{-1} were attributed to the C-O stretching vibration on the polysaccharide skeleton [41]. In comparison with SA, OSA had a new characteristic peak at 1734 cm^{-1}, which was assigned to the vibration absorption peak of the C=O bond on the aldehyde group. This peak was too weak to be detected for the hemiacetal formation of the free aldehyde groups [42]. In addition, the hydroxyl stretching vibration absorption peak at 3437 cm^{-1} in the spectrum of SA became blue-shifted to 3443 cm^{-1} in the spectrum of OSA, implying a decline in the amount of -OH groups of alginate. These results indicated that the adjacent hydroxyl groups at the C-2 and C-3 positions on alginate uronic acid monomer were oxidized to aldehyde groups by $NaIO_4$ [25,41]. Moreover, OSA displayed similar FT-IR spectroscopy results to the raw SA, which demonstrated the periodate ion only cleaved the C2–C3 linkage by the oxidation reaction, leading to the formation of a dialdehyde [24].

Figure 4b shows the ^1H NMR spectra of SA and OSA. The proton peaks of SA and OSA ranging from 5.0 to 3.5 ppm were assigned to the hydrogen atoms of native alginate backbone [40]. Compared with SA, the two new proton peaks were discovered at 5.3 and 5.6 ppm in the spectrum of OSA, which were attributed to a hemiacetalic proton generated from the hydroxyl groups of aldehyde and its neighbors, confirming the achievement of the oxidation with $NaIO_4$ [23]. Moreover, the proton peaks of SA appearing at 3.7~3.55 ppm that were assigned to the H2 of α-L-guluronic acid decreased and moved to the high field for cleavage of the C2–C3 bond of the uronic acid monomer [24]. The appearance of these new proton signal peaks and the change of the proton signal peaks also directly proved the successful oxidation of alginate with $NaIO_4$.

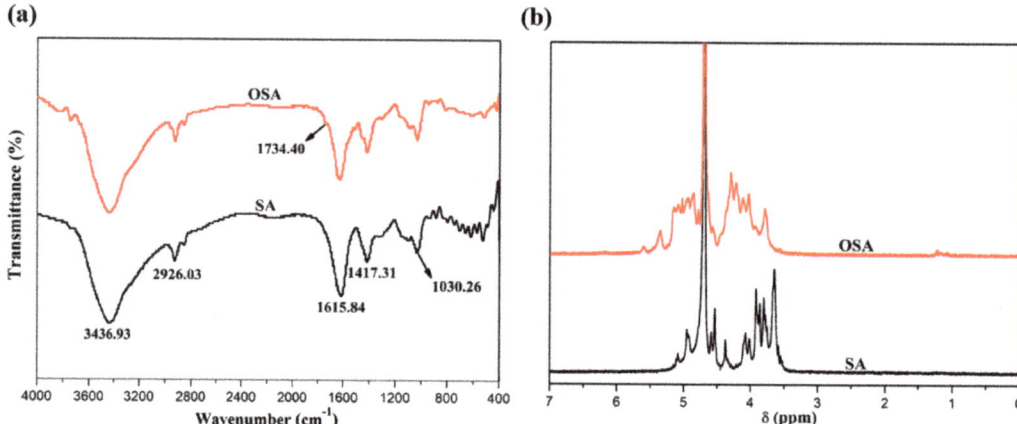

Figure 4. (a) FT-IR spectra and (b) ^{1}H NMR spectra of SA and OSA.

As a complementary technique, XPS was performed to further verify the surface composition of resultant OSA. As shown in Figure 5a, XPS spectra of SA and OSA confirmed the presence of the anticipated peaks for Na (1069.02 eV), O (530.23 eV), N (397.54 eV), and C (284.68 eV), with an O/Na ratio close to the theoretical value of 6 [43], indicating that partial oxidation of alginate by NaIO$_4$ preserved the main structure and functional groups of alginate. In addition, the peak assignment of XPS C1s narrow scans of SA and OSA that could elucidate the bonding atmosphere has been marked in the spectra in Figure 5b according to the previous report [44]. It can be observed that the C1s narrow scan revealed three peaks at 285.46 eV (O–C–O), 283.87 eV (C–O), and 282.20 eV (C–C), respectively. It is found that the peak ratio of the O–C–O species to the C–O species in the C1s narrow scan of OSA was significantly higher than that of SA, due to the partial oxidation of alginate by NaIO4, which was consistent with previous results reported by Jejurikar et al. [24]. Thus, these results further verified the successful partial oxidation of alginate by NaIO$_4$.

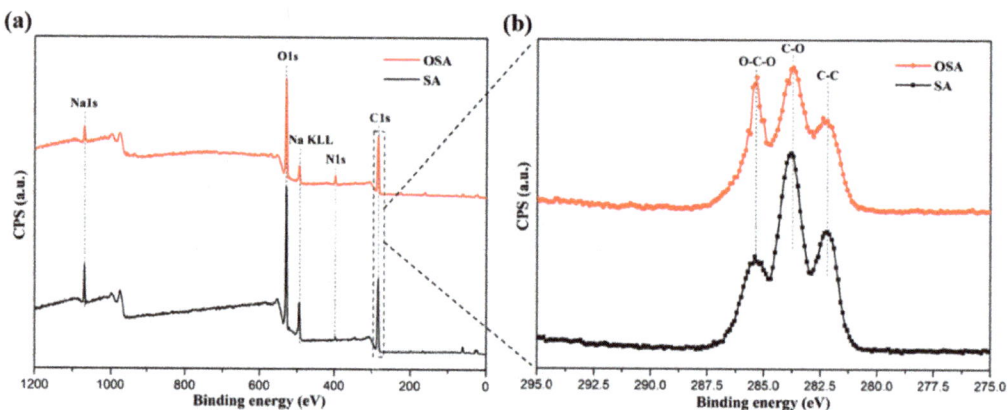

Figure 5. (a) XPS spectra and (b) XPS C1s narrow scans of SA and OSA.

The crystal structure of OSA was evaluated by XRD. As shown in Figure 6, the XRD patterns of SA and OSA were proved to be amorphous structures, consistent with previous reports [45]. The typical characteristic diffraction peaks of SA at 2θ = 14.5° and 22.0° represented the hydrated crystalline structure generated from the intramolecular hydrogen bonds of SA [46]. Compared with SA, the diffraction peaks of OSA were shifted to

2 θ = 14.9° and 23.0°, and the peak at 2 θ = 23.0° was a broader peak. These results implied that the structure of SA was changed during the oxidation process and the intermolecular hydrogen bonds were weakened with the decrease in molecular weight.

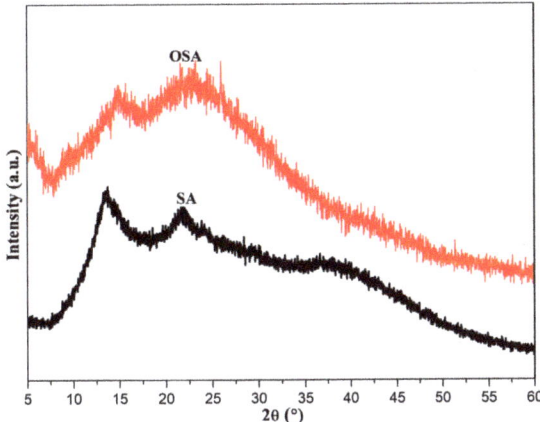

Figure 6. XRD patterns of SA and OSA.

Thermogravimetric analysis was the most effective method to characterize the different thermal stability of SA and OSA, which could indirectly reflect the change of molecular structures of alginate during the oxidation reaction. During the test, since the mass of the samples will change with the increase in temperature, the thermal stability of SA and OSA can be judged by comparing the initiating decomposition temperature and the final weight loss rate [47]. As shown in Figure 7a, the weight loss of the samples at the temperature below 100 °C corresponds to the evaporation of water in the samples. Major weight loss occurred at the temperature ranging from 200 °C to 300 °C, which resulted from dehydration of hydroxyl group along the alginate backbone and its thermal decomposition of hexuronic segments [48,49]. The total weight loss rates of SA and OSA at 800 °C were 74% and 77%, respectively. Compared with SA, OSA possessed lower residual weight, which may be ascribed to the oxidation of hydroxyl groups of SA. From the DTG curves of SA and OSA in Figure 7b, it can be observed that both SA and OSA displayed two main weight loss stages. The maximum rate of the first weight loss stage occurred at 100~150 °C, which was ascribed to the removal of physically adsorbed water and bound water from the polymer materials [27,28], while the second weight loss stage occurred at 200~300 °C and was attributed to the thermal decomposition of the polymer materials, which was gradually thermally cracked into CO, CO_2, and H_2O, resulting in a rapid decline in their weight [33]. The temperatures corresponding to the maximum rates of the thermal decomposition weight loss of SA and OSA were 249.6 and 232.2 °C, respectively. Obviously, the initiating thermal decomposition temperature of OSA was lower than that of raw SA, indicating that the thermal stability of OSA decreased after the oxidation reaction. This result was ascribed to the destruction of the intramolecular hydrogen bond for the cleavage of the C2–C3 bond of alginate uronic acid monomer during the oxidation process, thereby improving the molecular flexibility of alginate.

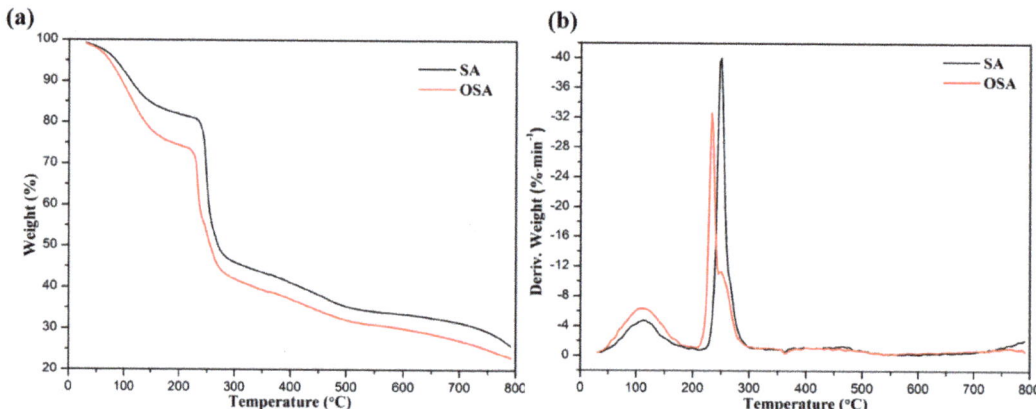

Figure 7. (**a**) TGA curves and (**b**) DTG curves of SA and OSA.

3.3. In Vitro Biodegradation Analysis

The weight-average molecular weights (M_W) of the water-soluble OSA were measured via GPC, which used 0.05% sodium azide as the mobile phase. Figure 8a depicts GPC chromatographs of SA and OSA with various ODs. It can be observed that the molecular weight of OSA decreased with an increase in the OD of OSA, similar to the previous results [50]. The SA and OSA with various ODs were immersed in the PBS buffer at 37 °C for in vitro biodegradation, and the biodegradation performance of the samples was studied by measuring their molecular weight at a specific time. As shown in Figure 8b, SA underwent less biodegradation because SA was hardly biodegradable in the body [9,10]. In contrast, OSA showed significant biodegradability, especially within 10 days of the onset of biodegradation, which could exhibit great potentials as the degradable hydrogel scaffolds or controlled release drug delivery system in biomedical field. Additionally, the M_W of OSA decreased very rapidly with the increase in their OD, and all of the OSA with various ODs revealed similar biodegradation trends. These behaviors were attributed to the cleavage of the C2–C3 bond of alginate uronic acid, as oxidation occurred because this was a free-radical-independent biodegradation process, which showed that the biodegradation depended directly on the concentration of $NaIO_4$ in the reaction solution [23]. Consequently, the biodegradation rate of alginate could be controlled by regulating the OD of OSA to meet different application requirements.

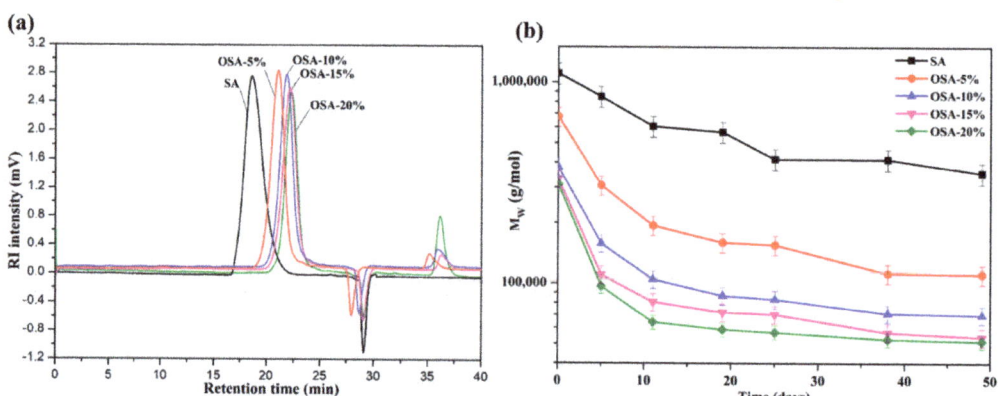

Figure 8. (**a**) GPC chromatographs of SA and OSA with various OD; (**b**) biodegradation curves of SA and OSA with various OD in PBS solution.

3.4. Gelation Study of OSA

OSA with oxidation degrees ranging from 5% to 20% was formulated into the corresponding concentration of homogeneous hydrogel, and their gelation ability was preliminarily determined by observing the results of the viscosity variation with the shear rate, as shown in Figure 9a. Since the formation time of hydrogel was related to the molecular weight of the oxidized alginate derivatives, high molecular weight of OSA could greatly reduce the formation time of hydrogel. It could be found that the viscosity (η) of the SA hydrogel exhibited a slight variation with the shear rate, which displayed typical Newtonian liquid characteristics. This result was mainly attributed to the highly stretched rigid molecular structure of SA [37]. On the contrary, the OSA hydrogel with various ODs displayed apparent shear-thinning behavior, which was ascribed to the destruction of the intramolecular hydrogen bond for cleavage of the C2–C3 bond of the alginate uronic acid monomer during the oxidation process. In comparison with SA hydrogel, the OSA hydrogel exhibited distinctly lower viscosity (η) at the same shear rate, indicating that the gelation ability of OSA significantly decreased due to the oxidation of the adjacent hydroxyl groups at C-2 and C-3 positions on alginate uronic acid monomer [24]. Simultaneously, the viscosity of OSA hydrogel decreased with the increase in OD at the same shear rate, which indicated that the hydrogel could be prepared by controlling the OD.

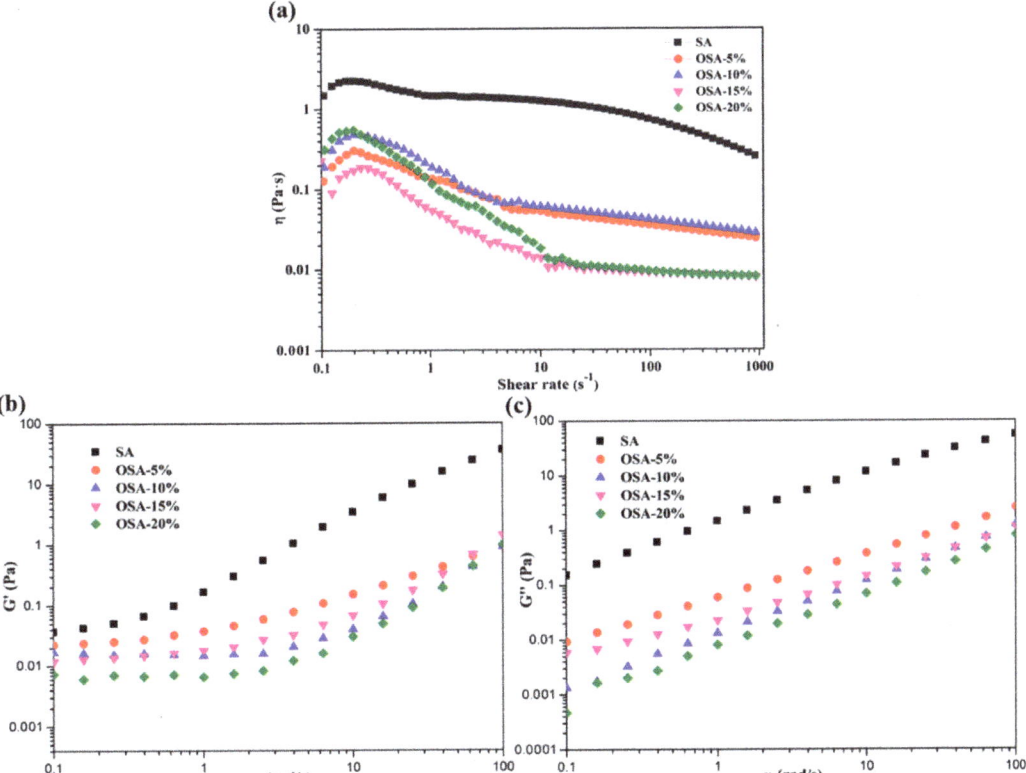

Figure 9. (a) Viscosity of SA hydrogel and OSA hydrogel with various ODs as a function of shear rate at 25 °C; (b) viscosity of SA hydrogel and OSA hydrogel with various ODs as a function of shear rate; (c) storage (G′) and loss (G″) moduli of SA hydrogel and OSA hydrogel with various ODs as functions of angular frequency (ω).

In addition, it can be also observed from Figure 9b,c that the SA hydrogel and OSA hydrogel with various ODs presented an upward trend for their storage modulus (G′) and loss modulus (G″) as the angular frequency (ω) increased. In general, the storage modulus (G′) and loss modulus (G″) were, respectively, related to the strength of the internal structure and dynamic viscosity. By contrast, the SA hydrogel exhibited relatively better mechanical properties as the storage (G′) and loss (G″) moduli of SA hydrogel were apparently higher than that of OSA hydrogel at the same angular frequency (ω). Furthermore, a reduction in both of the storage (G′) and loss (G″) moduli was found when the OD increased, which could be owing to the decrease of the molecular weight. Thus, it could be concluded that the oxidation reaction would cause the decrease in the molecular weight of the alginate, further affecting the formation of hydrogel. As presented in Figure 10, only OSA with ODs of 5% and 10% could be cross-linked by Ca^{2+} to form hydrogel, while no hydrogels were formed when the OD exceeded 10%. Therefore, the critical OD of OSA to form a hydrogel was 10%, which was consistent with the previous report [23]. Meanwhile, OSA with low molecular weight was beneficial to biomedical applications, such as regenerative medicine, 3D-printed/composite scaffolds, and tissue engineering, because it was found that, for $M_W \leq 50$ kDa, alginate could be removed from the human body [11,12].

Figure 10. The physical picture of OSA—5% and OSA—10% hydrogel.

3.5. Cytocompatibility of OSA Hydrogel

After 2 days and 5 days incubation, the in vitro cytotoxicity of OSA—10% hydrogel was measured using the CCK—8 assay kit. As shown in Figure 11, the MC3T3-E1 cells displayed better proliferative activity on the OSA—10% hydrogel than that on the control group, indicating that the MC3T3-E1 cells could survive on the OSA—10% hydrogel and grow in their 3D pore structure. These results further indicated that the OSA hydrogel had no cytotoxicity, presenting excellent biocompatibility. Therefore, the OSA hydrogel with excellent cytocompatibility could realize their extended biomedical applications in regenerative medicine, 3D-printed/composite scaffolds, and tissue engineering.

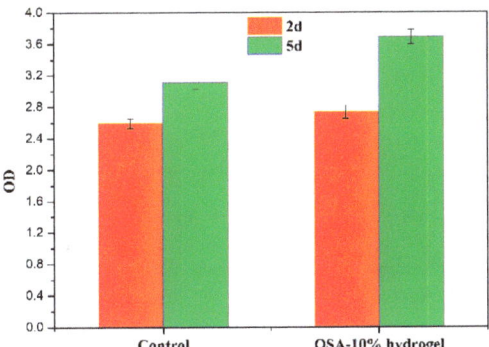

Figure 11. Cell viability of MC3T3-E1 cells cultured on the OSA—10% hydrogel for 2 and 5 days, respectively.

4. Conclusions

In this work, the oxidation of SA with various ODs was achieved using $NaIO_4$ as the oxidant, and the structure and properties of the resultant OSA were characterized by multiple testing methods. Meanwhile, the effect of OD on the biodegradability and gelation ability of OSA was also investigated. The adjacent hydroxyl groups at C-2 and C-3 position on alginate uronic acid monomer were oxidized to aldehyde groups by $NaIO_4$. On account of the cleavage of the C2–C3 bond of alginate uronic acid monomer during the oxidation process, the thermal stability of OSA was worse than that of SA. On the contrary, OSA possessed smaller molecular weight and better degradability in contrast to SA; the higher the OD of OSA, the better the degradability. Nevertheless, the gelation ability of OSA decreased with the increase in OD due to the decrease in molecular weight. It is worth noting that the critical OD of OSA to form a hydrogel was 10%. Thus, it could be concluded that both the formation and biodegradability of OSA hydrogel were controlled by the OD. This work aimed to improve the properties of alginate via the oxidation reaction to broaden its application in the biomedical field.

Author Contributions: Writing—original draft preparation, H.W.; methodology, X.C.; data curation, Y.W.; software, D.L.; conceptualization, X.S.; visualization, Z.L.; writing—review and editing, supervision, H.Y.; funding acquisition, Q.L. All authors have read and agreed to the published version of the manuscript.

Funding: This research was funded by the National Natural Science Foundation of China, 51963009; the Natural Science Foundation of Hainan Province, 220MS035; the Key Research and Development Project of Hainan Province, ZDYF2019018; the Open Fund for Innovation and Entrepreneurship of College Students of Hainan Normal University, S202111658065X; the Scientific Research Fund of Jiangxi Provincial Education Department, GJJ211509. The APC was funded by the National Natural Science Foundation of China, 51963009.

Institutional Review Board Statement: Not applicable.

Informed Consent Statement: Not applicable.

Data Availability Statement: The data presented in this study are available on request from the corresponding author. Samples of the compounds are available from the authors.

Conflicts of Interest: The authors declare no conflict of interest.

References

1. Draget, K.I.; Taylor, C. Chemical, physical and biological properties of alginates and their biomedical implications. *Food Hydrocoll.* **2011**, *25*, 251–256. [CrossRef]
2. Zia, K.M.; Zia, F.; Zuber, M.; Rehman, S.; Ahmad, M.N. Alginate based polyurethanes: A Review of Recent Advances and Perspective. *Int. J. Biol. Macromol.* **2015**, *79*, 377–387. [CrossRef]

3. Zheng, Z.; Qi, J.; Hu, L.; Ouyang, D.; Wang, H.; Sun, Q.; Lin, L.; You, L.; Tang, B. A cannabidiol-containing alginate based hydrogel as novel multifunctional wound dressing for promoting wound healing. *Mater. Sci. Eng. C Mater.* **2021**, *11*, 112560. [CrossRef]
4. Varaprasad, K.; Jayaramudu, T.; Kanikireddy, V.; Toro, C.; Sadiku, E.R. Alginate-based composite materials for wound dressing application: A Mini Review. *Carbohydr. Polym.* **2020**, *236*, 116025. [CrossRef] [PubMed]
5. Pina, S.; Oliveira, J.M.; Reis, R.L. Natural-based nanocomposites for bone tissue engineering and regenerative medicine: A Review. *Adv. Mater.* **2015**, *27*, 1143–1169. [CrossRef] [PubMed]
6. Pereira, L.; Cotas, J. Introductory chapter: Alginates—A General Overview. In *Alginates-Recent Uses of This Natural Polymer*; IntechOpen: London, UK, 2020.
7. Li, Q.; Liu, C.G.; Huang, Z.H.; Xue, F.F. Preparation and characterization of nanoparticles based on hydrophobic alginate derivative as carriers for sustained release of vitamin D3. *J. Agric. Food Chem.* **2011**, *59*, 1962–1967. [CrossRef]
8. Vallée, F.; Müller, C.; Durand, A.; Schimchowitsch, S.; Dellacherie, E.; Kelche, C.; Cassel, J.C.; Leonard, M. Synthesis and rheological properties of hydrogels based on amphiphilic alginate-amide derivatives. *Carbohydr. Res.* **2009**, *344*, 223–228. [CrossRef]
9. Montes, L.; Santamaria, M.; Garzon, R.; Rosell, C.M.; Moreira, R. Effect of the addition of different sodium alginates on viscoelastic, structural features and hydrolysis kinetics of corn starch gels. *Food Biosci.* **2022**, *47*, 101628. [CrossRef]
10. Hernández-González, A.C.; Téllez-Jurado, L.; Rodríguez-Lorenzo, L.M. Alginate hydrogels for bone tissue engineering, from injectables to bioprinting: A review. *Carbohydr. Polym.* **2020**, *229*, 115514. [CrossRef]
11. Kristiansen, K.A.; Tomren, H.B.; Christensen, B.E. Periodate oxidized alginates: Depolymerization kinetics. *Carbohydr. Polym.* **2011**, *86*, 1595–1601. [CrossRef]
12. Priddy, L.B.; Chaudhuri, O.; Stevens, H.Y.; Krishnan, L.; Uhrig, B.A.; Willett, N.J.; Guldberg, R.E. Oxidized alginate hydrogels for bone morphogenetic protein-2 delivery in long bone defects. *Acta Biomater.* **2014**, *10*, 4390–4399. [CrossRef] [PubMed]
13. Bonino, C.A.; Krebs, M.D.; Saquing, C.D.; Jeong, S.I.; Shearer, K.L.; Alsberg, E.; Khan, S.A. Electrospinning alginate-based nanofibers: From blends to crosslinked low molecular weight alginate-only systems. *Carbohydr. Polym.* **2011**, *85*, 111–119. [CrossRef]
14. Boontheekul, T.; Kong, H.J.; Mooney, D.J. Controlling alginate gel degradation utilizing partial oxidation and bimodal molecular weight distribution. *Biomaterials* **2005**, *26*, 2455–2465. [CrossRef] [PubMed]
15. Liang, Y.; Liu, W.; Han, B.; Yang, C.; Ma, Q.; Song, F.; Bi, Q. An in situ formed biodegradable hydrogel for reconstruction of the corneal endothelium. *Colloids Surf. B Biointerfaces* **2011**, *82*, 1–7. [CrossRef]
16. Kristiansen, K.A.; Potthast, A.; Christensen, B.E. Periodate oxidation of polysaccharides for modification of chemical and physical properties. *Carbohydr. Res.* **2010**, *345*, 1264–1271. [CrossRef]
17. Lee, K.Y.; Mooney, D.J. Alginate: Properties and Biomedical Applications. *Prog. Polym. Sci.* **2012**, *37*, 106–126. [CrossRef]
18. Lu, L.; Zhang, P.; Cao, Y.; Lin, Q.; Pang, S.; Wang, H. Study on partially oxidized sodium alginate with potassium permanganate as the oxidant. *J. Appl. Polym. Sci.* **2009**, *113*, 3585–3589. [CrossRef]
19. Yue, W.; Zhang, H.H.; Yang, Z.N.; Xie, Y. Preparation of low-molecular-weight sodium alginate by ozonation. *Carbohydr. Polym.* **2021**, *251*, 117104–117109. [CrossRef]
20. Mao, S.; Zhang, T.; Sun, W.; Ren, X. The depolymerization of sodium alginate by oxidative degradation. *Pharm. Dev. Technol.* **2012**, *17*, 763–769. [CrossRef]
21. Ghahramanpoor, M.K.; Najafabadi, S.A.H.; Abdouss, M.; Bagheri, F.; Eslaminejad, M.B. A hydrophobically-modified algi-nate gel system: Utility in the Repair of Articular Cartilage Defects. *J. Mater. Sci. Mater. Med.* **2011**, *22*, 2365–2375. [CrossRef]
22. Xu, Y.T.; Li, L.; Yu, X.X.; Gu, Z.P.; Zhang, X. Feasibility study of a novel crosslinking reagent (alginate dialdehyde) for bio-logical tissue fixation. *Carbohydr. Polym.* **2012**, *87*, 1589–1595. [CrossRef]
23. Gomez, C.G.; Rinaudo, M.; Villar, M.A. Oxidation of sodium alginate and characterization of the oxidized derivatives. *Carbohydr. Polym.* **2007**, *67*, 296–304. [CrossRef]
24. Jejurikar, A.; Seow, X.T.; Lawrie, G.; Martin, D.; Jayakrishnan, A.; Grøndahl, L. Degradable alginate hydrogels crosslinked by the macromolecular crosslinker alginate dialdehyde. *J. Mater. Chem.* **2012**, *22*, 9751–9758. [CrossRef]
25. Emami, Z.; Ehsani, M.; Zandi, M.; Foudazi, R. Controlling alginate oxidation conditions for making alginate-gelatin hydrogels. *Carbohydr. Polym.* **2018**, *198*, 509–517. [CrossRef]
26. Ding, F.; Shi, X.; Wu, S.; Liu, X.; Deng, H.; Du, Y.; Li, H. Flexible polysaccharide hydrogel with pH-regulated recovery of self-healing and mechanical properties. *Macromol. Mater. Eng.* **2017**, *302*, 1700221. [CrossRef]
27. Tallawi, M.; Germann, N. Self-crosslinked hydrogel with delivery carrier obtained by incorporation of oxidized alginate microspheres into gelatin matrix. *Mater. Lett.* **2020**, *263*, 127211. [CrossRef]
28. Wang, S.; Ju, J.; Wu, S.; Lin, M.; Sui, K.; Xia, Y.; Tan, Y. Electrospinning of biocompatible alginate-based nanofiber membranes via tailoring chain flexibility. *Carbohydr. Polym.* **2020**, *230*, 115665. [CrossRef]
29. Li, Y.; Rodrigues, J.; Tomas, H. Injectable and biodegradable hydrogels: Gelation, Biodegradation and Biomedical Applications. *Chem. Soc. Rev.* **2012**, *41*, 2193–2221. [CrossRef]
30. Ching, S.H.; Bansal, N.; Bhandari, B. Alginate gel particles—A review of production techniques and physical properties. *Crit. Rev. Food Sci. Nutr.* **2017**, *57*, 1133–1152. [CrossRef]
31. Yang, X.; Lu, Z.; Wu, H.; Li, W.; Zheng, L.; Zhao, J. Collagen-alginate as bioink for three-dimensional (3D) cell printing based cartilage tissue engineering. *Mater. Sci. Eng. C* **2018**, *83*, 195–201. [CrossRef]

32. Cerciello, A.; Del Gaudio, P.; Granata, V.; Sala, M.; Aquino, R.P.; Russo, P. Synergistic effect of divalent cations in improving technological properties of cross-linked alginate beads. *Int. J. Biol. Macromol.* **2017**, *101*, 100–106. [CrossRef] [PubMed]
33. Li, Z.Y.; Chen, X.Q.; Bao, C.L.; Liu, C.; Liu, C.Y.; Li, D.Z.; Yan, H.Q.; Lin, Q. Fabrication and Evaluation of Alginate/Bacterial Cellulose Nanocrystals-Chitosan-Gelatin Composite Scaffolds. *Molecules* **2021**, *26*, 5003. [CrossRef] [PubMed]
34. Bidarra, S.J.; Barrias, C.C.; Granja, P.L. Injectable alginate hydrogels for cell delivery in tissue engineering. *Acta Biomater.* **2014**, *10*, 1646–1662. [CrossRef] [PubMed]
35. Turco, G.; Marsich, E.; Bellomo, F.; Semeraro, S.; Donati, I.; Brun, F.; Grandolfo, M.; Accardo, A.; Paoletti, S. Alginate/hydroxyapatite biocomposite for bone ingrowth: A trabecular structure with high and isotropic connectivity. *Biomacromolecules* **2009**, *10*, 1575–1583. [CrossRef]
36. Yang, J.S.; Xie, Y.J.; He, W. Research progress on chemical modification of alginate: A review. *Carbohydr. Polym.* **2011**, *84*, 33–39. [CrossRef]
37. Nie, H.R.; He, A.H.; Zheng, J.F.; Xu, S.S.; Li, J.X.; Han, C.C. Effects of chain conformation and entanglement on the electrospinning of pure alginate. *Biomacromolecules* **2008**, *9*, 1362–1365. [CrossRef]
38. Vieira, E.F.S.; Cestari, A.R.; Airoldi, C.; Loh, W. Polysaccharide-based hydrogels: Preparation, Characterization, and Drug Interaction Behaviour. *Biomacromolecules* **2008**, *9*, 1195–1199. [CrossRef]
39. Singh, B.; Sharma, D.K.; Gupta, A. Controlled release of the fungicide thiram from starch-alginate-clay based formulation. *Appl. Clay Sci.* **2009**, *45*, 76–82. [CrossRef]
40. Yang, J.S.; Ren, H.B.; Xie, Y.J. Synthesis of amidic alginate derivatives and their application in microencapsulation of λ-cyhalothrin. *Biomacromolecules* **2011**, *12*, 2982–2987. [CrossRef]
41. Islam, M.S.; Karim, M.R. Fabrication and characterization of poly (vinylalcohol)/alginate blend nanofibers by electrospinning method. *Colloids Surf. A* **2010**, *366*, 135–140. [CrossRef]
42. Kang, H.A.; Shin, M.S.; Yang, J.W. Preparation and characterization of hydrophobically modified alginate. *Polym. Bull.* **2002**, *47*, 429–435. [CrossRef]
43. Lawrie, G.; Keen, I.; Drew, B.; Chandler-Temple, A.; Rintoul, L.; Fredericks, P.; Grøndahl, L. Interactions between alginate and chitosan biopolymers characterized using FTIR and XPS. *Biomacromolecules* **2007**, *8*, 2533–2541. [CrossRef] [PubMed]
44. Kovalenko, I.; Zdyrko, B.; Magasinski, A.; Hertzberg, B.; Milicev, Z.; Burtovyy, R.; Luzinov, I.; Yushin, G. A major constituent of brown algae for use in high-capacity Li-ion batteries. *Science* **2011**, *334*, 75–79. [CrossRef] [PubMed]
45. Chang, C.H.; Lin, Y.H.; Yeh, C.L.; Chen, Y.C.; Chiou, S.F.; Hsu, Y.M.; Chen, Y.S.; Wang, C.C. Nanoparticles incorporated in pH-sensitive hydrogels as amoxicillin delivery for eradication of Helicobacter pylori. *Biomacromolecules* **2010**, *11*, 133–142. [CrossRef] [PubMed]
46. Ionita, M.; Pandele, M.A.; Iovu, H. Sodium alginate/graphene oxide composite films with enhanced thermal and mechanical properties. *Carbohydr. Polym.* **2013**, *94*, 339–344. [CrossRef]
47. Liu, M.; Dai, L.; Shi, H.; Xiong, S.; Zhou, C. In vitro evaluation of alginate/halloysite nanotube composite scaffolds for tissue engineering. *Mater. Sci. Eng. C* **2015**, *49*, 700–712. [CrossRef]
48. Yan, H.Q.; Chen, X.Q.; Li, J.C.; Feng, Y.H.; Shi, Z.F.; Wang, X.H.; Lin, Q. Synthesis of alginate derivative via the Ugi reaction and its characterization. *Carbohydr. Polym.* **2016**, *136*, 757–763. [CrossRef]
49. Yan, H.Q.; Chen, X.Q.; Feng, M.X.; Shi, Z.F.; Zhang, W.; Wang, Y.; Ke, C.R.; Lin, Q. Entrapment of bacterial cellulose nanocrystals stabilized Pickering emulsions droplets in alginate beads for hydrophobic drug delivery. *Colloids Surf. B Biointerfaces* **2019**, *177*, 112–120. [CrossRef]
50. Dalheim, M.; Vanacker, J.; Najmi, M.A.; Aachmann, F.L.; Strand, B.L.; Christensen, B.E. Efficient functionalization of alginate biomaterials. *Biomaterials* **2016**, *80*, 146–156. [CrossRef]

Article

A New Decellularization Protocol of Porcine Aortic Valves Using Tergitol to Characterize the Scaffold with the Biocompatibility Profile Using Human Bone Marrow Mesenchymal Stem Cells

Marika Faggioli [1,2], Arianna Moro [2,3], Salman Butt [1,2], Martina Todesco [2,3], Deborah Sandrin [2,4], Giulia Borile [2,4], Andrea Bagno [2,3], Assunta Fabozzo [1,2], Filippo Romanato [2,4], Massimo Marchesan [5], Saima Imran [1,2,*] and Gino Gerosa [1,2]

1. Department of Cardiac, Thoracic, Vascular Sciences and Public Health, University of Padua, I-35128 Padua, Italy; marika.faggioli96@gmail.com (M.F.); salman.butt@studenti.unipd.it (S.B.); assunta.fabozzo@aopd.veneto.it (A.F.); gino.gerosa@unipd.it (G.G.)
2. L.I.F.E.L.A.B. Program, Consorzio per la Ricerca Sanitaria (CORIS), Veneto Region, I-35127 Padua, Italy; moroarianna97@gmail.com (A.M.); martina.todesco@unipd.it (M.T.); sandrin.deborah@gmail.com (D.S.); giulia.borile@unipd.it (G.B.); andrea.bagno@unipd.it (A.B.); filippo.romanato@unipd.it (F.R.)
3. Department of Industrial Engineering, University of Padua, I-35131 Padua, Italy
4. Department of Physics and Astronomy "G. Galilei", University of Padua, I-35131 Padua, Italy
5. Consultant of Animal and Food Welfare, 3500 Padova, Italy; massimo.marchesan@yahoo.it
* Correspondence: saima.imran@unipd.it

Abstract: The most common aortic valve diseases in adults are stenosis due to calcification and regurgitation. In pediatric patients, aortic pathologies are less common. When a native valve is surgically replaced by a prosthetic one, it is necessary to consider that the latter has a limited durability. In particular, current bioprosthetic valves have to be replaced after approximately 10 years; mechanical prostheses are more durable but require the administration of permanent anticoagulant therapy. With regard to pediatric patients, both mechanical and biological prosthetic valves have to be replaced due to their inability to follow patients' growth. An alternative surgical substitute can be represented by the acellular porcine aortic valve that exhibits less immunogenic risk and a longer lifespan. In the present study, an efficient protocol for the removal of cells by using detergents, enzyme inhibitors, and hyper- and hypotonic shocks is reported. A new detergent (Tergitol) was applied to replace TX-100 with the aim to reduce toxicity and maximize ECM preservation. The structural integrity and efficient removal of cells and nuclear components were assessed by means of histology, immunofluorescence, and protein quantification; biomechanical properties were also checked by tensile tests. After decellularization, the acellular scaffold was sterilized with a standard protocol and repopulated with bone marrow mesenchymal stem cells to analyze its biocompatibility profile.

Keywords: decellularization; Tergitol; valve bioprostheses; cardiac tissue engineering; mesenchymal stem cells; biocompatibility

1. Introduction

Cardiovascular diseases are the most common cause of mortality in high-income countries [1] and are estimated to be responsible for one in three deaths worldwide [2]. Amongst them, aortic stenosis (AS) is the most common heart valve lesion encountered in clinical practice and affects 2% to 5% of adults over the age of 65 years [3]. Furthermore, narrowing at various level of the left ventricle may also affect the pediatric population, which shows a prevalence of 0.22 per 1000 live births [4]. The currently available surgical intervention is the replacement of the diseased valve either with biological prostheses or mechanical devices. The major limitations associated with the former are short post-surgical lifespan (10 years) and low aptness in young patients [5,6]; the latter have longer durability but require lifelong anticoagulation therapy, which results in chronic complications.

Xenogeneic substitutes, i.e., chemically treated acellular valves, can be an alternative. Two main advantages are associated with these valves: first, their large availability; second, after implantation, they can stimulate mechanisms of tissue regeneration and healing, which favor specific cell types to grow and differentiate [7,8]. This is due to the presence of tissue-specific biochemical factors that are still preserved, even after decellularization. Indeed, the immunogenic responses triggered by biological tissues are still a hurdle that can be limited by improving decellularization protocols.

Decellularization involves the use of chemical/biological agents, physical stimulation, and/or a combination of both. Generally, decellularization protocols are applied through perfusion or under mechanical agitation to allow chemicals (mainly detergents, e.g., SDS, Triton, and sodium deoxycholate) or enzymes (e.g., trypsin) penetrating the tissue or the whole organ. Numerous studies demonstrated that decellularization is efficient for removing the DNA content, with an acceptable level of damage to the ECM composition, architecture, bioactivity, and mechanical properties [9,10]. The removal of nuclear materials is of primary importance since high levels of residual DNA can trigger severe immunological responses. Therefore, it is crucial to minimize the DNA content while avoiding potential structural damage to tissue/organ during decellularization [11].

In the present study, we propose a new efficient, low cytotoxic decellularization protocol (based on current data), which is based on a new detergent, Tergitol, which is eco friendly and highly degradable. The previously frequently applied Triton-100, is included in the candidate list by the European Chemical Agency (ECHA) because of its endocrine toxicity [12]; the alternative detergents were proposed by the same authorities, and we selected Tergitol due to its surfactant properties and rapid degradability. To our best of knowledge, the systemic toxicity of Tergitol has not been reported previously; however, it was reported to be toxic in the *C. elegans* model elsewhere [13]. The new decellularization protocol includes a cocktail of protease inhibitors, freezing and thawing cycles, and the application of Tergitol and sodium cholate in various steps along the whole process. Decellularization was applied to porcine aortic valves: the acellular matrices were characterized regarding their histological, immunological, biochemical, and biomechanical properties and shown to have significant perseverance.

ECMs are prone to cell attachment: this indicates scaffold biocompatibility and the possibility for tissue integration and remodeling in vivo. A well-preserved structure can provide a favorable environment for cell repopulation: the crosstalk between ECMs and cells can also promote the structural organization, directing cell functions. Understanding the process underlying the cross talk between cells and ECMs will lead to the design of the remodeling process of the engineered heart valves [14].

2. Materials and Methods

2.1. Tissue Procurement and Preservation

Fresh porcine hearts (20 in total) were harvested from 6–8 month-old adult pigs from the local abattoir (F.lli Guerriero S.r.l, Villafranca Padovana, Italy) within 2 h from death. The aortic roots were dissected from the heart and, after removing fat and blood remnants, washed with phosphate buffered saline (PBS, Sigma, Saint Louis, MO, USA), dried on blotting paper, and frozen in organ plastic bags at $-80\ °C$ until use.

2.2. Decellularization Procedure

The Tergitol-based decellularization protocol was evolved from the TRICOL protocol previously reported by our group [15]. Briefly, tissues were thawed at RT for 3 h, before initiating the decellularization procedure. Then, they were submerged in 500 mL sterile jars in agitation at $4\ °C$ with proteases inhibitors cocktail for 8 h, afterwards with 1% Tergitol (Sigma) at RT for the next 12 h; hypertonic/hypotonic shocks (8 h each cycle at RT) followed, and finally they were treated with 0.4% sodium cholate (Sigma) for 16 h in the dark at

RT. After decellularization, 2 cycles of washing with PBS (1X) were performed. Ethanol (4%, Carlo Erba, Cornaredo, Milano, Italy) and peracetic acid (0.1%, Sigma) solution was used for bioburden reduction. Finally, an endonuclease (Benzonase 25k U, Sigma, E1014) was applied to remove nucleic acids for 48 h at 37 °C.

2.3. DNA Extraction and Quantification

Native ($n = 4$) and decellularized tissues ($n = 4$) were lyophilized with the Speed vac SPD130DLX (Thermo Scientific). Dry tissue samples (5–10 mg each) were used for DNA extraction that was performed according to the procedures described in the DNeasy Blood & Tissue Kit (Qiagen®, 69506, Valencia, CA, USA). Briefly, tissues were incubated in proteinase K solution and the lysis buffer ATL was added overnight in thermomixer at 56 °C. DNA concentrations were measured by using Nanodrop One (Thermo Fisher Scientific, Waltham, MA, USA) and Qubit (Qubit™ 1X dsDNA HS Assay Kit, Q33231, Waltham, MA, USA) fluorescent-based assays.

2.4. Biochemical Assays

Biochemical assays were performed to determine the concentration of major structural proteins (collagen and elastin) and glycosaminoglycans.

2.4.1. Elastin Quantification

Native ($n = 3$) and decellularized ($n = 3$) aortic valves were dissected separately to investigate the myocardium, the aortic wall, and the cusps. Lyophilized samples (3–5 mg) were used for elastin quantification. Elastin concentration (mg/mg of dry weight of tissue) was measured according to the manufacturer's instructions (Elastin Assay-Fastin, Biocolor, F4000): the assay is reported in [16]. Absorbance (513 nm) was measured with the Spark spectrophotometer ((Spark 10M, Tecan, Manne Dorf, Switzerland).

2.4.2. Hydroxyproline Quantification

Samples (3–5 mg) of lyophilized native ($n = 3$) and decellularized ($n = 3$) valves were used for hydroxyproline quantification. The assay was carried out according to the manufacturer's instructions (MAK008, Sigma-Aldrich, Saint Louis, MO, USA). The protocol was based on the principle established in [17], and the end-point absorbance (560 nm) was measured spectrophotometrically.

2.4.3. Glycosaminoglycans (GAGs) Quantification

Lyophilized samples (3–5 mg) from myocardium, aortic wall and cusps ($n = 6$) were used for glycosaminoglycans quantification, according to the manufacturer's instructions (Blyscan Sulfated Glycosaminoglycan Assay, Biocolor, B1000, County Antrim, UK). This assay is based on the protocol established in [18]; the end point absorbance (656 nm) was measured spectrophotometrically.

2.5. Biomechanical Tests

The effects on the aortic arterial wall and leaflets due to the decellularization procedure were biomechanically assessed by uniaxial tensile tests: native and decellularized aortic valve responses to load were compared. A total of 3 native and 3 decellularized valves were analyzed.

To evaluate the anisotropy of the aortic wall, 3 specimens were cut from the aortic wall in the longitudinal direction and 3 specimens in the circumferential direction [19]. The leaflets were isolated and cut into dog-bone-shaped specimens, using a homemade cutter. The dimensions and shape of each specimen respect what is suggested by the ASTM D1708-13 standard concerning small-size tissues: the gauge length was 5 mm, and the width 2 mm. Sample thickness was measured using Mitutoyo digital caliber model ID-C112XB (Aurora, IL, USA).

Samples were biomechanically assessed by uniaxial tensile loading tests performed with a custom-made apparatus (IRS, Padova, Italy), which was operated by a dedicated LabVIEW software (National Instruments, Austin, TX, USA).

The axial force was measured by means of a 50 N load cell, and the displacement between two actuators was taken as a direct measurement of sample elongation. Tests were performed at room temperature; samples were preloaded up to 0.1 N, then elongated with a rate of 0.2 mm/s until rupture. Displacement and load were recorded with a sampling frequency of 1000 Hz.

Biomechanical data were analyzed by means of an in-house developed Matlab® script. For each sample, the stress–strain relationship was obtained, and the following parameters were calculated: engineering stress σ (MPa) as the tensile force measured by the loading cells (Newton) divided by the original cross-sectional area of the sample; strain ε (%) as the ratio between the grip displacement and the gauge length.

Since the stress–strain curve of soft tissues has a typical J-shape, it can be split into two regions with different stiffness. The first part of the curve is usually termed "elastin phase", while the second is named "collagen phase", to indicate the contribution of these proteins to the mechanical response to load. Tensile modulus E1 was calculated as the slope of the linear portion of the curve between 0% and 10% deformation. The modulus E2 was calculated between 60% and 100% deformation, to characterize the second portion of the curve.

The ultimate tensile strength (UTS) and failure strain (FS) values were also calculated, representing the maximum strength and maximum elongation of the sample, respectively. As a final parameter, toughness (I) was calculated: it represents the energy required for the sample to reach failure.

Statistical comparisons were performed using the t-test (GraphPad Prism Software, San Diego, CA, USA). Significance was set at $p < 0.05$.

2.6. Histology

Native ($n = 6$) and decellularized tissues ($n = 6$) were cut into pieces (5–8 mm) and embedded with OCT. Myocardium was fixed with 4% paraformaldehyde (PFA, Bioptica) followed by sucrose gradient (10% to 30%) before embedding with OCT under nitrogen vapors. All samples were frozen at $-80\ °C$ until analysis. Cryosections of each tissue (7–8 µm thick) were obtained using a cryostat (NX 70 HOMVPD Cryostar, Thermo Fisher Scientific, Waltham, MA, USA). Staining with hematoxylin and eosin (04-061010, BioOptica, Milan, Italy), Masson's trichrome (04-01082, Bio-Optica, Milan, Italy), Weigert Van Gieson (04-051802, Bio-Optica, Milan, Italy) were performed according to each kits' instructions. Images were acquired using EVOS XL Core Cell Imaging System (Thermo Fisher Scientific, Waltham, MA, USA) and processed with Fiji software (version 1.51n).

2.7. Immunofluorescence

Immunofluorescence was performed with antibodies (Table S3) to detect structural proteins in ECM. Primary antibodies included Collagen I, Collagen IV, Elastin, and Laminin; DAPI was applied for nuclear staining. Briefly, cryosections (7–8 µm thick) were fixed with 4% PFA, blocked with 1% of bovine serum albumin (BSA, Sigma) for 30 min at RT. Incubation with the primary antibodies at 4 °C in the dark in a humidified chamber was followed by secondary antibodies (Table S3) for 90 min at RT. After 3 washings with PBS (5 min each), samples were incubated with Nublu DAPI (Life Technologies, Thermo Fisher Scientific, Waltham, MA, USA) for 30 min at RT. After mounting the cryosections, images were taken with Leica CTR 6000. Post-imaging analysis was performed using Fiji software.

2.8. Determination of Alpha-Gal Epitope

Immunofluorescence was performed as described in the section material and method (immunofluorescence). Alpha-gal images were obtained with confocal microscope Axio Observer LSM 800 and processed with Fiji software.

2.9. Two Photon Microscopy

Native and decellularized samples (aortic wall, leaflets, and myocardium) were compared, acquiring the SHG signal to evaluate collagen I intensity and distribution and staining the tissues with DAPI (Life Technologies, Thermo Fisher Scientific, Waltham, MA, USA) to evaluate the presence of nuclei. Two-photon microscopy was performed by using a custom developed multiphoton microscope, previously described by Filippi et al. [20]. Briefly, an incident wavelength of 800 nm (~40 mW average laser power, under the microscope objective) was adopted to detect the collagen SHG signal at 400 nm, and DAPI was detected in the blue channel. Images were acquired at a fixed magnification through the Olympus 25X water immersion objective with 1.05 numerical aperture (1024 × 1024 pixels), averaged over 70 consecutive frames, with a pixel dwell time of 0.14 µs and a pixel width of 0.8 µm. First order analysis of collagen intensity and distribution was performed with dedicated FIJI plugins as reviewed in (https://www.mdpi.com/1422-0067/22/5/2657, accessed on 15 November 2021).

2.10. Scanning Electron Microscopy (SEM)

Patches of the aortic wall from native and decellularized valves were analyzed with SEM. Before the analysis, patches were stored in the fridge until the preparation for microscopy with an ascending scale of ethanol concentration.

Samples were analyzed by SEM in High Vacuum after fixing, dehydrating (with Ethanol), drying (with CPD Critical Point Dryer) and coating with Au.

2.11. Sterility Assessment

Sterilization of decellularized tissue was performed following the guidelines of the European Pharmacopoeia [21]. Decontamination of the decellularized samples was performed in two phases: first, samples were treated with 70% ethanol for 30 min at RT; afterwards they were treated with a cocktail of antibiotics and antimycotics (50 mg/L vancomycin hydrochloride (SBR00001, Merck, Kenilworth, NJ, USA), 8 mg/L gentamicin (G1397, Merck), 240 mg/L cefoxitin (C0688000, Merck) and 25 mg/L amphotericin B (A9528, Merck)) at 37 °C for 24 h.

For the sterility assessment, two specific liquid media (thioglycolate medium, Cat no. T9032, and soya-bean casein digest medium, Cat no. 22092, Sigma-Aldrich) were used, and turbidity was observed over time. The 8 mm punches of tissue samples (aortic wall, leaflets, and myocardium, each in duplicate) from native, decellularized, and eventually decontaminated valves were immersed into media to detect aerobic/anaerobic bacteria and fungal growth. Samples were incubated at 35 °C in the oven (Thioglycolate medium) and RT (Soya-bean casein digest medium) for 14 days. Images were taken during incubation at time intervals of 48 h.

2.12. Biocompatibility Test

Human bone marrow mesenchymal stem cells (hMSC-BM, 12974, PromoCell, Heidelberg, Germany) were cultured in bone marrow MSC media (C-28019, PromoCell). Cells were expanded and passaged up to 5 passages in T75 flasks in 5% CO_2 at 37 °C with 95% of humidity. Decellularized and sterilized tissue patches (8 mm) were placed in 48-well plate and equilibrated with the MSC media for 24 h prior to culturing cells. Cells were harvested and resuspended in media with a density of 1×10^6 cells/mL. Approximately 30,000 cells were plated over each tissue sample. The proliferation/toxicity analysis at each time point (24 h, 72 h, day 7, and day 14) was performed by the following protocols:

(a) Live and Dead assay: Cell viability was evaluated using the Live/Dead viability/cytotoxicity kit (MP 03224, Invitrogen, Thermo Fisher Scientific, Waltham, MA, USA). Briefly, cells were stained with Calcein AM and Ethidium homodimer-1 with final concentrations of 2 and 4 µM, respectively, by incubation at 37 °C for 45 min. Images were obtained using Olympus IX71 microscope.

(b) DNA extraction and quantification: Tissue patches were cultured with the hMSCs-BM and the amount of DNA was quantified to estimate cell proliferation. DNA extraction and quantification were performed as described in Section 2.3.
(c) WST-1 assay: The cell Proliferation and Cytotoxicity Assay kit (AR1159, Boster, Pleasanton, CA, USA) was used to assess the viability/proliferation of cells cultured on decellularized patches at each time point. The assay was performed according to the suppliers' instructions. Absorbance was measured at 450 nm using microplate reader (Spark 10M Tecan, Tecan). Cells plated on a multi-well plate served as a positive control.
(d) Scanning electron microscopy (SEM): SEM was performed after culturing hMSCs-BM for 14 days on the decellularized and decontaminated aortic tissue patches ($n = 3$).

2.13. Data Analysis

All data were expressed as mean ± SD. One way ANOVA was performed with GraphPad Prism 8 to analyze the groups of experiments and multiple comparisons within groups were performed if necessary. Differences were considered statistically significant when $p < 0.05$.

3. Results

3.1. Visual Inspection

After decellularization, porcine valves appeared white, except for the myocardium, exhibiting a light brown color (Figure 1a).

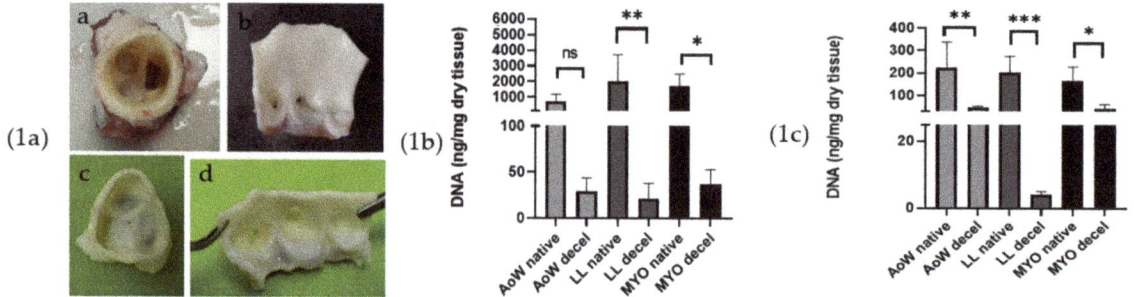

Figure 1. Macroscopic analysis of native (**a**,**b**) and decellularized (**c**,**d**) porcine aortic valve (**1a**). DNA quantification (ng/mg of dry tissue) comparative analysis on native and decellularized tissues performed by Nanodrop (**1b**) and Qubit assay (**1c**). Data were analyzed by one-way ANOVA ($p < 0.05$). There was a significant difference between native and corresponding decellularized tissues denoted by *, **, *** the level of significance from lower to highest p values, ns = not significant.

The tissues were subgrouped according to the analysis to be performed. Tissues were lyophilized for the DNA quantification and biochemical assays. For biomechanical tests, tissue was taken fresh after decellularization with the defrosted native aortic valves. For the rest of the analysis, tissues were frozen in liquid nitrogen after embedding into the optimal cutting temperature (OCT). The results obtained are presented in the following section.

3.2. DNA Extraction and Quantification

After decellularization, the DNA quantities resulted in being below the threshold value of 50 ng/mg dry tissue [22]. Samples from the aortic wall and myocardium showed a DNA content between 40 and 50 ng/mg of dry tissue, while it was found to be between 3 and 10 ng/mg of dry tissue in leaflets (Figure 1b,c).

3.3. Biochemical Assays

Biochemical assays showed a non-significant loss of GAGs (Figure 2a) and elastin (Figure 2b) after decellularization. On the other hand, the hydroxyproline concentration increased, particularly in leaflets, but the increase was not significant (Figure 2c).

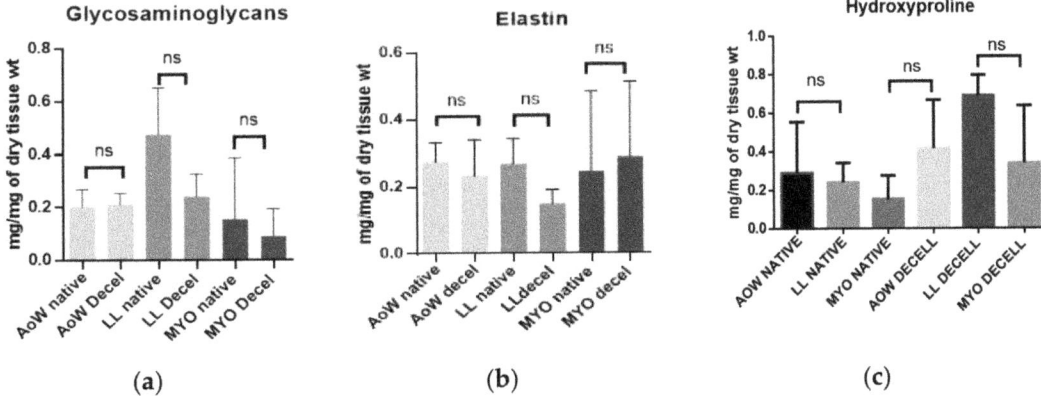

Figure 2. Biochemical analysis of (**a**) glycosaminoglycans, (**b**) elastin and (**c**) hydroxyproline shown. Native ($n = 3$) and decellularized ($n = 3$) samples were used for the analysis. Data were analyzed by one-way ANOVA ($p < 0.05$). There was a non-significant (ns) difference between native and corresponding decellularized tissues. Increasing values of hydroxyproline are not statistically significant.

3.4. Biomechanical Tests

The decellularization treatment caused a decrease in the thickness of the aortic wall and leaflet: from 1.8 ± 0.52 mm to 1.7 ± 0.35 mm (5.6%), and from 0.45 ± 0.25 to 0.38 ± 0.09 (15.6%), respectively. Differences were not significant (Figure 3a, Table 1).

Concerning the parameters that characterize the mechanical behavior of the investigated tissues upon loading (Figure 3), there was an increase in stiffness along the longitudinal direction, which is significant in the E2 module ($p = 0.0055$) that characterizes the collagen part of the stress–strain curve. Other significant differences along the longitudinal direction are found with respect to FS ($p = 0.0003$), UTS ($p = 0.0074$) and I value ($p = 0.0001$): decellularized tissues reached lower elongation and tension at break than native ones. Indeed, FS values range from $320.5 \pm 30.42\%$ for native tissues to $248.2 \pm 35.71\%$ for decellularized ones, while UTS values vary from 1.71 ± 0.55 to 1.12 ± 0.18, respectively; the parameter I decreases in decellularized tissues from 181.9 ± 47.47 MPa to 110.6 ± 24.62 MPa.

Table 1. Sample thickness measure in Native and Decellularized. Aortic Wall (AoW) and Native Leaflet (LL).

Sample	Thickness [mm]
AoW Native	1.8 ± 0.52
AoW Decellularized	1.7 ± 0.35
LL Native	0.45 ± 0.25
LL Decellularized	0.38 ± 0.09

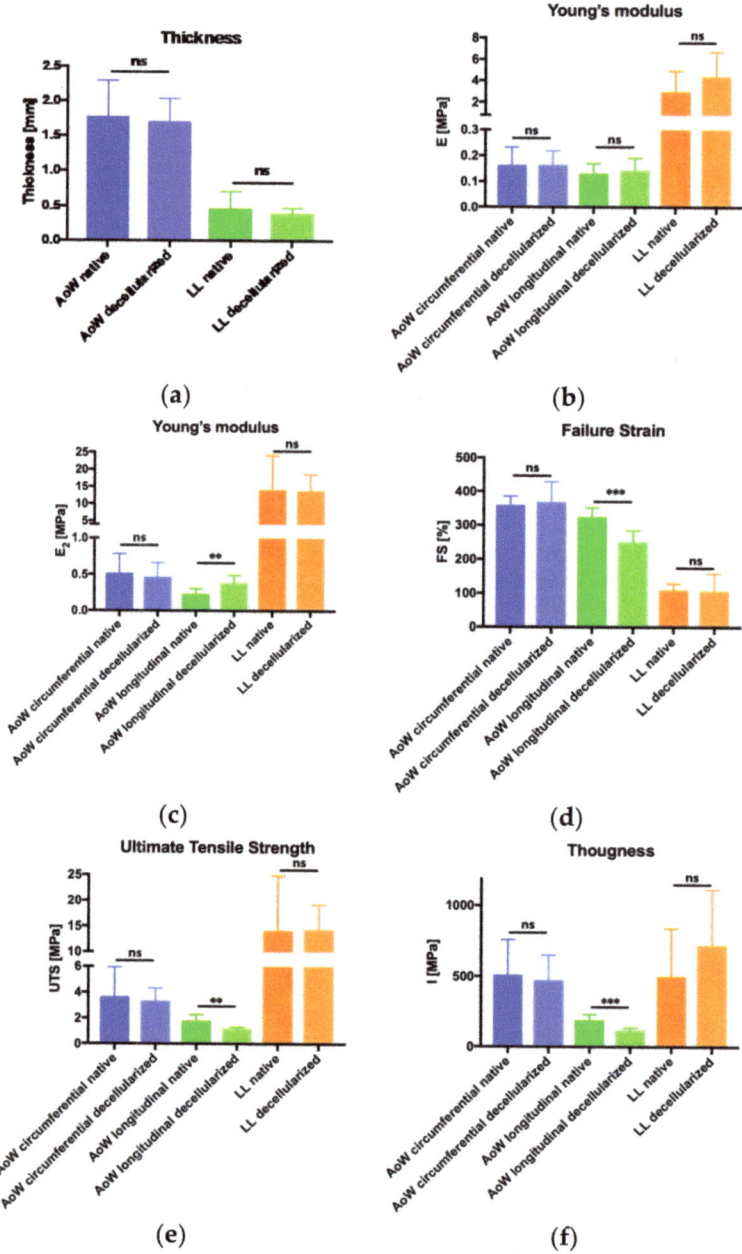

Figure 3. Histograms of mechanical tests. Sample thickness are not statistically different comparing native and decellularized samples (**a**). Modulus E1 values are not statistically different (**b**), whereas modulus E2 values are statistically different for the longitudinal aortic wall comparing samples before and after decellularization (**c**). These samples are statistically different also in terms of failure strain (**d**), ultimate tensile strength (**e**) and toughness (**f**) (ns: no significant different, ** $p < 0.001$, *** $p < 0.001$).

3.5. Histology

In general, histological analyses showed the integrity of the ECM, even after decellularization. In the representative pictures (Figure 4), the collagen and elastin arrangement were found to be intact, with a well-preserved fiber distribution that was confirmed by TEM; nuclei were effectively removed, and they were not more visible in the decellularized samples.

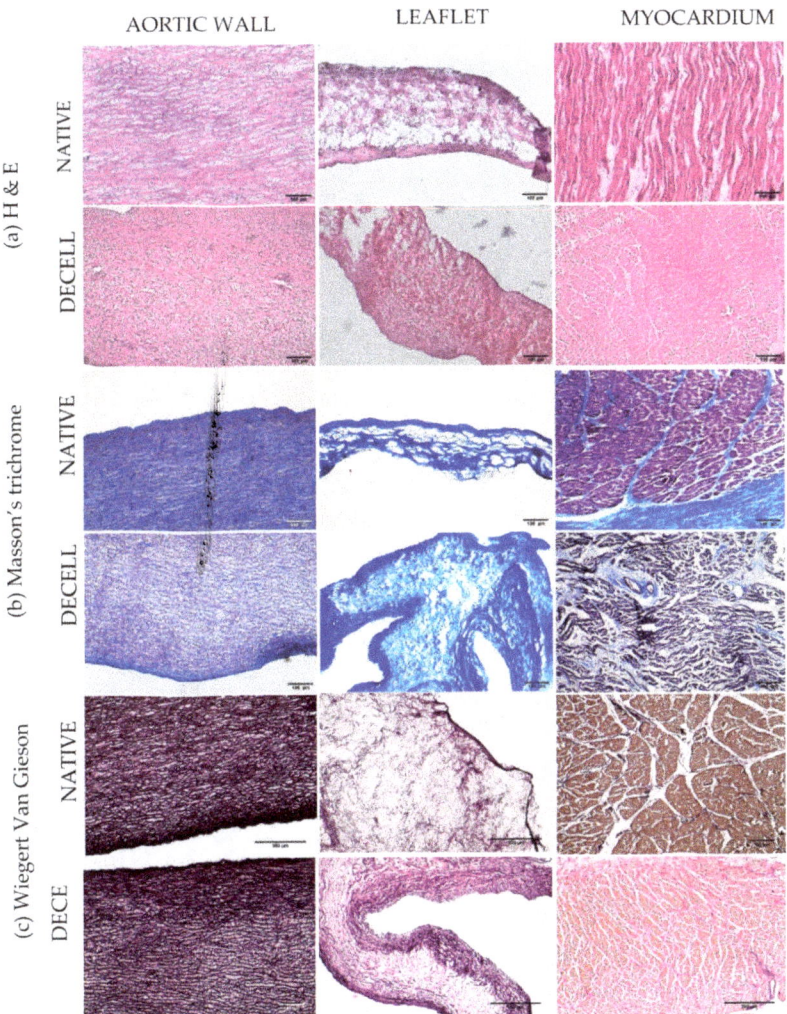

Figure 4. Histological analysis of native and decellularized tissues was performed on OCT embedded cryosections (7 μm). (**a**) Hematoxylin and Eosin staining showed the presence of nuclei in native aortic wall, leaflet, and myocardium and the removal of nuclei in the decellularized counterparts. (**b**) Masson's trichrome showed the nuclei in black staining aligned with collagen in blue. (**c**) Wiegert Van Gieson staining indicates the presence of nuclei in native tissues (elastin and collagen are in red) while nuclei are absent in the decellularized tissues.

3.6. Immunofluorescence

Immunofluorescence was used to evaluate ECM structure and decellularization effectiveness. DAPI staining showed the presence of nuclei in native samples while their absence was observed in decellularized tissues. Regarding the expression of collagen I, collagen IV, laminin and elastin, no significant difference was detected between native and decellularized aortic walls; in decellularized leaflets and myocardium, reduced signals of laminin and collagen IV were found (Figure 5).

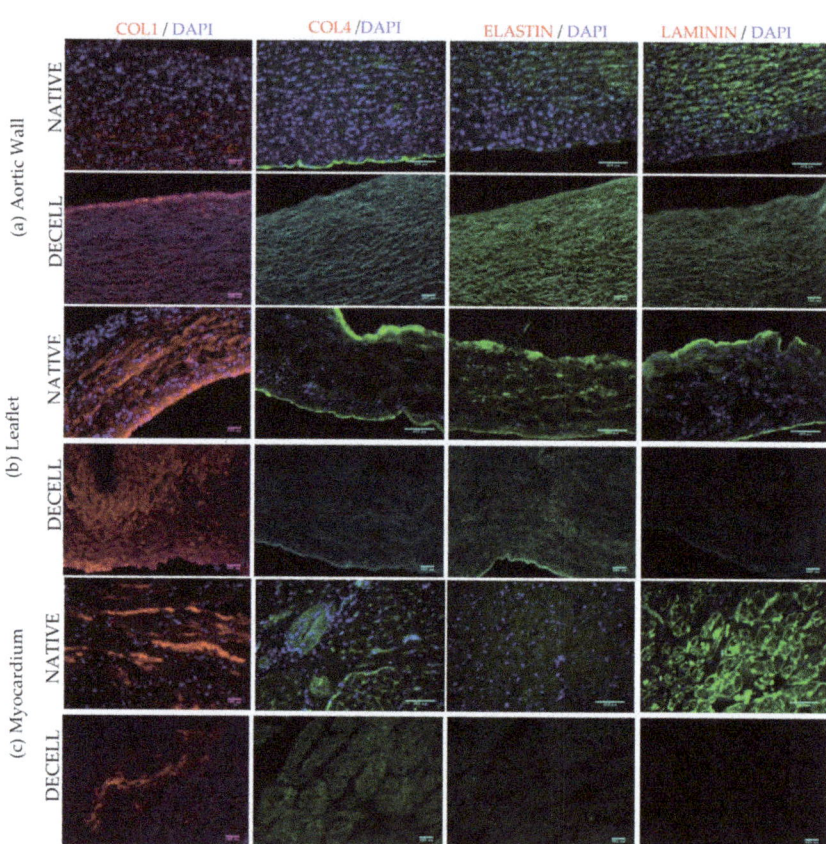

Figure 5. Immunofluorescence analysis showed the PFA fixed cryosections stained with primary and secondary antibodies and DAPI (blue). (**a**) aortic wall, (**b**) leaflets, and (**c**) myocardium. The representative images indicate the expression of collagen I (red), collagen IV (green), elastin (green) and laminin (green). Images were taken with the Leica imaging system at 20× magnification.

With regards to the alpha-gal epitope, which is major immunogenic factor in animal tissues, signals were significantly maintained comparing native and decellularized samples. However, detection with isolectin showed a reduction in the decellularized myocardium (Figure S1).

3.7. Two Photon Microscopy

To deepen the investigation of structural proteins, with a peculiar interest in collagen, we evaluated Collagen I in a semi-quantitative approach thanks to SHG imaging. Compared to immunofluorescence staining that requires a processing of the tissues, SHG offers a label-free approach to acquire the collagen I signal, proportional to the protein content. Moreover, samples were stained with DAPI.

In Figure 6, representative images of native and decellularized tissues from the aortic wall, leaflet, and myocardium are shown. The DAPI signal is completely absent in decellularized portions. This observation further confirms the effective nuclei removal, thanks to the decellularization protocols. The first-order analysis of the SHG signal is focused on the average intensity that is proportional to the protein content. We observed a decrease in signal intensity in the aortic wall and leaflets after decellularization, while no changes appeared in the myocardium samples. In line with this, a re-arrangement of the collagen fibers resulted from coherency analysis for the aortic wall and leaflets. Again, no differences were observed in myocardium when comparing native and decellularized tissue.

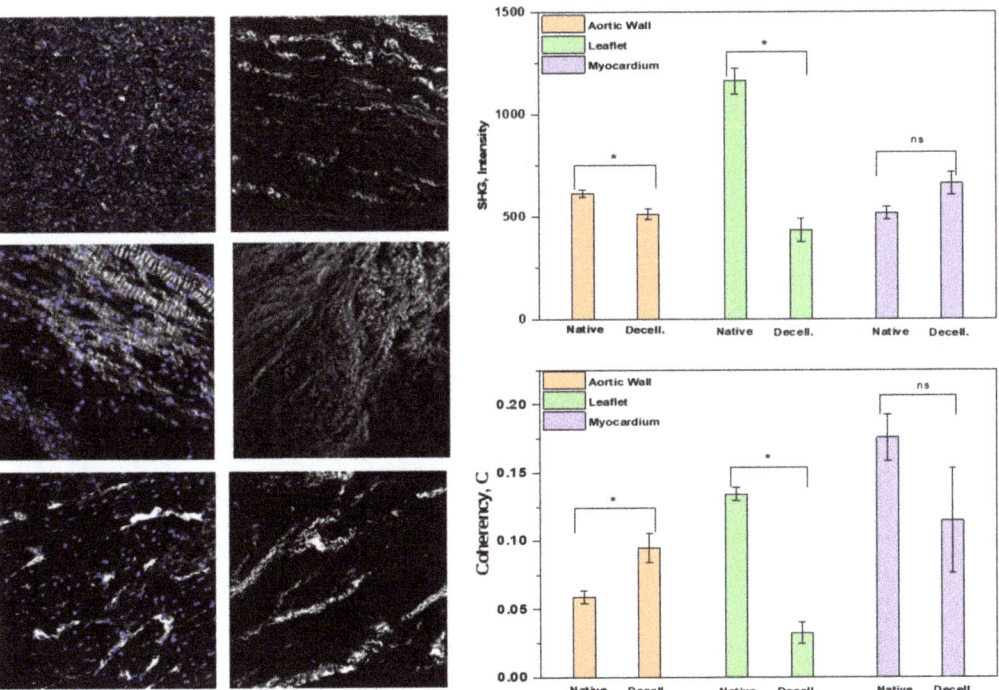

Figure 6. Pictures obtained by means of two photon microscope from native and decellularized aortic wall, leaflets and myocardium (ns = non significant, * denoted significance of difference between values of native and decellularized values).

3.8. Scanning Electron Microscopy (SEM)

Samples from the aortic root were analyzed at different steps all along the decellularization process: native, after Tergitol, sodium cholate, Benzonase, and after 14 days from cell seeding with hMSCs-BM (Figure 7). Images of native tissue showed the presence of cells and the typical structure of collagen fibers inside the matrix. After Tergitol treatment, cells appeared in a lower amount or completely absent, while the collagen structure was slightly disrupted. After sodium cholate treatment, pictures showed the presence of cellular debris while collagen fibers were maintained: less tissue disruption is caused during this step. After Benzonase, nuclear components are completely absent.

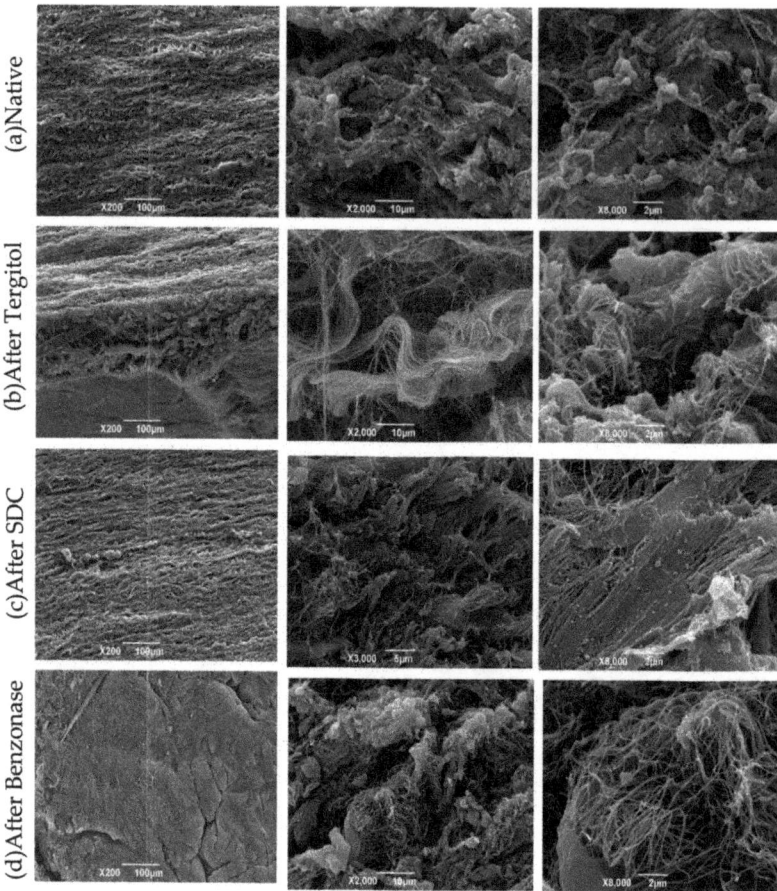

Figure 7. SEM images represent the gradual changes in morphological features of the native aortic wall tissue (**a**) and along the sequential steps of decellularization (**b**–**d**).

3.9. Sterility Assessment

From the visual inspection of native tissue samples (Figure S2), turbidity clearly indicated contamination within 24–72 h decellularized samples were eventually turbid on day 7. Sterilized samples showed the absence of turbidity during an incubation time of 14 days.

3.10. Biocompatibility Assessment

Live and dead staining showed that the cells were attached to the tissue surface within 24 h: they seemed to be more prone to attach at the edges of the aortic patches, whereas they spread over the entire surface of the leaflets (Figure 8). Cell proliferation over both tissues was observed progressively until day 7: their elongation and alignment were oriented by tissue structure. On day 14, the cell viability on the aortic wall was compromised significantly, while the proliferation on the leaflet was sustained (Figure 8).

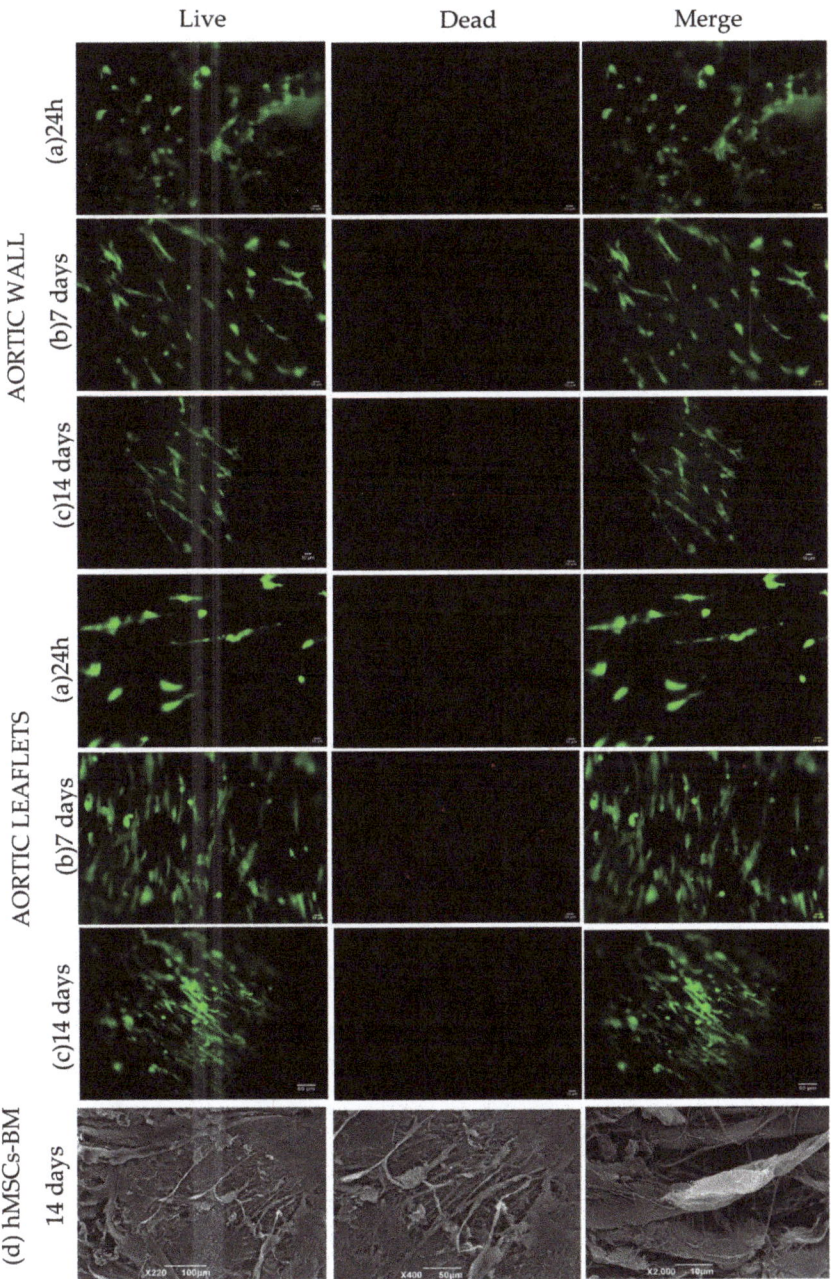

Figure 8. Live and dead staining analysis at (**a**) 24 h, (**b**) day 7 and (**c**) day 14 for aortic wall and leaflets samples. Images were taken with the Olympus IX71 microscope at 2× magnification. (**d**) SEM images of the decellularized tissue cultured with hMSCs-BM (day 14) represented with red arrows. Pictures are shown in different scale of magnification (see the scale bars).

DNA content in cell cultured tissues progressively increased over time in the case of aortic leaflets, while in the aortic wall, there was a significant decrease on day 14 (Figure 9).

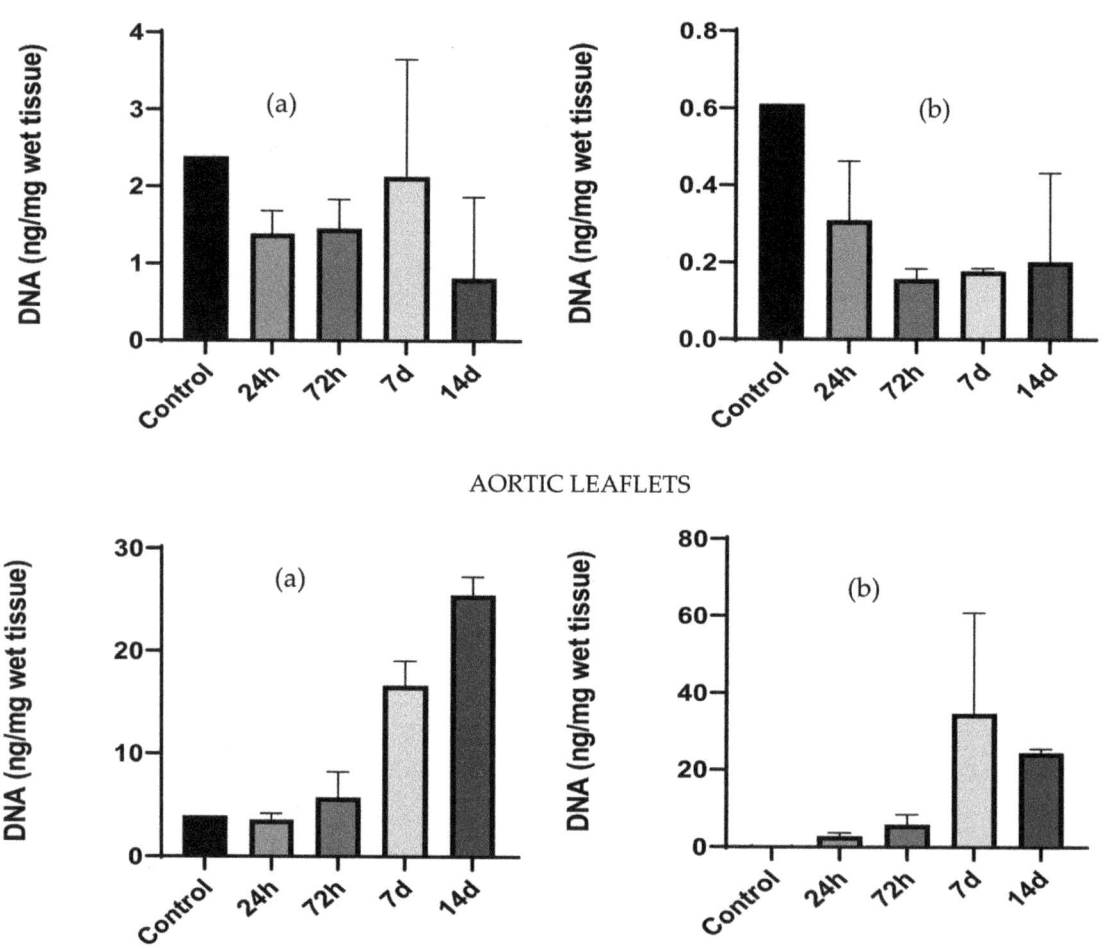

Figure 9. DNA quantification (ng/mg of dry tissue) comparative analysis between native and decellularized tissues performed by Nanodrop (**a**) and Qubit assay (**b**). Data were analyzed by one-way ANOVA ($p < 0.05$). There was a significant difference between native and corresponding decellularized tissues.

WST-1 assay is an indicator of the metabolic activity of living cells: the absorbance of the aortic wall samples was lower than the control (Figure 10). On the contrary, we observed a continuous absorbance increase over time in the leaflets (Figure 10), which was approximately equal to the positive control.

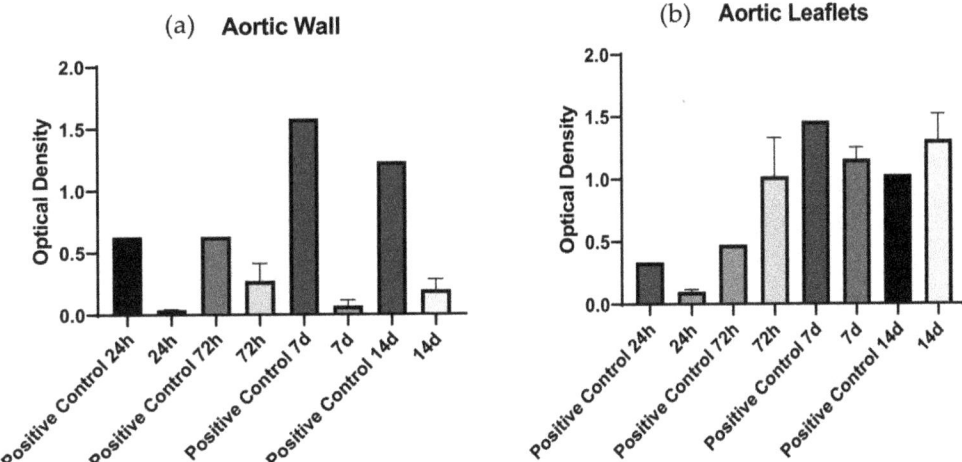

Figure 10. Analysis of metabolic activities of the cultured hMSC-BM cells at different time intervals was measured by WST-1 reduction assay. (**a**) aortic wall and (**b**) leaflets samples show significant differences in the proliferation capacity of the cells. Data were analyzed by one-way ANOVA ($p < 0.05$). There was a significant difference between native and corresponding decellularized samples for the aortic wall.

Scanning electron microscopy (SEM) was performed after culturing hMSCs-BM for 14 days, high magnification images depicted the cells attached to the surface of the aortic wall.

4. Discussion

Decellularized porcine heart valves can be exploited as substitutes for the surgical replacement of diseased valves: they present fundamental advantages over currently available mechanical and bioprosthetic devices since they do not evoke any adverse response and can be repopulated by circulating cells in vivo. The decellularization has to suppress the immunogenic potential of animal valves while preserving their structure and composition.

In the present paper, we illustrated a new decellularization protocol that is based on the previously reported one [15]. Major modifications include the use of a different detergent (Tergitol at room temperature instead of Triton X-100), freeze/thawing cycles, and reduced time points for certain inhibitor cocktail cycles. The Tergitol-based protocol was applied to porcine aortic valves (3 cm diameter), and it was demonstrated to be effective in removing cellular and nuclear components while preserving tissue structural integrity.

Histological analyses allowed excluding the presence of nuclei and confirmed substantial maintenance of collagen and elastin inside the ECM of decellularized samples. In particular, immunological assays revealed a minimal loss of collagen IV and laminin; microscopic observations showed that the nuclear material was efficiently removed with the persistence of collagen fibers.

With regard to the alpha-gal epitope, its removal seemed to be not effective: only in decellularized myocardium samples was it significantly less abundant than in native ones.

The determination of the DNA content in acellular biological scaffolds is of paramount importance to assess the effectiveness of decellularization protocol: previous studies have shown that residual DNA is still present in biological scaffolds already exploited for both research and commercial purposes [23,24]. In the present work, a maximal level of DNA was removed (≥ 95–97%) and the residual DNA amount was measured in valve leaflets

(~10 ng/mg). We also compared two quantification methods: Nanodrop spectrophotometer assay exhibited a sensitivity that is lower than fluorescent-based Qubit assay.

Hydroxyproline is a diagnostic amino acid for collagen: thus, its concentration was calculated as an indicator of collagen amount. No significant difference was found comparing native and decellularized tissues. On the other hand, a loss of elastin was observed in all tissues after decellularization, but it was not significant: this evidence is a positive indicator for tissue suitability. The little increase in the hydroxyproline content of the tissue after decellularization was mainly due to a relative increase in the ratio of these molecules to the total dry weight, due to the loss of soluble proteins and cell components. A non-significant reduction in GAGs was also observed after decellularization. The ECM proteins quantification data revealed that the ECM integrity was found to be in a highly acceptable range. This presented the aortic tissue suitability for further study for the animal implantation research trials.

With regard to the biomechanical characterization performed by means of tensile tests, the decellularization process did not cause significant differences with regard to the stiffness values of tissue samples from both longitudinal and circumferential aortic walls and from leaflets, as they were measured in the first part of the stress–strain curves (modulus E1). Moreover, in the second part of these curves, where the contribution of collagen fibers withstanding the mechanical deformation becomes prominent, a significant increase in the E2 modulus of the longitudinal aortic wall samples appeared: thus, the decellularization process makes the tissue stiffer along the longitudinal direction. This result is confirmed by the other statistical differences evidenced in Figure 3: they only regard the comparison of the aortic wall samples along the longitudinal direction, before and after decellularization. As the E2 modulus increases, the failure strain and the ultimate tensile strength decreases: the first represents the ability of the tissue to be elongated upon load; the second is the maximin value of the load applied until failure. Actually, the decellularization process is responsible for a significant improvement in the stiffness along the longitudinal direction of the aortic wall samples, which is followed by a significant reduction in their extensibility and ability to resist the application of a mechanical load. Toughness (I) is the last biomechanical parameter shown in Figure 3: it is calculated as the area below the stress–strain curve, and its physical meaning is related to the amount of energy absorbed by the material upon deformation. Once again, the decellularization process caused a reduction in toughness only for the samples withdrawn for the longitudinal direction of the aortic walls.

In terms of the biocompatibility profile, it is worthwhile to mention that two different surfaces showed a difference in the growth of hMSCs-BM: the leaflets surface was found to be the most appropriate to allow cell growth. The ECMs are prone to cell attachment, indicating scaffold biocompatibility and the possibility for tissue integration and remodeling in vivo. A well-preserved structure can provide a favorable environment for cell repopulation. There could be a possibility that the cross talk between the ECMs and the cells could also be promoting the structural organization, directing the cell function. Understanding the process underlying the cross talk between the cells and the ECMS will lead to designing the remodeling process of the engineered heart valves.

5. Conclusions

At present, acellular xenogeneic valve conduits are applied as scaffolds for heart valve regeneration, with promising results [25–27]. Due to the toxicity related to the detergent used in the previous protocols, it is necessary to find appropriate alternatives. In the present study, Tergitol was applied to decellularize aortic conduits, which were then characterized following standardized criteria. From the results obtained, we suggest the exploitation of acellular heart valve conduits for studies on the animal model. Preclinical studies will be a step ahead toward the application of decellularized heart valves in humans as an alternative to the currently available prosthetic devices.

Supplementary Materials: The following supporting information can be downloaded at: https://www.mdpi.com/article/10.3390/polym14061226/s1, Figure S1: Immunofluorescence data for the expression of α-gal; Figure S2: Sterility assessment; Tables S1 and S2: Biomechanical data; Table S3: List of antibodies.

Author Contributions: Conceptualization, G.G., A.F., A.B. and S.I.; investigation, experiments conducted and data analysis, M.F., A.M., M.T., S.B., D.S., G.B., F.R.; writing—original draft preparation, S.I., M.F., M.T., D.S., G.B.; resources, A.B., M.M., G.G.; writing—review and editing, A.F., S.I., A.B., G.G.; supervision, G.G., S.I.; project administration, G.G.; funding acquisition, G.G. All authors have read and agreed to the published version of the manuscript.

Funding: This work was supported by LifeLab Program of the Consorzio per la Ricerca Sanitaria (CORIS) of the Veneto Region, Italy (DGR1017, 17 July 2018). Funds for article submission are partially provided by y the "Centro Trapianti Cuore anno 2021—DGRV n.1237 del 14 September 2018".

Institutional Review Board Statement: Not applicable.

Informed Consent Statement: Not applicable.

Data Availability Statement: Not applicable.

Conflicts of Interest: The authors declare no conflict of interest.

References

1. Mahmood, S.S.; Levy, D.; Vasan, R.S.; Wang, T.J. The Framingham Heart Study and the epidemiology of cardiovascular disease: A historical perspective. *Lancet* **2014**, *383*, 999–1008. [CrossRef]
2. Zhao, D.; Liu, J.; Wang, M.; Zhang, X.; Zhou, M. Epidemiology of cardiovascular disease in China: Current features and implications. *Nat. Rev. Cardiol.* **2019**, *16*, 203–212. [CrossRef] [PubMed]
3. Lindman, B.R.; Clavel, M.A.; Mathieu, P. Calcific aortic stenosis. *Nat. Rev. Dis. Primers* **2016**, *2*, 16006. [CrossRef] [PubMed]
4. Van der Linde, D.; Konings, E.E.; Slager, M.A.; Witsenburg, M.; Helbing, W.A.; Takkenberg, J.J.; Roos-Hesselink, J.W. Birth prevalence of congenital heart disease worldwide: A systematic review and meta-analysis. *J. Am. Coll. Cardiol.* **2011**, *58*, 2241–2247. [CrossRef]
5. Rodriguez-Gabella, T.; Voisine, P.; Puri, R.; Pibarot, P.; Rodés-Cabau, J. Aortic Bioprosthetic Valve Durability Incidence, Mechanisms, Predictors, and Managment of Surgical and Transcatheter Valve Degeneration. *J. Am. Coll. Cardiol.* **2017**, *70*, 1013–1028. [CrossRef]
6. Brown, J.M.; O'Brien, S.M.; Wu, C.; Sikora, J.A.H.; Griffith, B.P.; Gammie, J.S. Isolated aortic valve replacement in North America comprising 108,678 patients i 10 years: Canges in risks, valve tyoes and outcomes in the Society of Thoracic Surgeons National Database. *J. Thorac. Cardiovasc. Surg.* **2008**, *137*, 82–90. [CrossRef] [PubMed]
7. Wilcox, H.E.; Korossis, S.A.; Booth, C.; Watterson, K.G.; Kearney, J.N.; Fisher, J.; Ingham, E. Biocompatibility and recellularization potential of an acellular porcine hearts valve matrix. *J. Hear. Valve Dis.* **2005**, *14*, 228.
8. Fioretta, E.S.; von Boehmer, L.; Motta, S.E.; Lintas, V.; Hoerstrup, S.P.; Emmert, M.Y. Cardiovascular tissue engineering: From basic science to clinical application. *Exp. Gerontol.* **2019**, *117*, 1–12. [CrossRef] [PubMed]
9. Sierad, L.N.; Shaw, E.L.; Bina, A.; Brazile, B.; Rierson, N.; Patnaik, S.S.; Kennamer, A.; Odum, R.; Cotoi, O.; Terezia, P.; et al. Functional heart valve scaffolds obtained by complete decellularization of porcine aortic roots in a novel differential pressure gradient perfusion system. *Tissue Eng. Part C Methods* **2015**, *21*, 1284–1296. [CrossRef]
10. Sánchez, P.L.; Fernández-Santos, M.E.; Costanza, S.; Climent, A.M.; Moscoso, I.; Gonzalez-Nicolas, M.A.; Sanz-Ruiz, R.; Rodríguez, H.; Kren, S.M.; Garrido, G.; et al. Acellular human heart matrix: A critical step toward whole hearts grafts. *Biomaterials* **2015**, *61*, 279–289. [CrossRef]
11. Spina, M.; Ortolani, F.; Messlemani, A.E.; Gandaglia, A.; Bujan, J.; Garcia-Honduvilla, N.; Vesely, I.; Gerosa, G.; Casarotto, D.; Petrelli, L.; et al. Isolation of intact aortic valve scaffolds for heart-valve bioprostheses: Extracellular matrix structure, prevention from calcification, and cell repopulation features. *J. Biomed. Mater. Res. Part A* **2003**, *67*, 1338–1350. [CrossRef] [PubMed]
12. Commission Regulation (EU) 2017/999—Of 13 June 2017—Amending Annex XIV to Regulation (EC) No 1907/2006 of the European Parliament and of the Council Concerning the Registration, Evaluation, Authorisation and Restriction of Chemicals. Available online: https://eur-lex.europa.eu/legal-content/EN/TXT/PDF/?uri=CELEX:32017R0999&from=EN (accessed on 17 February 2022).
13. De la Parra-Guerra, A.; Olivero-Verbel, J. Toxicity of nonylphenol and nonylphenol ethoxylate on Caenorhabditis elegans. *Ecotoxicol. Environ. Saf.* **2020**, *187*, 109709. [CrossRef] [PubMed]
14. Roderjan, J.G.; de Noronha, L.; Stimamiglio, M.A.; Correa, A.; Leitolis, A.; Bueno, R.R.L.; da Costa, F.D.A. Structural assessments in decellularized extracellular matrix of porcine semilunar heart valves: Evaluation of cell niches. *Xenotransplantation* **2019**, *26*, e12503. [CrossRef]

15. Gallo, M.; Naso, F.; Poser, H.; Rossi, A.; Franci, P.; Bianco, R.; Micciolo, M.; Zanella, F.; Cucchini, U.; Aresu, L.; et al. Physiological Performance of a Detergent Decellularized Heart Valve Implanted for 15 Months in Vietnamese Pigs: Surgical Procedure, Follow-up, and Explant Inspection. *Artif. Organs* **2012**, *36*, E138–E150. [CrossRef] [PubMed]
16. Winkelman, J. The Distribution of Tetraphenylporphinesulfonate in the Tumor-bearing Rat. *Cancer Res.* **1962**, *22*, 589–596. [PubMed]
17. Neuman, R.E.; Logan, M.A. The determination of hydroxyproline. *J. Biol. Chem.* **1950**, *184*, 299–306. [CrossRef]
18. Barbosa, I.; Garcia, S.; Barbier-Chassefière, V.; Caruelle, J.P.; Martelly, I.; Papy-García, D. Improved and simple micro assay for sulfated glycosaminoglycans quantification in biological extracts and its use in skin and muscle tissue studies. *Glycobiology* **2003**, *13*, 647–653. [CrossRef] [PubMed]
19. García-Herrera, C.M.; Atienza, J.M.; Rojo, F.J.; Claes, E.; Guinea, G.V.; Celentano, D.J.; García-Montero, C.; Burgos, R.L. Mechanical behaviour and rupture of normal and pathological human ascending aortic wall. *Med. Biol. Eng. Comput.* **2012**, *50*, 559–566. [CrossRef]
20. Filippi, A.; Dal Sasso, E.; Iop, L.; Armani, A.; Gintoli, M.; Sandri, M.; Gerosa, G.; Romanato, F.; Borile, G. Multimodal label-free ex vivo imaging using a dual-wavelength microscope with axial chromatic aberration compensation. *J. Biomed. Opt.* **2018**, *23*, 091403. [CrossRef]
21. Council of Europe. 2.6.1. Sterility. *Eur. Pharm.* **2005**, *5*, 145–149.
22. Crapo, P.M.; Gilbert, T.W.; Badylak, S.F. An overview of tissue and whole organ decellularization processes. *Biomaterials* **2011**, *32*, 3233–3243. [CrossRef] [PubMed]
23. Horke, A.; Tudorache, I.; Laufer, G.; Andreas, M.; Pomar, J.L.; Pereda, D.; Quintana, E.; Sitges, M.; Meyns, B.; Rega, F.; et al. Early results from a prospective, single-arm European trial on decellularized allografts for aortic valve replacement: The ARISE study and ARISE registry data. *Eur. J. Cardio-Thorac. Surg.* **2020**, *58*, 1045–1053. [CrossRef] [PubMed]
24. Simon, P.; Kasimir, M.T.; Seebacher, G.; Weigel, G.; Ullrich, R.; Salzer-Muhar, U.; Rieder, E.; Wolner, E. Early failure of the tissue engineered porcine heart valve SYNERGRAFT™ in pediatric patients. *Eur. J. Cardio-Thorac. Surg.* **2003**, *23*, 1002–1006. [CrossRef]
25. Baraki, H.; Tudorache, I.; Braun, M.; Höffler, K.; Görler, A.; Lichtenberg, A.; Bara, C.; Calistru, A.; Brandes, G.; Hewicker-Trautwein, M.; et al. Orthotopic replacement of the aortic valve with decellularized allograft in a sheep model. *Biomaterials* **2009**, *30*, 6240–6246. [CrossRef] [PubMed]
26. Tudorache, I.; Theodoridis, K.; Baraki, H.; Sarikouch, S.; Bara, C.; Meyer, T.; Höffler, K.; Hartung, D.; Hilfiker, A.; Haverich, A.; et al. Decellularized aortic allografts versus pulmonary autografts for aortic valve replacement in the growing sheep model: Haemodynamic and morphological results at 20 months after implantation. *Eur. J. Cardio-Thorac. Surg.* **2016**, *49*, 1228–1238. [CrossRef]
27. Da Costa, F.D.A.; Costa, A.C.B.A.; Prestes, R.; Domanski, A.C.; Balbi, E.M.; Ferreira, A.D.A.; Lopes, S.V. The early and midterm function of decellularized aortic valve allografts. *Ann. Thorac. Surg.* **2010**, *90*, 1854–1860. [CrossRef] [PubMed]

Article

Bioactive Low Molecular Weight Keratin Hydrolysates for Improving Skin Wound Healing

Laura Olariu [1,2], Brindusa Georgiana Dumitriu [1], Carmen Gaidau [3,*], Maria Stanca [3], Luiza Mariana Tanase [1], Manuela Diana Ene [1], Ioana-Rodica Stanculescu [4,5], and Cristina Tablet [5,6]

[1] SC Biotehnos SA, 3–5 Gorunului Street, 075100 Otopeni, Romania; lolariu@biotehnos.com (L.O.); dbrandusa@biotehnos.com (B.G.D.); luiza.craciun@biotehnos.com (L.M.T.); diana.ene@biotehnos.com (M.D.E.)
[2] Academy of Romanian Scientists, 3 Ilfov Street, 030167 Bucharest, Romania
[3] Leather Research Department, National Institute for Textiles and Leather Division Leather and Footwear Research Institute (ICPI), 93 Ion Minulescu Street, 031215 Bucharest, Romania; maria.stanca@icpi.ro
[4] Horia Hulubei National Institute of Research and Development for Physics and Nuclear Engineering, 30 Reactorului Str., 077125 Magurele, Romania; istanculescu@nipne.ro
[5] Department of Physical Chemistry, University of Bucharest, 4–12 Regina Elisabeta Bd., 030018 Bucharest, Romania; cristinatablet@yahoo.com
[6] Faculty of Pharmacy, Titu Maiorescu University, 16 Gh. Sincai Bd., 040317 Bucharest, Romania
* Correspondence: carmen.gaidau@icpi.ro

Citation: Olariu, L.; Dumitriu, B.G.; Gaidau, C.; Stanca, M.; Tanase, L.M.; Ene, M.D.; Stanculescu, I.-R.; Tablet, C. Bioactive Low Molecular Weight Keratin Hydrolysates for Improving Skin Wound Healing. *Polymers* 2022, 14, 1125. https://doi.org/10.3390/polym14061125

Academic Editors: Antonia Ressler, Inga Urlic and Jianxun Ding

Received: 20 January 2022
Accepted: 8 March 2022
Published: 11 March 2022

Publisher's Note: MDPI stays neutral with regard to jurisdictional claims in published maps and institutional affiliations.

Copyright: © 2022 by the authors. Licensee MDPI, Basel, Switzerland. This article is an open access article distributed under the terms and conditions of the Creative Commons Attribution (CC BY) license (https://creativecommons.org/licenses/by/4.0/).

Abstract: Keratin biomaterials with high molecular weights were intensively investigated but few are marketed due to complex methods of extraction and preparation and limited understanding of their influence on cells behavior. In this context the aim of this research was to elucidate decisive molecular factors for skin homeostasis restoration induced by two low molecular weight keratin hydrolysates extracted and conditioned through a simple and green method. Two keratin hydrolysates with molecular weights of 3758 and 12,400 Da were physico-chemically characterized and their structure was assessed by circular dichroism (CD) and FTIR spectroscopy in view of bioactive potential identification. Other investigations were focused on several molecular factors: $\alpha 1$, $\alpha 2$ and $\beta 1$ integrin mediated signals, cell cycle progression in pro-inflammatory conditions (TNFα/LPS stimulated keratinocytes and fibroblasts) and ICAM-1/VCAM-1 inhibition in human vascular endothelial cells. Flow cytometry techniques demonstrated a distinctive pattern of efficacy: keratin hydrolysates over-expressed $\alpha 1$ and $\alpha 2$ subunits, responsible for tight bounds between fibroblasts and collagen or laminin 1; both actives stimulated the epidermal turn-over and inhibited VCAM over-expression in pro-inflammatory conditions associated with bacterial infections. Our results offer mechanistic insights in wound healing signaling factors modulated by the two low molecular weight keratin hydrolysates which still preserve bioactive secondary structure.

Keywords: keratin hydrolysate; bioactive keratin; skin homeostasis restoration; skin wound healing

1. Introduction

Keratin can be extracted from keratin-rich materials using chemical (reduction, oxidation, hydrolysis, sulphitolysis), physical (steam explosion, microwave irradiation) or biological methods [1]. The properties of keratin such as amino acid content, molecular weight, thermal behavior and bioactivity depend on the extraction and preparation methods [2,3]. Biomaterials prepared from keratin extracted from wool or human hair showed excellent properties regarding biocompatibility and cellular proliferation abilities. These properties make wool keratin an excellent material for tissue engineering and drug delivery systems [4]. In vitro studies have shown that keratin has the ability to scavenge free radicals, similar to that of vitamin C and allows replacing cosmetic preservatives [5].

A study on the anti-inflammatory ability of keratin extracted from human hair showed that it has a more pronounced anti-inflammatory effect on monocytic cell line, than collagen or hyaluronic acid. The study showed that primary macrophage cells are altered when they are exposed to an immobilized keratin biomaterial surface and that these changes appear to target an anti-inflammatory phenotype. This phenomenon appears to be a function of the lower molecular weight of keratin, even if it is recognized that the mechanism of anti-inflammatory process is not fully understood [6]. Tachibana et al. showed also that due to the amino acid sequence with more carboxylic terminal groups, keratin is more effective than collagen in binding osteoblast cells and inducing osteogenesis [7]. Biomaterials based on keratin contain regulatory molecules that make them capable of functioning as synthetic extracellular matrices (ECM) and promoting nerve tissue regeneration due to their capacity to create fibronectin-like cell binding that facilitates cell adhesion [8,9]. Gao et al. conducted in vivo and in vitro studies on the ability of keratin to promote the regeneration of peripheral nerves. The in vitro study showed that it has neuroinducible activity and Schwann cells grown in keratin medium have a number of changes in terms of migration capacity, morphology and proliferation activity, with stimulating effect on neuronal axons extension. In vivo experiments have shown that keratin accelerates the regeneration of the axon in the early stages and has a beneficial effect for subsequent functional recovery [10].

Human keratin hair hydrogel efficiency for treatment of thermal or chemical wounds was proved by faster closure and no increase in wound size in the first four days of healing [11]. In the field of skin regeneration, there are several mechanisms working together to restore dramatically affected molecular processes. For example, cellular function is regulated during dermal repair flow by critical interactions between receptors and extracellular matrix proteins. Also, secretory immune cells release cytokines and growth factors (IL-1, IL-4, IL-6, IL-13, TGF-β, TNF) interrelated with structural proteins synthesis [12]. This kind of interactions leads to regenerative processes' activation, including angiogenesis and scar remodeling [13,14]. Integrins are heterodimeric proteins from plasmatic membrane, involved in cellular responses to growth factors, and direct regulation of transcriptional programs, signaling cascades and activation mechanisms [15]. Over-expression of specific integrins directs wound healing, fibrosis and scarring. They are transmembrane structures with α and β subunits assembled to form functional receptors: $\alpha 3\beta 1$ (receptor for laminin 332), $\alpha 6\beta 5$ (hemidesmosome component, receptor for laminin 332), $\alpha 2\beta 1$ (receptor for collagen and laminin, having a pivotal role for wound healing in vivo), $\alpha v\beta 5$ (receptor for vitronectin); $\alpha 9\beta 1$ (receptor for tenascin C) [16,17]. A correlation was reported between the level of $\beta 1$ integrins and proliferation in the interfollicular epidermis [18]. Integrins are also required for an essential step of skin regeneration, namely fibroblast infiltration into the wound clot. Normal fibroblasts and granulation tissue fibroblasts express many types of integrins: $\alpha 1\beta 1$, $\alpha 2\beta 1$, $\alpha 3\beta 1$, $\alpha 5\beta 1$, $\alpha 11\beta 1$, $\alpha v\beta 1$, $\alpha v\beta 3$, and $\alpha v\beta 5$, in order to bind collagens, fibronectins and other blood clot components. For example, fibroblasts interact with fibrillar collagens via $\alpha 1\beta 1$, $\alpha 2\beta 1$, and $\alpha 11\beta 1$ integrins, regulating matrix metalloproteinase (MMP) expression and collagen fibrillogenesis [19]. Another event accompanying skin injuries is inflammation, including cytokines (IL6, IL8, TNFα) signals progression and cellular factors over-expression. Key elements for vascular endothelial inflammation are proteins from CAM family, especially ICAM (intercellular adhesion molecule) and VCAM (vascular cell adhesion molecule) [20]. They play a different role, namely: ICAM-1 specifically participates in trafficking of inflammatory cells, in leukocyte effectors' functions, in adhesion of antigen-presenting cells to T lymphocytes, in microbial pathogenesis, and in signal transduction pathways through outside-in signaling events [21], and VCAM-1 that is one of the major regulators of leukocyte adhesion and transendothelial migration by interacting with $\alpha 4\beta 1$ integration, which in turn activates intracellular signaling that allows transendothelial migration of leukocytes [22].

In our paper we hypothesized that the wool keratin, an easily available byproduct, can be tuned by chemical-enzymatic hydrolyses in different keratin biomolecules so as to stimulate the main skin cell biochemical mechanisms for wound healing. As in many reported research studies, the human hair keratin was extracted and solubilized through complex reduction or oxidizing methods and prepared in hydrogel forms mainly with high molecular weights [23], the present research proposes alkaline-enzymatic hydrolysates with lower molecular weights, prepared by complete wool solubilization [24] and with preserved bioactive structured molecules.

It is recognized that few commercial products based on keratin reached the market of wound healing biomaterials as compared to other materials due to the complex methods of preparation and the insufficient understanding of cellular interaction with keratin [25,26]. In this regard the present research proposes keratin hydrolysates prepared by a green, reproducible and easy method of solubilization [24] and brings evidences in several molecular factors, decisive for restoring the skin homeostasis as new potential wound dressing active component.

If different keratin dressings with high molecular weights were recognized as activators of keratinocytes wound healing process [27], the present research brings mechanistic insights from damaged skin cells restitution processes induced by two low molecular bioactive keratin hydrolysates with still structured peptides.

Thus, the stimulation effect on β1 glycoprotein expression in pro-inflammatory conditions and α1 and α2 subunits generation in the basal state of fibroblasts were proved by low molecular keratins. The other anti-inflammatory factors involved in wound healing process, like significant down regulation of nonspecific stimulation—TNFα + PMA, and the ICAM expression were also influenced.

According to our knowledge no similar study was performed regarding the influence of low molecular wool keratin hydrolysates on skin cells homeostasis factors.

2. Materials and Methods

2.1. Materials

Raw wool was purchased from a local sheep farmer (Lumina, Constanta, Romania). Chemical reagents of analytical grade like sodium hydroxide (98%), ammonia (25%), sodium carbonate (99.7%), formic acid (85%) and sulfuric acid (92%) were purchased from Chimopar Trading SRL (Bucharest, Romania). Borron SE (ethoxylated alkyl derivatives with 65% concentration) was supplied by SC Triderma SRL (Bucharest, Romania). Esperase® 8.0 L, a serine endo-peptidase from *Bacillus lentus* with activity of 8 KNPU-E/g, working at elevated temperature and pH = 8–12.5, was purchased from Novozymes (Atasehir, Turkey). Valkerase®, a keratinase, serine protease from *B. licheniformis*, with activity of 80,549 U/g at pH = 5.5 and 55 °C, was supplied by BioResource International (Durham, NC, USA).

2.2. Keratin Hydrolysates Preparation

The alkaline hydrolysate was prepared by a method previously described [24]. Briefly, to obtain the alkaline hydrolysate, wool was pretreated by washing and degreasing using 4% w/w NH_4OH, 0.6% w/w Borron SE and 1% w/w Na_2CO_3 for 2 h at 40 °C, rinsed to neutral pH and minced with a bench grinder machine (La Minerva, Minerva Omega, Bologna, Italy). After this treatment, wool was mixed with 2.5% w/v sodium hydroxide solution in a stainless-steel vessel equipped with a mechanical stirrer and automatic temperature control (SC Caloris SA, Bucharest, Romania) at 80 °C, for 4 h. The enzymatic keratin hydrolysates were obtained from alkaline hydrolysate by enzymatic hydrolysis using 1% w/w concentration of two enzymes: Esperase or Valkerase under optimum activity conditions for 4 h. The enzymatic keratin hydrolysates were conditioned at pH = 7 with sulfuric acid. The enzymatic keratin hydrolysates were labeled Ker 1 and Ker 2. Keratin hydrolysates were centrifuged for 15 min at 6000 rpm (Eppendorf 5804, Wien, Austria) and lyophilized by freeze-drying of keratin dispersions in a DELTA 2-24 LSC Freeze-dryer, laboratory scale (Osterode am Harz, Germany).

2.3. Keratin Hydrolysates Characterisation

Physical-chemical characteristics were evaluated according to the standard in force or in house methods for: dry substance (SR EN ISO 4684:2006), ash (SR EN ISO 4047:2002), total nitrogen and protein content (SR EN ISO 5397:1996), aminic nitrogen (ICPI method), cysteine (SR 13208:1994), cystine sulphur (SR13208:1994) content, and pH (STAS 8619/3:1990). The results were expressed as the average of triplicate determinations with standard deviations. The molecular weights were determined by gel permeation chromatography (GPC) using an Agilent Technologies instrument (1260 model) (Agilent Technologies, Santa Clara, CA, USA) equipped with PL aqua gel-OH MIXED-H column (7.5 × 300 mm × 8 μm) and multi-detection unit. Optimum working conditions for GPC were flow rate of mobile phase containing 1 mL min^{-1}, injection volume of the sample 100 μL, and temperature of 35 °C for the detectors and column. Calculations of the Mw and number average molecular weight (Mn) were performed with the Agilent GPC/SECS software (Version 1.1, Agilent Technologies, Santa Clara, CA, USA). The free amino acid analysis was performed using a GC-MS (Thermo Scientific TRACE1310/TSQ 8000Evo, Waltham, MA, USA). Samples were derivatized with acetonitrile and BSTFA at 105 °C. The GC separation was achieved using a capillary column TG-5SILMS (5% diphenyl/95% dimethyl polysiloxane), 30 m × 0.25 mm × 0.25 μm. The operating conditions for GC were an initial temperature of 100 °C raised to 170 °C then to 190 °C (3 °C/min) and the final temperature was 280 °C. The total run time was 40 min. The temperature for the transfer line was set at 280 °C and the ionization source at 230 °C. Injection volume was 0.5 μL. The MS detector was operated in continuous scan mode with the m/z interval ranging from 40 to 600 amu. The data were analyzed using the Chromeleon v.7.2.7 software. For quantification of amino acids, a calibration curve was used.

Circular dichroism spectra were measured with a Jasco J-815 spectropolarimeter (Cremella, Italy) in 190–250 nm range, at room temperature in a 0.05 cm optical path quartz cuvette. The experimental parameters were set as follows: 1 nm bandwidth, 1 nm data pitch, 4 s response, 50 nm/min scanning speed, 3 accumulations. The sample solutions were prepared in a pH 7.4-buffer phosphate by diluting the stock solutions to 30 μM for Ker 1 and 121 μM for Ker 2. The normalized root mean square error deviations of deconvolutions were 0.02158 for Ker 1 and 0.02603 for Ker 2.

FTIR spectra were acquired in the 4000–400 cm^{-1} spectral range with a Bruker Vertex 70 instrument (Bruker, Ettlingen, Germany) with the following working parameters: 4 cm^{-1} resolution, 0.1 cm^{-1} wavenumber accuracy, 0.1% T photometric accuracy and 64 scans. A KBr beam splitter and a RTDLaTGS (Room Temperature Deuterated Lanthanum α Alanine doped TriGlycine Sulphate) detector were used. The measurements were performed by the technique of pastillation in KBr of spectroscopic purity (Merck reagent) at a KBr: sample weight ratio of 300:1 toward a 300 mg KBr reference disc. The discs were prepared by pressing at 10 t/cm^2, under vacuum, the fine powder obtained after grinding for 5 min in an agate mortar. Spectra were processed with the OPUS software (Bruker, Ettlingen, Germany) using atmospheric compensation, vector normalization, baseline correction with straight lines and one iteration additional concave rubber band correction and automatic peak peaking. The deconvolution parameters were settled in the spectral range of 1230–1330 cm^{-1} using Gaussian functions generated by ORIGIN 9 software. FTIR deconvolution data residual RMS error was less than 0.001.

The morphology of lyophilized enzymatic keratin hydrolysates was analyzed by scanning electron microscopy with a FEI Quanta 200 Scanning Electron Microscope (FEI, Eindhoven, The Netherlands) with a gaseous secondary electron GSED detector at an accelerating voltage of 12.5–20 kV.

2.4. Biocompatibility Tests

To establish the dose effect relation of lyophilized keratins, as well as the cytocompatibility conditions, standardized cell lines specific to the dermo-epidermal layer and vascular endothelium were used as follows: Human dermal Fibroblast (normal cell line HS27—ATCC® CRL-1634™) cultivated in DMEM (Dulbecco's Modified Eagle's Medium /Nutrient Mixture F-12 Ham, Sigma-Aldrich (Saint Louis, MO, USA) supplemented with 10% fetal bovine serum (Sigma-Aldrich) and 1% Antibiotic Antimycotic Solution (100×)

(Sigma Aldrich); Normal Human Keratinocytes (immortalized cell line HaCaT purchased from ThermoFisher Scientific) cultured in Dulbecco's Modified Eagle's Medium (DMEM) (ATCC®), with high content of glucose, supplemented with 10% fetal bovine serum and 1% Antibiotic Antimycotic Solution (100×) purchased from Sigma Aldrich; Human Umbilical Vein Endothelial Cells (HUVEC—CRL-1730™) cultured in RPMI-1640 media, supplemented with 10% fetal bovine serum, 1% L-glutamine, and 1% Antibiotic Antimycotic Solution (100×) purchased from Sigma-Aldrich. All cell lines were cultured or incubated at 37 °C with 5% CO_2.

Determination of the cytotoxic profile was performed correlating cell viability monitored by intracellular esterase activity (labeling with tetrazolium salt MTS using the CellTiter 96® AQueous One Solution Cell Proliferation Assay (Promega)) with release of lactate dehydrogenase in culture medium due to affected cell membrane permeability (LDH test using the CytoTox 96® Non-Radioactive Cytotoxicity Assay Kit (Promega)). The cells were allowed to adhere for 24 h (7000 cells/well) and treated for 48 h with the test substances. The staining was done according to the protocol for the specific reagent kit (MTS/LDH). All concentrations of active principle were tested in triplicate and the data were acquired using an ELISA TriStar 941 multimode plate reader from Berthold Technologies (Bad Wildbad, Germany).

For the specific, wound-healing activity investigation, several experimental series were processed, in the following cycle of cultivation: 24 h for adhesion + 48 h for physiological development and incubation with tested compounds. Some cells were grown under normal conditions, other cells were activated overnight either with 15 ng/mL TNFα and 0.1 µM PMA (phorbol myristate acetate), to mimic acute inflammation, associated with pro-oxidative conditions, or with 1 µM suspension of LPS (bacteria lipopolysaccharide) to mimic bacterial infection.

Cell cycle sequential analysis was performed by flow cytometry according to Cycle Test Plus DNA Reagent kit (BD Pharmingen) protocol, which involves dissolving the cell membrane lipids with a nonionic detergent, removing the cytoskeleton and nuclear proteins with trypsin, enzymatic digestion of cellular RNA and stabilization of nuclear chromatin with spermine. Propidium iodide is stoichiometrically bound to purified DNA and the complex formed is analyzed by flow cytometry [28].

The integrin detection technique by flow cytometry involves the use of monoclonal antibodies for α and β chains from BD Pharmingen (CD49a, fluorescent labeled for PE—corresponding to α2 integrin; CD49b, fluorescent labeled for FITC, corresponding to α1 integrin; and CD 29 fluorescent labeled for APC, corresponding to integrin β1). Results were compared with the specific isotype control (PE Mouse IgG1, k Isotype control, APC Mouse IgG1, k Isotype control, FITC Mouse IgG1, k Isotype control). Analysis of the vascular anti-inflammatory effect was performed by simultaneous fluorescent labeling with the appropriate antibodies for ICAM-1 and VCAM-1, respectively. The cell suspension is labeled with APC Mouse Anti-Human CD54 for ICAM-1 (intracellular-adhesion molecules) and with PE Mouse Anti-Human CD106 for highlighting VCAM-1 (vascular-cell-adhesion molecules). Flow cytometry results concerning membrane molecules stained with fluorescent antibodies were compared with the specific isotype controls (PE Mouse IgG1, k Isotype control, APC Mouse IgG1, k Isotype control, FITC Mouse IgG1, k Isotype control).

Acquisition and analysis were done with FACS CANTO II (Becton-Dickinson) and FCS Express software—DNA cell cycle module and DIVA 6.1 software. Figure 1 presents an example of flow-cytometry diagram concerning fluorescent antibody staining of membrane proteins.

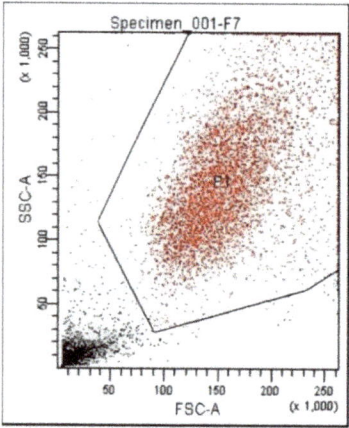 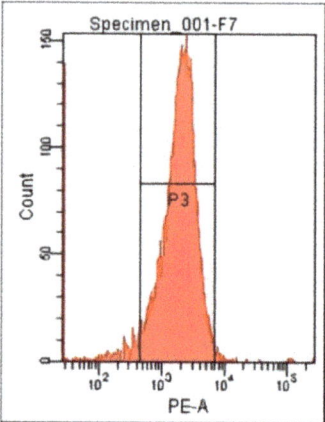 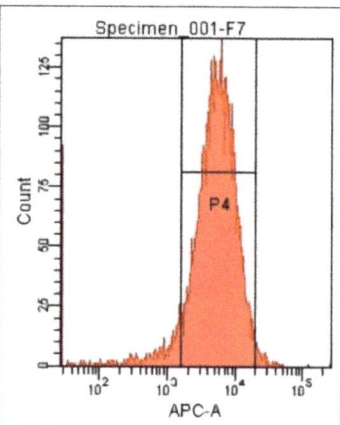

Figure 1. Dot-plot of cellular population and histograms corresponding to ICAM-1 (APC-A) and VCAM-1 (PE-A) signals.

2.5. Statistical Analysis

The statistical processing was done using analysis of variance (ANOVA) (95% significant level) on each pair of interest and differences at $p < 0.05$ were considered statistically significant.

3. Results

3.1. Keratin Hydrolysates Characteristics

The main physical-chemical characteristics of the two keratin hydrolysates in powder form are presented in Table 1. It can be seen that Ker 1 has lower molecular weight in correlation with higher aminic nitrogen concentration and lower cysteine and cystinic sulphur due to the higher breaking degree of keratin molecules. The polydispersity and free aminoacids concentrations are very similar. Molecular weights of 10,000–30,000 Da are known to be considered low molecular weights for keratin extracts [29] and according to our knowledge, the behaviour of keratin extracts under these values related to their structure and properties in interaction with different cell lines was not reported.

Table 1. Physical-chemical characteristics of lyophilized keratin hydrolysates.

Characterization	Samples	
	Ker 1	Ker 2
Dry substance, %	96.91 ± 0.35	94.95 ± 0.32
Total ash, %	15.07 ± 0.24	14.26 ± 0.20
Total nitrogen, %	12.38 ± 0.34	11.68 ± 0.30
Protein substance, %	75.04 ± 0.34	70.78 ± 0.32
pH, pH units	7.27 ± 0.10	6.80 ± 0.10
Aminic nitrogen, %	0.89 ± 0.05	0.92 ± 0.05
Cysteine, %	9.03 ± 0.03	7.88 ± 0.03
Cystinic sulphur,%	2.41 ± 0.03	2.10 ± 0.03
Mw, Da	12,400	3758
Polydispersity index	1	1.05
Free amino acids,%	4.9	4.3

The morphology of lyophilized keratins is presented in Figure 2 and showed that the powders have a porous, interconnected structure.

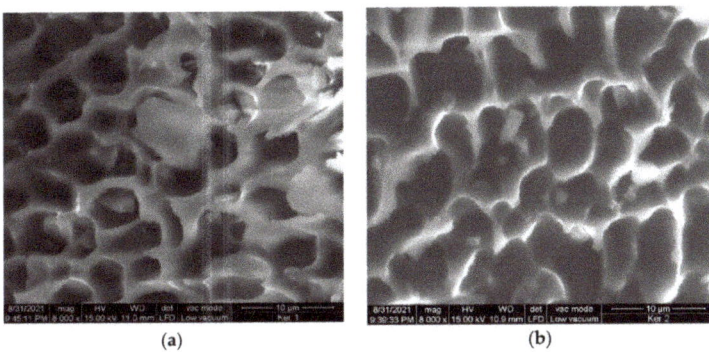

Figure 2. SEM images of lyophilized keratins: (**a**) Ker 1 and (**b**) Ker 2.

3.2. Circular Dichroism Spectroscopy Analyses (CD)

Circular dichroism (CD) is widely used to gain information about a protein's secondary structure. Thus, the presence of the α-helix in the protein structure is reflected in the CD spectra by the existence of three bands: a negative band at 208 nm and a positive one at ~195 nm due to π-π* transitions and a 220 nm negative band due to n-π* transitions. A quick look at the spectra from Figure 3 suggests low α-helix content for both samples. For a quantitative analysis of the secondary structure, we have used the BeStSel webserver [30,31]. The analysis results are displayed in Table 2.

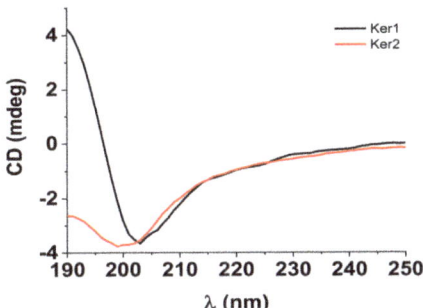

Figure 3. Circular dichroism spectra of lyophilized keratins.

Table 2. Secondary structure percent composition obtained with BeStSel webserver for keratin samples.

Sample	α-Helix	ß Sheets		Turn	Others
		Antiparallel	Parallel		
		%			
Ker 1	5.8	27.5	0.0	15.4	51.3
Ker 2	4.7	25.5	7.4	15.2	47.2

3.3. Fourier Transform Infrared Spectroscopy (FTIR) Analysis

Due to the variation of the protein content, molecular weight and conformation of Ker 1 and Ker 2, important changes can be observed in the FTIR spectra highlighted in the Figure 4a, regarding the main protein bands, Amide A, Amide I, Amide II and Amide III, their shape and intensity [32].

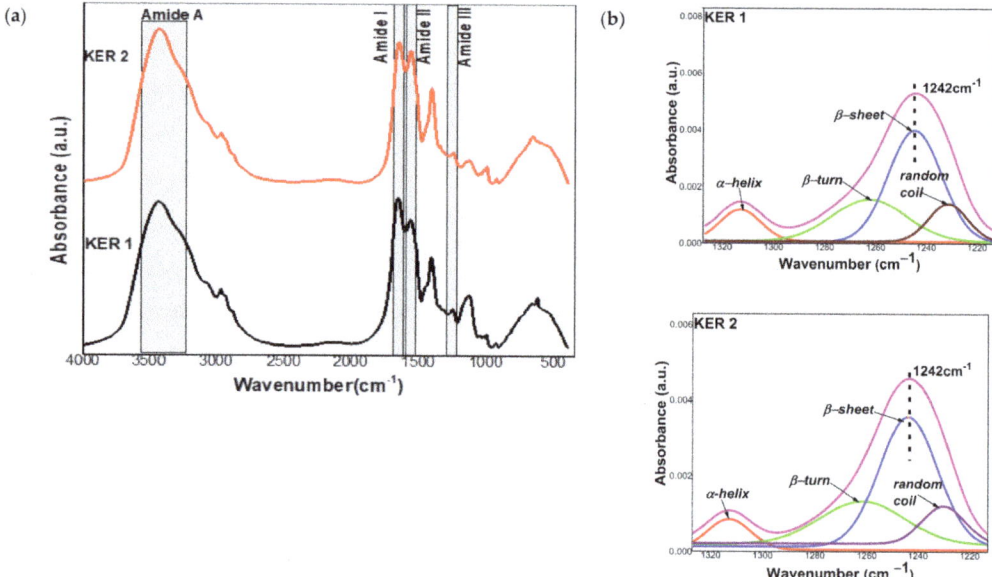

Figure 4. (a) FTIR spectra of Ker1 and Ker2; (b) Amide III band deconvolution of Ker1 and Ker 2.

The Amide I band is due to the stretching vibrations of the C=O and C–N groups in the peptide bonds and is correlated with the backbone conformation. The Amide II band is associated with N–H plane deformation vibrations and C–N and C–C tensile vibrations and is conformationally sensitive. The Amide I/Amide II increased ratio of Ker 1 is correlated with an increase in total ordered structures content as seen in Table 3 [33,34]. Ker 1 has an intense band between 1105 and 1140 cm^{-1} which indicates the presence of oxidized cysteine and Bunte salt [32]. The Amide A band is correlated with N–H stretching vibrations and is an indicator of the hydrogen bonds formed between polypeptide chains leading to twisted sheet structures. The shift of the Amide A band of Ker2 to a higher wavenumber indicates a weaker hydrogen bond network in correlation with lower molecular weight (Table 1). Data from the literature show that the deconvolution of the Amide III band (Figure 4b) allows the reliable quantification of the secondary structures using the characteristic peaks of α-helix (1330–1295 cm^{-1}), β-sheet (1250–1220 cm^{-1}), β-turn (1295–1270 cm^{-1}) and random coil (1270–1250 cm^{-1}) [35,36]. Thus, the secondary structure of the Ker 1 additive consists of 49.8% ordered structures, while the Ker 2 additive has a slightly lower proportion of ordered structures, respectively 46.2% (Table 3).

Table 3. Secondary structure percent composition obtained from deconvolution of Amide III band of keratin powders.

Sample	α-Helix	ß Sheets	ß Turn	Random Coil
		%		
Ker 1	11.0	12.3	26.5	50.2
Ker 2	9.2	11.2	25.8	53.8

3.4. Cytotoxicity Determination

The effect of keratin hydrolysates on three cell lines viability (MTS test) and their cytotoxicity (LDH test), respectively, was plotted as the ratio of test sample concentrations to the corresponding control and is presented in Figure 5.

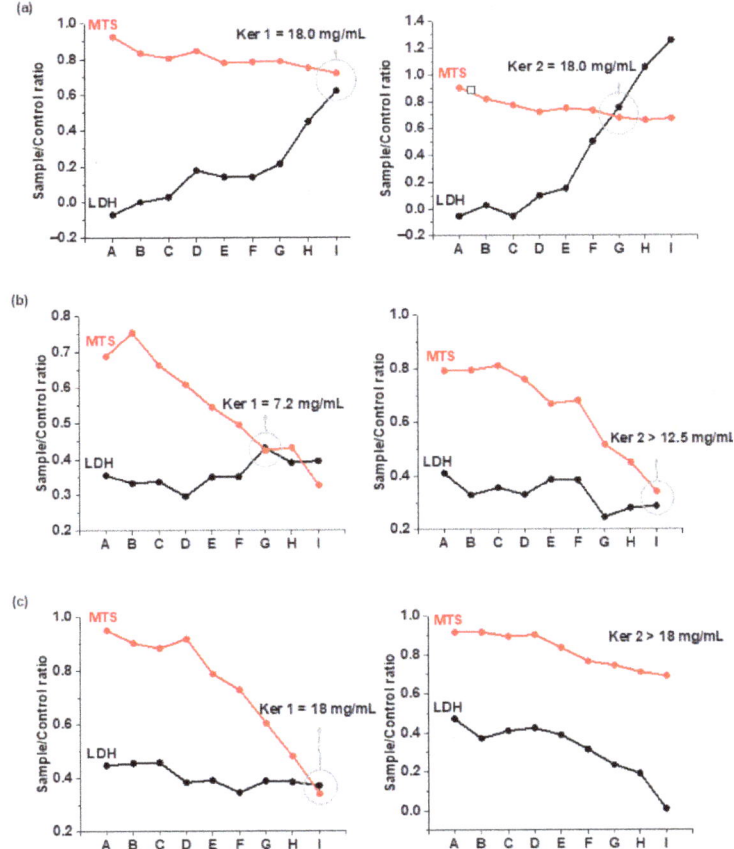

Figure 5. Cytotoxic profile of Ker 1 and Ker 2 on: (**a**) dermal fibroblasts (HS27); (**b**) keratinocytes (HaCaT); (**c**) endothelial cells (HUVEC). The sample concentrations were: A: 0.6 mg/mL, B: 3.0 mg/mL, C: 6.0 mg/mL, D: 9.0 mg/mL, E: 10.2 mg/mL, F: 12.0 mg/mL, G: 13.2 mg/mL, H: 15.0 mg/mL, I: 18.0 mg/mL.

In Figure 5 can be seen that the intersection point between the curves of the two analyzed parameters (LDH and MTS), represents the dose of the signal for the amount of lactate dehydrogenase released and exceeds the signal for the amount of MTS transformed into formazan, being equivalent to the dose at which cell viability is significantly affected. In Table 4 the maximum doses allowed for Ker 1 and Ker 2 are presented for different cell lines. It can be seen that Ker 2 has higher maximum doses for two cell lines as compared to Ker 1 which can be attributed to lower molecular weight with more available carboxylic groups [7].

Table 4. The maximum dose allowed for Ker 1 and Ker 2 depending on the cell line tested.

No. crt	Sample	Cell Lines, mg mL^{-1}		
		HS27	HaCaT	HUVEC
1	Ker 1	18	7.5	18
2	Ker 2	13.2	>12	>18

3.5. Specific Activity of Lyophilized Keratins on Epidermal, Dermal and Vascular Endothelial Cells

In the healing of skin wounds the combination of the two mechanisms concurred: regeneration (affected cells are "de novo" replaced) and repair (the scar tissue reconstruction). The process by which fibroblasts and keratinocytes migrate to the site of the wound is replicated by mitosis to form a thin film of cells interconnected by adhesion molecules, growth factors and structural proteins, playing a key role in proper healing, as we previously described. This study's investigations focused on the analysis of several molecular factors with a decisive role in restoring skin homeostasis: the dermo-epidermic proliferative status on HS27 and HaCaT cells, α1, α2 and β1 integrin mediated signals, and ICAM-1/VCAM-1 inhibition in human vascular endothelial cells (HUVEC). In order to simulate the inflammation associated with skin lesions, the unstimulated cells were compared with the pro-inflammatory state induced by TNFα + PMA or LPS stimulation, respectively. The positive control was considered dexamethasone, a potent anti-inflammatory drug. Results regarding the proliferative status of keratinocytes and fibroblasts, expressed as cell cycle progression are presented in Tables 5 and 6 and Figure 6.

Table 5. HS27 proliferative status expressed as cell cycle progression in normal and pro-inflammatory conditions.

Tested Material	Unstimulated, %			TNF-α +PMA Stimulated,%		
	G0/G1	S	G2/M	G0/G1	S	G2/M
Control cells	69.22 ± 4.8	2.57 ± 0.2	29.69 ± 3.03	55.64 ± 7.4	18.99 ± 2.1	25.34 ± 5.3
Ker 1 (0.5 mg/mL)	70.20 ± 3.9 ***	2.10 ± 0.1 ***	27.70 ± 4.01	57.13 ± 4.8 ***	15.38 ± 3.7 ***	27.48 ± 1.1 **
Ker 2 (0.5 mg/mL)	70.55 ± 5.3 ***	2.60 ± 0.9 ***	26.18 ± 7.3 **	54.09 ± 1.1 ***	18.87 ± 1.4 **	27.03 ± 2.6 **
Dexamethasone	63.37 ± 1.5	5.90 ± 1.0	30.71 ± 2.5	48.40 ± 1.7	27.30 ± 1.3	24.27 ± 0.3

** $p < 0.01$, *** $p < 0.001$, using Repeated Measures ANOVA, Dunnett's Multiple Comparison Test.

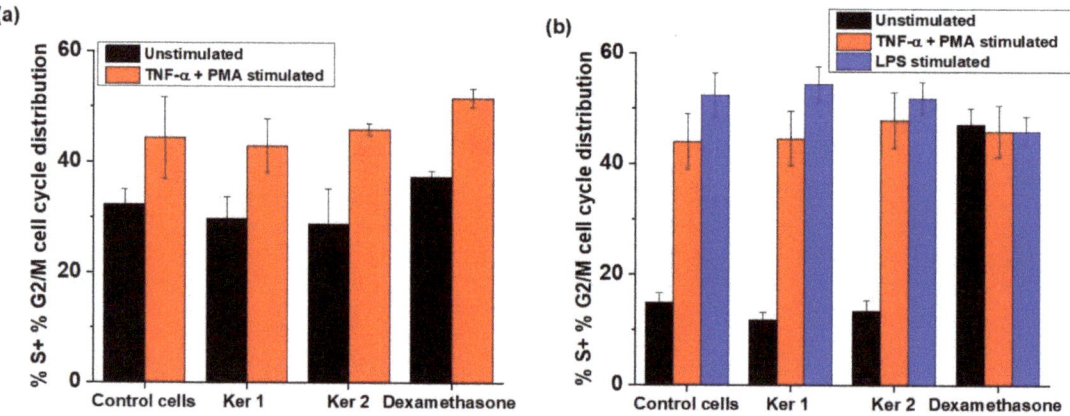

Figure 6. Mitotic phases' evolution modulated by the lyophilized keratins: (**a**) HS27, (**b**) HaCaT (differences between cells treated with Ker1, Ker2 and positive control were done using ANOVA test— 95% significant level on each pair of interest and differences at $p < 0.05$ were considered statistically significant).

Table 6. HaCaT proliferative status expressed as cell cycle progression in normal and pro-inflammatory conditions.

Tested Material	Unstimulated, %		
	G0/G1	S	G2/M
Control cells	49.03 ± 2.3	4.86 ± 0.12	10.12 ± 1.01
Ker 1 (0.5 mg/mL)	48.14 ± 3.1	2.05 ± 0.6 ****	9.81 ± 0.8 ****
Ker 2 (0.5 mg/mL)	46.49 ± 1.6	3.42 ± 0.3 ****	10.09 ± 0.4 ****
Dexamethasone	52.93 ± 4.2	31.7 ± 0.4	15.37 ± 0.9
	TNFα +PMA stimulated,%		
	G0/G1	S	G2/M
Control cells	56 ± 0.92	32.59 ± 1.05	11.42 ± 0.12
Ker 1 (0.5 mg/mL)	55.4 ± 1.1	33.99 ± 0.9 ****	10.60 ± 0.6 **
Ker 2 (0.5 mg/mL)	52.05 ± 1.3	34.86 ± 1.9 **	13.09 ± 0.9
Dexamethasone	54.05 ± 1.6	30.99 ± 1.2	14.95 ± 0.7
	LPS stimulated, %		
	G0/G1	S	G2/M
Control cells	47.59 ± 4.7	39.56 ± 2.09	12.85 ± 1.6
Ker 1 (0.5 mg/mL)	45.53 ± 5.03 ***	43.02 ± 3.21 ****	11.44 ± 0.9
Ker 2 (0.5 mg/mL)	48.14 ± 3.21	41.90 ± 1.23 ****	9.92 ± 1.06
Dexamethasone	54.08 ± 4.78	34.47 ± 2.21	11.46 ± 0.75

** $p < 0.01$, *** $p < 0.001$, **** $p < 0.0001$, using Repeated Measures ANOVA, Dunnett's Multiple Comparison Test.

The fibroblasts' mitotic phases' evolution has no significant change after keratin hydrolysates' treatment, but Ker 1 and Ker 2 increased the keratinocytes' percent of nuclei in S-phase (DNA replication), Ker 2 being more active than Ker 1.

Cell adhesion, essential for tissue repair process and mediated by specific interactions of integrins with extracellular matrix, was investigated through membrane over-expression of α1, α2 and β1 integrins in fibroblasts. Flow cytometry data presented in Table 7 and Figure 7 revealed a significant contribution of both types of keratins on stimulation of β1 glycoprotein expression in pro-inflammatory conditions, and higher presence of α1 and α2 subunits in the basal state of fibroblasts induced by Ker1 and Ker2. The influence of low molecular keratin on cell communication and function was revealed for hair keratin gel in the first stage of wound healing process when stronger regulatory ability was highlighted [29]. The same study stated the relation between keratin molecular weight and wound healing properties is still unknown.

Cell adhesion, essential for tissue repair process and mediated by specific interactions of integrins with extracellular matrix, was investigated through membrane over-expression of α1, α2 and β1 integrins in fibroblasts. Flow cytometry data presented in Table 7 and Figure 7 revealed a significant contribution of both types of keratins on stimulation of β1 glycoprotein expression in pro-inflammatory conditions, and higher presence of α1 and α2 subunits in the basal state of fibroblasts induced by Ker 1 and Ker 2.

Table 7. Membrane-specific fluorescence quantified through flow-cytometry showing integrins α1, α2 and β1expressed by fibroblast cells (HS27).

Tested Material	Integrin α1 (FITC-Mean-Relative Fluorescence Units)		Integrin α2 (PE-Mean-Relative Fluorescence Units)		Integrin β1 (APC-Mean-Relative Fluorescence Units)	
	Unstimulated	TNF α + PMA	Unstimulated	TNF α + PMA	Unstimulated	TNF α + PMA
Control cells	6611.0 ± 70.8	11,227.3 ± 716.2	2723.7 ± 215.5	12,599.3 ± 890.6	4831.7 ± 677.3	5427.3 ± 377.9
Ker 1 (0.5 mg/mL)	9601.0 ± 112.5 **	12553.0 ± 273.4 ***	3507.7 ± 152.3 ****	14,892.0 ± 1265.1 ****	6162.0 ± 923.4 ****	9677.7 ± 309.0 **
Ker 2 (0.5 mg/mL)	10,271.7 ± 1054.6 **	12,498.3 ± 647.7	3645.3 ± 398.6 ***	12,375.3 ± 1213.5	4543.7 ± 292.2 ***	9439.0 ± 204.2 **
Dexamethasone	8872.3 ± 413.7	8951.0 ± 875.0	1843.0 ± 46.9	6715.0 ± 829.1	2361.0 ± 560.8	8400.7 ± 571.2

** $p < 0.01$, *** $p < 0.001$, **** $p < 0.0001$, using Repeated Measures ANOVA, Dunnett's Multiple Comparison Test.

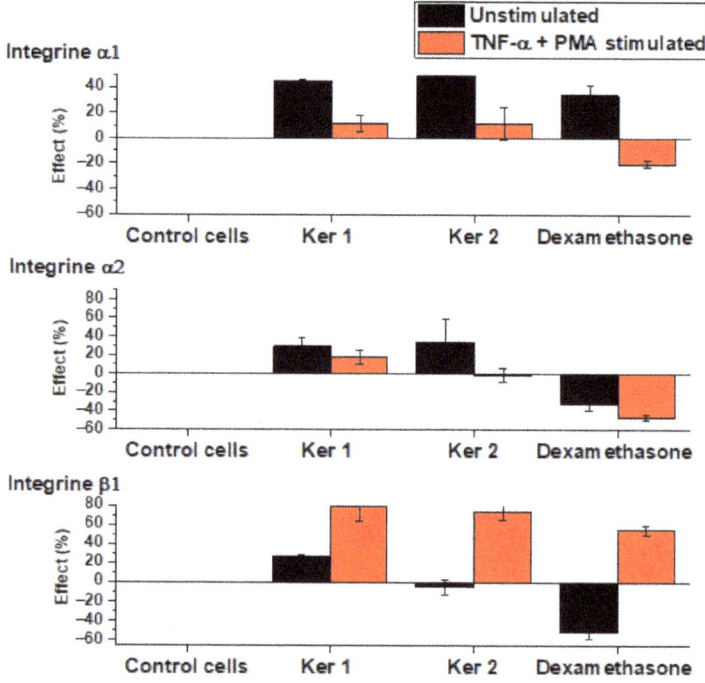

Figure 7. Percentage variation of integrins in fibroblasts compared to control cells for integrins α1, α2 and β1 over-expression modulated by Ker 1 and Ker 2.

Inflammation associated with skin wounds reflected the expressions of CAM-family proteins at vascular endothelial cells' level (VCAM and ICAM), with role in monocytes' adhesion to endothelium. The experimental results obtained on pro-inflammatory conditions simulating acute systemic disregulation (TNFα + PMA) and bacteria aggression (LPS) are presented in the Table 8 and Figure 8. Data reflects the vascular endothelium membrane expression of ICAM-1 and VCAM-1 in fixed cells stained with fluorescent specific antibodies and acquired by flow–cytometry.

The bacterial attack at the endothelial level (stimulation with LPS) is counteracted by the Ker1 and Ker2 which significantly reduce the expression of ICAM (particularity for small blood vessels), as well as VCAM (specific to large blood vessel). Ker 2 showed increased efficiency in ICAM and VCAM expression reduction for LPS stimulation. In the case of nonspecific stimulation—TNFα + PMA, the ICAM expression is significantly down-regulated compared to the positive control (dexamethasone) for both keratins.

Table 8. Adhesion molecules' modulation by keratin hydrolysates in activated vascular endothelial cells (HUVEC).

Tested Material	VCAM (PE-A Mean-Relative Fluorescence Units)		ICAM (APC-A Mean-Relative Fluorescence Units)	
	LPS Stimulation	(TNF α+ PMA) Stimulation	LPS Stimulation	(TNFα+ PMA) Stimulation
Control cells	7428 ± 180.0	3516 ± 194.5	18,249 ± 507.7	13367 ± 751.1
Ker 1 (0.5 mg/mL)	6834 ± 88.5 ***	3448 ± 323 ****	15,168 ± 1768.7 ***	4612 ± 189.6 **
Ker 2 (0.5 mg/mL)	5677 ± 456.2 ***	3454 ± 327.14 ****	11,833 ± 130.9 ***	5122 ± 544.4
Dexamethasone	4921 ± 220.5	2912 ± 107.17	10,670 ± 533.5	5684 ± 330.5

** $p < 0.01$, *** $p < 0.001$, **** $p < 0.0001$, using Repeated Measures ANOVA, Dunnett's Multiple Comparison Test.

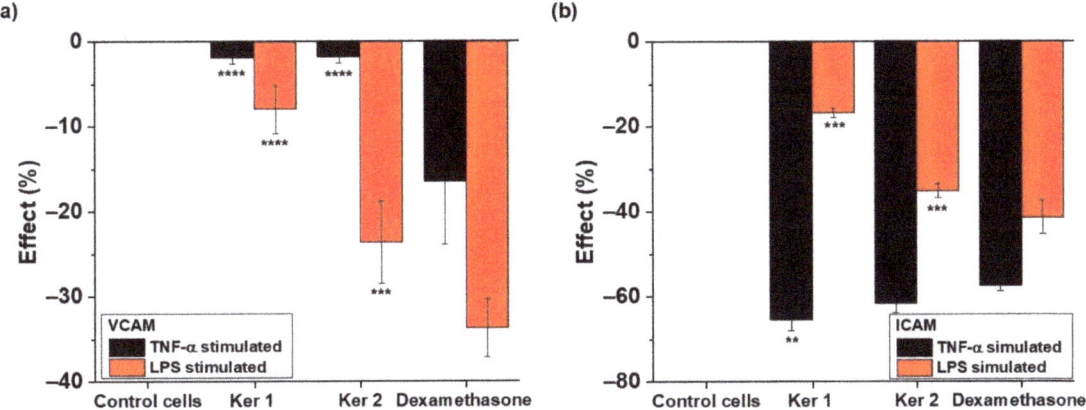

Figure 8. Percentage variation of (a) VCAM and (b) ICAM expression modulated by Ker1 and Ker2 compared to the cell control. ** $p < 0.01$, *** $p < 0.001$, **** $p < 0.0001$.

4. Discussion

Keratin extracted from sheep wool or human hair showed a high potential for wound healing biomaterials design [37–39], but the methods for keratin solubilization are very sophisticated and the mechanisms of cellular response are not completely understood [26]. The present research proposes the keratin hydrolysates with low molecular weights as wound healing additive, prepared by complete solubilization of sheep wool and by enzymatic refinery of molecular weight with still preserved secondary structure and showing bioactivity in integrin signaling, cell cycle modulation and vascular-cell-adhesion molecules inhibition.

The morphology of lyophilized keratin powders shows porous and interconnected architecture that represents a premise for cell proliferation, adhesiveness favoring the cell migration and wound healing [40]. The porosity of biomaterials is considered a critical property for cell migration due to the potential of facilitating oxygen, nutrition and metabolism exchange [41].

Circular dichroism investigations demonstrated that the solutions of keratin hydrolysates Ker 1 and Ker 2, with low molecular weights, still preserve the secondary structures of native keratin which represent the premise for their biocompatibility and cell biostimulation of wound healing mechanisms [42].

The high content in cysteine which generates very stable cystine bridges make keratin one of the most stable proteins with low solubility, which is difficult to extract. As compared to the reported methods for keratin hydrolysate extraction known as harsh methods that lead to the loss of natural function of protein, the present research showed that native structure is still preserved with effect on keratin immunogenicity which is much reduced as compared to collagen [43]. Recently reported research showed that the enzymatic route of wool keratin preparation is a new approach for obtaining high molecular keratin, with self-assembling properties for tissue engineering, but with disordered molecular structures [40], as compared to the present research which approaches methods based on chemical-enzymatic hydrolyses allowing to develop still organized structures with bioactivity in skin cell regeneration.

Ker 2 in solution has a larger percent of β-sheet organized molecules than Ker 1, meanwhile Ker 2 shows to have a parallel β-sheet fraction of 7.4%, probably due to the self-assembly ability of proteins with more ionic side groups and lower molecular weight [2]. The higher ratio of β sheet/β turn of 2.964 for Ker 2 as compared to Ker 1 with 1.785, showed improved biocompatibility [44], in correlation with higher cytotoxicity concentration for 2 cell lines (Table 3), more efficiency in keratinocytes increase and higher activity in ICAM and VCAM reduction in LPS stimulation test.

Ker 1 with a superior molecular weight and higher concentration of α-helix secondary structure is more active than Ker 2 on the binding site of integrins, both in normal and pro-inflammatory conditions, restoring the dermal strength through fibroblasts—collagen/laminin/fibronectin interactions.

FTIR spectra showed the shift to higher wavenumber of the Amide A band of Ker 2 indicating a weaker network of hydrogen bonds due to the presence of more side functional groups and lower molecular weight. Amide III deconvolution of keratin powders has revealed higher proportion of α-helix, 11% and 9.2% respectively, as compared to keratin solutions with 5.8% and 4.7% respectively, calculated from circular dichroism analyses which suggest a potentially slightly higher biocompatibility [45] in powder form. The results are similar with reported results for human hair keratin hydrolysate, with 12.18% α-helix attributed to the extraction conditions which favoured the conversion of native α-helix into β-sheet structures [46].

The behaviour of keratin hydrolysates in solution showed a tendency of developing more β-sheet and β-turn organized structures (41.9% and 48.1%, respectively) as compared to powder forms (38.8% and 37%, respectively). The β sheet/β turn ratio values for keratin powders has very similar values of 0.434 for Ker 1 and 0.464 for Ker 2, respectively. Similar values were reported in the case of wool keratin extracted by oxidative methods and with molecular weight of 13–31 kDa [47]. It can be concluded that keratin hydrolysates in solution showed a higher ability to organize and to generate β sheet and β turn secondary structures with a higher ability for lowest molecular weight Ker 2, as compared to powder form for which, α-helix ratio is higher for Ker 1, with higher molecular weight.

The biocompatibility studies performed through correlative methods for metabolic activity and cellular toxicity revealed a very good skin tolerance for higher doses of keratin preparations, expressed at all significant cellular exponents for wound healing: standardized cell lines HaCaT—human immortalized keratinocytes, HS27—human normal dermal fibroblasts and HUVEC—primary vascular endothelium from human umbilical vein.

Our studies also demonstrated a homeostasis restoring role of hydrolyzed keratin forms on skin inflammatory disruptions, injury associated. Ker 1 and Ker 2 presented a differentiated action at epidermal, dermal and vascular level, regulating significant molecular targets of the healing process. Inflammation was induced by external stimuli such as molecules from bacterial infection—lipopolysaccharide (LPS), originating from outer membrane of Gram-negative bacteria but also by molecules produced in several signaling cascades [48]. We chose tumor necrosis factor alpha (TNFα), a pro-inflammatory cytokine that triggers the expression of inflammatory molecules, including cell adhesion molecules—intercellular cell adhesion molecule-1 (ICAM-1) and vascular cell adhesion molecule-1 (VCAM-1) [22]. Compared with unstimulated series, the in vitro experimentally-induced pro-inflammatory signals were effective in key factors modulation and appropriate for the investigation of compounds effects in the wound healing context. Integrin-mediated adhesion is a bidirectional event in order to modulate the extracellular matrix structure and composition as well as to decide cell fate during wound healing through complex pathways: proliferation, differentiation, migration [19]. One of the most important integrins is β1 integrin (fibronectin receptor), which mediates the cellular response such as adhesion, migration, extracellular matrix assembly and signal transduction [49]. The presented screening showed a marked influence of keratin hydrolysates, Ker1 being stronger than Ker2, on β1 integrin, especially in pro-inflammatory conditions. Results gain thus relevance for the rapid relief of skin injuries after keratin treatment. In addition, Ker1 and Ker2 have a strong affinity for α1 and α2 subunits in the basal state of fibroblasts, preserving the essential cell—collagen interactions. The endothelium plays an important role in inflammation by regulating vascular permeability for macromolecules and leukocytes, vascular tone and hemostasis, and by producing and binding inflammatory mediators [50]. Skin injury activates adhesion molecules VCAM-1 and ICAM-1 in endothelial cells, and triggers the subsequent release of chemokines and inflammatory cell infiltration. Ker1 and Ker2 are involved in this factor's membrane expression, after their up-regulation during endothelial activation. We have to mention the complex but unspecific action on both ICAM-1 and VCAM-1 when the inflammatory phenotype was induced by bacterial LPS and the selective effect on ICAM-1 when the systemic cytokines signals (TNF α) are involved.

Further research will be carried out for new formulations based on Ker1 or Ker2 and in vivo tests of their efficiency in wound healing.

5. Conclusions

The complex screening focused on essential mechanisms in delayed and difficult wound healing has demonstrated the involvement of two keratin hydrolysates with low molecular weights in skin regeneration processes.

The main conclusions of the study are that the lowest molecular weight keratin hydrolysate has a better cytotoxicity profile in dermo-epidermal cells and an anti-inflammatory effect on endothelial cells stimulated with LPS concerning the monocytes—endothelium adhesion molecules inhibition.

The study highlights correlations between the structural profile and the molecular weight of the two types of keratins, Ker1 and Ker2, and the modulation of certain signaling pathways in activating the post-traumatic recovery of the dermo-epidermal tissue. Thus, Ker1 (with superior molecular weight and higher concentration of α-helix secondary structure) is more active than Ker2 on the binding site of integrins, both in normal and pro-inflammatory conditions, restoring the dermal strength through fibroblasts—collagen/laminin/fibronectin interactions. In turn, Ker 2, better than Ker 1, accelerates the epidermal turnover, improving reepithelization probably due to the more available surface contact and biocompatibility induced by the higher ratio of β sheet/β turn structure. The slight impact of keratins on the proliferative status of fibroblasts has to be mentioned, which brings another therapeutical advantage, preventing tissue fibrosis and imperfect scarring. The obtained results contribute to the knowledge regarding mechanistic insight in wound repair, directing future medical applications based on low molecular hydrolyzed keratin preparations.

Author Contributions: Conceptualization, L.O., B.G.D. and C.G.; methodology, M.S., L.M.T., M.D.E., I.-R.S. and C.T.; software, C.T.; validation, L.O., B.G.D., C.G., I.-R.S. and C.T.; formal analysis, L.O., B.G.D., C.G., M.S., I.-R.S. and C.T.; investigation, B.G.D., L.M.T., M.D.E., I.-R.S., C.T. and M.S.; resources, L.O. and C.G.; data curation, L.O., B.G.D. and C.G.; writing—original draft preparation, C.G., M.S., B.G.D. and L.O.; writing—review and editing, C.G., B.G.D., L.O. and M.S.; visualization, C.G. and M.S.; supervision, L.O., B.G.D. and C.G.; project administration, L.O. and C.G.; funding acquisition, L.O. and C.G. All authors have read and agreed to the published version of the manuscript.

Funding: The research was supported by grants funded by the Romanian Ministry of Research, Innovation and Digitalization, CCCDI-UEFISCDI, project PN III P2.1 PTE 2019 0655, contract 5PTE/2020,

BIOTEHKER and project number 4N/2019–PN 19 17 01 02/2021, CREATIV_PIEL, National Program Nucleus.

Institutional Review Board Statement: Not applicable.

Informed Consent Statement: Not applicable.

Data Availability Statement: Data are contained within the article.

Conflicts of Interest: The authors declare no conflict of interest. The funders had no role in the design of the study; in the collection, analyses, or interpretation of data; in the writing of the manuscript, or in the decision to publish the results.

References

1. Perta-Crisan, S.; Ursachi, C.S.; Gavrilas, S.; Oancea, F.; Munteanu, F.-D. Closing the loop with keratin-rich fibrous materials. *Polymers* **2021**, *13*, 1896. [CrossRef] [PubMed]
2. Gaidau, C.; Stanca, M.; Niculescu, M.-D.; Alexe, C.-A.; Becheritu, M.; Horoias, R.; Cioineag, C.; Râpă, M.; Stanculescu, I.R. Wool Keratin Hydrolysates for Bioactive Additives Preparation. *Materials* **2021**, *14*, 4696. [CrossRef] [PubMed]
3. Râpă, M.; Gaidău, C.; Stefan, L.M.; Matei, E.; Niculescu, M.; Berechet, M.D.; Stanca, M.; Tablet, C.; Tudorache, M.; Gavrilă, R.; et al. New Nanofibers based on Protein By-Products with Bioactive Potential for Tissue Engineering. *Materials* **2020**, *13*, 3149. [CrossRef] [PubMed]
4. Idrees, H.; Zaidi, S.Z.J.; Sabir, A.; Khan, R.U.; Zhang, X.; Hassan, S.U. A review of biodegradable natural polymer-based nanoparticles for drug delivery applications. *Nanomaterials* **2020**, *10*, 1970. [CrossRef]
5. Matyašovsky, J.; Sedliačik, J.; Valachova, K.; Novak, I.; Jurkovič, P.; Duchovič, P.; Mičušik, M.; Kleinova, A.; Šoltes, L. Antioxidant Effects of Keratin Hydrolysates. *J. Am. Leather Chem.* **2017**, *112*, 327–337.
6. Waters, M.; VandeVord, P.; Van Dyke, M. Keratin biomaterials augment anti-inflammatory macrophage phenotype in vitro. *Acta Biomater.* **2018**, *66*, 213–223. [CrossRef]
7. Tachibana, A.; Nishikawa, Y.; Nishino, M.; Kaneko, S.; Tanabe, T.; Yamauchi, K. Modified Keratin Sponge: Binding of Bone Morphogenetic Protein-2 and Osteoblast Differentiation. *J. Biosci. Bioeng.* **2006**, *102*, 425–429. [CrossRef]
8. Sierpinski, P.; Garrett, J.; Ma, J.; Apel, P.; Klorig, D.; Smith, T.; Koman, L.A.; Atala, A.; Van Dyke, M. The use of keratin biomaterials derived from human hair for the promotion of rapid regeneration of peripheral nerves. *Biomaterials* **2008**, *29*, 118–128. [CrossRef]
9. Bordeleau, F.; Bessard, J.; Sheng, Y.; Marceau, N. Keratin contribution to cellular mechanical stress response at focal adhesions as assayed by laser tweezers. *Biochem. Cell Biol.* **2008**, *86*, 352–359. [CrossRef]
10. Gao, J.; Zhang, L.; Wei, Y.; Chen, T.; Ji, X.; Ye, K.; Yu, J.; Tang, B.; Sun, X.; Hu, J. Human hair keratins promote the regeneration of peripheral nerves in a rat sciatic nerve crush model. *J. Mater. Sci. Mater. Med.* **2019**, *30*, 82. [CrossRef]
11. Poranki, D.; Whitener, W.; Howse, S.; Mesen, T.; Howse, E.; Burnell, J.; Greengauz-Roberts, O.; Molnar, J.; Van Dyke, M. Evaluation of skin regeneration after burns in vivo and rescue of cells after thermal stress in vitro following treatment with a keratin biomaterial. *J. Biomater. Appl.* **2013**, *29*, 26–35. [CrossRef]
12. Pfisterer, K.; Shaw, L.-E.; Symmank, D.; Weninger, W. The Extracellular Matrix in Skin Inflammation and Infection. *Front. Cell Dev. Biol.* **2021**, *9*, 682414. [CrossRef]
13. Eckes, B.; Nischt, R.; Krieg, T. Cell-matrix interactions in dermal repair and scarring. *Fibrogenes. Tissue Repair* **2010**, *3*, 4. [CrossRef]
14. Olachi, J.M.-N.; Maheshwari, A. The role of integrins in inflammation and angiogenesis. *Pediatr. Res.* **2021**, *89*, 1619–1626.
15. Samarth, H.; Srikala, R.A. Skin-depth Analysis of Integrins: Role of the Integrin Network in Health and Disease. *Cell Commun. Adhes.* **2013**, *20*, 155–169.
16. Ford, A.J.; Rajagopalan, P. Extracellular matrix remodeling in 3D: Implications in tissue homeostasis and disease progression. *WIREs Nanomed. Nanobiotechnol.* **2017**, *10*, e1503. [CrossRef]
17. Zweers, M.C.; Davidson, J.M.; Pozzi, A.; Hallinger, R.; Janz, K.; Quondamatteo, F.; Leutgeb, B.; Krieg, T.; Eckes, B. Integrin α2β1 Is Required for Regulation of Murine Wound Angiogenesis but Is Dispensable for Reepithelialization. *J. Investig. Dermatol.* **2007**, *127*, 467–478. [CrossRef]
18. Lopez-Rovira, T.; Silva-Vargas, V.; Watt, F.M. Different consequences of beta1 integrin deletion in neonatal and adult mouse epidermis reveal a context-dependent role of integrins in regulating proliferation, differentiation, and intercellular communication. *J. Investig. Dermatol.* **2005**, *125*, 1215–1227. [CrossRef]
19. Koivisto, L.; Heino, J.; Hakkinen, L.; HannuLarjava, H. Integrins in Wound Healing. *Adv. Wound Care* **2014**, *3*, 12. [CrossRef]
20. Van Buul, J.D.; van Rijssel, J.; van Alphen, F.-P.J.; van Stalborch, A.-M.; Mul, E.P.J.; Hordijk, P.L. ICAM-1 Clustering on Endothelial Cells Recruits VCAM-1. *J. Biomed. Biotechnol.* **2010**, *2010*, 120328. [CrossRef]
21. Hua, S. Targeting sites of inflammation: Intercellular adhesion molecule-1 as a target for novel inflammatory therapies. *Front. Pharmacol.* **2013**, *4*, 127. [CrossRef] [PubMed]
22. Kong, D.-H.; Kim, Y.K.; Kim, M.R.; Jang, J.H.; Lee, S. Emerging Roles of Vascular Cell Adhesion Molecule-1 (VCAM-1) in Immunological Disorders and Cancer. *Int. J. Mol. Sci.* **2018**, *19*, 1057. [CrossRef] [PubMed]

23. Mohamed, J.M.; Alqahtani, A.; Fatease, A.A.; Alqahtani, T.; Khan, B.A.; Ashmitha, B.; Vijaya, R. Human Hair Keratin Composite Scaffold: Characterisation and Biocompatibility Study on NIH3T3 Fibroblast Cells. *Pharmaceuticals* **2021**, *14*, 781. [CrossRef] [PubMed]
24. Gaidau, C.; Epure, D.-G.; Enascuta, C.E.; Carsote, C.; Sendrea, C.; Proietti, N.; Chen, W.; Gu, H. Wool keratin total solubilisation for recovery and reintegration—An ecological approach. *J. Clean. Prod.* **2019**, *236*, 117586. [CrossRef]
25. Feroz, S.; Muhammad, N.; Ratnayake, J.; Dias, G. Keratin-Based materials for biomedical applications. *Bioact. Mater.* **2020**, *5*, 496–509. [CrossRef]
26. Lazarus, B.S.; Chadha, C.; Velasco-Hogan, A.; Barbosa, J.D.V.; Jasiuk, I.; Meyers, M.A. Engineering with keratin: A functional material and a source of bioinspiration. *iScience* **2021**, *24*, 102798. Available online: https://www.sciencedirect.com/science/article/pii/S2589004221007665 (accessed on 1 December 2021). [CrossRef]
27. Pechter, P.M.; Gil, J.; Valder, J.; Tomic-c, M.; Pastar, J.; Stojadinovic, O.; Kirsner, R.S.; Davis, S.C. Keratin dressings speed epithelisation of deep partial-thickness wounds. *Wound Repair Regen.* **2012**, *20*, 236–242. [CrossRef]
28. Vindelov, L.L.; Christensen, I.J.; Nissen, N.I. A detergent-trypsin method for the preparation of nuclei for flow cytometric DNA analysis. *Cytometry* **1983**, *3*, 323–327. [CrossRef]
29. Gao, F.; Li, W.; Ding, Y.; Wang, Y.; Deng, J.; Qing, R.; Wang, B.; Hao, S. Insight into the Regulatory Function of Human Hair Keratins in Wound Healing Using Proteomics. *Adv. Biosyst.* **2020**, 1900235. [CrossRef]
30. Micsonai, A.; Wien, F.; Bulyaki, E.; Kun, J.; Moussong, E.; Lee, Y.H.; Goto, Y.; Refregiers, M.; Kardos, J. BeStSel: A web server for accurate protein secondary structure prediction and fold recognition from the circular dichroism spectra. *Nucleic Acids Res.* **2018**, *46*, W315–W322. [CrossRef]
31. Micsonai, A.; Wien, F.; Kernya, L.; Lee, Y.H.; Goto, Y.; Refregiers, M.; Kardos, J. Accurate secondary structure prediction and fold recognition for circular dichroism spectroscopy. *Proc. Natl. Acad. Sci. USA* **2015**, *112*, E3095–E3103. [CrossRef]
32. Kissi, N.; Curran, K.; Vlachou-Mogire, C.; Fearn, T.; McCullough, L. Developing a non-invasive tool to assess the impact of oxidation on the structural integrity of historic wool in Tudor tapestries. *Herit. Sci.* **2017**, *5*, 1–13. [CrossRef]
33. Roach, P.; Farrar, D.; Perry, C.C. Interpretation of protein adsorption: Surface-induced conformational changes. *J. Am. Chem. Soc.* **2005**, *127*, 8168–8173. [CrossRef]
34. Peng, D.; Zhao, J.; Mashayekhi, H.; and Xing, B. Adsorption of bovine serum albumin and lysozyme on functionalized carbon nanotubes. *J. Phys. Chem. C* **2014**, *118*, 22249–22257.
35. Rajabinejad, H.; Zoccola, M.; Patrucco, A.; Montarsolo, A.; Rovero, G.; Tonin, C. Physicochemical properties of keratin extracted from wool by various methods. *Text. Res. J.* **2018**, *88*, 2415–2424. [CrossRef]
36. Chiaramaria, S.; Vaccari, L.; Mitri, E.; Birarda, G. FTIR investigation of the secondary structure of type I collagen: New insight into the amide III band. *Spectrochim. Acta Part A Mol. Biomol. Spectrosc.* **2020**, *229*, 118006. [CrossRef]
37. Knop, M.; Rybka, M.; Drapda, A. Keratin Biomaterials in Skin Wound Healing, an Old Player in Modern Medicine: A Mini Review. *Pharmaceutics* **2021**, *13*, 2029. [CrossRef]
38. Tang, A.; Li, Y.; Yang, X.; Cao, Z.; Nie, H.; Yang, G. Injectable keratin hydrogels as hemostatic and wound dressing materials. *Biomater. Sci.* **2021**. [CrossRef]
39. Chen, Y.; Li, Y.; Yang, X.; Cao, Z.; Nie, H.; Brian, Y.; Yang, G. Glucose-triggered in situ forming keratin hydrogel for the treatment of diabetic wounds. *Acta Biomater.* **2021**, *125*, 208–2018. [CrossRef]
40. Su, C.; Gong, J.-S.; Ye, J.-P.; He, J.-M.; Li, R.-Y.; Jiang, M.; Geng, Y.; Zhang, Y.; Chen, J.-H.; Xu, Z.-H.; et al. Enzymatic Extraction of Bioactive and Self-Assembling Wool Keratin for Biomedical Applications. *Macromol. Biosci.* **2020**, 2000073. [CrossRef]
41. Sundaram, J.; Durance, T.D.; Wang, R. Porous scaffold of gelatin-starch with nanohydroxyapatite composite processed via novel microwave vacuum drying. *Acta Biomater.* **2008**, *4*, 932–942. [CrossRef]
42. Ito, H.; Miyamoto, T.; Inagaki, H.; Noishiki, Y. Biocompatibility of Denatured Keratins from Wool. *Kobunshi Ronbunshu* **1982**, *39*, 249–256. [CrossRef]
43. Bianchera, A.; Catanzano, O.; Boateng, J.; Elviri, L. *The Place of Biomaterials in Wound Healing*, 1st ed.; Boateng, J., Ed.; John Wiley & Sons Ltd.: Hoboken, NJ, USA, 2020.
44. Vanea, E.; Magyari, K.; Simion, V. Protein attachment on alumino silicates surface studied by XPS and FTIR spectroscopy. *J. Optoelectron. Adv. Mater.* **2010**, *12*, 1206–1212.
45. Ramirez, D.O.S.; Cruz-Maya, I.; Vineis, C.; Guarino, V.; Tonetti, C.; Varesano, A. Wool Keratin-Based Nanofibres—In Vitro Validation. *Bioengineering* **2021**, *8*, 224. [CrossRef]
46. Ko, J.; Nguyen, L.T.H.; Surendran, A.; Tan, B.Y.; Ng, K.W.; Leong, W.L. Human Hair Keratin for Biocompatible Flexible and Transient Electronic Devices. *ACS Appl. Mater. Interfaces* **2017**, *9*, 43004–43012. [CrossRef]
47. Fernández-d'Arlas, B. Tough and Functional Cross-linked Bioplastics from Sheep Wool Keratin. *Sci. Rep.* **2019**, *9*, 14810. [CrossRef]
48. Nam, U.; Kim, S.; Park, J.; Jeon, J.S. Lipopolysaccharide-Induced Vascular Inflammation Model on Microfluidic Chip. *Micromachines* **2020**, *11*, 747. [CrossRef]
49. Liu, S.; Xu, S.; Blumbach, K.; Eastwood, M.; Denton, C.P.; Eckes, B.; Krieg, T.; Abraham, D.J.; Leask, A. Expression of integrin beta1 by fibroblasts is required for tissue repair in vivo. *J. Cell Sci.* **2010**, *123 Pt 21*, 3674–3682. [CrossRef]
50. Videm, V.; Albrigtsen, M. Soluble ICAM-1 and VCAM-1 as Markers of Endothelial Activation. *Scand. J. Immunol.* **2008**, *67*, 523–531. [CrossRef]

MDPI
St. Alban-Anlage 66
4052 Basel
Switzerland
Tel. +41 61 683 77 34
Fax +41 61 302 89 18
www.mdpi.com

Polymers Editorial Office
E-mail: polymers@mdpi.com
www.mdpi.com/journal/polymers

www.ingramcontent.com/pod-product-compliance
Lightning Source LLC
LaVergne TN
LVHW070235100526
838202LV00015B/2132